T5-DIJ-677

Photosensitive Metal–Organic Systems

ADVANCES IN CHEMISTRY SERIES 238

Photosensitive Metal–Organic Systems

Mechanistic Principles and Applications

Charles Kutal, EDITOR
University of Georgia

Nick Serpone, EDITOR
Concordia University

Developed from a symposium sponsored
by the Division of Inorganic Chemistry
of the American Chemical Society

American Chemical Society, Washington, DC 1993

Library of Congress Cataloging-in-Publication Data

Photosensitive metal–organic systems : mechanistic principles and applications / Charles Kutal, editor, Nick Serpone, editor.

p. cm.—(Advances in Chemistry Series, ISSN 0065–2393; 238).

"Developed from a symposium sponsored by the Division of Inorganic Chemistry at the Fourth Chemical Congress of North America (202nd National Meeting of the American Chemical Society), New York, New York, August 25–30, 1991."

Includes bibliographical references and index.

ISBN 0–8412–2527–3

1. Photocatalysis—Congresses. 2. Organic compounds—Congresses. 3. Organometallic compounds—Congresses.

I. Kutal, Charles. II. Serpone, Nick, 1939– . III. American Chemical Society. Division of Inorganic Chemistry. IV. Chemical Congress of North America (4th : 1991 : New York, NY). V. Series

QD1.A355 no. 238
[QD716.P45]
540 s—dc20
[547.1'35]

93–22823
CIP

The paper used in this publication meets the minimum requirements of American National Standard for Information Sciences—Permanence of Paper for Printed Library Materials, ANSI Z39.48–1984. ∞

PRINTED IN THE UNITED STATES OF AMERICA

QD
1
.A355
no. 238

FOREWORD

The ADVANCES IN CHEMISTRY SERIES was founded in 1949 by the American Chemical Society as an outlet for symposia and collections of data in special areas of topical interest that could not be accommodated in the Society's journals. It provides a medium for symposia that would otherwise be fragmented because their papers would be distributed among several journals or not published at all.

Papers are reviewed critically according to ACS editorial standards and receive the careful attention and processing characteristic of ACS publications. Volumes in the ADVANCES IN CHEMISTRY SERIES maintain the integrity of the symposia on which they are based; however, verbatim reproductions of previously published papers are not accepted. Papers may include reports of research as well as reviews, because symposia may embrace both types of presentation.

ABOUT THE EDITORS

CHARLES KUTAL is a professor and Head of the Department of Chemistry at the University of Georgia. He received his Ph.D. and M.S. from the University of Illinois and his B.S. from Knox College. He is the recipient of the Undergraduate Teaching Award and the Chemist of the Year Award of the Northeast Georgia section of the American Chemical Society. Kutal has authored more than 70 articles and chapters dealing with the photochemistry of inorganic and organometallic compounds, solar energy conversion and storage, and photosensitive polymer materials. He is coeditor of the books *Solar Energy: Chemical Conversion and Storage* and *Electron Transfer in Biology and the Solid State.*

NICK SERPONE is a professor and Director of the Laboratory of Pure and Applied Studies in Catalysis, Environment, and Materials in the Department of Chemistry and Biochemistry at Concordia University. He received his Ph.D. from Cornell University and his B.Sc. from Sir George Williams University. He was the founder and the first Director of the Canadian Center for Picosecond Laser Spectroscopy (1981–1987). He is the holder of one patent and has authored more than 180 journal articles and review papers, coedited three books, and jointly translated one book (French to English). He was visiting professor at the University of Bologna, Italy (1975–1976), an invited professor at the Ecole Polytechnique Federale of Lausanne, Switzerland (1983–1984), and a visiting professor and Directeur de Recherche at the Ecole Centrale de Lyon, France (1990–1991). His principal research interests are in heterogeneous photocatalysis, environmental and inorganic photochemistry, picosecond laser spectroscopy, and imaging science.

CONTENTS

INDEXES

PREFACE

NUMEROUS CLASSES OF ORGANIC REAGENTS undergo useful and, quite often, novel chemistry when irradiated in the presence of a metal-containing coordination compound or particulate semiconductor. Unlike thermal catalytic processes, which occur with all reactants, intermediates, and products residing in their electronic ground states, this photocatalysis of organic reactions occurs by pathways that, at some point, involve the participation of an electronic excited state. Not surprisingly, then, considerable effort has been expended to elucidate the roles played by excited states and other photogenerated species in the overall reaction sequence. Apart from its fundamental value, such mechanistic information provides guidance to those interested in the practical uses of photosensitive metal–organic systems. Promising applications of these systems ranging from the photocuring of coatings to the remediation of polluted waters are under active investigation.

The purpose of the symposium upon which this book is based was to bring together scientists with diverse backgrounds and interests for a comprehensive discussion of the conceptual and practical advances that have occurred in the burgeoning area of photosensitive metal–organic systems. Of the 23 symposium presentations, 21 are included in this volume. Each author was asked to provide sufficient review and tutorial material to afford the reader a sound introduction. In addition, each chapter describes research results and, where appropriate, indicates promising future directions. Given the interdisciplinary scope of the material covered, and the treatment of both basic and applied topics, we feel that this volume will prove to be of value to a broad spectrum of scientists.

We are grateful to the Division of Inorganic Chemistry of the American Chemical Society for their sponsorship of the symposium and to the Petroleum Research Fund, administered by the American Chemical Society; Ciba-Geigy; the 3M Company; and Loctite Corporation for their generous financial support of the symposium speakers.

CHARLES KUTAL
University of Georgia
Athens, GA 30602

NICK SERPONE
Concordia University
Montreal, Quebec
Canada H3G 1M8

May 1993

1

Photosensitive Metal–Organic Systems

An Overview

Charles Kutal

Department of Chemistry, University of Georgia, Athens, GA 30602

Organic molecules undergo a rich assortment of transformations when irradiated in the presence of a metal-containing coordination compound or particulate semiconductor. This chapter provides an overview of the mechanistic principles needed to understand the chemistry occurring in these photosensitive metal–organic systems. The important concept of photocatalysis is defined, and processes that satisfy this definition are divided into two operationally distinct categories: photogenerated catalysis and catalyzed photolysis. Photocatalysis proceeds via mechanisms that, at some point, involve an electronic excited state, and therefore a brief description is provided of the various excited states that arise in coordination compounds and semiconductors. Finally, some specific examples of photogenerated catalysis and catalyzed photolysis are presented.

METAL-CATALYZED REACTIONS of organic substrates have figured prominently in the development of modern chemical science and technology. This type of transformation can be represented generically by equation 1:

$$\mathrm{O} \xrightarrow{\mathrm{M}} \mathrm{P} \qquad (1)$$

where O and P denote the organic reactant and product, respectively, and

0065–2393/93/0238–0001$07.50/0

M is the metal-containing catalyst or catalyst precursor. Examples of commercially important metal-catalyzed processes include the polymerization of ethylene and propylene, the metathesis of olefins, the hydrocyanation of butadiene to adiponitrile, the hydroformylation of olefins to alcohols, and the syntheses of acetic acid and acetic anhydride (*1*). Typically, such processes occur via a succession of steps, at least one of which involves an interaction between M and O that lowers the activation enthalpy and thereby accelerates the reaction rate.

Reactions that conform to equation 1 are thermally activated and, quite often, require elevated temperatures to generate the active catalyst from an inactive precursor. Regardless of the exact sequence of steps leading from reactants to products, all species (e.g., M, O, and reactive intermediates) participating in the catalyzed transformation reside in their electronic ground states. An alternative strategy for activating the system, described by equation 2, involves irradiation with ultraviolet or visible light.

$$\mathrm{O} \xrightarrow[h\nu]{\mathrm{M}} \mathrm{P} \tag{2}$$

This photoactivation process generates electronic excited states whose physical properties (e.g., energy and geometry) and, most importantly, chemical properties (e.g., ease of bond making and breaking, redox potential, and acid–base character) can differ substantially from those of the ground state. In a very real sense, light adds another dimension to catalytic behavior by introducing new reaction pathways that, at some point, include an excited-state species. Irradiation of a system frequently yields the same set of products formed in the corresponding thermal reaction, but the product distribution may differ because of a change in chemoselectivity. Moreover, because light rather than heat activates the system, thermally sensitive products can be isolated by irradiating at low temperatures. Photoactivation also provides a possible route to highly energetic (contrathermodynamic) products that cannot be formed via a ground-state process.

Table I contains a sampling of photosensitive metal–organic systems discussed later in this chapter. The diverse assortment of organic substrates, metal-containing compounds, and reaction classes attests to the generality of equation 2. Detailed mechanistic studies of such systems over the past 10–15 years have yielded important insights concerning the various interactions that can occur between M, O, and light. Beside being of fundamental interest, such knowledge facilitates the rational design of practical devices and processes that incorporate this chemistry as an essential component. Promising applications of equation 2 have been reported

Table I. Examples of Photosensitive Metal–Organic Systems

Organic Substrate(s)	*Metal-Containing Compound*	*Transformation*	*Ref.*
1-Pentene	$Fe(CO)_5$	double-bond migration and *cis–trans* isomerization	25–27
Epoxides	$Fe(\eta^5\text{-}C_5H_5)(\eta^6\text{-arene})^+$	cationic polymerization	29–31
α-Cyanoacrylate	*trans*-$Cr(NH_3)_2(NCS)_4^-$	anionic polymerization	32
Acrylonitrile	$Co(C_2O_4)_3^{3-}$	radical polymerization	36
Quadricyclene	$PdCl_2(\eta^4\text{-NBD})$[a]	valence isomerization	43, 44
Dimethyl fumarate + BNAH[b]	$Ru(bipy)_3^{2+}$	two-electron reduction of olefin	46, 47
Methanol[c]	$MoO(OCH_3)$ (porphyrin)	production of H_2O_2 and formaldehyde	48
α,β-Unsaturated ketones[c]	$CuSO_4$	formation of dimeric lactones	49
Norbornadiene	CuCl	valence isomerization	50, 51

[a] NBD is norbornadiene.

[b] BNAH is 1-benzyl-1,4-dihydronicotinamide.

[c] O_2 is required.

in such diverse fields as chemical synthesis, catalysis, imaging and coating technologies, biology and medicine, photochemical energy storage, and materials design.

This chapter provides an overview of the mechanistic principles that govern the behavior of photosensitive metal–organic systems. In the following section we define the important concept of photocatalysis and describe several possible pathways by which it can occur. Next, we consider the chemical consequences of populating the various electronic excited states of metal-containing compounds. Finally, we employ the preceding information as a guide in discussing the chemistry of a few exemplary systems.

Definition and Classification of Photocatalysis

Reactions that proceed according to equation 2 require both light and the presence of a catalyst. Given the generality (*see* Table I) and complexity (i.e., the tripartite interactions of M, O, and light) of this process, it is perhaps not surprising that a variety of labels have been coined to describe it. This often confusing and still contentious situation has been discussed by several authors (*2–9*). Rather than restating the pros and cons of various viewpoints, we shall adopt what appears to be the consensus opinion concerning terminology in this area. For sake of completeness, however, alternate labels still in use will be mentioned as appropriate.

Kisch (*8*) suggested that the generic term *photocatalysis* be applied to equation 2. Specifically, photocatalysis is defined as the "acceleration of a photoreaction by the presence of a catalyst." Furthermore, "the catalyst may accelerate the photoreaction by interaction with the substrate in its ground or excited state and/or with a primary photoproduct, depending upon the mechanism of the photoreaction." For the systems treated in this volume, the catalyst or catalyst precursor will be a metal-containing species such as a mononuclear or dinuclear coordination compound, polynuclear cluster, or colloidal (<1000-Å radius) or macroparticulate (>1000-Å radius) semiconductor.

Salomon (*2*) proposed that reactions satisfying the rather broad definition of photocatalysis be divided into two operationally distinct classes. The first, termed *photogenerated catalysis* (also photoinitiated or photoinduced catalytic reactions), involves the light-induced generation of a ground-state catalyst, C, from M (eq 3) and/or O (eq 4). In one or more subsequent reactions, C catalyzes the conversion of organic substrate to product (eq 5).

$$M \xrightarrow{h\nu} C \tag{3}$$

$$M + O \xrightarrow{h\nu} C \tag{4}$$

$$C + O \longrightarrow C + P \tag{5}$$

Salomon (*2*) and Chanon and Chanon (*9*) enumerated several generalized mechanisms by which photogenerated catalysis can occur. Scheme 1, for example, involves the direct photochemical conversion of M to the active catalyst, and Scheme 2 requires the participation of both M and O in catalyst formation. Schemes 3 and 4, which feature the transformation of a catalyst–substrate complex, C•O, to a catalyst–product complex, C•P, differ in the mode of regenerating C.

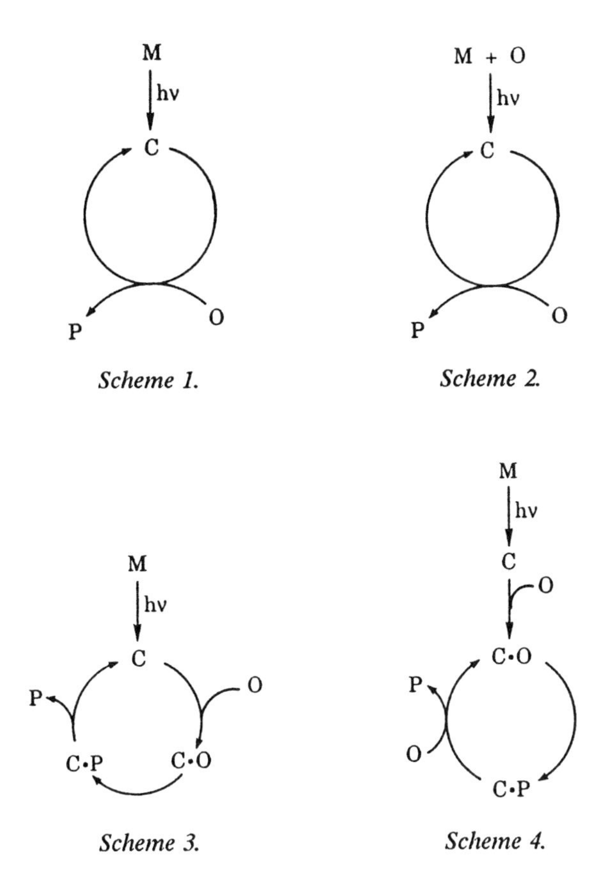

Scheme 1.

Scheme 2.

Scheme 3.

Scheme 4.

Closely related to photogenerated catalysis is the process of photoinitiation (*10*), which involves the photochemical production of an initiator, I, from M and/or O (replace C by I in eqs 3 and 4). Unlike catalyst C, which is continuously regenerated (eq 5), I is consumed while initiating a chain reaction (eq 6).

$$\mathrm{I} + \mathrm{O} \longrightarrow \mathrm{P}_1 \xrightarrow{\mathrm{O}} \mathrm{P}_2 \xrightarrow{\mathrm{O}} \text{etc.} \tag{6}$$

In either case, however, the reactive species (C or I) produced by the action of a single photon can cause the conversion of several substrate molecules to product. This multiplicative response constitutes chemical amplification (*11*) of the initial photochemical act and affords a means of designing systems that exhibit high photosensitivity. Thus the observed quantum efficiency, Φ_{ob}, will be the product of the true quantum efficiency of the photochemical step (eq 3 or 4) and the turnover number (i.e., number of catalytic events) or the average chain length. Values of Φ_{ob} greater than unity are clearly diagnostic of photogenerated catalysis or photoinitiation. Other characteristics of these processes are an induction period, during which the active species is produced, and the continuation of thermal chemistry after irradiation has ceased.

The other major class of reactions satisfying the definition of photocatalysis has been labeled *catalyzed photolysis* (also catalyzed photochemistry) (*2*). This type of process begins with the absorption of light by M, O, or a pre-formed M–O complex (eqs 7–9; an asterisk denotes an electronic excited state).

$$\mathrm{M} \xrightarrow{h\nu} \mathrm{M}^* \xrightarrow{\mathrm{O}} \mathrm{P} + \mathrm{M} \tag{7}$$

$$\mathrm{O} \xrightarrow{h\nu} \mathrm{O}^* \xrightarrow{\mathrm{M}} \mathrm{P} + \mathrm{M} \tag{8}$$

$$\mathrm{M} + \mathrm{O} \rightleftarrows \mathrm{M{-}O} \xrightarrow{h\nu} (\mathrm{M{-}O})^* \longrightarrow \mathrm{P} + \mathrm{M} \tag{9}$$

Transformation of organic substrate to final product occurs during the course of the subsequent photochemical event. Even though the overall process may be catalytic with respect to M, at least one photon is required per product molecule formed and, consequently, Φ_{ob} never exceeds unity. Moreover, neither an induction period nor a postirradiation reaction is observed.

Catalyzed photolysis can occur by a variety of mechanisms (*2, 9*) because, depending upon the system, any one (or more) of three species may absorb the incident radiation (eqs 7–9) and then yield a reactive excited state or primary photoproduct. Scheme 5 illustrates the process, commonly termed *photosensitization* (*12, 13*), in which the interaction between electronically excited M and ground-state substrate activates the latter and regenerates M. Alternatively, in Scheme 6 the reaction of M* produces a ground-state species, C′, which assists the transformation of substrate to product and then reverts to M. C′ has been called a photoassistor (also pseudocatalyst), and the overall mechanism has been termed photoassistance (also stoichiometric photogenerated catalysis) (*4*).

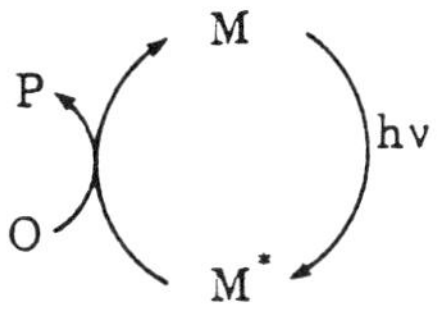

Scheme 5.

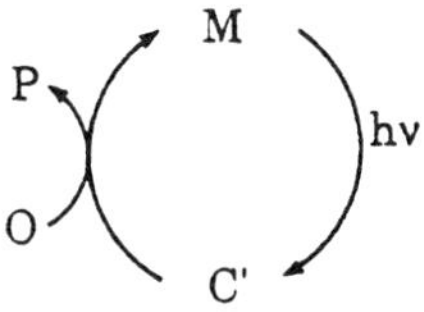

Scheme 6.

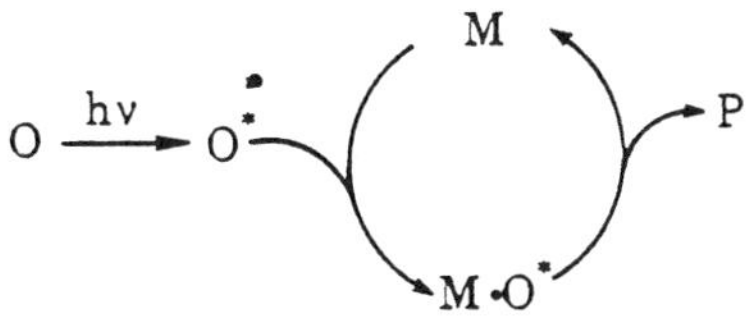

Scheme 7.

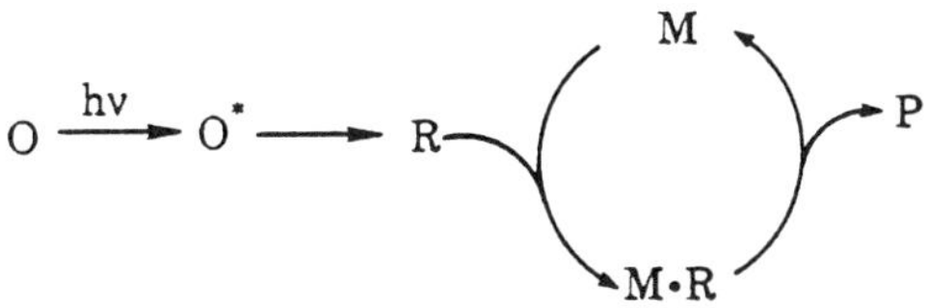

Scheme 8.

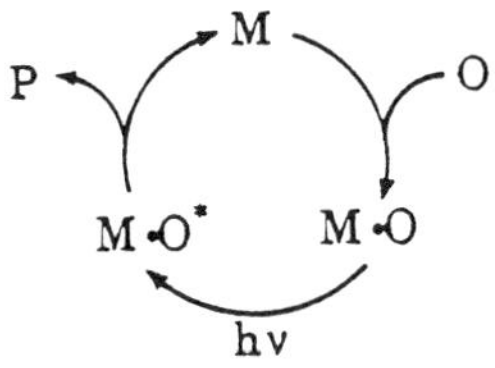

Scheme 9.

Scheme 7 describes the case in which M catalyzes the reaction of an electronically excited organic substrate via formation of an exciplex (excited complex), M•O*. Another possibility, shown in Scheme 8, involves a metal-catalyzed reaction of a primary photoproduct, R. Finally, Scheme 9 illustrates a transformation that results from irradiation of a ground-state M–O complex. Typically, complex formation shifts the absorption spectrum of the system to longer wavelengths, and may introduce a sterically or electronically favorable path to the product that is inaccessible to the uncoordinated substrate.

Excited-State Properties of Coordination Compounds and Semiconductors

As discussed, photocatalysis occurs by pathways that, at some point, involve the participation of an electronic excited state produced by the absorption of light. Consequently, it will be useful to review the properties of the various excited states that can arise in metal–organic systems. Coordination compounds and semiconductors serve as the principal light-absorbing species in the vast majority of cases studied to date, so we shall focus on their excited states.

Coordination Compounds. Each electronic state in a coordination compound can be conveniently classified in terms of its dominant molecular orbital configuration. Transitions between states are then labeled according to the orbitals that undergo a change in electron occupancy. This simple formalism can be illustrated with the aid of the qualitative orbital energy diagram in Figure 1, which shows the molecular orbi-

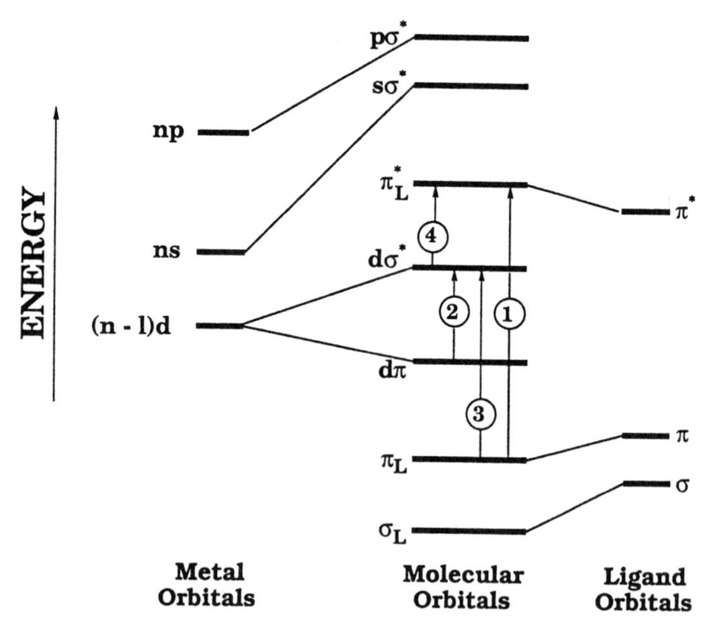

Figure 1. Qualitative energy level diagram of the molecular orbitals and electronic transitions in an octahedral coordination compound. Transition types are ①, intraligand; ②, ligand field; ③, ligand-to-metal charge transfer (LMCT); and ④, metal-to-ligand charge transfer (MLCT). For clarity, all orbitals of a given type are represented by a single energy level. Orbitals designated as dπ may be bonding (as shown here), nonbonding, or antibonding, depending upon the particular metal–ligand combination.

tals that arise from combining the valence orbitals of a transition metal with the appropriate symmetry-adapted orbitals of six ligands located at the vertices of an octahedron. Under the assumption of weak metal–ligand covalency, the molecular orbitals σ_L, π_L, and π_L* are localized predominantly on the ligands, whereas dπ, dσ*, sσ*, and pσ* are mainly metal in character (when part of an orbital label, an asterisk denotes antibonding character; the letters d, s, and p identify the specific metal orbital involved). The various electronic transitions and resulting excited states that can occur in coordination compounds will now be considered. More detailed treatments of this topic can be found in several of the references (*14–18*).

Intraligand excited states arise from electronic transitions between orbitals localized primarily on a coordinated ligand (① in Figure 1). Typically, the metal causes a relatively minor (<1000 cm^{-1}) perturbation in the transition energy and this characteristic simplifies the task of spectral assignment. It appears reasonable to expect intraligand states to undergo ligand-centered reactions, but the influence of the metal on such processes can be appreciable and can result in photochemistry different from that of the free ligand. Accordingly, no useful generalizations concerning reactivity can be made.

Ligand-field excited states result from transitions between valence d orbitals formally localized on the metal. For example, the dπ → dσ* transition (② in Figure 1) involves the promotion of an electron from a d orbital that undergoes a π-bonding interaction with the ligands to a higher lying d orbital that is strongly σ-antibonding with respect to the metal–ligand bonds. This angular redistribution of electron density does not alter the oxidation state of the metal, but it weakens the metal–ligand bonding in the complex and thereby enhances the likelihood of ligand loss. Various rearrangement processes such as geometrical isomerization and racemization also may occur.

Charge-transfer excited states arise from the radial redistribution of electron density between the components (metal and ligands) of a coordination compound or between the compound and the surrounding medium. Thus the π_L → dσ* transition (③ in Figure 1) results in the transfer of an electron from a ligand-centered orbital to a metal d orbital and generates a ligand-to-metal charge-transfer (LMCT) excited state. Electron flow in the opposite direction, as occurs in the dσ* → π_L* transition (④ in Figure 1), produces a metal-to-ligand charge-transfer (MLCT) excited state. A transition that causes the transfer of electron density from one coordinated ligand to another affords a ligand-to-ligand charge-transfer (LLCT) excited state (Figure 2) (*18*). A transition that results in the movement of electron density from the coordination compound to the surrounding solvent produces a charge-transfer-to-solvent (CTTS) excited state, and electron transfer to an ion-paired partner yields an ion-pair charge-transfer

(IPCT) excited state (*4*). All of these transitions occur with a change in the oxidation state or charge of the species involved (metal, ligands, solvent, or counterion). Consequently, charge-transfer excited states are prone to oxidation–reduction reactions and, in some cases, to accompanying ligand dissociation (i.e., charge transfer may create a substitutionally labile metal center). Moreover, changes in the electronic distribution about a coordinated ligand can enhance its reactivity toward external reagents.

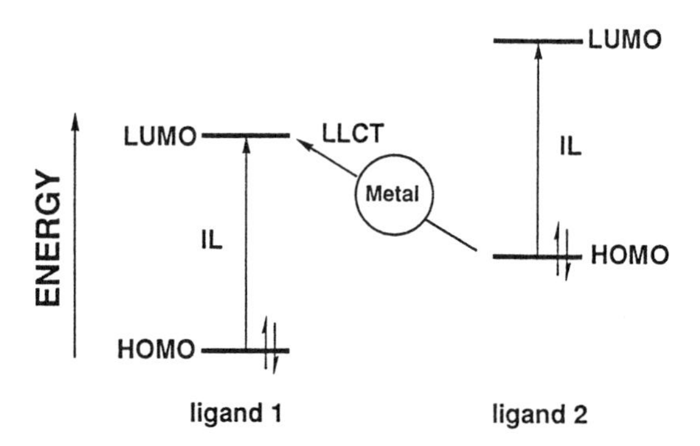

Figure 2. Orbital representation of a ligand-to-ligand charge-transfer (LLCT) transition in a coordination compound. HOMO and LUMO denote the highest occupied and lowest unoccupied molecular orbitals, respectively, and IL identifies an intraligand transition.

Sigma-a_π *excited states* arise in complexes containing a filled metal–E σ-bonding orbital (E denotes an element) and an empty, ligand-based antibonding orbital of π origin (denoted a_π). In the example depicted in Figure 3, the excitation of an electron from a metal–P σ orbital to an a_π orbital situated on a phenyl ring affords a σ–a_π state. Although the net effect of the excitation process could be viewed as a kind of metal-to-ligand charge transfer, the σ–a_π label is more explicit in identifying the key orbi-

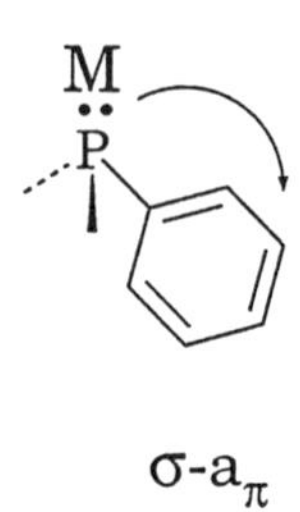

σ-a_π

Figure 3. Pictorial representation of the σ–a_π transition in metal–arylphosphine complexes.

tal components involved. This type of excited state has been observed in arylphosphine complexes of d^{10} metals (*18*) and in complexes containing a Re–Ge, Re–Sn, or Rh–Si bond and a π-acceptor ligand such as 1,10-phenanthroline (*19, 20*). Little is presently known about the photochemistry of σ–a_π excited states, although it seems reasonable to expect labilization of the metal–E bond resulting from loss of an electron from the corresponding σ orbital (*19*).

Metal–metal bonded excited states occur in multinuclear (containing two or more metal atoms) coordination compounds containing at least one direct metal–metal bond. These states arise from electronic transitions between orbitals delocalized over the metal framework. For the simplest case of a dinuclear complex, exemplified by $Mn_2(CO)_{10}$ in Figure 4 (*21*), mutual overlap of the d_{z^2} orbital on each metal yields a stable σ-bonding orbital, σ_z, occupied by two electrons, and a higher-energy empty σ-antibonding orbital, σ_z*. This model predicts that the identical halves of the complex are held together by a single metal–metal bond. The $\sigma_z \rightarrow \sigma_z^*$ transition effectively destroys this bond and thereby facilitates the formation of separated $Mn(CO)_5$ fragments. A less pronounced bond-weakening effect should result from the $d\pi \rightarrow \sigma_z^*$ transition.

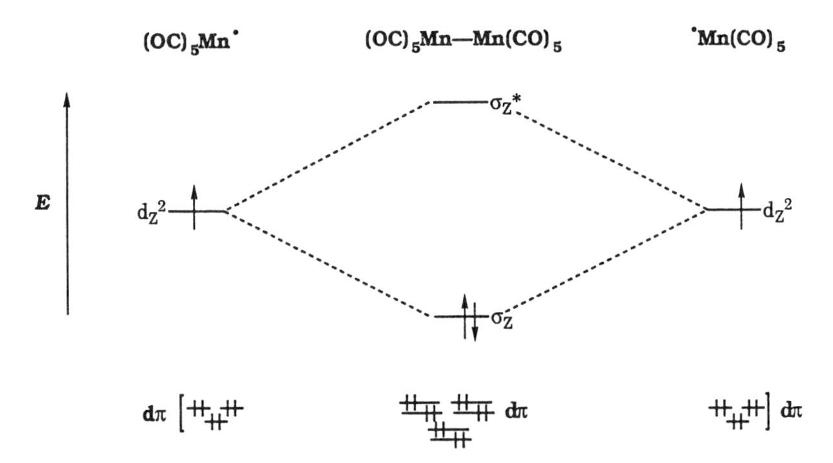

Figure 4. Simplified representation of the metal–metal bonding in $Mn_2(CO)_{10}$. (Reproduced from reference 21. Copyright 1975 American Chemical Society.)

Thus far the discussion of excited-state behavior has emphasized the intramolecular reactions of coordination compounds. Sufficiently long-lived excited states also can undergo a variety of intermolecular processes such as protonation, exciplex formation, energy transfer, and oxidation–reduction. Occurrence of intermolecular photoredox chemistry reflects the fact that the photoexcitation of an electron from an occupied molecular orbital to a higher lying empty orbital creates an excited state that is both a stronger reductant and a stronger oxidant than the ground state by an amount that corresponds to the excitation energy ΔE (neglecting structural reorganization effects) (*12, 17*). As depicted in Figure 5, the presence of an electron in the upper orbital facilitates oxidation of the coordination compound, and the vacancy in the lower orbital provides a ready pathway for reduction. This photoenhancement of the driving force for intermolecular electron transfer is a general phenomenon that occurs for all types of electronic excited states.

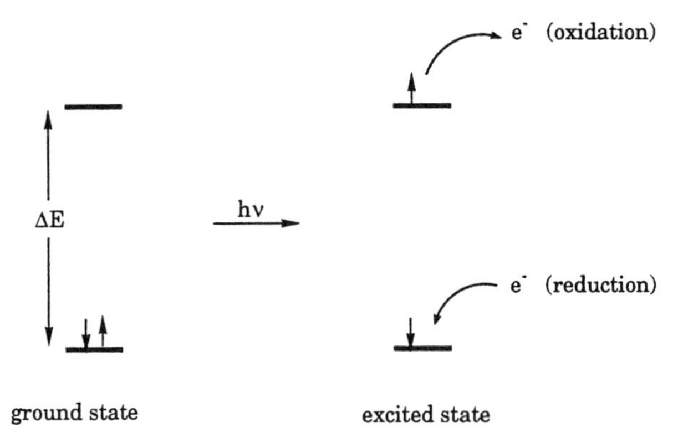

Figure 5. Alteration of the redox properties of a coordination compound upon photoexcitation. The excited state undergoes oxidation (loss of the electron from the higher lying orbital) and reduction (addition of an electron to the partially filled lower orbital) more readily than the ground state.

Semiconductors. Coordination compounds containing a small number (e.g., 20–30) of metal atoms are discrete molecular units whose ground-state and excited-state properties can be understood within the conceptual framework of molecular orbital theory. As the number of interacting metal atoms in a cluster increases, however, the properties approach those of a bulk solid, and an alternative bonding model, termed *band theory* (22), becomes more appropriate. To illustrate this model, consider a linear chain of N lithium atoms, each contributing a 2s orbital to the bonding of the assembly. As depicted in Figure 6, these N atomic orbitals combine to form N molecular orbitals that extend over the entire structure. The $N = 2$ system yields bonding and antibonding orbitals; for $N = 3$, a third orbital having nonbonding character is added. Energy differences between the molecular orbitals decrease as N increases until, at large N values, the collection of orbitals essentially forms a continuous energy *band*. Each lithium atom contributes one valence electron, and each orbital can accommodate two electrons with spins paired; therefore only one-half of this band will be occupied. More generally, a partially occupied band is a characteristic feature of the bonding in metallic lattices (Figure 7a). Electrons lying near the top of the filled portion of this band require little energy to be promoted to nearby vacant orbitals. Such electrons are mobile and move relatively freely through the solid, and this property accounts for the high electrical conductivity of metals.

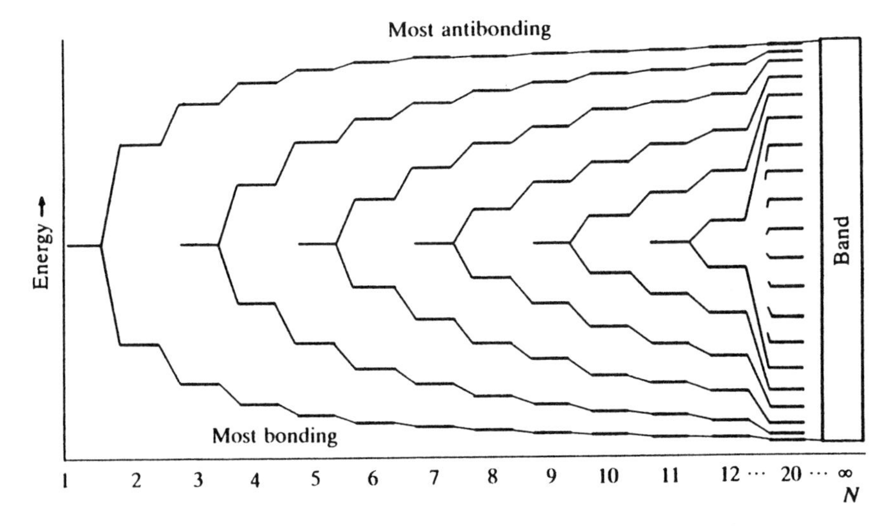

Figure 6. Molecular orbitals formed upon combining the 2s atomic orbitals of N *lithium atoms arranged in a linear array. As* N *approaches ∞, the orbitals are so closely spaced energetically that they form a continuous band. (Reproduced with permission from reference 22. Copyright 1990 W. H. Freeman and Company.)*

Semiconductors are materials that contain a fully occupied valence band separated from an empty conduction band by an energy gap (Figure 7b). In TiO_2, for example, this gap is greater than 3 eV, so that at ordinary temperatures few electrons are thermally excited to the conduction band. Consequently, the electrical conductivity of semiconductors falls well below that of metals. Illuminating the semiconductor with light of energy equal to or greater than the band gap increases conductivity because of the promotion of electrons to the conduction band and the creation of positive holes (h^+) in the valence band. More importantly for our purposes, this separation of charge can be exploited to effect useful chemistry. Figure 8 illustrates this possibility for an irradiated semiconductor particle. Migration of the hole to the surface generates an oxidizing site whose redox potential is defined by the energy of the valence band, whereas the promoted electron constitutes a reducing equivalent at the potential of the conduction band (this situation is similar to that depicted in Figure 5). Scavenging of h^+ by an appropriate electron donor D and/or of e^- by an electron acceptor A prevents unproductive e^-–h^+ recombination and, in so doing, effects redox chemistry in the surrounding medium. A recent monograph (*23*) contains detailed discussions of the various physical and chemical processes that can occur at the surface of an illuminated semiconductor.

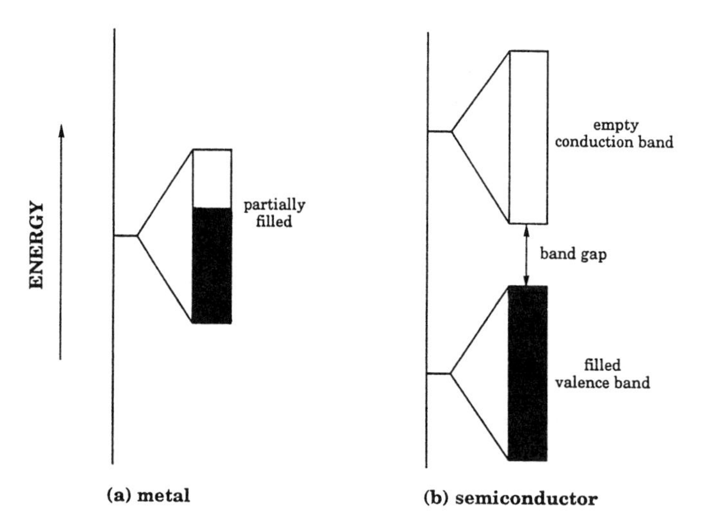

Figure 7. Distinguishing features of the band structure of (a) a metal and (b) a semiconductor. Darkened areas are occupied by electrons.

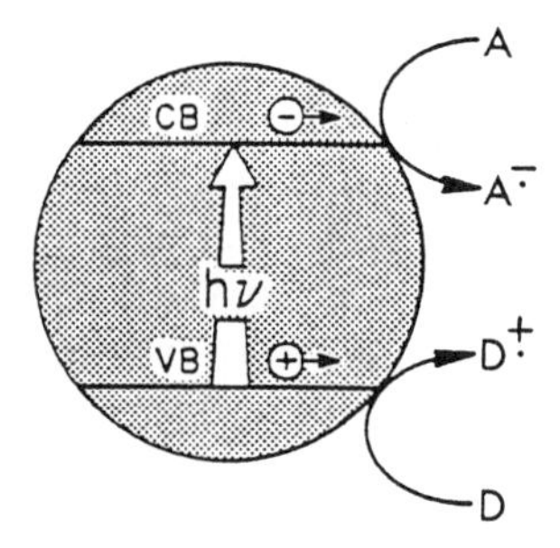

Figure 8. Simplified depiction of the photochemistry occurring at an illuminated semiconductor particle. Photoexcitation of an electron from the valence band (VB) to the conduction band (CB) can result in reduction of electron acceptor A or oxidation of electron donor D.

Examples of Photogenerated Catalysis and Photoinitiation

Photochemical generation of a ground-state catalyst (eq 3 or 4) or initiator (eq 6) from a thermally robust coordination compound can occur by a number of different routes. A very common pathway involves the photoinduced loss of one or more ligands to produce a coordinatively unsaturated metal center that serves as a template for binding the organic substrate and other potential reagents (e.g., H_2, O_2, and CO_2) present in the system. Further reaction of these bound species then occurs within the coordination sphere of the metal to yield the final product(s).

Mononuclear metal carbonyls such as $Cr(CO)_6$ and $Fe(CO)_5$ typify this behavior (*24*). Upon population of a ligand-field excited state, these compounds undergo highly efficient dissociation of CO to generate coordinatively unsaturated catalysts for a variety of reactions including geometric isomerization, double-bond migration, hydrogenation, hydrosilation, hydroformylation, and cycloaddition. For example, irradiation of $Fe(CO)_5$ in the presence of linear pentenes induces double-bond migration and *cis–trans* isomerization to yield a mixture of isomeric olefin products close to the thermodynamic ratio (*25–27*). The characteristics of this transformation are clearly diagnostic of photogenerated catalysis: occurrence of an induction period, quantum yields of substrate conversion above unity, and continuation of reaction after the cessation of photolysis. Scheme 10 presents a catalytic cycle that can accommodate this behavior (this cycle corresponds to the generic mechanism in Scheme 2). Successive photochemical loss of two CO ligands from the $Fe(CO)_5$ precursor produces the active thermal catalyst, $Fe(CO)_3$(pentene), which then undergoes a sequence of well-precedented organometallic reactions to form product. Comparable chemistry occurs when $Fe(CO)_5$ is replaced by $Fe_3(CO)_{12}$ or $Ru_3(CO)_{12}$ (*28*), a result suggesting that photodeclusterification of these trinuclear species produces catalytically active mononuclear fragments that follow a pathway similar to that outlined in Scheme 10.

Photoinduced ligand loss from a ligand-field excited state also provides a convenient route to thermal initiators. Visible-light irradiation of $FeCp(\eta^6\text{-arene})^+$ complexes (Cp is $\eta^5\text{-}C_5H_5$) has been shown (*29, 30*) to produce an initiator for the cationic polymerization of epoxides. The proposed mechanism (*31*), summarized in Scheme 11, involves photolabilization of the arene group to yield the ring-slipped intermediate, I, which then undergoes substitution of the η^4-arene by epoxide to form $FeCp(epoxide)_3{}^+$. Subsequent thermal activation of $FeCp(epoxide)_3{}^+$ causes ring opening of a coordinated epoxide to produce II, the active initiator for polymerization.

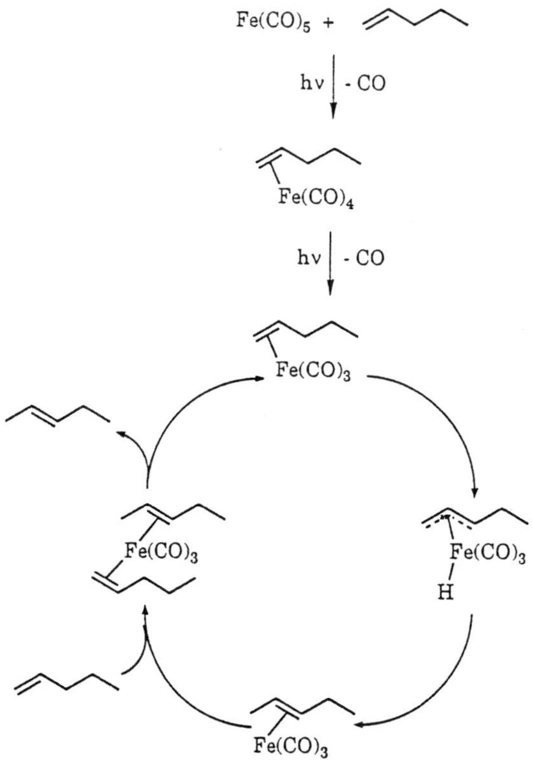

$Fe(CO)_5$ +

hν | - CO

$Fe(CO)_4$

hν | - CO

$Fe(CO)_3$

$Fe(CO)_3$

$Fe(CO)_3$

H

$Fe(CO)_3$

Scheme 10.

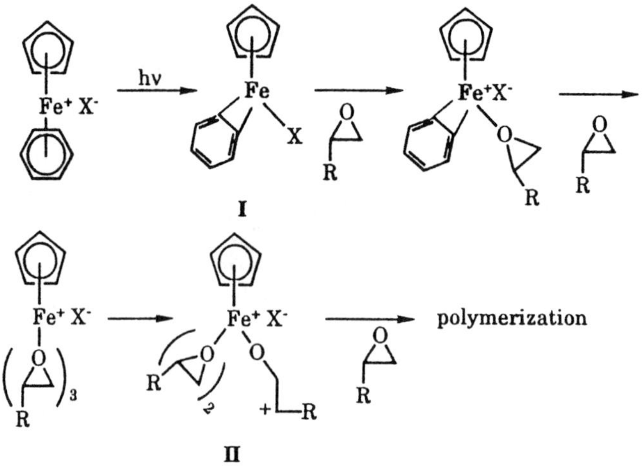

hν

$Fe^{+}X^{-}$

Fe

X

O

R

I

$Fe^{+}X^{-}$

O

R

O

R

$Fe^{+}X^{-}$

O

R

3

$Fe^{+}X^{-}$

O

O

R

2

+ R

O

R

polymerization

II

$X^{-} = PF_6^{-}; CF_3SO_3^{-}; SbF_6^{-}$

Scheme 11.

Scheme 12 depicts a novel example of photoinitiated anionic polymerization that occurs upon irradiating *trans*-$Cr(NH_3)_2(NCS)_4^-$ (the anion of Reinecke's salt, abbreviated R^-) dissolved in α-cyanoacrylate (*32*). The primary photochemical step is release of NCS^- from a ligand-field excited state of R^-. Initiation of polymerization results from the addition of NCS^- to the carbon–carbon double bond of the acrylate monomer, which contains electron-withdrawing substituents to stabilize the negative charge.

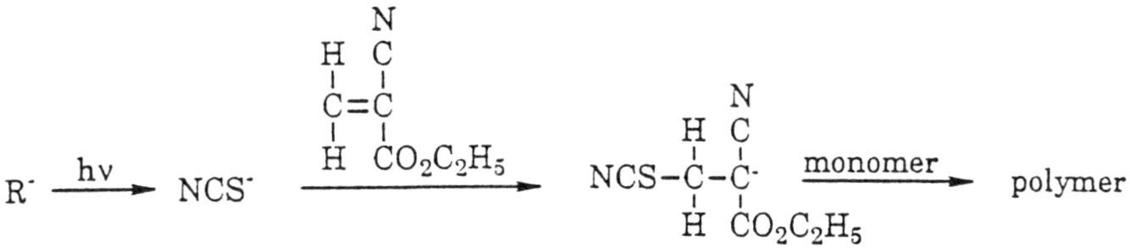

$$R^- \xrightarrow{h\nu} NCS^- \xrightarrow{HC(H)=C(CN)CO_2C_2H_5} NCS{-}CH_2{-}\dot{C}(CN)(CO_2C_2H_5) \xrightarrow{\text{monomer}} \text{polymer}$$

Scheme 12.

Photolysis of dinuclear metal carbonyl complexes such as $Mn_2(CO)_{10}$ and $Re_2(CO)_{10}$ occurs via two competing pathways (*33, 34*). The first is loss of a CO ligand without disruption of the metal–metal bond (eq 10a) and yields a coordinatively unsaturated metal center that, in principle, can function as a catalyst by pathways available to mononuclear carbonyls, as already discussed. The other primary process, homolytic cleavage of the metal–metal bond (eq 10b), originates from a σ_z–σ_z* or $d\pi$–σ_z* excited state (Figure 4) and produces two 17-electron metal-centered radicals.

$$Mn_2(CO)_{10} \xrightarrow{h\nu} Mn_2(CO)_9 + CO \quad (10a)$$

$$Mn_2(CO)_{10} \xrightarrow{h\nu} 2{}^{\bullet}Mn(CO)_5 \xrightarrow[CCl_4]{} 2Mn(CO)_5Cl + 2{}^{\bullet}CCl_3 \quad (10b)$$

These reactive species either recombine or, in the presence of a halogenated compound such as CCl_4, undergo atom abstraction to form carbon-centered radicals that can initiate vinyl polymerization. Other initiation strategies based upon the photohomolysis of metal–metal bonds have been described (*10*).

Generation of a thermally active catalyst or initiator also can result from photoinduced electron-transfer reactions. Trisoxalato complexes of cobalt(III) and iron(III), for example, undergo efficient intramolecular redox chemistry from LMCT excited states to yield the corresponding divalent metal complex and an oxalate radical anion (eq 11).

$$Fe(C_2O_4)_3^{3-} \xrightarrow{h\nu} Fe(C_2O_4)_2^{2-} + {}^{\bullet}C_2O_4^{-} \quad (\text{or } {}^{\bullet}CO_2^{-} + CO_2) \tag{11}$$

The oxalate radical anion species, or its ${}^{\bullet}CO_2^{-}$ radical anion daughter, can function as an initiator for radical polymerization (*35, 36*). Redox decomposition of acidopentaam(m)inecobalt(III) complexes produces divalent cobalt, multiequivalents of a Lewis base, and a radical (eq 12) (*37, 38*).

$$Co(NH_3)_5Br^{2+} \xrightarrow{h\nu} Co^{2+} + 5NH_3 + Br^{\bullet} \tag{12}$$

Any one or combination of these species could catalyze or initiate useful chemistry in a suitably designed system. Thus Co^{2+} catalyzes the oxidative decolorization of the red dye alizarin S by hydrogen peroxide in aqueous solution (*39*), and photoreleased base initiates cross-linking in thin films of an epoxide-containing photoresist (*40–42*).

Irradiation of the square-planar complex $PdCl_2(\eta^4\text{-NBD})$ (*see* structure) in the presence of quadricyclene, Q, causes valence isomerization of this highly strained molecule to norbornadiene, NBD, with quantum yields that can exceed 10^2 (eq 13) (*43, 44*).

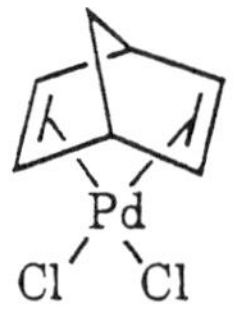

Pd
Cl Cl

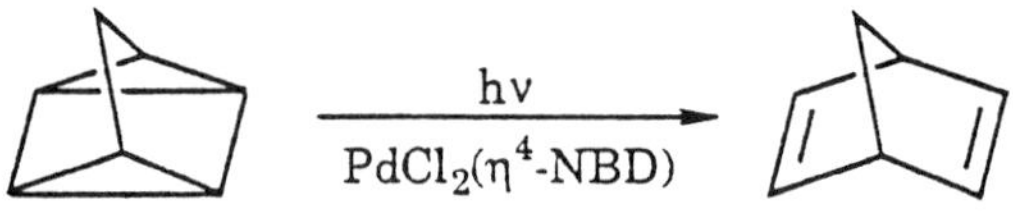

$$Q \xrightarrow[PdCl_2(\eta^4\text{-NBD})]{h\nu} NBD \tag{13}$$

This behavior results from an intermolecular redox process in which Q reductively quenches a MLCT excited state of the palladium complex. As summarized in Scheme 13, the resulting quadricyclene radical cation rearranges to the more stable isomer, $NBD^{\bullet +}$, which then oxidizes another Q molecule to restart the cycle.

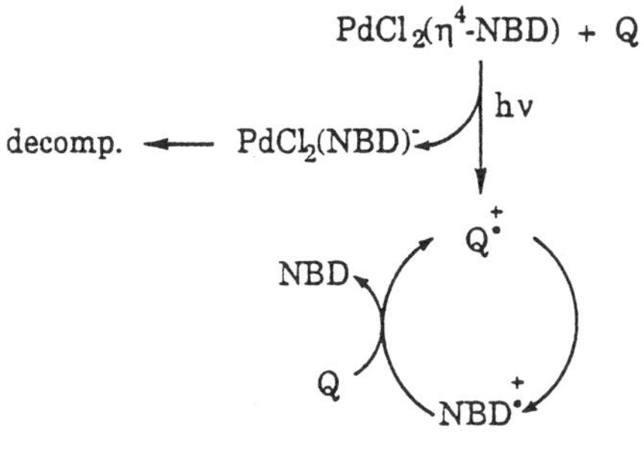

$PdCl_2(\eta^4\text{-NBD}) + Q$

decomp. ← $PdCl_2(NBD)^{\cdot-}$ ← hν

$Q^{\cdot+}$ NBD Q $NBD^{\cdot+}$

Scheme 13.

Examples of Catalyzed Photolysis

Absorption of light by M (eq 7) produces an electronically excited species that, if sufficiently long-lived, can photosensitize an organic substrate via energy-transfer or electron-transfer pathways (Scheme 5) (*12, 13*). The isomerization of *cis*- and *trans*-piperylene (eq 14), for example, can be photosensitized with high quantum efficiency by $Cu(diphos)BH_4$ (diphos is 1,2-bis(diphenylphosphino)ethane) and $Cu(prophos)BH_4$ (prophos is 1,3-bis(diphenylphosphino)propane) (*45*).

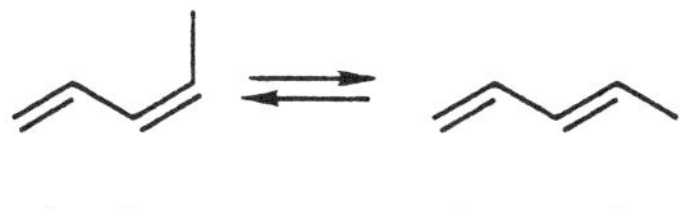

cis-pip ⇌ *trans*-pip (14)

Sensitization is accompanied by quenching of the emissive $^3(\sigma\text{–}a_\pi)$ excited state in each copper compound (the left superscript designates spin multiplicity), and the identical kinetics of the two bimolecular processes suggest that both originate from this state. As illustrated in Scheme 14, energy transfer from $^3(\sigma\text{–}a_\pi)$ to the lowest $^3(\pi\text{–}\pi^*)$ state of the diene has been assigned as the mechanism of sensitization.

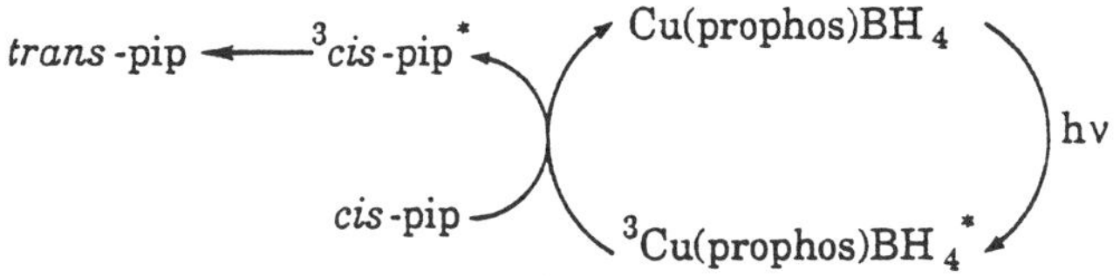

trans-pip ← ^{3}*cis*-pip* ; *cis*-pip ; $Cu(prophos)BH_4$ —hν→ $^3Cu(prophos)BH_4{}^*$

Scheme 14.

The other common sensitization mechanism involves electron transfer between photoexcited M and ground-state substrate. In the example outlined in Scheme 15 (*46, 47*), visible-light irradiation of the photosensitizer $Ru(bipy)_3^{2+}$ (bipy is 2,2′-bipyridine) populates a MLCT excited state, which then undergoes reductive quenching by 1-benzyl-1,4-dihydronicotinamide, BNAH. One of the primary photoproducts, $BNAH^{\bullet +}$, loses a proton to yield the $BNA^{\bullet}$ radical. Oxidation of the other photoproduct, $Ru(bipy)_3^{+}$, by an olefin such as dimethyl fumarate or dimethyl maleate regenerates the photosensitizer with concomitant production of the olefin radical anion, which protonates to form a radical. Reduction of this radical by $BNA^{\bullet}$, followed by protonation, affords the fully reduced product. Overall, this complicated sequence of events corresponds to the $Ru(bipy)_3^{2+}$-photosensitized two-electron reduction of an olefin by BNAH.

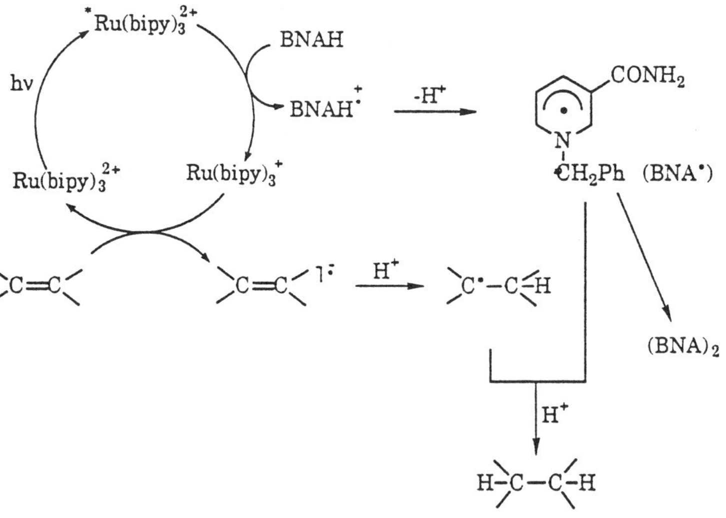

Scheme 15.

Scheme 16 depicts the proposed mechanism for the oxidation of methanol to formaldehyde and hydrogen peroxide in the presence of O_2, light, and a tetraphenylporphyrin(oxomethoxy)molybdenum(V) complex (*48*). Irradiation into the intraligand Soret absorption band of the complex causes homolytic cleavage of the Mo–OCH_3 bond. In subsequent thermal reactions, the methoxy radical affords formaldehyde, and the reduced metalloporphyrin undergoes oxidation by O_2 to produce H_2O_2 and the original complex. The entire cycle can be repeated upon absorption of another photon, and this repetition accounts for the production of up to 56 mol of H_2O_2 per mol of complex. The active (reduced) form of the complex must be regenerated in each cycle; thus Scheme 16 constitutes an example of photoassistance (cf. Scheme 6).

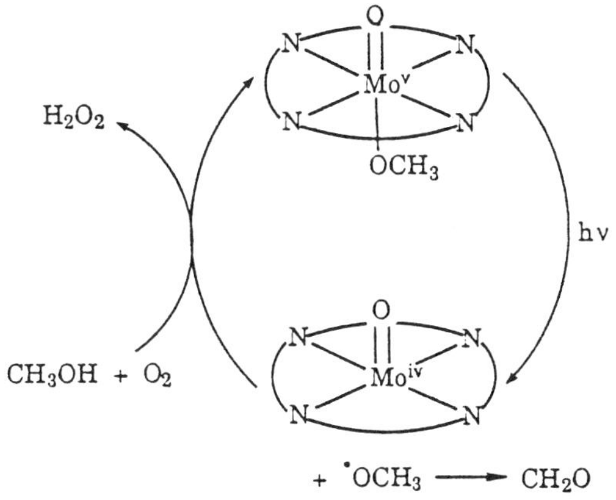

H_2O_2 ... $CH_3OH + O_2$ ... Mo^{V}–OCH_3 ... hν ... Mo^{iv} ... $+ \,^{\bullet}OCH_3 \longrightarrow CH_2O$

Scheme 16.

Catalyzed photolysis occurs via equation 8 only in relatively few cases because in most metal–organic systems examined to date, competitive absorption by M or an M–O complex effectively precludes direct photoexcitation of the uncomplexed organic substrate. An exception is the Cu^{2+}-catalyzed photooxidation of α,β-unsaturated ketones to dimeric lactones (eq 15; Ar is phenyl, *p*-tolyl, or *p*-bromophenyl) (*49*).

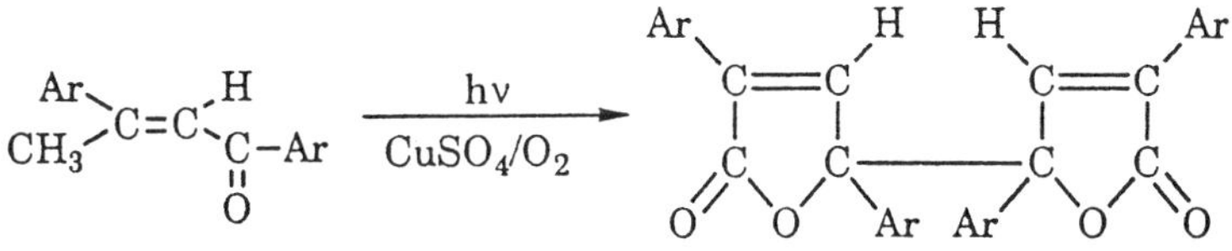

$$\text{Ar(CH}_3\text{)C=CH–C(=O)–Ar} \xrightarrow[CuSO_4/O_2]{h\nu} \text{dimeric lactone} \quad (15)$$

The key role played by the metal ion in this transformation is underscored by the observation that irradiating the ketone in the absence of Cu^{2+} leads to completely different chemistry, that of *cis–trans* isomerization about the carbon–carbon double bond. The lack of new spectral features upon mixing Cu^{2+} and the ketone argues against ground-state complex formation between these components; in fact, absorption by added Cu^{2+} acts as an inner filter that retards the photooxidation reaction. Collectively, these observations support a mechanism in which the metal ion and O_2 interact with an excited state (e.g., Scheme 7) or a primary photoproduct (e.g., Scheme 8) of the organic substrate.

Transformations that result from the irradiation of a pre-formed M–O complex (eq 9) are quite common (*5*). This complex is a distinct chemical species whose excited-state characteristics differ from those of its progenitors. Figure 9 dramatically illustrates this point for the 1:1 complex formed between CuCl and NBD (*50, 51*). The intense, long-wavelength absorption band that appears upon mixing the components arises from a transition to a low-energy Cu-to-NBD charge-transfer excited state. Irradiation at wavelengths that selectively populate this MLCT state predisposes the coordinated diene to rearrange to Q (Scheme 17). Q exhibits little affinity for CuCl, so the photoactive complex can be regenerated and the cycle repeated.

Concluding Remarks

Classifying the transformations that occur in photosensitive metal–organic systems as either photogenerated catalysis or catalyzed photolysis conveys useful information about the mechanisms involved. These two classes can be distinguished on the basis of experimental information such as reaction quantum yields, observation of an induction period or post-irradiation chemistry, luminescence quenching kinetics, and the detection of reaction intermediates. In some cases, however, insufficient or ambiguous data preclude a confident assignment of mechanism. Kisch (*8*) noted that this problem arises frequently in heterogeneous systems, where turnover number is defined as the moles of product formed per mole of active sites on the illuminated catalyst surface. The latter quantity often is unknown or only roughly estimated; therefore, a distinction between photogenerated catalysis and catalyzed photolysis becomes difficult if the observed quantum efficiency is below unity. A case in point is the oxidatively induced valence isomerization of quadricyclene to norbornadiene that occurs in the presence of an illuminated CdS semiconductor powder (*52*). Because the number of active surface sites was not determined, the low quantum efficiency for product formation ($\sim 10^{-2}$) does not, by itself, rule out the operation of a photogenerated catalytic cycle involving $Q^{\cdot +}$ and $NBD^{\cdot +}$ (refer to Scheme 13). Consequently, in this and other systems where mechanistic uncertainties exist, it is appropriate to apply the general label photocatalysis to the transformation in question.

Irradiation of an organic substrate in the presence of a transition metal compound or semiconductor has proven to be a convenient, mild, and often highly selective route to a wide variety of products. In this overview I have attempted to construct a mechanistic framework within which such transformations can be understood. Succeeding chapters in this volume will explore more fully the scope of this interesting chemistry and its potential applications.

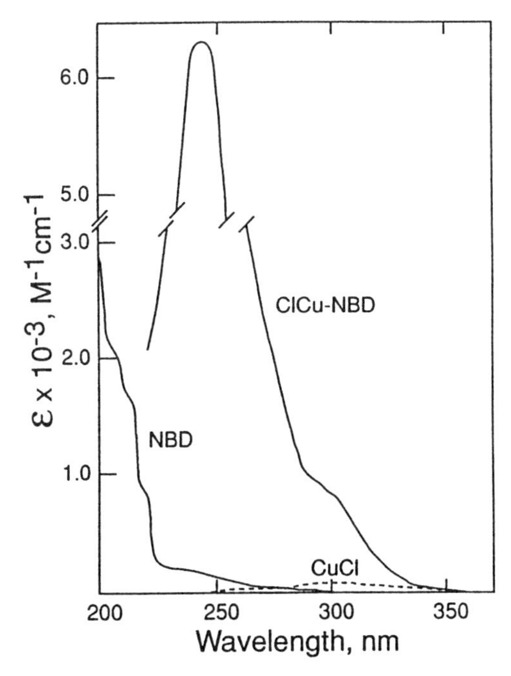

Figure 9. Spectral manifestation of complex formation between CuCl and NBD in ethanol. (Reproduced from reference 50. Copyright 1977 American Chemical Society.)

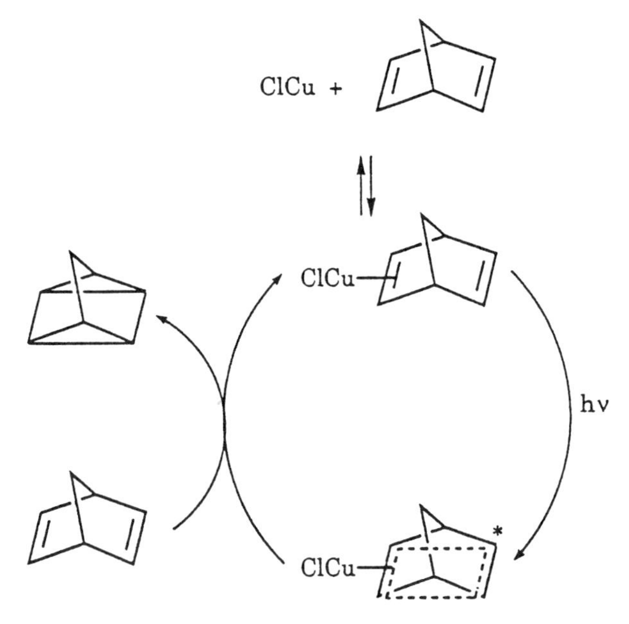

Scheme 17.

Acknowledgment

I thank the National Science Foundation (Grant No. DMR–8715635) for financial support of my recent work in the area of photosensitive metal–organic systems.

References

1. *J. Chem. Educ.* **1986,** *63,* 188–225.
2. Salomon, R. G. *Tetrahedron* **1983,** *39,* 485.
3. Wubbels, G. G. *Acc. Chem. Res.* **1983,** *16,* 285.
4. Hennig, H.; Rehorek, D.; Archer, R. D. *Coord. Chem. Rev.* **1985,** *61,* 1.
5. Kutal, C. *Coord. Chem. Rev.* **1985,** *64,* 191.
6. Albini, A. *J. Chem. Educ.* **1986,** *63,* 383.
7. Chanon, M.; Eberson, L. In *Photoinduced Electron Transfer;* Fox, M. A.; Chanon, M., Eds.; Elsevier: Amsterdam, 1988; Vol. A, Chapter 1.11.
8. Kisch, H. In *Photocatalysis;* Serpone, N.; Pelizzetti, E., Eds.; Wiley: New York, 1989; Chapter 1.
9. Chanon, F.; Chanon, M. In *Photocatalysis;* Serpone, N.; Pelizzetti, E., Eds.; Wiley: New York, 1989; Chapter 15.
10. Yang, D. B.; Kutal, C. In *Radiation Curing: Science and Technology;* Pappas, S. P., Ed.; Plenum: New York, 1992; Chapter 2.
11. Reichmanis, E.; Houlihan, F. M.; Nalamasu, O.; Neenan, T. X. *Chem. Mater.* **1991,** *3,* 394.
12. Balzani, V.; Bolletta, F.; Ciano, M.; Maestri, M. *J. Chem. Educ.* **1983,** *60,* 447.
13. Scandola, F.; Balzani, V. *J. Chem. Educ.* **1983,** *60,* 814.
14. Balzani, V.; Carassiti, V. *Photochemistry of Coordination Compounds;* Academic Press: New York, 1970; Chapter 5.
15. Forster, L. S. In *Concepts of Inorganic Photochemistry;* Adamson, A. W.; Fleischauer, P., Eds.; Wiley-Interscience: New York, 1975; Chapter 1.
16. Crosby, G. A. *J. Chem. Educ.* **1983,** *60,* 791.
17. Scandola, F.; Balzani, V. In *Photocatalysis;* Serpone, N.; Pelizzetti, E., Eds.; Wiley: New York, 1989; Chapter 2.
18. Kutal, C. *Coord. Chem. Rev.* **1990,** *99,* 213.
19. Luong, J. C.; Faltynek, R. A.; Wrighton, M. S. *J. Am. Chem. Soc.* **1980,** *102,* 7892.
20. Djurovich, P. I.; Safir, A.; Keder, N.; Watts, R. J. *Coord. Chem. Rev.* **1991,** *111,* 201.
21. Levenson, R. A.; Gray, H. B. *J. Am. Chem. Soc.* **1975,** *97,* 6042.
22. Shriver, D. F.; Atkins, P. W.; Langford, C. H. *Inorganic Chemistry;* W. H. Freeman: New York, 1990; Chapter 3.
23. *Photocatalysis;* Serpone, N.; Pelizzetti, E., Eds.; Wiley: New York, 1989.
24. Geoffroy, G. L.; Wrighton, M. S. *Organometallic Photochemistry;* Academic Press: New York, 1979; Chapter 2.

25. Schroeder, M. A.; Wrighton, M. S. *J. Am. Chem. Soc.* **1976,** *98,* 551.
26. Whetten, R. L.; Fu, K.-J.; Grant, E. R. *J. Am. Chem. Soc.* **1982,** *104,* 4270.
27. Miller, M. E.; Grant, E. R. *J. Am. Chem. Soc.* **1987,** *109,* 7951.
28. Austin, R. G.; Paonessa, R. S.; Giordano, P. J.; Wrighton, M. S. In *Inorganic and Organometallic Photochemistry;* Wrighton, M. S., Ed.; ACS Advances in Chemistry Series No. 168; American Chemical Society: Washington, DC, 1978; Chapter 12.
29. Roloff, A.; Meier, K.; Riediker, M. *Pure Appl. Chem.* **1986,** *58,* 1267.
30. Meier, K.; Zweifel, H. *J. Imaging Sci.* **1986,** *30,* 174.
31. Park, K. M.; Schuster, G. B. *J. Organomet. Chem.* **1991,** *402,* 355.
32. Kutal, C.; Grutsch, P. A.; Yang, D. B. *Macromolecules* **1991,** *24,* 6872.
33. Yasufuku, K.; Noda, H.; Iwai, J.-I.; Ohtani, H.; Hoshino, M.; Kobayashi, T. *Organometallics* **1985,** *4,* 2174.
34. Kobayashi, T.; Ohtani, H.; Noda, H.; Teratani, S.; Yamazaki, H.; Yasufuku, K. *Organometallics* **1986,** *5,* 110.
35. Sahul, K.; Natarajan, L. V.; Anwaruddin, Q. *J. Polym. Sci., Part B, Polym. Lett. Ed.* **1977,** *15,* 605.
36. Mahaboob, S.; Natarajan, L. V.; Anwaruddin, Q. *J. Macromol. Sci., Chem.* **1978,** *12,* 971.
37. Endicott, J. F. In *Concepts of Inorganic Photochemistry;* Adamson, A. W.; Fleischauer, P., Eds.; Wiley-Interscience: New York, 1975; Chapter 3.
38. Weit, S. K.; Kutal, C. *Inorg. Chem.* **1990,** *29,* 1455.
39. Varfolomeev, S. D.; Zaitsev, S. V.; Vasil'eva, T. E.; Berezin, I. V. *Dokl. Akad. Nauk SSSR* **1974,** *219,* 895.
40. Kutal, C.; Willson, C. G. *J. Electrochem. Soc.* **1987,** *134,* 2280.
41. Kutal, C.; Weit, S. K.; MacDonald, S. A.; Willson, C. G. *J. Coat. Technol.* **1990,** *62,* 63.
42. Weit, S. K.; Kutal, C.; Allen, R. D. *Chem. Mater.* **1992,** *4,* 453.
43. Borsub, N.; Kutal, C. *J. Am. Chem. Soc.* **1984,** *106,* 4826.
44. Kelley, C. K.; Kutal, C. *Organometallics* **1985,** *4,* 1351.
45. Liaw, B.; Orchard, S. W.; Kutal, C. *Inorg. Chem.* **1988,** *27,* 1311.
46. Pac, C.; Ihama, M.; Yasuda, M.; Miyauchi, Y.; Sakurai, H. *J. Am. Chem. Soc.* **1981,** *103,* 6495.
47. Pac, C.; Miyauchi, Y.; Ishitani, O.; Ihama, M.; Yasuda, M.; Sakurai, H. *J. Org. Chem.* **1984,** *49,* 26.
48. Ledon, H. J.; Bonnet, M. *J. Am. Chem. Soc.* **1981,** *103,* 6209.
49. Sato, T.; Tamura, K.; Maruyama, K.; Ogawa, O. *Tetrahedron Lett.* **1973,** *43,* 4221.
50. Schwendiman, D. P.; Kutal, C. *Inorg. Chem.* **1977,** *16,* 719.
51. Schwendiman, D. P.; Kutal, C. *J. Am. Chem. Soc.* **1977,** *99,* 5677.
52. Ikezawa, H.; Kutal, C. *J. Org. Chem.* **1987,** *52,* 3299.

RECEIVED for review February 10, 1992. ACCEPTED revised manuscript May 15, 1992.

2

Reactive Intermediates in the Carbonylation of Metal–Alkyl Bonds

Time-Resolved Infrared Spectral Techniques

Peter C. Ford, David W. Ryba, and Simon T. Belt

Department of Chemistry, University of California, Santa Barbara, CA 93106

In this chapter, we describe flash photolysis experiments using time-resolved infrared (TRIR) detection techniques to probe the reactivities of key intermediates proposed for the mechanisms of the thermal "migratory insertion" of CO into a metal–alkyl bond. The intermediates studied were generated by the photodissociation of CO from the metal complexes $CpFe(CO)L(COCH_3)$ *(Cp is* η^5-C_5H_5*, L is CO or phosphine) and* $Mn(CO)_5(COCH_3)$*. Kinetic and spectroscopic evidence points to the formation of a solvated species in each case, which undergoes migration of the acyl methyl group to the metal center, competitive with trapping by addition of a ligand. In cyclohexane, the CpFe intermediate* $CpFe(CO)(sol)(COCH_3)$ *(sol is solvent) undergoes methyl migration at a rate* ($k_1 = 5.6 \times 10^4\ M^{-1}\ s^{-1}$) *several orders of magnitude faster than does the remarkably slow manganese analog* ($k_1 = 6.0\ M^{-1}\ s^{-1}$)*. Comparisons are also made to the reactivity of the unsaturated intermediate* $Mn(CO)_4(sol)(CH_3)$ *formed by CO photodissociation from* $Mn(CO)_5(CH_3)$*.*

A THOROUGH UNDERSTANDING OF THE MECHANISMS of the photoreactions of organometallic compounds requires far more than the measurements of quantum yields and the determinations of photoproduct

0065–2393/93/0238–0027$06.00/0

identities and distributions. Most photochemical products are the result of a sequence of photophysical and photochemical events. For example, photoexcitation into some initial excited state (ES) is often followed by vibrational relaxation, internal conversion, or intersystem crossing to lower energy excited state(s) from which the actual chemical events leading to products are initiated. Reactive deactivation of such an ES may lead directly to the eventual photoproducts; commonly, however, the first chemical species formed from the reaction of the ES is itself a high-energy, reactive species, which undergoes further rapid thermal reactions to give the eventual photoproducts. Such species are generally formed only in (very) low concentrations under continuous photolysis. Thus direct evidence for their presence often requires flash photolysis techniques for which a broad range of time regimes down to femtoseconds have become accessible.

Flash photolysis studies of organometallic species in this laboratory were stimulated in part by our interest in fundamental photoreaction mechanisms. However, a second consideration was the opportunity to interrogate the quantitative reactivities of organometallic intermediates often proposed in homogeneous catalytic schemes for the activation of various organic and other small-molecule substrates (e.g., refs. 1–4). Thus, flash photolysis techniques allow one to generate nonequilibrium concentrations of organometallic intermediates from stable precursors that can be interrogated kinetically and spectroscopically. This reaction is illustrated qualitatively by Figure 1. Such intermediates may be coordinatively unsaturated products from ligand dissociation or reductive elimination, oxidized or reduced complexes resulting from ES electron transfer, radical products of homolytic bond cleavage, or high-energy isomers.

With organometallic compounds a major problem is that the UV–visible spectra of reactants and products generally are poorly defined and often provide little information addressing structural properties of key transient species. In this context, studies in this laboratory followed the pioneering work by other researchers (*5–12*) in developing methodologies to employ probe and detection systems with the flash photolysis excitation designed to obtain time-resolved infrared (TRIR) spectra. Under favorable circumstances the TRIR spectral characterizations of reactive intermediates are aided by comparison with results from low-temperature matrix experiments, in which normally highly reactive transient species may be trapped indefinitely and studied by using a full range of spectroscopic methods (*13*). In this chapter are described several flash photolysis studies of the photodecarbonylation of metal–acetyl complexes studied by both IR and UV–visible spectroscopic techniques. Also reviewed briefly is the apparatus used to obtain TRIR spectral data for the reactive organometallic intermediates.

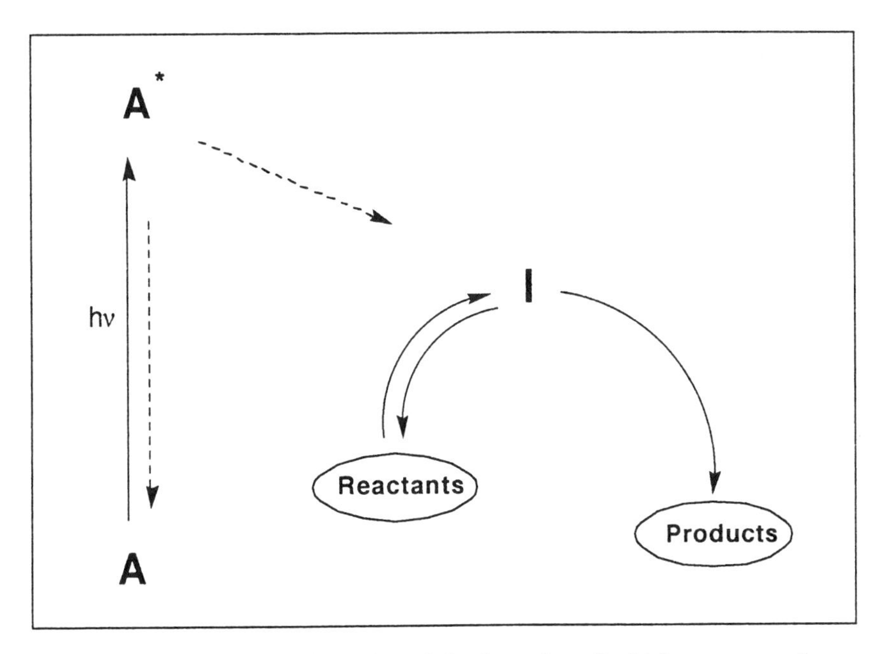

Figure 1. Qualitative illustration of the formation of a high-energy reactive intermediate I by the flash photolysis of a stable precursor A.

The carbonylation of metal–alkyl bonds is a key step in homogeneously catalyzed activation of carbon monoxide in processes such as alkene hydroformylations, alcohol carbonylations, carboxylic acid homologations, and reductive polymerization of CO (*14*). One fundamental organometallic reaction commonly proposed in schemes for such catalytic cycles is the "migratory insertion" of CO into a metal–alkyl bond (*15*), e.g.,

$$L_nM(R)-CO + L' \longrightarrow L_nM(L')-C(=O)R \qquad (1)$$

Migratory insertion mechanisms have been extensively investigated for the model compounds $Mn(CO)_5(CH_3)$ and $CpFe(CO)L(CH_3)$ (Cp is η^5-C_5H_5, L is CO or phosphine) (*16–24*). These studies have concluded that the rate-limiting initial step involves a methyl migration to CO, that is, equation 2, rather than CO insertion into the metal–alkyl bond, and have noted that donor solvents have marked effects on the reaction dynamics (*21, 22*).

$$L_nM(R)-CO \rightleftharpoons \underset{(\mathrm{I})}{L_nM-C(=O)R} \xrightarrow{+L'} L_n(L')M-C(=O)R \qquad (2)$$

However, evidence regarding the nature and kinetic behavior of the key intermediate I has remained indirect.

The reverse process, namely the decarbonylation of metal–acetyls, can often be effected photochemically *(24)*, the likely mechanism being CO photodissociation to give the intermediate I, which undergoes subsequent methyl migration to the metal center. The carbon monoxide lost is a terminally bound CO rather than the carbonyl from the acetyl ligand *(25)*. This chapter will present an overview of ongoing studies using a photochemical strategy to probe the reactivities of intermediates such as I in the decarbonylations of $CpFe(CO)_2(COCH_3)$ (*see* eq 3) and $Mn(CO)_5$-$(COCH_3)$ (*see* eq 4).

$$CpFe(CO)_2(C(=O)CH_3) \xrightarrow{h\nu} CpFe(CO)_2(CH_3) + CO \qquad (3)$$

$$Mn(CO)_5(C(=O)CH_3) \xrightarrow{h\nu,\ -CO} Mn(CO)_5(CH_3) \qquad (4)$$

These experiments provide both spectroscopic and kinetic parameters relevant to the identity and reactivity of key intermediates in the thermal migratory insertion mechanisms of the respective metal–alkyl complex. Also described are investigations of the reactivities of the "unsaturated" intermediate formed by photolysis of the analogous methyl complex $Mn(CO)_5(CH_3)$.

The TRIR Spectroscopy Method: Flash Photolysis of $Ru_3(CO)_{12}$

The TRIR flash photolysis apparatus, illustrated in Figure 2, was described in detail in reference 12. Key features of this system are the use of a Laser Analytics lead salt diode IR laser as the probe source and a SBRC photovoltaic Hg–Cd–Te fast rise-time detector. These features allow the manual tuning of the observation frequency with high resolution over a wide range, which depends on the diode lasers installed in the cryogenically cooled laser head. (In the current configuration this range is 1550 to 2200 cm^{-1}.) The excitation source is a Lambda Physik XeCl excimer laser (308 nm) or excimer laser–dye laser combination. With this apparatus, an IR spectrum for a transient with lifetime as short as 100 ns can be recorded. Sample solutions prepared under the desired gas mixtures were passed through the photolysis cells using a simple flow

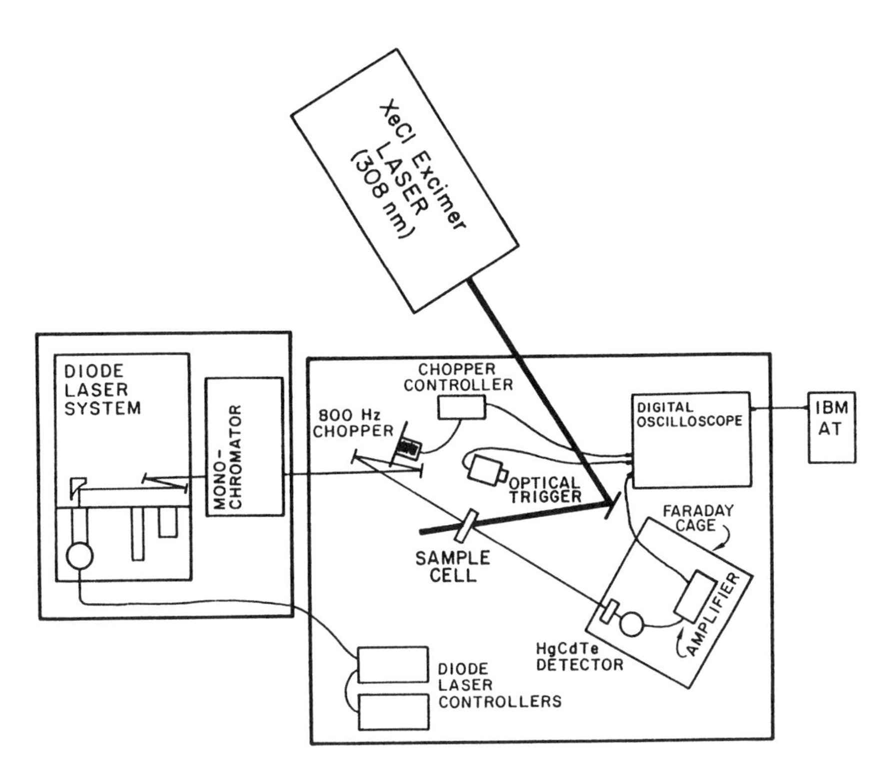

Figure 2. Diagram of TRIR apparatus. (Reproduced from reference 12. Copyright 1989 American Chemical Society.)

apparatus that allowed for multiple-pulse data collection and averaging experiments. The photolysis cell consisted of a modified McCarthy IR cell with CaF_2 windows and a Teflon spacer (0.5 or 1.0 mm). Stainless steel cannula were silver-soldered to the brass cell body for transfer of solutions into and from the reservoirs of the flow apparatus.

These techniques were employed (*12*) to investigate the TRIR spectra and reaction dynamics of the coordinatively unsaturated triruthenium cluster $Ru_3(CO)_{11}$, the type of intermediate proposed for photoassisted hydrogenation of alkenes by metal carbonyl clusters (*4*). A key observation (*12*) in this case is that short-wavelength photoexcitation of $Ru_3(CO)_{12}$ in isooctane solution leads to formation of a species with the same IR spectrum (Figure 3) as that seen in studies carried out in low-temperature hydrocarbon glasses (*26*). The first spectrum in Figure 3 was recorded 200 ns after the flash. In ambient-temperature solution this reactive intermediate is trapped competitively by CO or by a donor ligand such as tetrahydrofuran (THF) at rates approaching diffusion limits (1.3×10^{10} M^{-1} s^{-1}) in the isooctane medium (*12*). These rates are somewhat surprising given the observation that certain mononuclear coordinatively unsaturated species such as $Cr(CO)_5$ form much stronger adducts with

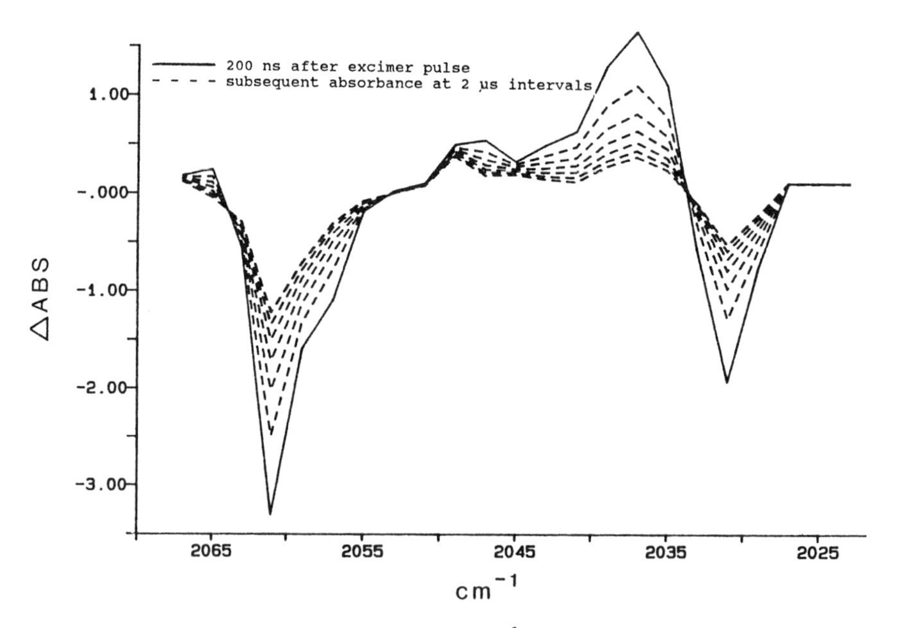

Figure 3. IR difference spectrum (2-cm^{-1} resolution) of the transients formed by 308-nm flash photolysis of $Ru_3(CO)_{12}$ in isooctane solution at ambient temperature 200 ns after flash. Subsequent curves are recorded at regular intervals of 2.0 µs. (Reproduced from reference 12. Copyright 1989 American Chemical Society.)

hydrocarbon solvents, hence are considerably less reactive (*27–33*). The adduct $Ru_3(CO)_{11}$•THF itself is quite labile and reacts with CO by an apparent dissociative mechanism to re-form the starting cluster. These observations are summarized in Scheme I.

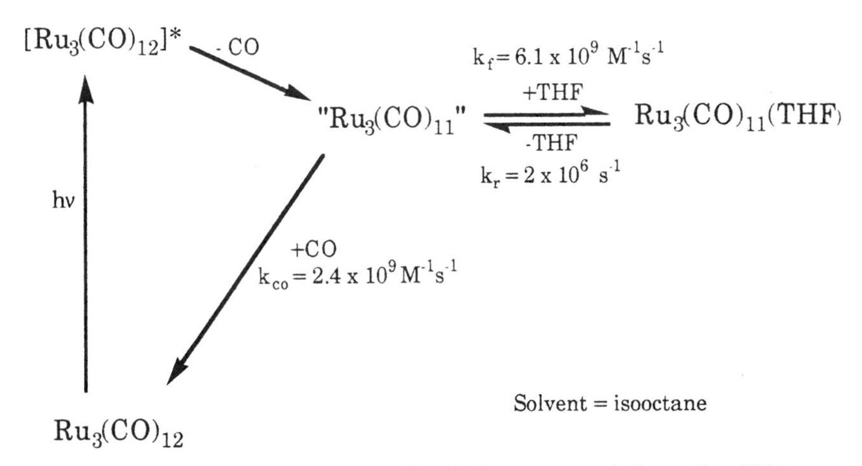

Scheme I. Reactions of the coordinatively unsaturated cluster $Ru_3(CO)_{11}$.

Pentacarbonyl(methyl)manganese

Flash photolysis studies of $Mn(CO)_5(CH_3)$ were initiated with the goal of providing a model for the pertinent spectroscopic and kinetic data relevant to unsaturated Mn(I) acyl intermediates (discussed later). Laser flash photolysis (λ_{ir} = 308 nm) of $Mn(CO)_5(CH_3)$ in cyclohexane or isooctane solution resulted in a 100-μs TRIR spectrum in which the depletion of starting material is evident with the negative absorbance changes (Δabs) corresponding to the ν_{CO} modes at 2014 and 1991 cm^{-1} (ν is frequency) and the formation of a transient species (J) is evidenced by new absorptions at ν_{CO} 1986, 1974, and 1940 cm^{-1} (*34*). These TRIR properties and a transient λ_{max} at 410 nm in the UV–visible spectrum are close to those attributed to *cis*-$Mn(CO)_4(CH_3)$•CH_4 formed by CO photodissociation from $Mn(CO)_5(CH_3)$ in a methane matrix (eq 5) (*35*).

CH_3, OC, CO, Mn, OC, CO, CO — hν, - CO → CH_3, OC, Sol, Mn, OC, CO, CO (J) (5)

We did not observe transients resulting from either *trans*-CO labilization or homolytic metal–alkyl bond cleavage, although prolonged irradiation does lead to the appearance of visible and IR absorbances indicating the production of $Mn_2(CO)_{10}$.

Under argon, the decay of **J** follows second-order kinetics, but under CO, both the rates of decay of **J** and the re-formation of $Mn(CO)_5(CH_3)$ (eq 6) are accelerated and follow pseudo-first-order kinetics (Figure 4).

$$\textit{cis}\text{-}Mn(CO)_4(CH_3)(Sol) \xrightarrow{+\ CO,\ k_2} Mn(CO)_5(CH_3) \qquad (6)$$

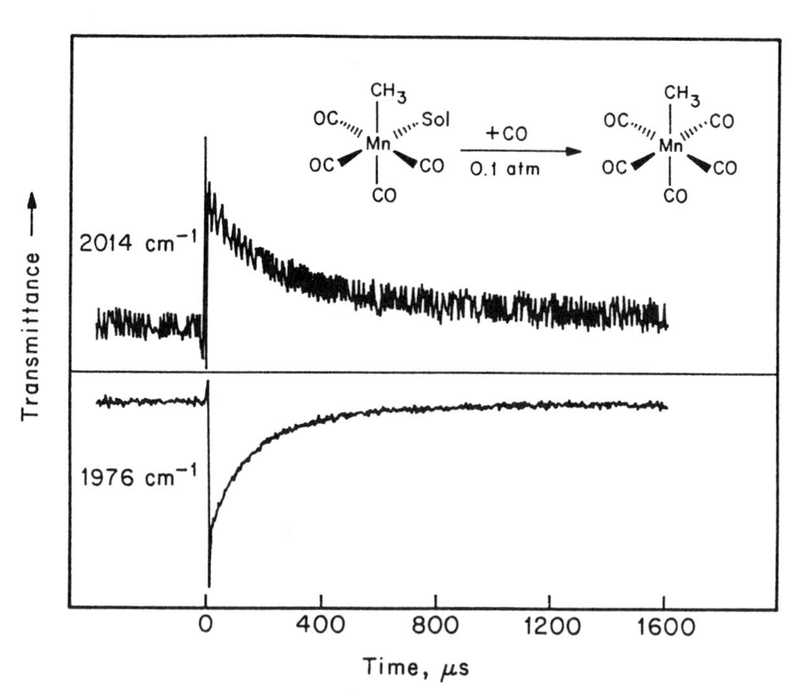

Figure 4. Kinetic traces showing the decay of cis-*$Mn(CO)_4(CH_3)$sol (sol is cyclohexane) at 1976 cm^{-1} and the re-formation of $Mn(CO)_5(CH_3)$ at 2014 cm^{-1} following laser flash photolysis of $Mn(CO)_5(CH_3)$ in cyclohexane under 0.1 atm CO. IR changes are shown in transmittance mode. (Reproduced from reference 34. Copyright 1990 American Chemical Society.)*

The second-order rate constant for the reaction with CO ($2.1 \pm 0.1 \times 10^6$ M^{-1} s^{-1}) showed excellent agreement between the IR and UV–visible detection methods and lies in the same range as that found for other weakly bound solvento–carbonylmetal intermediates, such as $Cr(CO)_5sol$ (sol is solvent), measured by flash photolysis techniques (*27*). In THF solution, the reaction of **J** with CO (as studied by UV–visible detection), is 4 orders of magnitude slower ($k_2 = 1.4 \times 10^2$ M^{-1} s^{-1}), consistent with the increased donor strength of THF. Preliminary flash photolysis studies with the analogous metal–alkyl complexes $Mn(CO)_5(CF_3)$ and $Re(CO)_5(CH_3)$ in cyclohexane demonstrated that the back reaction of the respective intermediates $Mn(CO)_4(CF_3)sol$ and $Re(CO)_4(CH_3)sol$ (sol is cyclohexane) with CO displayed second-order rate constants ($k_2 = 1.4 \times 10^7$ and 1.3×10^6 M^{-1} s^{-1}, respectively, at 298 K) (*36*), similar to that observed for **J** in cyclohexane.

Pentacarbonyl(acetyl)manganese

Laser flash photolysis (λ_{ir} = 308 nm) of the acyl complex $Mn(CO)_5(CH_3CO)$ in cyclohexane causes CO photodissociation to give a transient **X** that displays the TRIR difference absorption spectrum at 100 μs displayed in Figure 5 (*36*). The spectrum of the new transient species

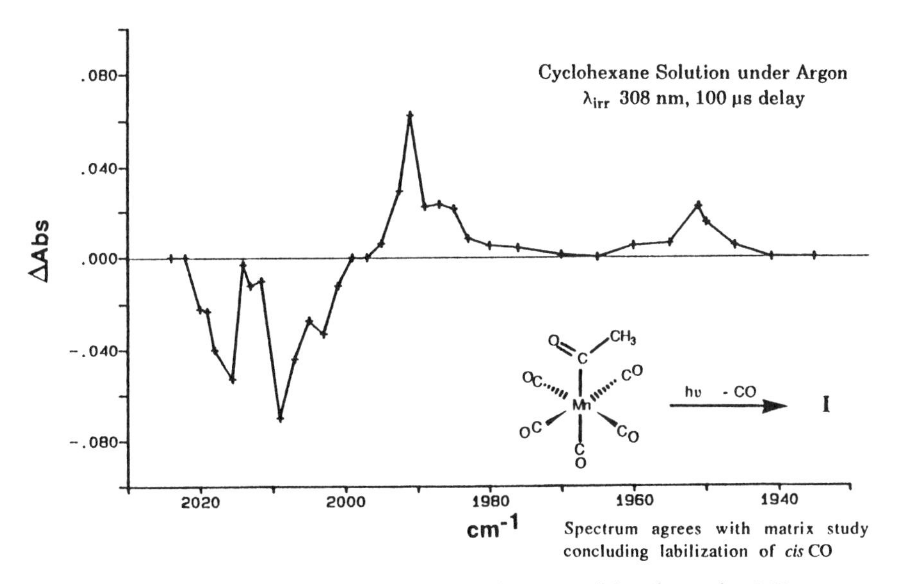

Figure 5. TRIR difference spectrum (100 μs) resulting from the 308-nm flash photolysis of $Mn(CO)_5(CH_3CO)$ in cyclohexane solution.

is quite close to that seen for the product of CO photodissociation from $Mn(CO)_5(CH_3CO)$ in a methane matrix (*25*). In ambient cyclohexane, **X** decays by rearrangement to the alkyl complex competitive with trapping by CO to regenerate the starting complex (eq 7), the rates of both processes being solvent-dependent.

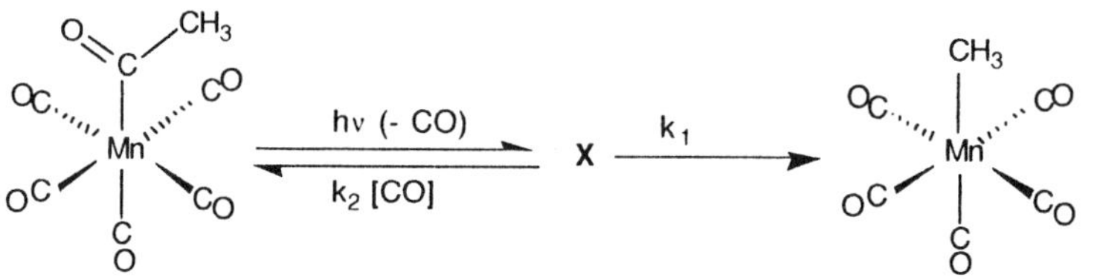

$$Mn(CO)_5(C(O)CH_3) \underset{k_2\,[CO]}{\overset{h\nu\,(-\,CO)}{\rightleftharpoons}} X \xrightarrow{k_1} Mn(CO)_5(CH_3) \quad (7)$$

Kinetics studies (*36*) using UV–visible detection have shown the decay to follow the rate law

$$-\frac{d[X]}{dt} = k_{obs}[X] = (k_1 + k_2[CO])[X]$$

and have determined values for k_1 and k_2 in cyclohexane as 6.0 s^{-1} and 5.6 $\times 10^3$ M^{-1} s^{-1}, respectively, and in THF solution 7×10^{-2} s^{-1} and 3×10^2 M^{-1} s^{-1}, respectively. The remarkable feature of these results is the relative unreactivity of **X** in cyclohexane. For example, the back reaction of **X** with CO (eq 7) is nearly 3 orders of magnitude slower than that for **J** (eq 6) described for cyclohexane solutions. In contrast the values of k_2 in THF solution for **X** and **J** are comparable.

Structures **U**, **C**, **S**, and **B** can be proposed for the intermediate **X** (terminally bound COs are not shown).

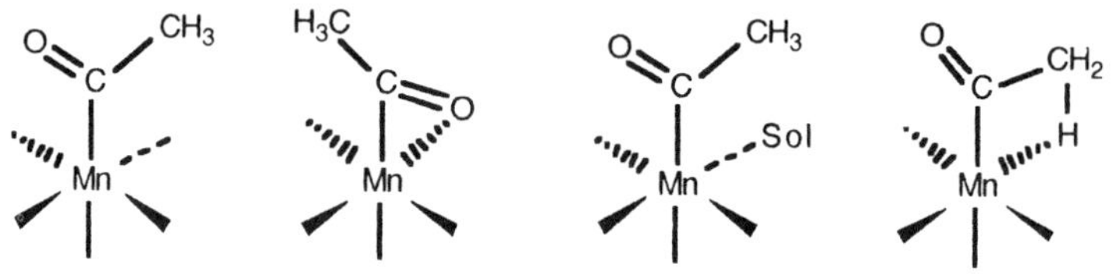

U C S B

Of these the unsaturated intermediate **U** seems the least likely, given the known tendency of coordinatively unsaturated metal carbonyls to bind alkane solvents with stabilities ~10 kcal/mol (*37*). On the other hand, the chelated intermediate **C**, with an η^2-carbonyl of the acetyl group, has been predicted by theoretical calculations (*38, 39*) to be the most stable structure for $Mn(CO)_4(CH_3CO)$ in the absence of solvent interactions. An alternative bidentate acetyl structure would be **B** with an agostic interaction between a methyl C–H and the metal. This alternative is especially attractive as a likely precursor for methyl migration to the metal. However, preliminary TRIR experiments (*40*) show the terminal ν_{CO} frequencies to differ for **X** in cyclohexane (1951 cm^{-1}) from those for **X** in THF (1931 cm^{-1}), and one can conclude that the structure of this intermediate is different in the two solvents. The simplest explanation of these differences is that **X** has the solvated structure **S**, the variation of ν_{CO} being attributable in part to the different donor strengths of cyclohexane and THF (*40*). Alternatively, in THF **X** may be present as the solvated species **S**, while in cyclohexane it may be present in either the **C** or **B** bidentate acetyl configuration. Such a possibility may explain the relative passivity of **X** in its reactions with CO in cyclohexane. These questions are the focus of ongoing investigations of the photodecarbonylation reactions of $Mn(CO)_5(CH_3CO)$ and related manganese acyls.

Photodecarbonylation of $CpFe(CO)_2(COCH_3)$

Laser flash photolysis (308 nm) of $CpFe(CO)_2(COCH_3)$ in cyclohexane (10^{-3} mol/L) under Ar was shown (*41*) to lead to the rapid and permanent depletion of the parent compound as monitored by TRIR detection of the terminal CO stretching bands at 2021 and 1965 cm^{-1} and the acetyl ν_{CO} at 1669 cm^{-1}. Within 100 μs a new species was observed to grow in; maximum absorbance changes of this new species appeared at 2012 and 1959 cm^{-1}, indicative of the formation of the final product, $CpFe(CO)_2CH_3$ (Figure 6a). On a shorter time scale (1 μs), the TRIR difference spectrum showed only the transient metal carbonyl absorbance at 1949 cm^{-1} (Figure 6b) and none of the absorptions attributable to the final product. The position of the 1949-cm^{-1} band is in close agreement with that found for the purported intermediate $CpFe(CO)(COCH_3)$ (ν_{CO} = 1948 cm^{-1}) produced by photolysis of $CpFe(CO)_2(COCH_3)$ in a methane matrix at 12 K (*42, 43*).

At 1949 cm^{-1} the decay of this transient (**M**) followed first-order kinetics (k_{obs} = 5.7×10^4 s^{-1}) identical to those for formation of $CpFe(CO)_2CH_3$ as measured at either 2010 or 1958 cm^{-1} (*41*). The rate of decay proved to be independent of CO pressure (P_{CO}), in agreement with the behavior of overall quantum yields for the photodecarbonylation ($\Phi = 0.64 \pm 0.02$ mol/einstein, independent of P_{CO} to 1 atm and of solvent). These results imply that **M** is not measurably trapped by added CO in competition with methyl migration to the metal center under these conditions.

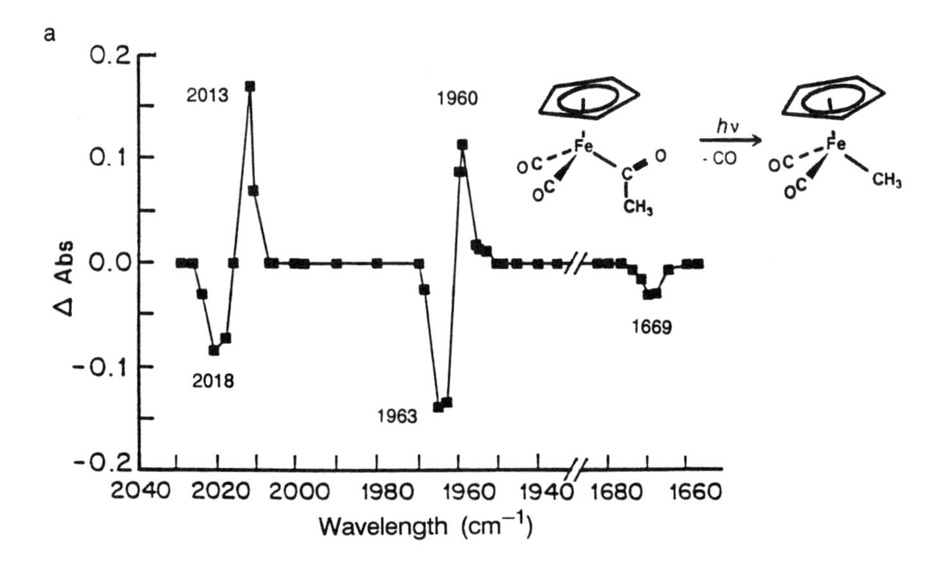

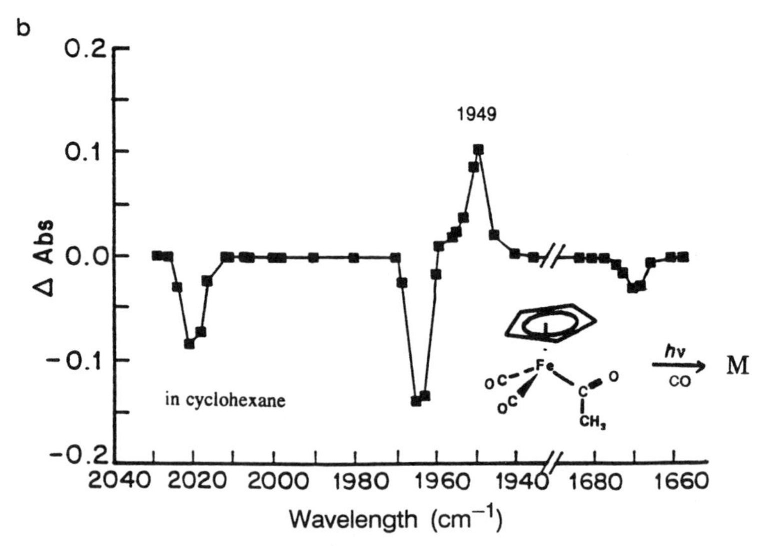

Figure 6. Part a: Transient difference spectrum of $CpFe(CO)_2(COCH_3)$ in cyclohexane taken 100 μs after a 308-nm laser pulse. Part b: Transient difference spectrum of $CpFe(CO)_2(COCH_3)$ in cyclohexane taken 1 μs after a 308-nm laser pulse.

In contrast, when photolyses of $CpFe(CO)_2(COCH_3)$ were carried out in the presence of added PPh_3, the phosphine complex $CpFe(CO)PPh_3(COCH_3)$ (ν_{CO} = 1924 cm^{-1}) was also formed as one product of the photoreaction, in agreement with a previous, qualitative, report (*44*). The decay of **M** was accelerated under these conditions, and a plot of k_{obs} versus $[PPh_3]$ proved to be linear (slope = 2.4×10^6 L/mol·s) with a nonzero intercept (5.6×10^4 s^{-1}) in agreement with the first-order rate constant determined under a CO atmosphere. These observations can be summarized in terms of the reactions displayed in Scheme II.

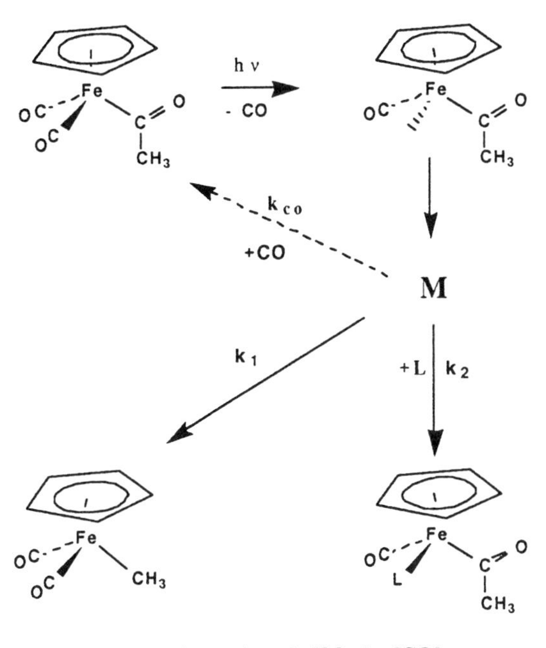

$$k_{obs} = k_1 + k_2[L] + k_{co}[CO]$$

Scheme II. Reactions of transient(s) formed by the laser flash photolysis of $CpFe(CO)_2(COCH_3)$.

Photolysis of $CpFe(CO)_2(COCH_3)$ leads to CO photodissociation and the formation of an intermediate, **M**, that undergoes first-order methyl migration to give $CpFe(CO)_2(CH_3)$ in competition with second-order trapping by PPh_3 or (in principle) by CO. The failure to observe kinetic effects of CO on k_{obs} implies that k_{CO} would have an upper limit of $\sim 6 \times 10^5$ M^{-1} s^{-1} given a CO concentration of ~0.01 M in such solutions.

Laser flash photolysis of $CpFe(CO)_2(COCH_3)$ in *n*-heptane or isooctane under Ar or CO resulted in the formation of transient species whose TRIR spectra and rates of reaction are close to those found for cyclohexane solution. However, in THF, the single ν_{CO} for **M** was observed at 1921 cm^{-1}, and the first-order decay was somewhat slower ($k_1 = 1.3 \times 10^4$ s^{-1}). Once again, the decay of this intermediate was accompanied by the formation of $CpFe(CO)_2CH_3$, as shown by ν_{CO} maxima at 2010 and 1952 cm^{-1}. When CH_3CN was used as solvent, the intermediate that was formed has a broad ν_{CO} mode centered at 1926 cm^{-1} and a lifetime longer than 1 ms. A difference Fourier transform IR spectrum recorded several minutes after irradiation revealed that the intermediate had indeed isomerized to $CpFe(CO)_2(CH_3)$, as observed in all other solvents.

In none of these solvents were we able to observe for the intermediate **M** the acyl ν_{CO} mode, which occurs at 1669 cm^{-1} in the IR spectrum of $CpFe(CO)_2(COCH_3)$. Two possible explanations come to mind: a shift of this band in **M** to frequencies below the 1550-cm^{-1} limit of the current instrumentation, or a decrease in extinction coefficient for this weak band to a value in **M** too low to be detected in the present experiment. The latter appears to be a likely explanation given that the intensities of the acyl bands in the monocarbonyl complexes $CpFe(CO)(PPh_3)$-$(COCH_3)$ and $CpFe(CO)(Xe)(COCH_3)$ are much smaller than in the dicarbonyl analog. The TRIR experiment was unable to detect the expected acyl carbonyl frequency of **M** in ambient temperature solution over the range 1680–1550 cm^{-1} despite clear observation of the terminal ν_{CO} mode. Therefore, preliminary experiments were carried out (*45*) to examine the IR spectra of intermediates in the analogous flash photolysis of $CpFe(CO)_2(COCH_3)$ in liquid Xe (193 K). A weak acyl band at 1582 cm^{-1} was detected using FTIR methods (the terminal carbonyl ν_{CO} was detected at 1938 cm^{-1}). Spectral and rate data obtained in different solvents are summarized in Table I.

Again four possibilities analogous to the structures **U**, **C**, **S**, and **B** described for the manganese complexes can be proposed for the nature of the intermediate **M**. Clearly, the data in Table I point to the nearly identical spectral and kinetic properties of this intermediate in the three alkane solvents. The effect on the kinetics of using THF as solvent is to decrease k_1 by only a factor of ~4, although the effect of using CH_3CN is considerably more dramatic, as might be expected if **M** were to have the solvento complex structure analogous to **S**. Notably, solvent has a significant effect on the frequency of the lone terminal ν_{CO} of **M** (Table I). For **S**, the frequency of the metal carbonyl ν_{CO} band would be expected to shift to lower values as the donor strength of the solvent is increased, a prediction certainly consistent with the observation that upon changing the solvent from hydrocarbons to THF, the position of the ν_{CO} mode for **M** shifted 26

Table I. IR Spectral Data for the Starting Material and for Transients Formed, Quantum Yields for Product Formation, and Rate Constants for the Decay of Transients from the Photolysis of $CpFe(CO)_2(COCH_3)$ in Various Solvents

Solvent	η_{CO} *(parent)* (cm^{-1})	η_{CO} *(transient)* (cm^{-1})	Φ_{313}[a]	k_1[b] (s^{-1})
Cyclohexane	2018, 1963, 1669	1949	0.64 ± 0.02	5.7×10^4
Isooctane	2018, 1962, 1670	1949	0.64 ± 0.02	4.0×10^4
Hexane	2020, 1965, 1670	1949	0.64 ± 0.02	4.0×10^4
THF	2015, 1955, 1658	1921	0.62 ± 0.04	1.3×10^4
CH_3CN	2018, 1958, 1650	1926	0.62×0.04	<1.0
Xe (l)[b]	2025, 2019, 1965, 1673	1938, 1582		

[a]Quantum yield for continuous wave photolysis at 313 nm. Experimental uncertainties are calculated for five or more duplicate runs.

[b]First-order rate constant for disappearance of the intermediate **M** at ambient temperature. Experimental uncertainties are estimated conservatively at $<\pm 10\%$ based upon five or more duplicate runs.

cm^{-1} to lower frequency (from 1949 to 1923 cm^{-1}). Shifts from the alkane solution values were also seen for CH_3CN solution and even in liquid xenon solution (*45*).

Thus, if a single structure is appropriate to describe **M**, a solvated species such as **S** with the η^1-acyl configuration is more consistent with the observations. One of the chelated configurations similar to **C** or **B** may certainly be present in hydrocarbon solutions, and **S** is the dominant species in more strongly donating solvents; however, the observation that **M** has a significantly lower frequency (ν_{CO}) in liquid xenon than in hydrocarbons (Table I) may argue for the presence of an S-type configuration even in this medium.

Concluding Remarks

In summary, the ongoing laser flash photolysis experiments described here have allowed the detection and characterization of the reaction kinetics of the short-lived (<100 μs in hydrocarbon solutions at ambient temperature) transient species relevant to the mechanism of the migratory insertion reaction. Comparisons of reaction rates and of the TRIR spectra for these reactive intermediates clearly point to the formation of a solvent coordinated species analogous to **S** when the solvent is a relatively good donor such as THF or acetonitrile. Greater ambiguity exists for those experiments carried out in more weakly donating alkane solvents in which the occupation of the empty coordination site by the oxygen of the acyl group to give an η^2-carbonyl functionality (e.g., **C**) may be competitive with alkane coordination. The flash photolysis dynamics of other carbonylmanganese and CpFe acyl complexes are being scrutinized with the goal of providing additional insight into these matters.

Acknowledgments

This research was sponsored by a grant (DE–FG03–85ER13317) from the Division of Chemical Sciences, Office of Basic Energy Sciences, U.S. Department of Energy. The TRIR instrumentation used was constructed from components purchased with funds from the National Science Foundation (Grants CHE–87–22561 and CHE–84–113020), the University of California—Santa Barbara (UCSB) Faculty Research Committee and the UCSB Quantum Institute and from components donated by the Newport Corporation and the Amoco Technology Company. John A. DiBenedetto is the individual primarily responsible for the design and construction of the TRIR apparatus. S. T. Belt acknowledges support from a NATO Fellowship awarded through the SERC (Science and Education Research Council, United Kingdom).

References

1. Wink, D. A.; Ford, P. C. *J. Am. Chem. Soc.* **1987,** *109,* 436–442.
2. Desrosiers, M. F.; Wink, D. A.; Trautman, R.; Friedman, A. E.; Ford, P. C. *J. Am. Chem. Soc.* **1986,** *108,* 1917–1927.
3. Ford, P. C.; Netzel, T. L.; Spillett, C. T.; Pourreau, D. B. *Pure Appl. Chem.* **1990,** *62,* 1091–1094.
4. Ford, P. C.; Friedman, A. In *Photocatalysis: Fundamentals and Applications;* Serpone, N., Ed.; John Wiley and Sons: New York, 1989; pp 541–564, Chapter 16.
5. Moore, B. D.; Simpson, M. B.; Poliakoff, M.; Turner, J. J. *J. Chem. Soc., Chem. Commun.* **1984,** 972.
6. Church, S. P.; Herman, H.; Grevels, F.-W.; Schaffner, K. *J. Chem. Soc., Chem. Commun.* **1984,** 785–786.
7. Church, S. P.; Herman, H.; Grevels, F.-W.; Schaffner, K. *Inorg. Chem.* **1985,** *24,* 418–422.
8. Poliakoff, M.; Weitz, E. *Adv. Organomet. Chem.* **1986,** *25,* 277–316.
9. Dixon, A. J.; Healy, M. A.; Hodges, P. M.; Moore, B. D.; Poliakoff, M.; Simpson, M. B.; Turner, J. J.; West, M. A. *J. Chem. Soc., Faraday Trans. 2* **1986,** *82,* 2083–2092.
10. Dobson, G. R.; Hodges, P. M.; Healy, M. A.; Poliakoff, M.; Turner, J. J.; Firth, S.; Asali, K. J. *J. Am Chem. Soc.* **1987,** *109,* 4218–4224.
11. Weiller, B. H.; Wasserman, E. P.; Bergman, R. G.; Moore, C. B.; Pimentel, G. C. *J. Am Chem. Soc.* **1989,** *111,* 8288–8289.

12. DiBenedetto, J. A.; Ryba, D. W.; Ford, P. C. *Inorg. Chem.* **1989,** *28,* 3503–3507.
13. Perutz, R. N. *R. Soc. Chem., Annu. Rep., Sect. C,* **1985,** 157.
14. Henrici-Olivé, G.; Olivé, S. *Catalyzed Hydrogenation of Carbon Monoxide;* Springer-Verlag: Berlin, 1984.
15. Collman, J. P.; Hegedus, L. S.; Norton, J. R.; Finke, R. G. *Principles and Applications of Organotransition Metal Chemistry;* University Science Books: Menlo Park, CA, 1987; Chapter 6.
16. Wojcicki, A. *Adv. Organomet. Chem.* **1973,** *11,* 87–145.
17. Calderazzo, F. *Angew. Chem., Int. Ed. Engl.* **1977,** *16,* 299–311.
18. Flood, T. C. *Top. Stereochem.* **1981,** *12,* 37–118.
19. Flood, T. C.; Jensen, J. E.; Statler, J. A. *J. Am. Chem. Soc.* **1981,** *103,* 4410.
20. Butler, I.; Basolo, F.; Pearson, R. G. *Inorg. Chem.* **1967,** *6,* 2074.
21. Cawse, J. N.; Fiato, R. A.; Pruett, R. *J. Organomet. Chem.* **1979,** *172,* 405.
22. Wax, M. J.; Bergman, R. G. *J. Am. Chem. Soc.* **1981,** *103,* 7028–7030.
23. Webb, S.; Giandomenico, C.; Halpern, J. *J. Am. Chem. Soc.* **1986,** *108,* 345.
24. King, R. B.; Bisnette, M. B. *J. Organomet. Chem.* **1964,** *2,* 15–37.
25. McHugh, T. M.; Rest, A. J. *J. Chem. Soc., Dalton Trans.* **1980,** 2323–2332.
26. Bentsen, J. G.; Wrighton, M. S. *J. Am. Chem. Soc.* **1987,** *109,* 4530.
27. Kelly, J. M.; Bonneau, R. *J. Am. Chem. Soc.* **1980,** *102,* 1220–1221.
28. Rothberg, L. J.; Cooper, N. J.; Peters, K. S.; Vaida, V. *J. Am. Chem. Soc.* **1982,** *104,* 3536–3537.
29. Simon, J. D.; Xie, X. *J. Phys. Chem.* **1986,** *90,* 6751–6753.
30. Joly, A. G.; Nelson, K. A. *J. Phys. Chem.* **1989,** *93,* 2876–2878.
31. Wang, L.; Zhu, X.; Spears, K. G. *J. Phys. Chem.* **1989,** *93,* 2–4.
32. Lee, M.; Harris, C. B. *J. Am. Chem. Soc.* **1989,** *111,* 8963–8965.
33. Yu, S.-C.; Xu, X.; Lingle, R.; Hopkins, J. B. *J. Am. Chem. Soc.* **1990,** *112,* 3668–3669.
34. Belt, S. T.; Ryba D. W.; Ford, P. C. *Inorg. Chem.* **1990,** *29,* 3633–3634.
35. Horton-Mastin, A.; Poliakoff, M.; Turner, J. J. *Organometallics* **1986,** *5,* 405–408.
36. Ryba D. W., Ph.D. dissertation, University of California, Santa Barbara, CA, 1991.
37. Morse, J.; Parker, G.; Burkey, T. J. *Organometallics,* **1989,** *7,* 2471.
38. Ziegler, T.; Versluis, L.; Tschinke, V. *J. Am. Chem. Soc.* **1986,** *108,* 612.
39. Axe, F. U.; Marynick, D. S. *J. Am. Chem. Soc.* **1988,** *110,* 3728–3734.
40. Gutmann, V. *Coord. Chem. Rev.* **1976,** *18,* 225–255.
41. Belt, S. T.; Ryba D. W.; Ford, P. C. *J. Am. Chem. Soc.* **1991,** *113,* 9524–9528.
42. Fettes, D. J.; Narayanaswamy, R.; Rest, A. J. *J. Chem. Soc., Dalton Trans.* **1981,** 2311–2316.
43. Hitam, R. B.; Narayanaswamy, R.; Rest, A. J. *J. Chem. Soc., Dalton Trans.* **1983,** 615–619.
44. Alexander, J. J. *J. Am. Chem. Soc.* **1975,** *97,* 1729–1732.
45. The experiments herein reported were carried out by Belt, S. T.; Ryba, D. W.; and Kyle, K. R. in the laboratories of Bergman, R. G., and Moore, C. B., at the University of California, Berkeley, CA.

RECEIVED for review November 7, 1991. ACCEPTED revised manuscript April 21, 1992.

3

Photochemical Reactions of Organometallic Molecules on Surfaces

CO Substitution Chemistry of Surface-Confined Derivatives of $(\eta^5\text{-}C_5H_5)Mn(CO)_3$

Doris Kang, Eric W. Wollman, and Mark S. Wrighton*

Department of Chemistry, Massachusetts Institute of Technology, Cambridge, MA 02139

Derivatives of $(\eta^5\text{-}C_5H_5)Mn(CO)_3$ attached to SiO_2, Si, or Au surfaces undergo photoreactions that allow the surface to be tailored in a rational manner. Photosubstitution of functionalized phosphines for CO occurs on all substrates, although the scope of the reaction is more limited for the surface-confined species than for the analogous complexes in solution. Flat surfaces modified with the derivatives of $(\eta^5\text{-}C_5H_5)Mn(CO)_3$ can be patterned photochemically, because no thermal CO substitution occurs.

THE PHOTOCHEMISTRY OF METAL CARBONYLS on surfaces is of potentially practical and fundamental importance. Possible applications include microelectronic device fabrication and photoimaging. Recent relevant studies include the formation of Fe thin films on GaAs (*1*) and Si (*2*) by UV photolysis of adsorbed $Fe(CO)_5$ and the assembly of a reversible photoimaging system based on poly$[(vbpy)Re(CO)_3]_2$ (vbpy is 4-methyl-4′-vinyl-2,2′-bipyridine) (*3*). These systems exploit photoinduced CO loss from $Fe(CO)_5$ and metal–metal bond cleavage in the photoexcited Re dimer.

*Corresponding author

0065–2393/93/0238–0045$06.50/0

Fundamental studies of surface-confined metal carbonyls may lead to new photoreactions or the elucidation of reaction mechanisms. Photoprocesses of metal carbonyls on solid substrates have been investigated in many systems (*1–23*). Although the chemical and physical characteristics of the surface may influence the reactivity of the adsorbed species, the primary photochemical events of surface-confined metal carbonyls are often identical to those of the analogous solution complexes.

For a variety of mononuclear metal carbonyls in solution, CO loss occurs as the primary photoprocess to generate a coordinatively unsaturated intermediate that can be trapped by another $2e^-$ donor L (*24–26*):

$$M(CO)_n \xrightarrow[-CO]{h\nu} M(CO)_{n-1} \xrightarrow{L} M(CO)_{n-1}L \qquad (1)$$

This substitution process has been observed for several surface-supported metal carbonyls and provides a means of functionalizing a surface with L. Importantly, many highly photosensitive metal carbonyls are quite thermally inert (*24*).

This chapter describes the photochemistry of derivatives of (η^5-C_5H_5)$Mn(CO)_3$ covalently bound to high-surface-area SiO_2, single-crystal Si, and Au. Photosubstitution of functionalized phosphines for CO is observed on all modified surfaces. Photochemical patterning of flat substrates can be achieved because the (η^5-C_5H_5)$Mn(CO)_3$ derivatives are inert toward thermal CO substitution.

Experimental Section

Procedures describing general spectroscopic and photochemical methods; handling of reagents; preparation of manganese compounds; and modification of high-surface-area SiO_2, single-crystal Si, and Au thin films were published in detail elsewhere (*27*).

11-Diphenylphosphinoundecylferrocene. Triethylsilane (1.8 mL, 11 mmol) was added to a solution of 11-bromoundecanoylferrocene (*28*) (2.2 g, 5.0 mmol) dissolved in 4 mL of trifluoroacetic acid (Aldrich) under Ar. The mixture was stirred for 48 h and then diluted with water. The organic product was extracted with Et_2O, washed with aqueous $NaHCO_3$, and dried over $MgSO_4$. The residue obtained upon removal of solvent was chromatographed on silica gel with hexane to give pure 11-bromoundecylferrocene in 62% yield. ^1H NMR (250 MHz, $CDCl_3$): δ 4.09 (s, 5H), 4.05 (m, 4H), 3.39 (t, 2H), 2.30 (t, 2H), 1.85 (m, 2H), and 1.17–1.52 (m, 16H) ppm. (NMR results are reported as chemical shifts (δ) in parts per million downfield from tetramethylsilane. Abbreviations used are s, singlet; m, multiplet; and t, triplet.)

$LiPPh_2$ was generated by addition of one equivalent of *n*-BuLi to a solution of PPh_2H (Aldrich, 0.4 g, 2.4 mmol) in 17 mL of dry tetrahydrofuran (THF) under Ar at −78 °C. 11-Bromoundecylferrocene (1.0 g, 2.4 mmol) dissolved in 4 mL of dry THF was added to the solution, which was then allowed to warm to room temperature. After 1.5 h of additional stirring, 10 mL of aqueous saturated NH_4Cl was added to the reaction mixture. The organic layer was collected and dried over $MgSO_4$. The crude material obtained upon solvent evaporation was chromatographed on silica gel with 9:1 hexane–CH_2Cl_2 to give pure 11-diphenylphosphinoundecylferrocene as a red–orange oil in 60% yield. ^1H NMR (300 MHz, $CDCl_3$): δ 7.29–7.48 (m, 10H), 4.09 (s, 5H), 4.05 (m, 4H), 2.32 (t, 2H), 2.05 (t, 2H), and 1.17–1.52 (m, 18H) ppm.

Preparation of Modified Au Electrodes. Au electrodes were made from 2000 Å of Au (99.999%) evaporated onto 100-mm Si wafers with a 100-Å adhesion layer of Cr. The Au-coated wafers were cut into pieces approximately 0.5 × 1.0 cm. The pieces were rinsed with hexane and then functionalized by immersing in a 1 mM solution of $HS(CH_2)_{11}(\eta^5\text{-}C_5H_4)Mn(CO)_3$ in hexane overnight. The modified Au was rinsed with hexane upon removal from solution.

Electrochemical Methods. Electrochemical measurements were carried out with a Pine Instruments model RDE-4 bipotentiostat. Voltammetric traces were recorded with a Kipp and Zonen model BD 91 *XY* recorder. Linear sweep cyclic voltammetry was performed in CH_3CN–0.1 M $[n\text{-}Bu_4N]PF_6$ at 298 K under Ar. Pt gauze was the counterelectrode, and oxidized Ag wire was the quasi-reference.

Studies of Derivatives of $(\eta^5\text{-}C_5H_5)Mn(CO)_3$ *in Solution*

$(CH_3)_3Si(\eta^5\text{-}C_5H_4)Mn(CO)_3$ and $HS(CH_2)_{11}(\eta^5\text{-}C_5H_4)Mn(CO)_3$ exhibit electronic absorption spectra similar to those reported for $(\eta^5\text{-}C_5H_5)Mn(CO)_3$ and $(\eta^5\text{-}C_5H_4CH_3)Mn(CO)_3$ (Table I). The 330-nm bands are assigned to Mn → $(\eta^5\text{-}C_5H_4R)$ charge-transfer (CT) transitions, which obscure ligand-field (LF) transitions also present in the same energy region (*24*). The 330-nm absorption is reported to have some Mn → COπ^* CT character as well (*24*).

The complexes $(\eta^5\text{-}C_5H_4R)Mn(CO)_3$ (where R = $(CH_3)_3Si$– or $-(CH_2)_{11}SH$) undergo efficient photoinduced CO substitution at 298 K upon near-UV irradiation in the presence of excess free phosphine L in alkane solution under Ar. Figures 1A, 2, and 3A show IR difference spectra recorded during irradiations of these molecules in solutions containing L = PPh_2(*n*-octyl), $PPh_2(CH_2)_{11}Fc$, or PPh_2Fc (Fc is ferrocenyl).

Upon irradiation of the tricarbonyl complexes, initially only

Table I. Spectroscopic Data for Relevant Compounds

Compound	*IR* ν_{CO} [a]	*UV–Visible* [b]
$(\eta^5\text{-}C_5H_5)Mn(CO)_3$ [c]		330 (1100), 216 (12,000)
$(\eta^5\text{-}C_5H_4CH_3)Mn(CO)_3$ [c]		330 (–)
$(CH_3)_3Si(\eta^5\text{-}C_5H_4)Mn(CO)_3$	2029 (7100), 1947 (10,500)	330 (940)
$(CH_3)_3Si(\eta^5\text{-}C_5H_4)Mn(CO)_2L$ [d]	1938 (7200), 1877 (7200)	355 (1010), 292(1960)
$(CH_3)_3Si(\eta^5\text{-}C_5H_4)Mn(CO)_2L'$ [e]	1934, 1873	
$(CH_3)_3Si(\eta^5\text{-}C_5H_4)Mn(CO)_2L''$ [f]	1932, 1873	
$(CH_3)_3Si(\eta^5\text{-}C_5H_4)Mn(CO)L_2$	1837	
$HS(CH_2)_{11}(\eta^5\text{-}C_5H_4)Mn(CO)_3$	2022 (1.0), 1942 (1.4)	330 (1040), 228 (–)
$HS(CH_2)_{11}(\eta^5\text{-}C_5H_4)Mn(CO)_2L$	1931 (1.3), 1871 (1.0)	
$HS(CH_2)_{11}(\eta^5\text{-}C_5H_4)Mn(CO)L_2$	1827	
$HS(CH_2)_{11}(\eta^5\text{-}C_5H_4)Mn(CO)L_2''$	1830	

$CH_3OSi(CH_3)_2$-(η^5-C_5H_4)$Mn(CO)_3$	1026 (1.0), 1946 (1.5)
Cl_3Si(η^5-C_5H_4)$Mn(CO)_3$	2036 (1.0), 1960 (1.2)
[SiO_2]–$OSi(CH_3)_2$-(η^5-C_5H_4)$Mn(CO)_3$	2023 (1.0), 1938 (1.0)
[SiO_2]–$OSi(CH_3)_2$-(η^5-C_5H_4)$Mn(CO)_2L$	1931 (1.1), 1861 (1.0)
[SiO_2]–$OSi(CH_3)_2$-(η^5-C_5H_4)$Mn(CO)_2L'$	1922, 1870
[Si]–OSi(η^5-C_5H_4)$Mn(CO)_3$	2025 (1.0), 1939 (1.2)
[Si]–OSi(η^5-C_5H_4)$Mn(CO)_2L$	1939 (1.0), 1867 (2.7)
[Au]–$S(CH_2)_{11}$(η^5-C_5H_4)$Mn(CO)_3$	2015 (1.0), 1925 (1.3)
[Au]–$S(CH_2)_{11}$(η^5-C_5H_4)$Mn(CO)_2L$	1922 (1.0), 1852 (1.0)
[Au]–$S(CH_2)_{11}$(η^5-C_5H_4)$Mn(CO)_2L''$	1920 (1.0), 1850 (1.0)

NOTE: All data were recorded at 298 K. All data for solution species were recorded in alkane solution. For SiO_2-supported species of high surface area, IR spectra were recorded as Nujol mulls. Characteristic frequencies for Si wafer and Au-supported species were obtained from transmission IR spectra.

[a]Molar absorptivity (ϵ) or relative optical density (OD). Band positions are given in reciprocal centimeters and extinction coefficients, in parentheses, are in reciprocal (centimeters Molar).

[b]Molar absorptivity. Band positions are given in nanometers and extinction coefficients, in parentheses, are in reciprocal (centimeters Molar).

[c]Reference 24.

[d]L is PPh_2(*n*-octyl).

[e]L′ is PPh_2Fc.

[f]L′′ is $PPh_2(CH_2)_{11}Fc$.

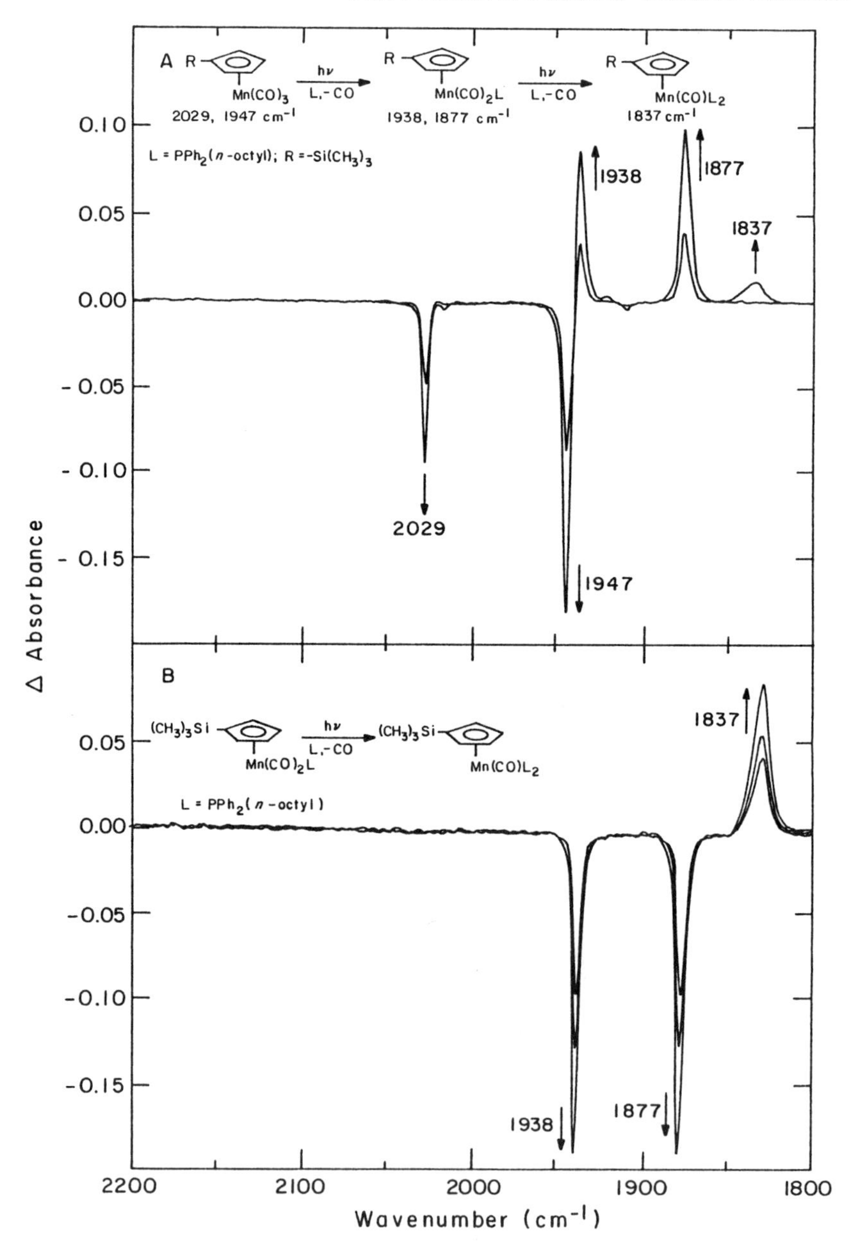

Figure 1. Part A: IR difference spectra accompanying photoreactions of $(CH_3)_3Si$-$(\eta^5$-$C_5H_4)Mn(CO)_3$ at 2029 and 1947 cm^{-1}, with L = PPh_2(n-octyl) in n-hexane at 25 °C to give $(CH_3)_3Si(\eta^5$-$C_5H_4)Mn(CO)_2L$ (absorptions at 1938 and 1877 cm^{-1}) and $(CH_3)_3Si(\eta^5$-$C_5H_4)Mn(CO)L_2$ (absorption at 1837 cm^{-1}). Part B: IR difference spectra accompanying photoreaction of $(CH_3)_3Si(\eta^5$-$C_5H_4)Mn(CO)_2$ at 1938 and 1877 cm^{-1}, with L = PPh_2(n-octyl) in n-hexane at 25 °C to give $(CH_3)_3Si(\eta^5$-$C_5H_4)Mn(CO)L_2$ (absorption at 1837 cm^{-1}).

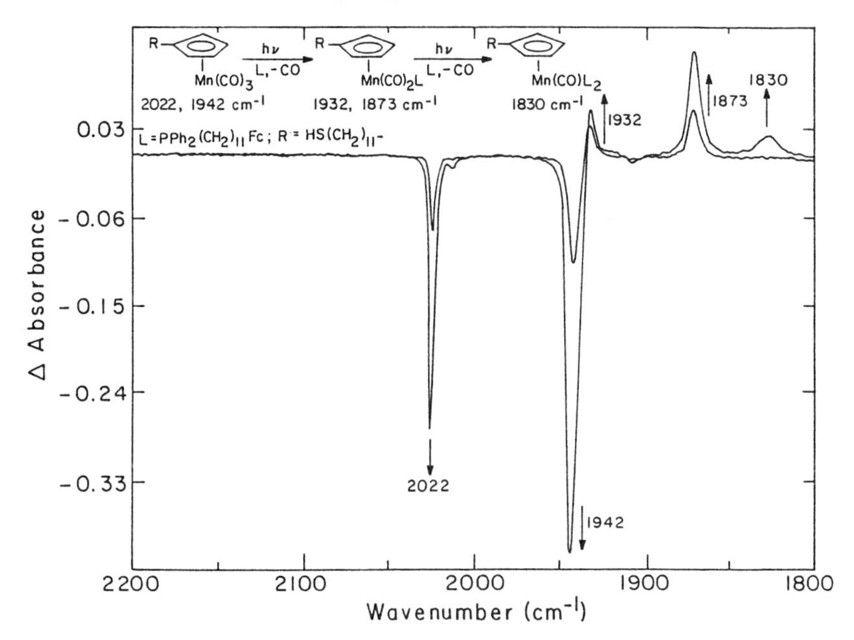

Figure 2. IR difference spectrum accompanying photoreaction of $HS(CH_2)_{11}(\eta^5\text{-}C_5H_4)Mn(CO)_3$ at 2022 and 1942 cm^{-1}, with L = $PPh_2(CH_2)_{11}Fc$ in methylcyclohexane (MCH) at 25 °C to give $HS(CH_2)_{11}(\eta^5\text{-}C_5H_4)Mn(CO)_2L$ (absorptions at 1932 and 1873 cm^{-1}) and $HS(CH_2)_{11}(\eta^5\text{-}C_5H_4)Mn(CO)L_2$ (absorption at 1830 cm^{-1}).

monosubstituted products, $(\eta^5\text{-}C_5H_4R)Mn(CO)_2L$, form. These result from CO loss from $(\eta^5\text{-}C_5H_4R)Mn(CO)_3$:

$$(\eta^5\text{-}C_5H_4R)Mn(CO)_3 \xrightarrow[L]{h\nu} (\eta^5\text{-}C_5H_4R)Mn(CO)_2L + CO \quad (2)$$

Further irradiation of $(\eta^5\text{-}C_5H_4R)Mn(CO)_2L$, where L is PPh_2(*n*-octyl) or $PPh_2(CH_2)_{11}Fc$, leads to the formation of disubstituted products, $(\eta^5\text{-}C_5H_4R)Mn(CO)L_2$, formed from substitution of L for CO:

$$(\eta^5\text{-}C_5H_4R)Mn(CO)_2L \xrightarrow[L]{h\nu} (\eta^5\text{-}C_5H_4R)Mn(CO)L_2 + CO \quad (3)$$

No IR spectral features indicating the formation of $(\eta^5\text{-}C_5H_4R)\text{-}Mn(CO)L_2$, where L is PPh_2Fc, were observed. The formation of $(CH_3)_3Si(\eta^5\text{-}C_5H_4)Mn(CO)L_2$ upon irradiation of pure $(CH_3)_3Si(\eta^5\text{-}C_5H_4)Mn(CO)_2L$ in an alkane solution containing L was also observed (Figure 1B).

The initial quantum yields at 366 nm for the processes in equations 2 and 3 are 0.63 ± 0.06 and 0.25 ± 0.03, respectively, for R = $(CH_3)_3Si$– and L = PPh_2(*n*-octyl). For L = $PPh_2(CH_2)_{11}Fc$ good quantum yields are difficult to obtain because of adsorption of the ferrocenyl center at 366 nm. The lower efficiency of the second substitution may result from possible photoinduced phosphine extrusion, (equation 4), which would be competitive with CO loss.

$$Cp'Mn(CO)_2L \xrightarrow[-L]{h\nu} Cp'Mn(CO)_2 \xrightarrow{L'} Cp'Mn(CO)_2L' \qquad (4)$$

where Cp′ is $(CH_3)_3Si(\eta^5\text{-}C_5H_4)$, L is PPh_2(*n*-octyl), and L′ is THF. $Cp'Mn(CO)_2L$ irradiated in THF yields $Cp'Mn(CO)_2$(THF), and $(\eta^5\text{-}C_5H_5)Mn(CO)_2PPh_3$ irradiated in alkane solution containing CO forms

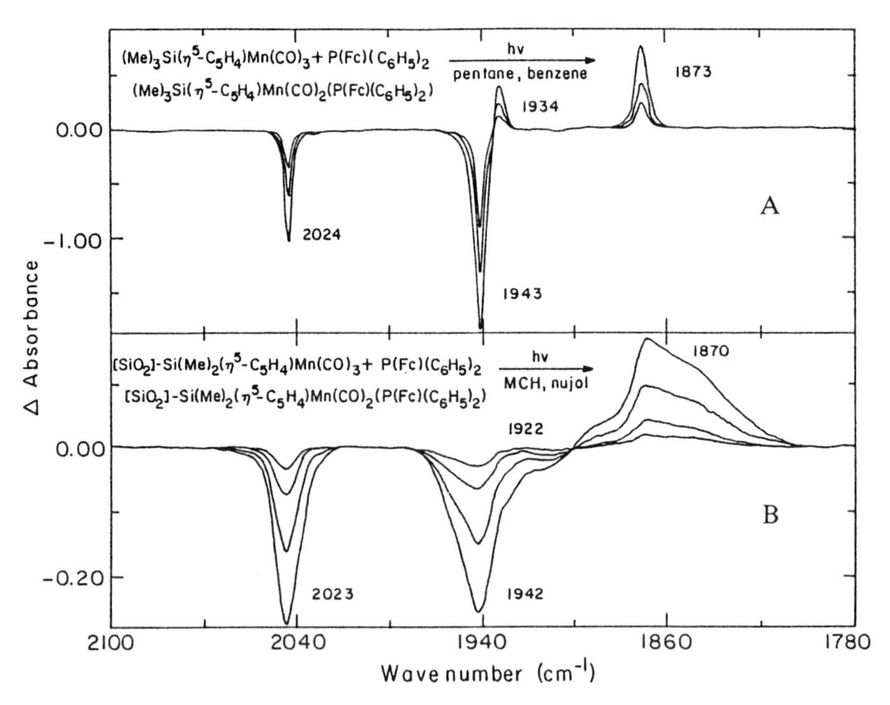

Figure 3. Part A: IR spectrum accompanying photoreaction of $(CH_3)_3Si(\eta^5\text{-}C_5H_4)Mn(CO)_3$ at 2024 and 1943 cm^{-1}, with L = PPh_2Fc in pentane–benzene at 25 °C to give $(CH_3)_3Si(\eta^5\text{-}C_5H_4)Mn(CO)_2L$ (absorption at 1934 and 1873 cm^{-1}). Part B: IR spectrum accompanying photoreaction of $[SiO_2]\text{–}(CH_3)_2Si(\eta^5\text{-}C_5H_4)Mn(CO)_3$ with L = PPh_2Fc in MCH–Nujol to give $[SiO_2]\text{–}(CH_3)_2Si(\eta^5\text{-}C_5H_4)Mn(CO)_2L$ (absorptions at 1922 and 1870 cm^{-1}).

$(\eta^5\text{-}C_5H_5)Mn(CO)_3$. Interestingly, however, irradiation of $(\eta^5\text{-}C_5H_4CH_3)Mn(CO)_2PPh_3$ in the presence of CO fails to yield the tricarbonyl (*29*).

These results show that the CO photosubstitution chemistry of the complexes $(\eta^5\text{-}C_5H_4R)Mn(CO)_3$ (R is $(CH_3)_3Si$– or $HS(CH_2)_{11}$– is analogous to that observed upon irradiation of $(\eta^5\text{-}C_5H_5)Mn(CO)_3$ in the presence of phosphines (*24–26*).

Photochemistry of Modified Surfaces

Modified surfaces studied were

- $[SiO_2]\text{–}OSi(CH_3)_2(\eta^5\text{-}C_5H_4)Mn(CO)_3$
- high-surface-area SiO_2 treated with $CH_3OSi(CH_3)_2(\eta^5\text{-}C_5H_4)Mn(CO)_3$
- $[Si]\text{–}OSi(\eta^5\text{-}C_5H_4)Mn(CO)_3$, single crystal Si (n- or p-type, 100 face) treated with $Cl_3Si(\eta^5\text{-}C_5H_4)Mn(CO)_3$
- $[Au]\text{–}S(CH_2)_{11}(\eta^5\text{-}C_5H_4)Mn(CO)_3$, Au-coated Si wafers treated with $HS(CH_2)_{11}(\eta^5\text{-}C_5H_4)Mn(CO)_3$

Preparation of these functionalized surfaces follows chemistry previously used in the modification of oxide and Au surfaces with organometallic reagents (Scheme I). In the modification of the oxide surfaces, the molecules with $\text{–}Si(CH_3)_2(OCH_3)$ and $\text{–}SiCl_3$ functionalities can react with surface –SiOH groups to covalently bind the metal carbonyl to the surface (*30–32*). The formation of monolayers on Au by adsorption of organometallic species with long-chain alkylthiol linkages was demonstrated previously (*33–35*), and the species formed from the chemisorption

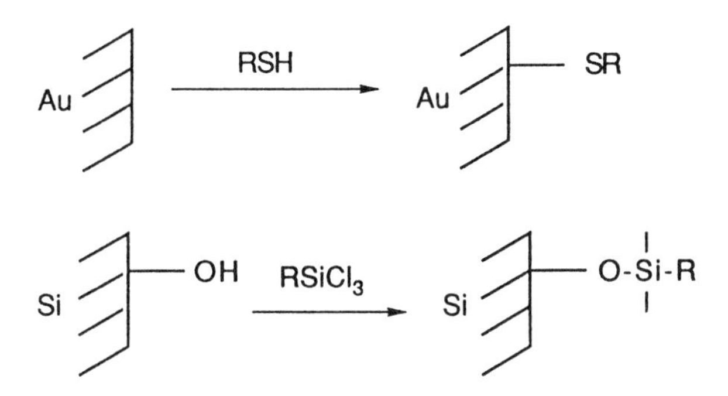

Scheme I. Modification of a Au surface with a thiol and a Si surface with a trichlorosilane.

of a thiol RSH onto Au is believed to be a thiolate ($RS^{-}Au(I)$) (*35*). Coverages from the silane and thiol reagents are typically in the range of one monolayer.

$[SiO_2]–OSi(CH_3)_2(\eta^5\text{-}C_5H_4)Mn(CO)_3$ suspended in a mineral oil (Nujol) mull can be easily characterized by transmission Fourier transform infrared (FTIR) spectroscopy (Figure 4A). Although the IR band positions of the surface-confined species agree fairly well with those of the related solution species, the bands of the surface-confined species are much broader and slightly red-shifted. Coverage of the metal carbonyl was established by elemental analysis for C and Mn and is approximately 4×10^{-11} mol/cm^2.

Irradiation of $[SiO_2]–OSi(CH_3)_2(\eta^5\text{-}C_5H_4)Mn(CO)_3$ (at 2023 and 1938 cm^{-1}) in a Nujol mull containing excess L = PPh_2(*n*-octyl) yields the monosubstituted product $[SiO_2]–OSi(CH_3)_2(\eta^5\text{-}C_5H_4)Mn(CO)_2L$ (absorptions at 1931 and 1861 cm^{-1}). Figure 4B shows the IR difference spectrum accompanying the irradiation. The band at 2023 cm^{-1} is associated with $[SiO_2]–OSi(CH_3)_2(\eta^5\text{-}C_5H_4)Mn(CO)_3$, and the band at 1946 cm^{-1} results from the disappearance of the 1938-cm^{-1} band of $[SiO_2]–OSi(CH_3)_2(\eta^5\text{-}C_5H_4)Mn(CO)_3$ and the growth of the lower energy band of $[SiO_2]–OSi(CH_3)_2(\eta^5\text{-}C_5H_4)Mn(CO)_2L$. The actual position of the lower energy IR band of the photosubstitution product is ~1931 cm^{-1} and was obtained from an absorption spectrum recorded after nearly all of $[SiO_2]–OSi(CH_3)_2(\eta^5\text{-}C_5H_4)Mn(CO)_3$ was converted to $[SiO_2]–OSi(CH_3)_2(\eta^5\text{-}C_5H_4)Mn(CO)_2L$. The band at 1861 cm^{-1} is assigned to $[SiO_2]–OSi(CH_3)_2(\eta^5\text{-}C_5H_4)Mn(CO)_2L$. Similar photochemistry occurs with L is PPh_2Fc (Figure 3B).

$[Si]–OSi(\eta^5\text{-}C_5H_4)Mn(CO)_3$ can also be characterized by transmission FTIR spectroscopy. Irradiation of $[Si]–OSi(\eta^5\text{-}C_5H_4)Mn(CO)_3$ (at 2025 and 1939 cm^{-1}) in air results solely in the decomposition of the metal carbonyl (Figure 5A). Coverage of the metal carbonyl can be determined from the magnitude of the negative peaks of the IR difference spectrum corresponding to complete decomposition of the surface-bound species. If the extinction coefficients of the solution species and of the corresponding Si-confined complexes are assumed to be the same, coverages on the order of one monolayer are obtained. As has previously been shown (*36*), Cl_3SiR reagents yield about one monolayer of the functional group R attached to a Si–SiO_x surface.

Irradiation of $[Si]–OSi(\eta^5\text{-}C_5H_4)Mn(CO)_3$ (at 2025 and 1939 cm^{-1}) in neat PPh_2(*n*-octyl) results in the loss of the tricarbonyl and formation of $[Si]–OSi(\eta^5\text{-}C_5H_4)Mn(CO)_2L$ (absorptions at 1939 and 1867 cm^{-1}) (Figure 5B). This substitution product is generated in about 75% yield. Irradiations performed in dry alkane or ether solutions containing the phosphine in concentrations of ~0.1 M yielded only photodecomposition of the metal carbonyl. These results are identical for either n- or p-type Si-confined species.

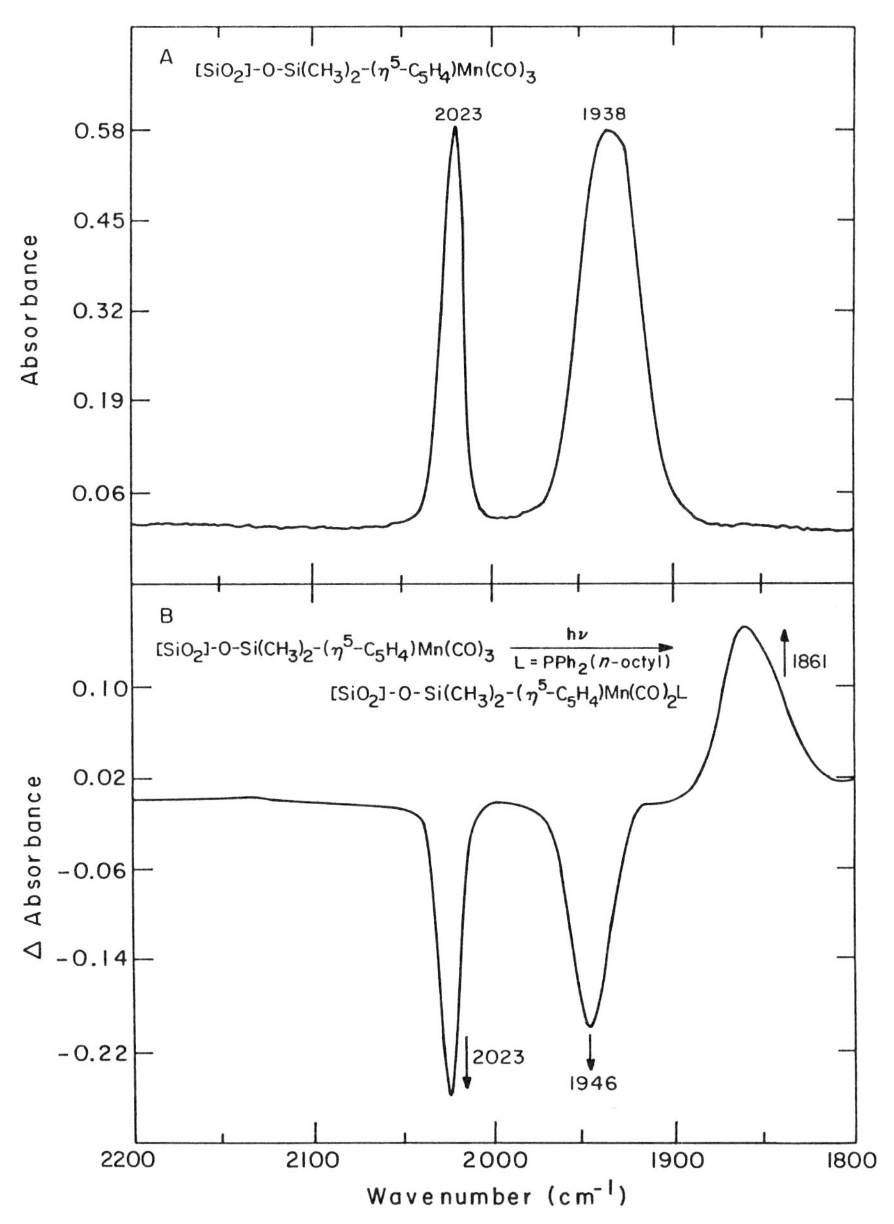

Figure 4. Part A: IR spectrum of $[SiO_2]$–$OSi(CH_3)_2(\eta^5$-$C_5H_4)Mn(CO)_3$ (absorptions at 2023 and 1938 cm^{-1}) in a Nujol mull at 25 °C. Part B: IR spectral changes accompanying photoreaction of $[SiO_2]$–$OSi(CH_3)_2(\eta^5$-$C_5H_4)$-$Mn(CO)_3$ with L = PPh_2(n-octyl) at 25 °C to give $[SiO_2]$–$OSi(CH_3)_2(\eta^5$-$C_5H_4)Mn(CO)_2L$. Bands at 2023 and 1861 cm^{-1} are assigned to $[SiO_2]$–OSi-$(CH_3)_2(\eta^5$-$C_5H_4)Mn(CO)_3$ and $[SiO_2]$–$OSi(CH_3)_2(\eta^5$-$C_5H_4)Mn(CO)_2L$, respectively. The band at 1946 cm^{-1} is a result of the disappearance of starting material and the formation of $[SiO_2]$–$OSi(CH_3)_2(\eta^5$-$C_5H_4)Mn(CO)_2L$.

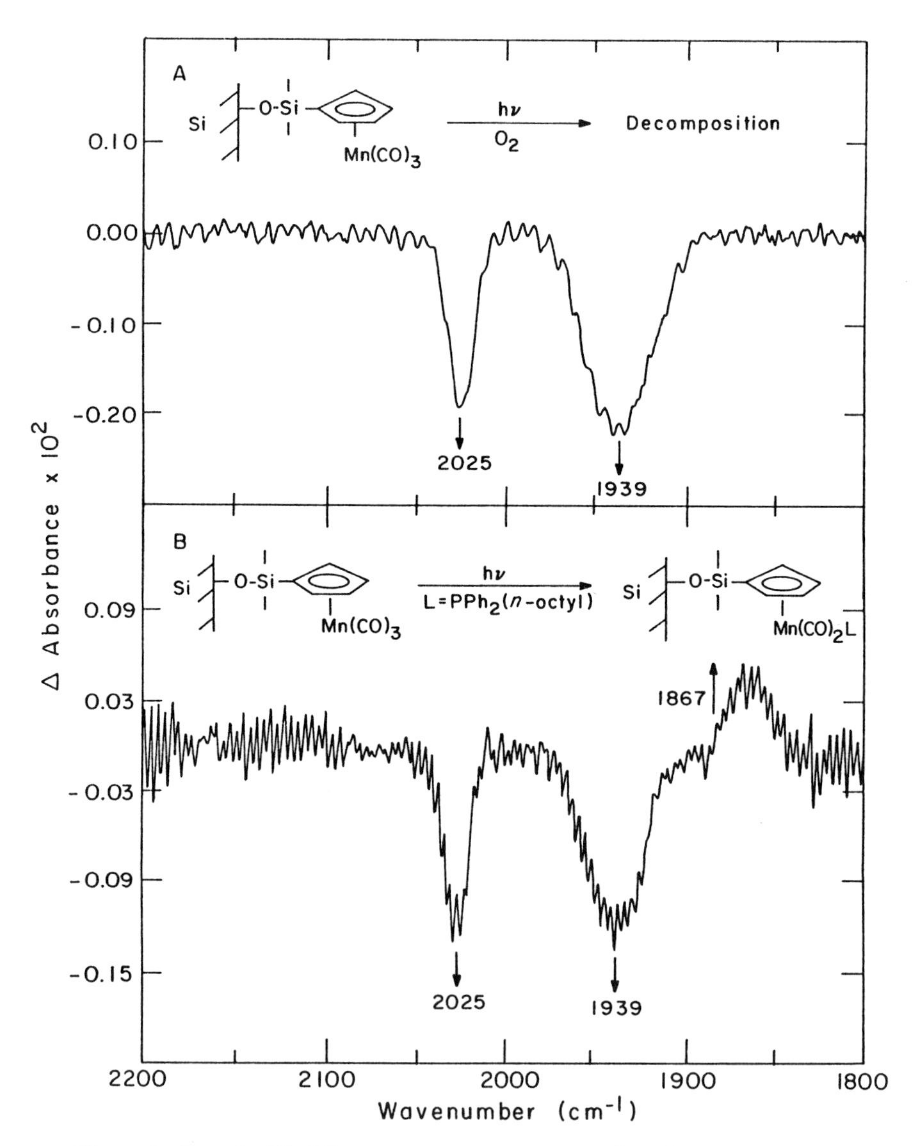

Figure 5. Part A: IR difference spectrum accompanying photodecomposition of [Si]–OSi(η^5-C_5H_4)Mn(CO)$_3$ (absorptions at 2025 and 1939 cm^{-1}) upon irradiation in air at 25 °C. Part B: IR difference spectrum accompanying photoreaction of [Si]–OSi(η^5-C_5H_4)Mn(CO)$_3$ at 25 °C in the presence of L = PPh_2(n-octyl) to give [Si]–OSi(η^5-C_5H_4)Mn(CO)$_2$L (absorptions at 1939 and 1867 cm^{-1}).

The thiol-modified Au thin film (~100 Å) on Si–[Au]–$S(CH_2)_{11}(\eta^5$-$C_5H_4)Mn(CO)_3$ has the transmission IR spectrum shown in Figure 6A. In contrast to the Si-confined species, successful CO photosubstitution in [Au]–$S(CH_2)_{11}(\eta^5$-$C_5H_4)Mn(CO)_3$ can be effected with lower concentrations of phosphine. Irradiation of [Au]–$S(CH_2)_{11}(\eta^5$-$C_5H_4)Mn(CO)_3$ (at 2015 and 1925 cm^{-1}) in 0.05–0.1 M L = PPh_2(*n*-octyl) in dry hexane under Ar results in loss of tricarbonyl and formation of [Au]–S-$(CH_2)_{11}(\eta^5$-$C_5H_4)Mn(CO)_2L$ (absorptions at 1922 and 1852 cm^{-1}). The difference spectrum for this process is shown in Figure 6B. Similar results are obtained upon irradiation of [Au]–$S(CH_2)_{11}(\eta^5$-$C_5H_4)Mn(CO)_3$ in a solution containing 0.05 M $PPh_2(CH_2)_{11}Fc$ in hexane under Ar (Figure 7).

These results show that these derivatives of $(\eta^5$-$C_5H_5)Mn(CO)_3$ bound to Si or Au surfaces are photosensitive with respect to CO loss. Coordination of a functionalized phosphine to the metal center can occur following photoejection of CO. Photodissociation of the surface-confined metal carbonyl apparently is competitive with quenching by energy transfer to the metal substrate (*37–41*).

Comparison of Photochemistry of Solution and Surface-Confined Species

In contrast to the solution species, very high phosphine concentrations (3.3 M for neat PPh_2-*n*-octyl) are required for efficient CO photosubstitution in [Si]–$OSi(\eta^5$-$C_5H_4)Mn(CO)_3$. For metal carbonyls in solution, entering ligand concentrations of <0.01 M are often adequate to efficiently trap coordinatively unsaturated intermediates formed by CO loss from the parent molecule. For example, near-UV irradiation of $(\eta^5$-$C_5H_5)Mn(CO)_3$ in hexane containing 0.01 M PPh_2(*n*-octyl) gives efficient formation of substitution products.

One possibility is that silanol groups present on the SiO_x surface compete with the phosphine for reaction with the Mn center. The formation of a derivative of dicarbonyl(cyclopentadienyl)manganese coordinated to a phosphinol group was observed upon irradiation of the tricarbonyl derivative intercalated into α-$Zr(HPO_4)_2 \cdot H_2O$ (*42*). Likewise, irradiation of $W(CO)_6$ adsorbed on porous Vycor glass yields a $W(CO)_5L$ complex, where L is chemisorbed water or a surface silanol group (*14*). Although successful photosubstitution of phosphine for CO in [Au]–$S(CH_2)_{11}(\eta^5$-$C_5H_4)Mn(CO)_3$ at lower phosphine concentrations would support this contention, no IR bands that could be assigned to [Si]–$OSi(\eta^5$-$C_5H_4)Mn(CO)_2(ROH)$ could be observed.

Another difference between the photochemistry of the surface-confined complexes and that of the solution species is that no disubsti-

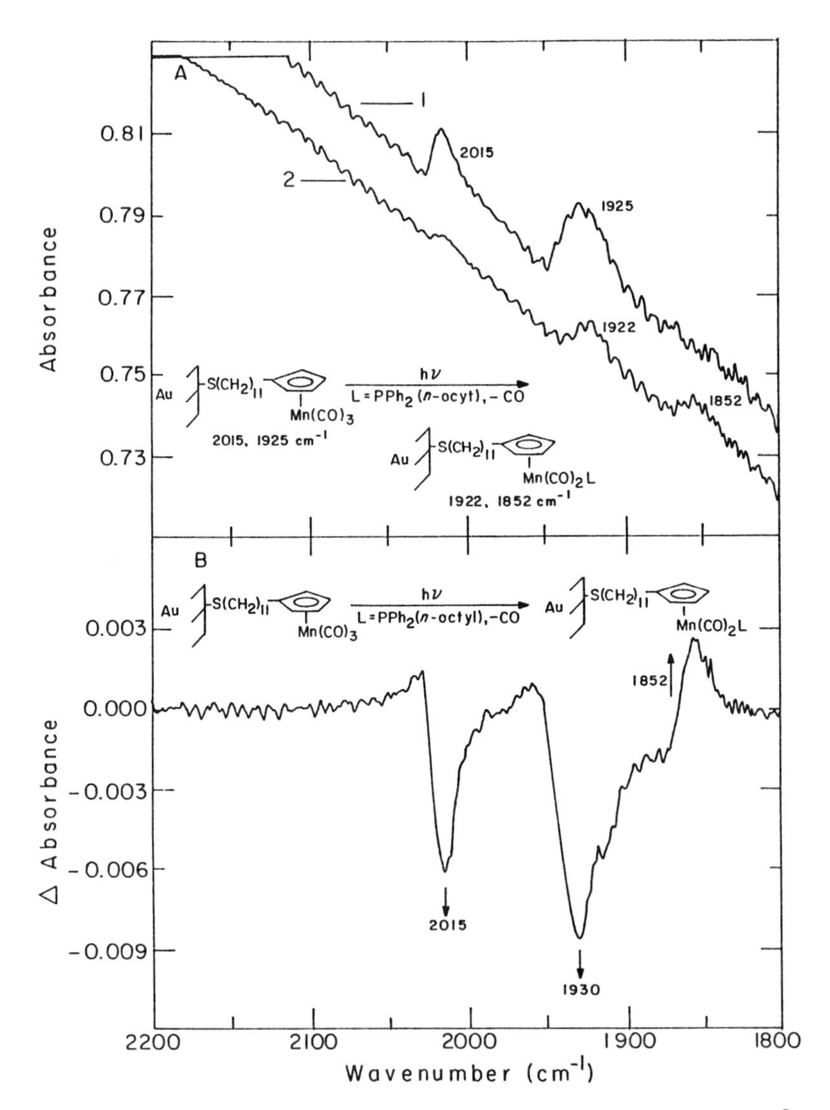

Figure 6. Part A: Initial IR spectrum (1) of [Au]–S$(CH_2)_{11}(\eta^5$-$C_5H_4)Mn(CO)_3$ (absorptions at 2015 and 1925 cm^{-1}) at 25 °C in air and final IR spectrum (2) after photoreaction with L = PPh_2(n-octyl) to give [Au]–S$(CH_2)_{11}(\eta^5$-$C_5H_4)Mn(CO)_2L$ (absorptions at 1922 and 1852 cm^{-1}). Part B: Difference IR spectrum of 1 and 2.

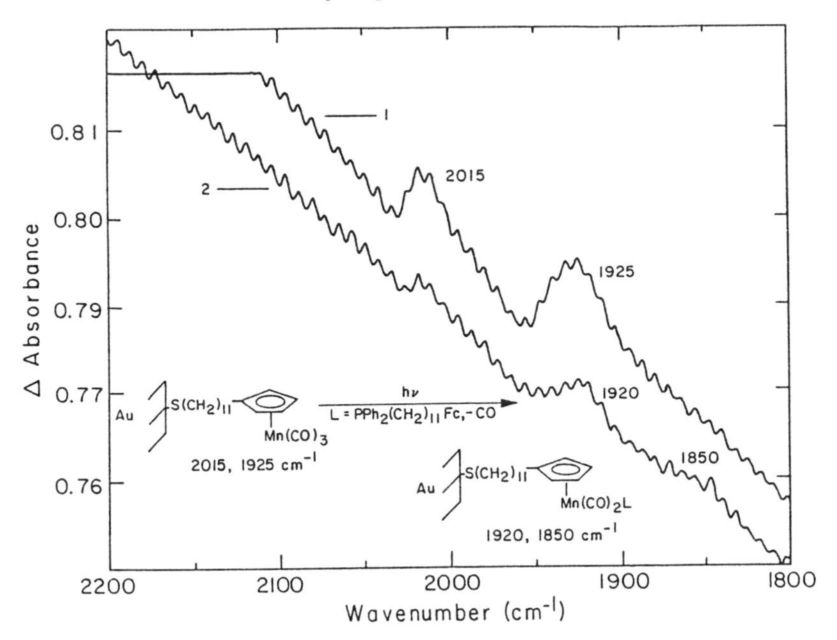

Figure 7. Initial IR spectrum (1) of [Au]–S$(CH_2)_{11}(\eta^5$-$C_5H_4)Mn(CO)_3$ (absorptions at 2025 and 1925 cm^{-1}) at 25 °C in air and final IR spectrum (2) after photoreaction with L = $PPh_2(CH_2)_{11}Fc$ to give [Au]–S$(CH_2)_{11}(\eta^5$-$C_5H_4)Mn(CO)_2L$ (absorptions at 1920 and 1850 cm^{-1}).

tuted species are formed on the surface. Prolonged irradiation only leads to decomposition of the monosubstituted product. Because the Mn center in the SiO_2 and Si-confined species is close to the surface, formation of the bisphosphine adduct could be sterically disfavored. For the Au-confined complex, close packing of the metal carbonyl head groups and of the hydrocarbon chains could render the formation of the disubstitution product sterically unfavorable. The fact that no substitution product forms upon irradiation of [Au]–S$(CH_2)_{11}(\eta^5$-$C_5H_4)Mn(CO)_3$ in the presence of PPh_2Fc indicates that steric crowding near the metal center may be greater for the molecule in the adsorbed monolayer than for free $(\eta^5$-$C_5H_5)Mn(CO)_3$ in solution, where substitution of PPh_2Fc occurs readily.

If close packing does exist, the metal carbonyl head group of [Au]–S$(CH_2)_{11}(\eta^5$-$C_5H_4)Mn(CO)_2L$ could behave in the monolayer as it would in a solid matrix. $(\eta^5$-$C_5H_5)Mn(CO)_2PPh_3$ irradiated in methylcyclohexane, 3-methylpentane, or 2-methyl-THF matrices at 110 K fails to undergo ligand loss (CO or PPh_3), as no IR spectral changes can be observed. In contrast, irradiation of $(CH_3)_3Si(\eta^5$-$C_5H_4)Mn(CO)_3$ in a methylcyclohexane matrix induces CO loss to give the 16e$^-$ species

$(CH_3)_3Si(\eta^5\text{-}C_5H_4)Mn(CO)_2$, analogous to results for related complexes $(\eta^5\text{-}C_5Me_5)Mn(CO)_3$ and $(\eta^5\text{-}C_5Cl_5)Mn(CO)_3$ (*43, 44*). These low-temperature matrix experiments suggest that dissociative CO loss from the monosubstituted species is less efficient for the molecule in a rigid matrix than in a fluid solution. The manganese carbonyl centers of [Au]–$S(CH_2)_{11}(\eta^5\text{-}C_5H_4)Mn(CO)_2L$ may behave as such centers do in a rigid matrix.

Thus, the CO photosubstitution chemistry of the surface-confined species is considerably more limited than that observed for the corresponding species in solution. Only monosubstitution photoproducts are formed from the surface-confined tricarbonyl, whereas disubstitution products form readily in solution once the monosubstitution species is formed. In addition, the formation of monosubstitution products is more difficult on a surface than in solution. The difficulty is manifested by the need for high entering ligand concentrations and the inability to achieve significant chemical yields of the surface-confined monosubstitution product with very bulky ligands, for example, PPh_2Fc.

Photochemical Patterning of Flat Substrates

Photopatterning of the Si and Au surfaces at relatively low resolution is readily accomplished by immersing ~16- × 32-mm pieces of Si derivatized with $Cl_3Si(\eta^5\text{-}C_5H_4)Mn(CO)_3$ and Au derivatized with $HS(CH_2)_{11}(\eta^5\text{-}C_5H_4)Mn(CO)_3$ in the appropriate phosphine solutions and irradiating only a portion of the surface. By IR the dicarbonyl phosphine forms only on the irradiated portion, and the tricarbonyl remains on the nonirradiated portion.

These results demonstrate that the properties of a surface probably can be photochemically tailored in a spatially selective manner by substitution of suitable $2e^-$ donor ligands having unique functional groups. For example, irradiation of [Au]–$S(CH_2)_{11}(\eta^5\text{-}C_5H_4)Mn(CO)_3$ in the presence of L = $PPh_2(CH_2)_{11}Fc$ introduces the redox active ferrocenyl group intact onto the surface of Au by formation of the substitution product [Au]–$S(CH_2)_{11}(\eta^5\text{-}C_5H_4)Mn(CO)_2L$. A cyclic voltammogram of the irradiated electrode shows the presence of redox active centers (Figure 8B). The waves at $E_{1/2}$ of +0.36 and +0.66 versus a Ag quasi-reference are assigned to the surface-confined ferrocene and substituted Mn centers, respectively. This assignment was made after comparison with the cyclic voltammograms of Au electrodes modified with $HS(CH_2)_{11}(\eta^5\text{-}C_5H_4)\text{-}Mn(CO)_2L$ (L = $PPh_2(CH_2)_{11}Fc$ or PPh_2(*n*-octyl), Figure 9) synthesized independently in solution. The tricarbonyl is electrochemically silent at potentials negative of +1.3 V where it is irreversibly oxidized. Examination of electrodes by specular reflectance FTIR spectroscopy and electrochemistry shows that photoconversion of two-thirds of a monolayer of the

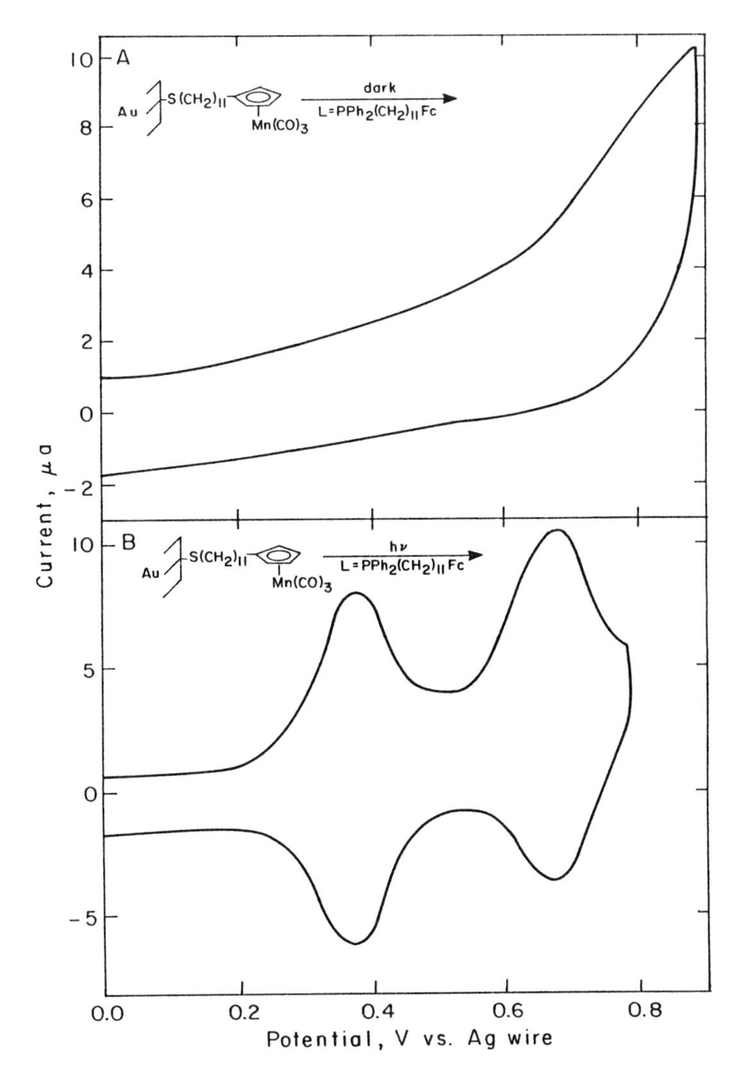

Figure 8. Part A: Cyclic voltammetry (CH_3CN–0.1 M [n-Bu_4N]PF_6, 200 mV/s) of a Au electrode derivatized with $HS(CH_2)_{11}(\eta^5\text{-}C_5H_4)Mn(CO)_3$ and treated first with 0.05 M $PPh_2(CH_2)_{11}Fc$ in hexane under Ar in the dark and then 20% benzyl bromide in EtOH in the dark. Part B: Cyclic voltammetry of a Au electrode derivatized with $HS(CH_2)_{11}(\eta^5\text{-}C_5H_4)Mn(CO)_3$, irradiated in 0.05 M $PPh_2(CH_2)_{11}Fc$ in hexane under Ar, and then treated with 20% benzyl bromide in EtOH in the dark.

tricarbonyl yields one-third to one-half of a monolayer of substitution product. Thus, the substitution product forms in only 50–75% yield.

Photopatterning of the electrodes can be accomplished, because no [Au]–$S(CH_2)_{11}(\eta^5\text{-}C_5H_4)Mn(CO)_2PPh_2(CH_2)_{11}Fc$ forms on nonirradiated portions of the electrode treated with the phosphine, as determined by

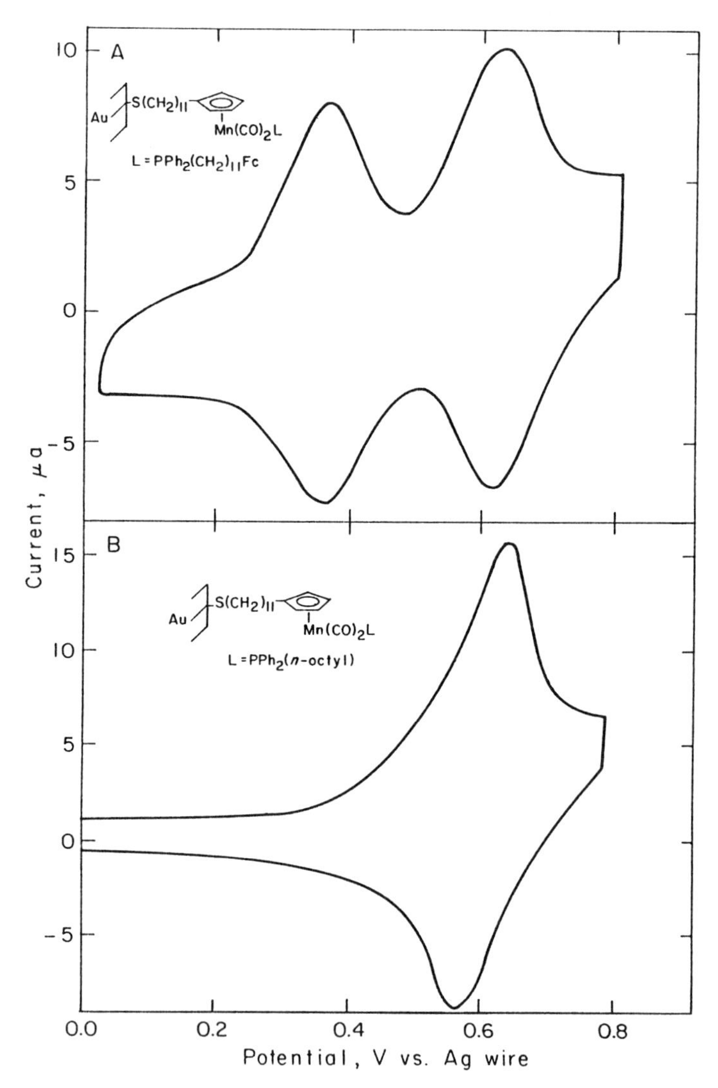

Figure 9. Part A: Cyclic voltammetry (CH_3CN–0.1 M [n-Bu_4N]PF_6, 200 mV/s) of a Au electrode derivatized in a 0.5 mM solution of $HS(CH_2)_{11}(\eta^5\text{-}C_5H_4)Mn(CO)_2PPh_2(CH_2)_{11}Fc$ in hexane under Ar overnight. Part B: Cyclic voltammetry (CH_3CN–0.1 M [n-Bu_4N]PF_6, 300 mV/s) of a Au electrode derivatized in a 1 mM solution of $HS(CH_2)_{11}(\eta^5\text{-}C_5H_4)Mn(CO)_2PPh_2$(n-octyl) in hexane under Ar overnight.

cyclic voltammetry. Residual $PPh_2(CH_2)_{11}Fc$ remaining on surfaces treated in the dark can be removed by treatment with 20% benzyl bromide in EtOH at 25 °C for 20 min. A cyclic voltammogram of a modified Au electrode $[Au]-S(CH_2)_{11}(\eta^5\text{-}C_5H_4)Mn(CO)_3$ treated in the dark with $PPh_2(CH_2)_{11}Fc$ and then benzyl bromide is shown in Figure 8A.

Because the surface-confined derivatives of $(\eta^5\text{-}C_5H_5)Mn(CO)_3$ are, like the parent molecule in solution, photosensitive with respect to CO loss but thermally inert toward CO substitution, photochemical patterning of flat surfaces modified with these complexes can be achieved by selectively irradiating portions of the surface exposed to the incoming ligand. Through this process, the properties of the entering ligand can be introduced onto the surface. In the case presented here, we show that redox centers can be introduced onto Au.

Conclusions

The results of our research show that monolayers of derivatives of $(\eta^5\text{-}C_5H_5)Mn(CO)_3$ confined to Au or Si can undergo photosubstitution of phosphine for CO. This process allows the introduction of unique functional groups present in the phosphine onto the surface.

Photosubstitution of bulky ligands for CO is limited to a single substitution, whereas multiple substitutions occur for the molecules in solution. Steric constraints created by the oxide surface or the monolayer may be responsible for the limited uptake of the entering ligand.

The fact that we find excited-state CO loss from the molecules confined to Au is consistent with reported results showing the dissociative loss of CO from metal carbonyls to be exceedingly fast and therefore competitive with the expected fast quenching processes associated with the metallic substrate (*37–41*). $Fe(CO)_5$ on Ag (110) dissociated upon irradiation at 337 nm with approximately the same yield as on Al_2O_3 (*19*). The rate constant for quenching by energy transfer to the surface was estimated to be $\sim 10^{12}\ s^{-1}$. The lower limit placed on the rate of photodissociation of $Fe(CO)_5$ was $\sim 3 \times 10^{12}\ s^{-1}$.

Because Si surfaces modified with $Cl_3Si(\eta^5\text{-}C_5H_4)Mn(CO)_3$ and Au surfaces modified with $HS(CH_2)_{11}(\eta^5\text{-}C_5H_4)Mn(CO)_3$ are inert toward thermal CO substitution, they can be patterned with respect to their photoproducts with $PPh_2(n\text{-octyl})$ or $PPh_2(CH_2)_{11}Fc$. Substitution of $PPh_2(CH_2)_{11}Fc$ yields redox active ferrocene on the surface. This method complements other reported procedures used to pattern flat surfaces by selective irradiation of adsorbed monolayers (*45, 46*) and should be useful in tailoring microfabricated structures.

Acknowledgment

We thank the Office of Naval Research and the National Science Foundation for partial support of this research.

References

1. Bottka, N.; Walsh, P. J.; Dalbey, R. Z. *J. Appl. Phys.* **1983,** *54,* 1104–1109.
2. Foord, J. S.; Jackman, R. B. *Chem. Phys. Lett.* **1984,** *112,* 190–194.
3. O'Toole, T. R.; Sullivan, B. P.; Meyer, T. J. *J. Am. Chem. Soc.* **1989,** *111,* 5699–5706.
4. Kinney, J. B.; Staley, R. H.; Reichel, C. L.; Wrighton, M. S. *J. Am. Chem. Soc.* **1981,** *103,* 4273–4275.
5. Reichel, C. L.; Wrighton, M. S. *J. Am. Chem. Soc.* **1981,** *103,* 7180–7189.
6. Liu, D. K.; Wrighton, M. S. *J. Am. Chem. Soc.* **1982,** *104,* 898–901.
7. Liu, D. K.; Wrighton, M. S.; McKay, D. R.; Maciel, G. E. *Inorg. Chem.* **1984,** *23,* 212–220.
8. Klein, B.; Kazlauskas, R. J.; Wrighton, M. S. *Organometallics* **1982,** *1,* 1338–1350.
9. Bentsen, J. G.; Wrighton, M. S. *Inorg. Chem.* **1984,** *23,* 512–515.
10. Trusheim, M. R.; Jackson, R. L. *J. Phys. Chem.* **1983,** *87,* 1910–1916.
11. Jackson, R. L.; Trusheim, M. R. *J. Am. Chem. Soc.* **1982,** *104,* 6590–6596.
12. Wild, F. R. W. P.; Gubitosa, G.; Brintzinger, H. H. *J. Organomet. Chem.* **1978,** *148,* 73–80.
13. Simon, R.; Gafney, H. D.; Morse, D. L. *Inorg. Chem.* **1983,** *22,* 573–574.
14. Simon, R. C.; Gafney, H. D.; Morse, D. L. *Inorg. Chem.* **1985,** *24,* 2565–2570.
15. Darsillo, M. S.; Gafney, H. D.; Paquette, M. S. *Inorg. Chem.* **1988,** *27,* 2815–2819.
16. Darsillo, M. S.; Gafney, H. D.; Paquette, M. S. *J. Am. Chem. Soc.* **1987,** *109,* 3275–3286.
17. Dieter, T.; Gafney, H. D. *Inorg. Chem.* **1988,** *27,* 1730–1736.
18. Germer, T. A.; Ho, W. *J. Vac. Sci. Technnol. A* **1989,** *7,* 1878–1881.
19. Celii, F. G.; Whitmore, P. M.; Janda, K. C. *Chem. Phys. Lett.* **1987,** *138,* 257–260.
20. Celii, F. G.; Whitmore, P. M.; Janda, K. C. *J. Phys. Chem.* **1988,** *92,* 1604–1612.
21. Creighton, J. R. *J. Appl. Phys.* **1986,** *59,* 410–414.
22. Gluck, N. S.; Ying, Z.; Bartosch, C. E.; Ho, W. *J. Chem. Phys.* **1987,** *86,* 4957–4978.
23. Swanson, J. R.; Friend, C. M.; Chabal, Y. J. *J. Chem. Phys.* **1987,** *87,* 5028–5037.
24. Geoffroy, G. L.; Wrighton, M. S. *Organometallic Photochemistry;* Academic: New York, 1979; p 35.
25. Wrighton, M. *Chem. Rev.* **1974,** *4,* 401–430.

26. Cox, A. *Photochemistry* **1983,** *14,* 158–178.
27. Kang, D.; Wrighton, M. S. *Langmuir* **1991,** *7,* 2169–2174.
28. Okahata, Y.; Enna, G.; Takenouchi, K. *J. Chem. Soc., Perkin Trans. 2* **1989,** 835–843.
29. Teixeira, G.; Aviles, T.; Dias, A. R.; Pina, F. *J. Organomet. Chem.* **1988,** *353,* 83–91.
30. Arkles, B. *CHEMTECH* **1977,** *7,* 766–778.
31. Bailey, D. C.; Langer, S. H. *Chem. Rev.* **1981,** *81,* 109–148.
32. Hartley, F. R.; Verey, P. N. *Adv. Organomet. Chem.* **1977,** *15,* 189–235.
33. Hickman, J. J.; Ofer, D.; Zou, C.; Wrighton, M. S.; Laibinis, P. E.; Whitesides, G. M. *J. Am. Chem. Soc.* **1991,** *113,* 1128–1132.
34. Chidsey, C. E. D.; Bertozzi, C. R.; Putvinski, T. M.; Mujsce, A. M. *J. Am. Chem. Soc.* **1990,** *112,* 4301–4306.
35. Whitesides, G. M.; Laibinis, P. E. *Langmuir* **1990,** *6,* 87–96.
36. Fischer, A. B.; Bruce, J. A.; McKay, D. R.; Maciel, G. E.; Wrighton, M. S. *Inorg. Chem.* **1982,** *21,* 1766–1771.
37. Chance, R. R.; Prock, A.; Silbey, R. *Adv. Chem. Phys.* **1978,** *37,* 1–65.
38. Campion, A.; Gallo, A. R.; Harris, C. B.; Robota, H. J.; Whitmore, P. M. *Chem. Phys. Lett.* **1980,** *73,* 447–450.
39. Whitmore, P. M.; Robota, H. J.; Harris, C. B. *J. Chem. Phys.* **1982,** *77,* 1560–1568.
40. Whitmore, P. M.; Robota, H. J.; Harris, C. B. *J. Chem. Phys.* **1982,** *76,* 740–741.
41. Alivisatos, A. P.; Waldeck, D. H.; Harris, C. B. *J. Chem. Phys.* **1985,** *82,* 541–547.
42. Lee, C. F.; Thompson, M. E. *Inorg. Chem.* **1991,** *30,* 4–5.
43. Hill, R. H.; Wrighton, M. S. *Organometallics* **1987,** *6,* 632–638.
44. Young, K. M.; Wrighton, M. S. *Organometallics* **1989,** *8,* 1063–1066.
45. Fodor, S. P. A.; Read, J. L.; Pirrung, M. C.; Stryer, L.; Lu, A. T.; Solas, D. *Science* **1991,** *251,* 767–773.
46. Dulcey, C. S.; Georger, J. H., Jr.; Krauthamer, V.; Stenger, D. A.; Fare, T. L.; Calvert, J. M. *Science* **1991,** *252,* 551–554.

RECEIVED for review November 7, 1991. ACCEPTED revised manuscript May 27, 1992.

4

Photocatalytic Behavior of Tungsten, Iron, and Ruthenium Carbonyls on Porous Glass

Shu-Ping Xu and Harry D. Gafney

Department of Chemistry and Biochemistry, Queens College, City University of New York, Flushing, NY 11367

Although the primary photoprocess of metal carbonyls physisorbed onto porous Vycor glass is similar to that in fluid solution, the secondary chemistry is quite different. UV photolysis of adsorbed $W(CO)_6$ leads to CH_4 evolution, but photolysis of $Ru_3(CO)_{12}$ yields a surface-grafted oxidative-addition product, $(\mu\text{-}H)Ru_3(CO)_{10}(\mu\text{-}OSi)$. The catalytic activity of these hybrid systems was examined with respect to methane evolution and olefin isomerization. Isotope labeling experiments showed that UV photolysis of $W(CO)_6$ leads to WO_3, and WO_3 photocatalyzes the conversion of $^{13}CO_2$ to $^{13}CH_4$. Spectroscopic data point to a reactive site composed of the metal oxide and Lewis acid site in the glass matrix. The addition of 1-pentene to $(\mu\text{-}H)Ru_3(CO)_{10}(\mu\text{-}OSi)$ disrupts the multicentered bonds, and photolysis of the adduct leads to cis- *and* trans-*2-pentene. The* trans–cis *ratio increases during photolysis but is significantly smaller than the thermodynamic ratio. Deuterating the oxidative addition product yields deuterated olefins, and the product distribution, 90% as 2-D-2- and 3-D-2-pentenes and 10% 1-D-1-pentene, suggests an excited state similar to a pi-allyl complex.*

THE EXCITED STATE differs from the ground state in energy, electron distribution, and nuclear configuration. It is not surprising then that its chemistry can be quite different. Certainly with classical transition metal complexes and metal carbonyls, studies of their spectroscopy, photophysics, and photochemistry over the past 30 years have elucidated a chemis-

0065–2393/93/0238–0067$06.00/0

try of the different excited states as well as the intramolecular processes that partition the energy among these states. Nevertheless, it remains a chemistry of high-energy, short-lived intermediates, and developing the means to control their reactivity or promote a specific chemistry is a significant experimental objective. This objective is readily apparent in the catalytic activity of photoactivated metal carbonyls. Although the specific nature of the catalytic intermediate remains controversial, it is generally agreed that light generates a coordinatively unsaturated intermediate, and this species, or a species derived from it, promotes the conversion of the organic substrate.

One approach being examined is to bind or adsorb the precursor onto a support (*1*), that is, to assemble a hybrid catalyst (*2*). A support provides a unique microenvironment, in which the photochemistry of the precursor and its subsequent catalytic activity reflect the chemical nature and dimensionality of the support surface. Most inorganic oxide supports, including porous glasses, possess hydroxylated surfaces on which the surface functionality, typically a hydroxyl group, is capable of acting as a scavenging nucleophile. Coordination to a surface functionality stabilizes the primary photoproduct, influences its surface mobility, and changes its optical absorption characteristics. The morphology of the surface, that is, its topology or fractal dimensionality, can also influence reactivity (*3*). In zeolites, in which cage dimensions approach molecular dimensions, encapsulation of a reagent influences its molecular dynamics (*4*). Amorphous substrates, such as porous glasses, do not possess the crystalline regularity of zeolites. In fact, porous glasses derived from the base-catalyzed polymerization of alkylorthosilicates (xerogels) frequently exhibit two realms of porosity. A microporosity on the order of tens of angstroms exists within the silicate clusters, and a mesoporosity that can range from tens or hundreds of angstroms to micrometers exists between the clusters (*5*). Nevertheless, the morphology of these amorphous materials restricts adsorbate mobility and, at least in one case, curtails the fragmentation of a photoactivated cluster (*6*). As a result, the intermediates generated on a support do not necessarily possess a direct analog among those generated in homogeneous solution and need not exhibit an equivalent photoactivated catalytic chemistry.

Our interest in photoactivated hybrid catalysis stems from studies of the photochemistry of metal carbonyls adsorbed onto Corning's code 7930 porous Vycor glass (PVG) (*7*). In many cases, the intermediates generated on this support and their subsequent chemistry differ from that in fluid solution. These changes arise from the participation of the glass in the secondary thermal and photochemical reactions of the adsorbates, but the choice of PVG as a reaction medium does not stem from a specific advantage with respect to catalysis. Rather, PVG and porous glasses in general offer a unique combination of rigidity, transparency, and porosity.

Transparency offers spectroscopic access, and in turn, an amenability to fast reaction techniques. Porosity offers chemical access; access not only in the sense of intercepting a short-lived intermediate, but also in the synthetic sense of utilizing the chemical nature, rigidity, and morphology of the porous matrix to modify adsorbate chemistry. Certainly in our case, the underlying strategy is to take advantage of the microstructure and microenvironment of the support to impose some control on a photoactivated reaction system. In this chapter, we summarize the properties of PVG and the photocatalytic activity of $W(CO)_6$, $Fe(CO)_5$, and $Ru_3(CO)_{12}$ physisorbed onto this glass.

Porous Vycor Glass (PVG)

PVG (*8–11*) is a 96% SiO_2, 3% B_2O_3, and 1% Na_2O and Al_2O_3 glass (*9*). When the borosilicate melt is cooled below its phase-transition temperature, the silica phase separates from the boron oxide–alkali oxide phase, and acid leaching of the oxide phase yields a random, three-dimensional network of interconnected pores throughout the glass. Pore size and surface area are determined by the extent of phase separation in the melt and acid leaching. Pore sizes ranging from 20 to 2500 Å are currently available, but larger pore sizes reduce optical transparency. The glass used in our experiments has an average pore diameter of 100 ± 10 Å and a surface area of 183 ± 15 m^2/g (*12*).

Scanning electron microscope (SEM) analyses of calcined (650 °C) samples reveal a nodular surface composed of silicate nodules with intervening crevices that contain the openings into the interior pore structure (*12*). The intervening spaces, which in total correspond to a void volume of ~35%, range from 40 to 100 Å (*12*). This substrate is clearly amorphous, and although the term "amorphous" appears to have a negative connotation with respect to organizing a reaction system, it is a length-dependent term; its significance depends on the dimensions of the events under consideration. Small-angle X-ray scattering (SAXS) and small-angle neutron scattering (SANS) yield a correlation length, that is, a length of uniform density, of 242 ± 8 Å (*13, 14*). Although it does not possess the geometric regularity of a zeolite, with respect to the dimensions of an adsorbate and the distance over which its chemistry occurs, this amorphous glass is a relatively uniform substrate.

With any heterogeneous medium, the active region is the surface. Diffuse reflectance Fourier transform (DRIFT) infrared spectra of the calcined glass reveal a surface composed of free 3744-cm^{-1} and associated 3655-cm^{-1} silanol groups (*7, 11*). The number of silanol groups depends on the sample's thermal history, but studies of a variety of hydroxylated silicas yield silanol numbers of 4–7 per 100 Å^2 with the highest density within the pores (*15, 16*). Trace amounts of chemisorbed water are also present in calcined (650 °C) samples (*17*). PVG is often likened to silica gel, but the two materials are not chemically equivalent. In addition to the silanol groups, which function as weak Brønsted acids, PVG also possesses B_2O_3 Lewis acid sites. As a result of its method of manufacture, these sites are dispersed on surfaces throughout the glass matrix. X-ray photoelectron spectroscopic (XPS) analyses indicate that the amount of B present in the first 50 Å of the samples used in our experiments is 2.6 ± 0.1% (*12*).

Impregnation is accomplished by conventional solution adsorption or vapor deposition techniques (*17, 18*). Regardless of initial loading, however, neither technique yields a uniform distribution of the organometallic compound throughout the pore volume. Impregnation of the bulk is limited by the narrow, tortuous passes connecting the pores, and typical exposure times of ≤24 h result in impregnation of the outermost volumes of glass. Nevertheless, within these volumes, optical spectra confirm a uniform distribution of the complexes (*17, 18*).

Photocatalyzed Methanation of CO_2

$W(CO)_6$ physisorbs on PVG without disruption of its primary coordination sphere. In contrast to its photochemistry in fluid solution, 300- or 254-nm photolyses of the adsorbed complex lead to CO loss followed by CH_4 evolution (*17*). Similar results occur in the thermal activation of $W(CO)_6$ physisorbed onto silica gel (*19, 20*). As noted, PVG is similar to silica gel, but two fundamental differences exist in the observed chemistry.

First, in the thermally activated system, methane is attributed to the hydrogenation of the coordinated CO (*19, 20*). However, stoichiometric measurements and ^{13}C-labeling experiments confirm that the carbon source in the photochemical system is not coordinated CO (*17*). In this case, photolysis of the precursor leads to CO loss, but when the adsorbed complex achieves an average molecularity of $W(CO)_4$, CH_4 evolution accompanies metal oxidation. CH_4 is the sole hydrocarbon evolved, and the amount evolved falls within known impurity levels; therefore, CH_4 is attributed to the hydrogenation of a carbonaceous impurity thought to be a C_1 oxide within the glass matrix (*17*).

Second, the photoactivated system becomes catalytic, whereas thermal activation yields a stoichiometric reaction. Although photoinduced CH_4 evolution initially occurs with concurrent oxidation of the tungsten, photoactivation of the resultant oxide continues CH_4 evolution (*17*), whereas thermally activated CH_4 evolution is limited to a stoichiometric reaction (*19, 20*). With less than monolayer coverage and 254-nm excitation, the glass, rather than the metal carbonyl or resultant metal oxide, is the dominant absorbing species. However, CH_4 evolution does not occur in the absence of the metal. UV–visible spectra indicate that the resultant oxide is WO_3, and on the basis of the amount of W present, the photoactivated oxide exhibits turnover numbers ranging from ~5:1 to 12:1 in the $W(CO)_6$–PVG systems examined (*17*).

These results raised two questions that are the focus of the current experiments. If the carbon source is a C_1 oxide, is this hybrid system capable of photocatalyzing the hydrogenation of an external carbon source, specifically CO_2? Also, does the glass matrix provide specific site(s) that promote, in concert with the metal oxide, the hydrogenation of a C_1 oxide?

When a sample of PVG containing 10^{-6} mol of $W(CO)_6$(ads)/g (ads denotes an adsorbed species) is exposed to 25 torr (3332 Pa) of $^{13}CO_2$ (2×10^{-4} mol), the gas rapidly equilibrates between the adsorbed and gas phases. However, a 254-nm photolysis does not result in immediate $^{13}CH_4$ evolution. Instead, $^{12}CH_4$ initially accompanies ^{12}CO evolution, but as the photolysis proceeds, the rate of $^{12}CH_4$ evolution declines and is replaced by $^{13}CH_4$ and ^{13}CO evolution. The 1986-cm^{-1} band of the hexacarbonyl disappears during $^{12}CH_4$ evolution, and when $^{13}CH_4$ evolution occurs, the electronic spectrum of the adsorbate consists of a strong UV absorbance (50% transmission (T) at 232 nm) with a weak shoulder at 330 nm. As noted, the spectrum of the adsorbate is essentially equivalent to the spectrum of PVG impregnated with WO_3 (*17*), and repeating the experiments with WO_3-impregnated samples (Table I) yields similar results. A 254-nm photolysis of a PVG sample containing 2×10^{-6} mol of WO_3/g under 25 torr (3332 Pa) of $^{13}CO_2$ leads to $^{12}CH_4$ and ^{12}CO evolution followed by $^{13}CH_4$ and ^{13}CO evolution.

These results implicate WO_3, or a species derived from it, as the photocatalytic reagent. At low loadings ($\leq$15% by weight) on Al_2O_3, WO_3 exists as the tetrahedral ion WO_4^{2-} (*21*). WO_3 is less dispersed on SiO_2, and at monolayer coverage exists as a mixture of bulk WO_3 and a tetrahedral species bound to SiO_2 through W–O–Si bonds (*22–24*), the latter being suggested by a corresponding reduction in the intensity of the SiO–H proton resonance (*24*). Although the WO_3 loadings in our experiments correspond to surface coverages ($\leq$0.4%) that are well below monolayer coverage, the spectral data suggest that both species are present in PVG. When WO_3 is adsorbed onto PVG, a corresponding decline occurs in the 3744-cm^{-1} SiOH band, whereas the UV–visible spectrum of the impregnated sample corresponds to that of the agglomerated metal oxide. However, bulk WO_3 is the dominant light-adsorbing species present during the photocatalyzed evolution of $^{13}CH_4$.

Table I. Average Rates of CO and CH_4 Evolution during Photolysis of $W(CO)_6$(ads) and WO_3(ads)

		Amount (mol)			*Rate (mol/h)*			
Exp.	*Impregnate*	*Adsorbed*[a] $\times 10^{-6}$	$^{13}CO_2$ $\times 10^{-4}$	^{12}CO $\times 10^{-8}$	$^{12}CH_4$ $\times 10^{-7}$	^{13}CO $\times 10^{-8}$	$^{13}CH_4$ $\times 10^{-8}$	*Turnover Frequency*[b] $\times 10^{-6}$
1	$W(CO)_6$	2.8 (1.9)	2.0	5.6	1.5	0.85	0.32	0.32
2	$W(CO)_6$[c]	4.0 (2.7)	2.0	8.9	6.0	7.0	1.8	1.3
3	WO_3[c]	1.4 (1.1)	1.6	3.1	1.2	14.0	1.6	3.0
4	WO_3	2.8 (1.9)	1.6	2.3	1.8	4.8	3.3	3.2
5	WO_3[d]	2.6 (1.9)	1.6	2.3	0.20	3.8	1.8	1.9
6	WO_3[e]	2.2 (1.9)	1.6	0.30	0.25	11.2	0.23	0.30

[a] Total moles adsorbed; numbers in parentheses are the number of moles adsorbed per gram of PVG.

[b] For $^{13}CH_4$ evolution in reciprocal seconds.

[c] $^{13}CO_2$ was added after $^{12}CH_4$ evolution (*see* text).

[d] Evacuated sample ($P \leq 10^{-5}$ torr or 13.33×10^{-6} Pa) was equilibrated with water vapor.

[e] Extensively dehydrated by evacuation at 650 °C and $P \leq 10^{-4}$ (13.33×10^{-5} Pa) for 15 h.

Whether present as an individual molecular entity or an aggregate, formation of the metal oxide, in itself, does not account for methane formation. The immediate appearance of $^{12}CH_4$ in all experiments with both $W(CO)_6$ and WO_3, regardless of the initial $^{13}CO_2$ pressure, implies that the carbonaceous impurity is more easily hydrogenated than $^{13}CO_2$. Comparisons of individual reaction rates of hybrid systems are tenuous, but in all experiments (Table I), the rate of $^{13}CH_4$ is smaller than the rate of $^{12}CH_4$ evolution. This result suggests that an equally important parameter is the nature of the site on the glass itself.

Regardless of whether the reaction is initiated with $W(CO)_6$ or WO_3, the first events are the conversion of the C_1 oxide impurity to $^{12}CH_4$. As the impurity sites are depleted, $^{13}CO_2$ begins to adsorb onto the vacated sites, and continued photolysis of WO_3 leads to $^{13}CH_4$ evolution. In fact, if the impurity sites are depleted prior to exposure to $^{13}CO_2$, photolysis leads to immediate $^{13}CH_4$ evolution. For example, sample 2 (Table I) was photolyzed until the rate of $^{12}CH_4$ evolution was below that detectable by gas chromatography. At that point, the sample was evacuated [pressure $P \leq 10^{-5}$ torr (133.3×10^{-5} Pa)], charged with 25 torr (3332 Pa) of $^{13}CO_2$, and photolyzed. Periodic gas chromatographic–mass spectrometric (GC–MS) analyses of the surrounding gas phase confirmed immediate $^{13}CH_4$ formation.

Photolyses on partially deuterated PVG indicate that the hydrogen source is either the silanol group or chemisorbed water (*17*). In these experiments, calcined powdered or plate samples of the glass were refluxed in neat D_2O ($\geq$99%) for 8–12 h. After drying under vacuum at room temperature, the reduced intensities of the 3744- and 3655-cm^{-1} SiO–H bands and corresponding growth of the 2760- and 2641-cm^{-1} SiO–D bands indicated that 33–50% of the Si–OH groups were converted to Si–OD. A 310- or 254-nm photolysis of $W(CO)_6$ adsorbed onto deuterated PVG results in the evolution of CH_4, CH_3D, and small amounts of CH_2D_2. The amount of deuterated products increases with the extent of deuteration, but never exceeds 25% of the total amount of CH_4 evolved. This finding might be taken as prima facie evidence that the glass acts as the hydrogen source, but water is tenaciously held on PVG. Thermal gravimetric analyses show a slow evolution of water from samples calcined at 650 °C for 48 h. A weak high-frequency shoulder on the 2760-cm^{-1} band indicates the presence of small amounts of D_2O in these samples (*17*). Rapid exchange between the hydroxyl moieties is expected, so the data designate only the silanol group and chemisorbed water as the hydrogen source.

We suspect that water is both the hydrogen source and ultimately the source of the reducing equivalents. However, methane evolution exhibits a complex dependence on the amount of adsorbed water. Extensive dehydration of a sample containing 2.2×10^{-6} mol of WO_3/g under vacuum [P

$\leq 6 \times 10^{-5}$ torr (799.8×10^{-5} Pa)] at 650 °C (Table I, sample 6) reduces the rate of $^{13}CH_4$ evolution by an order of magnitude during subsequent photolysis under 25 torr (3332 mPa) of $^{13}CO_2$. On the other hand, increasing the water content by exposing a sample containing 2.6×10^{-6} mol of WO_3/g to water vapor (Table I, sample 5) also decreases the rate of $^{13}CH_4$ evolution relative to an unexposed sample. Consistent with it acting as the source of the reducing equivalents, O_2 is detected as a reaction product, although further experiments are needed to establish the stoichiometry of the reaction. In excess, however, water may reduce the amount of $^{13}CO_2$ adsorbed onto the glass or compete with $^{13}CO_2$ for adsorption onto the active site.

Exposing PVG samples containing WO_3 to CO_2 does not result in IR bands indicative of an interaction between the surface oxide and CO_2. Consistent with a WO_3 surface coverage of $\leq 0.4\%$, the spectral changes are dominated by those observed when unimpregnated PVG is exposed to CO_2. The intense, sharp, SiOH band at 3744 cm^{-1} is replaced by a series of bands at 3730, 3632, and 3599 cm^{-1} and bands indicative of physisorbed CO_2 at 2360, 2236, and 671 cm^{-1}. However, a series of weak bands also appear in the 1700–1200-cm^{-1} region and have been assigned to carbonate and bicarbonate (*25, 26*). The intensities of these bands are independent of the pressure of CO_2 and do not disappear on evacuation, whereas the bands indicative of physisorbed CO_2 and the 3730-, 3632-, and 3599-cm^{-1} silanol bands disappear immediately. This behavior suggests a limited number of adsorption sites, and on adsorption onto these sites, CO_2 is converted to species similar to carbonate and bicarbonate. On irradiation, the bands in the 3730–3599- and 1700–1200-cm^{-1} regions disappear concurrent with CH_4 evolution.

Of the bands that appear in the 1700–1200-cm^{-1} region, however, only the weak band at 1655 cm^{-1} agrees with the bands observed when CO_2 is adsorbed onto SiO_2. The other bands, which appear at 1704, 1516, 1467, and 1293 cm^{-1}, are similar to bands that appear when CO_2 is adsorbed onto Al_2O_3 (*21*). In addition to SiOH Brønsted acid sites, PVG also possesses electron-deficient Lewis acid sites in the form of surface B_2O_3 sites. The amount of B present in the glass, $2.6 \pm 0.1\%$ in the first 50 Å (*12*), is consistent with a limited number of active sites, and the spectrum CO_2 adsorbed onto these sites would be more analogous to that adsorbed onto Al_2O_3 than that adsorbed onto SiO_2. Furthermore, adsorption onto a Lewis acid site would be expected to increase the susceptibility of CO_2 to reduction, and recent experiments show that tying up the Lewis acid sites through a reaction with NH_3 (*27*) reduces the rate of CH_4 evolution substantially.

Further experiments are in progress, but the data gathered to date clearly show that this hybrid system photocatalyzes the conversion of CO_2

to CH_4. The data point to an active site composed of either a surface-bound tetrahedral tungsten oxide or agglomerated WO_3 and a B_2O_3 Lewis acid site.

Photocatalyzed Isomerization of 1-Pentene

$Fe(CO)_5$. $Fe(CO)_5$ physisorbs onto PVG without disruption of its primary coordination sphere. IR and electronic spectra of the adsorbed complex are essentially equivalent to fluid solution spectra. UV photolysis leads to CO loss, and the tetracarbonyl rapidly reacts with the silanol groups or chemisorbed water to form the oxidative addition products $H–Fe(CO)_4–OSi$ and $H–Fe(CO)_4OH$ (*18*). Photolysis of $Fe(CO)_5$ under a 1-pentene atmosphere, however, leads to quantitative formation of $Fe(CO)_4(1\text{-}C_5H_{10})$(ads), which is then photochemically converted to the shorter-lived $Fe(CO)_3(1\text{-}C_5H_{10})$(ads) and $HFe(CO)_4(1\text{-}C_5H_9)$(ads) (*28*). The latter species are thought to be crucial intermediates in the catalyzed reaction in homogeneous solution (*29, 30*), and the quantum yield of isomerization obtained with the hybrid system, 152 $\pm$ 23 mol/einstein, suggests a similar reaction sequence (*31*). Light generates the active species, which then thermally catalyzes alkene isomerization. However, the *trans–cis* product ratio differs from the thermodynamic ratio, 4.82, and varies with irradiation time. The ratio increases from 1.6 $\pm$ 0.2 after 10 min of photolysis to 3.7 $\pm$ 0.2 after 60 min. Similar results have been obtained with $Fe(CO)_5$ adsorbed onto the outer surfaces of small-pore zeolites, in the supercages of large-pore zeolites (*32*), and with phosphine derivatives anchored to a styrene microporous resin (*33*). In each of these hybrid systems, the *trans–cis* product ratio varies with irradiation time and differs from the thermodynamic ratio, 4.82.

$Ru_3(CO)_{12}$. Ruthenium complexes, in general, catalyze a number of transformations, and thermal activation of $Ru_3(CO)_{12}$ on high-surface-area supports yields catalysts of enhanced activity and selectivity. $Ru_3(CO)_{12}$ physisorbs onto PVG without a change in its molecular integrity (*6*). In contrast to the substitution and fragmentation reactions that occur in fluid solution and on functionalized silica gel (*34, 35*), however, UV photolysis of $Ru_3(CO)_{12}$ physisorbed onto PVG yields the surface-grafted, oxidative-addition product $(\mu\text{-H})Ru_3(CO)_{10}(\mu\text{-OSi})$ (*6*) (*see* structure on page 76). Oxygen binds the complex to the glass surface and acts as a three-electron donor, whereas hydrogen binds through a two-electron, three-centered bond. Although stable for weeks in vacuo, the grafted complex remains highly reactive.

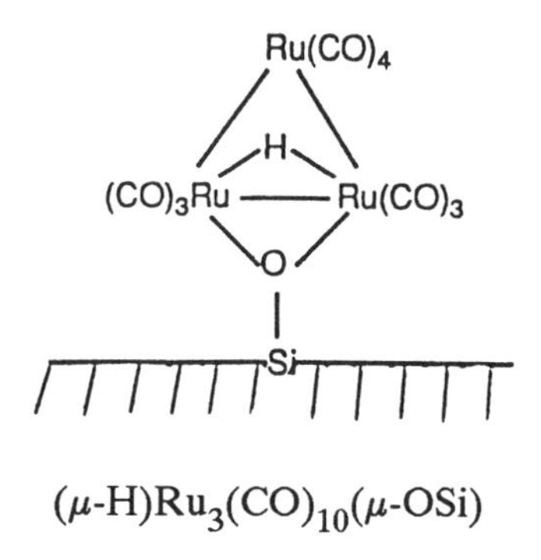

(μ-H)$Ru_3(CO)_{10}$(μ-OSi)

Exposing (μ-H)$Ru_3(CO)_{10}$(μ-OSi) to 400 torr (53.32 kPa) of 1-pentene results in an immediate reaction. A decline in its characteristic 330-nm absorption is accompanied by a broad, nondescript increase in absorbance in the 300–450-nm region with a weak shoulder at ~310 nm. (μ-H)$Ru_3(CO)_{10}$(μ-OSi) exhibits a weak band at 2109 cm^{-1}, relatively intense bands at 2078 and 2068 cm^{-1}, and a broad band at 2035 cm^{-1} with shoulders at 2017 and 1999 cm^{-1}. On exposure to 1-pentene (Figure 1), the bands at 2109 and 2078 cm^{-1} decline, the 2068-cm^{-1} band declines slightly and shifts to 2066 cm^{-1}, and the 2035-cm^{-1} band shifts to 2028 cm^{-1}. Concurrent with these changes, new bands appear at 2102 and 1830 cm^{-1}.

Evacuating the cell to a pressure of 4×10^{-4} torr (53.32 mPa) reverses the spectral changes and regenerates the IR spectrum of the oxidative addition product. Although regenerating the starting material requires hours, the increase in absorbance at 330 nm indicates ≥90% recovery of (μ-H)$Ru_3(CO)_{10}$(μ-OSi), and GC analysis of the effluent from the reactor confirms that the reaction with 1-pentene does not result in CO loss. The ability to quantitatively cycle the system and recover $Ru_3(CO)_{12}$(ads) in ≥90% yield when (μ-H)$Ru_3(CO)_{10}$(μ-OSi) is exposed to CO (1 atm) strongly suggests that the metal trimer remains intact during the reaction sequence. As expected, addition of 1-pentene disrupts the multicentered bonds binding the trimer to the glass surface.

Three reaction products are possible: disruption of an Ru–O bond (I), an Ru–H bond (II), or both (III). Basset and co-workers (*36*) proposed a structure analogous to **I** in $Os_3(CO)_{12}SiO_2$-catalyzed olefin hydrogenation. Although IR spectra recorded in our experiments do not exhibit a distinct absorption in the 1960–2060-cm^{-1} region that could be assigned to a terminal Ru–H vibration (*37*), we tentatively assign the product to **III**. This assignment is based solely on the stability of the (μ-H)$Ru_3(CO)_{10}$(μ-OSi)-olefin product. The adduct persists for hours at room temperature, and the observed stability seems more consistent with a species in which the Ru atoms are formally 18-electron species. Conse-

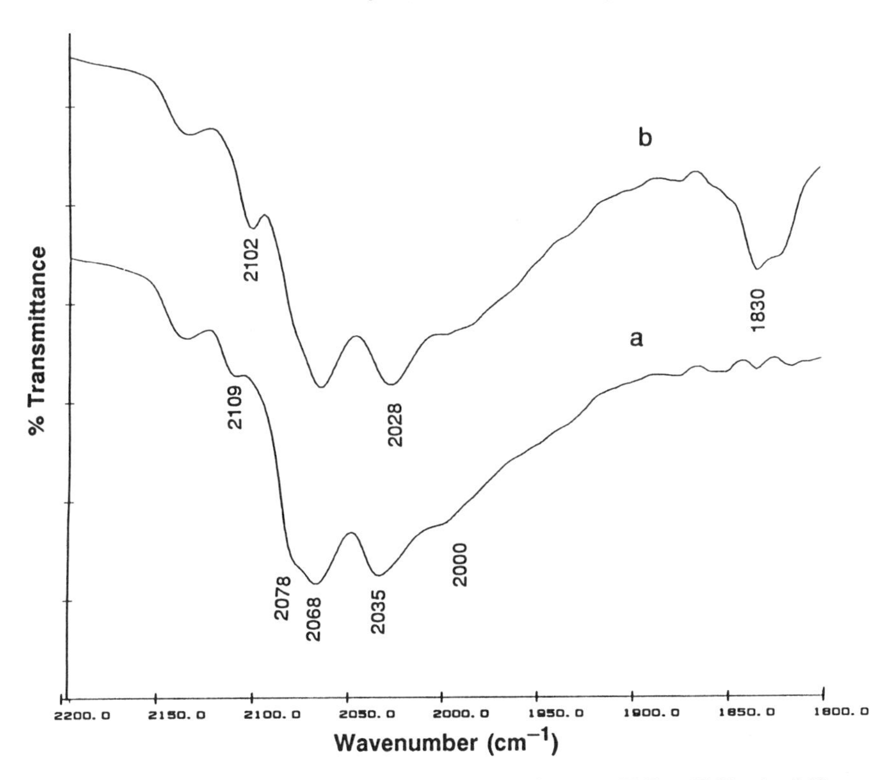

Figure 1. Diffuse reflectance FTIR spectra of (a) (μ-H)$Ru_3(CO)_{10}$(μ-OSi) and (b) (μ-H)$Ru_3(CO)_{10}$(μ-OSi) under 400 torr (53.32 kPa) of 1-pentene.

quently, the bands at 2109, 2078, 2066, 2028, and 2102 cm^{-1} are assigned to III. Although shifted, the band at 1830 cm^{-1} is assigned to 1-pentene adsorbed on the glass, because 1-pentene adsorbed onto calcined PVG exhibits a relatively intense band at 1850 cm^{-1} (*28*), and the 1830-cm^{-1} band disappears under vacuum more rapidly than those assigned to III.

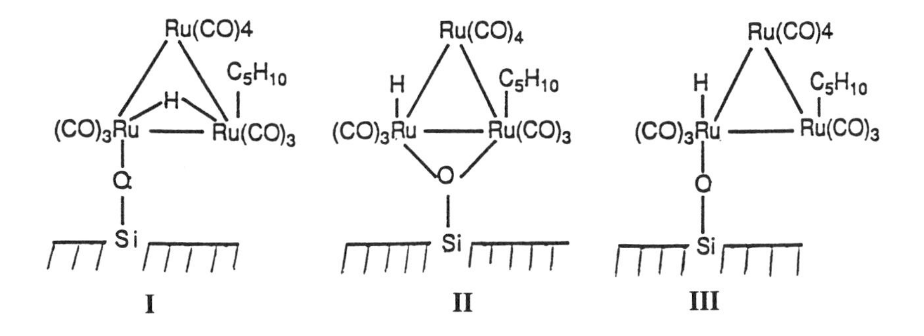

Electronic and DRIFT spectra recorded during UV photolysis of **III** under 400 torr (53.32 kPa) of 1-pentene show relatively little change. A slight decline in the 2102-cm^{-1} band, characteristic of **III**, is accompanied by a corresponding increase in the 2078-cm^{-1} band characteristic of (μ-H)$Ru_3(CO)_{10}$(μ-OSi). Nevertheless, periodic CG analyses of the surrounding vapor phase indicate the conversion of 1-pentene to *cis*- and *trans*-2-pentene.

The *trans–cis* ratio is initially 1.5 ± 0.1, and increases to a relatively constant value of 2.0 ± 0.1 as the photolysis proceeds (Figure 2). The ratio is smaller than the thermodynamic ratio, 4.82, and smaller than that obtained with $Fe(CO)_5$(ads), where the ratio increases from 1.6 ± 0.2 to 3.7 ± 0.2 during photolysis (*28*).

CO evolution does not occur during photolysis, and introducing H_2 into the reactor neither increases the rate of isomerization nor results in hydrogenation of the olefin. However, generating (μ-D)$Ru_3(CO)_{10}$(μ-OSi) by photolysis of $Ru_3(CO)_{12}$ adsorbed onto deuterated PVG (*6*) and using the deuterated analog yields deuterated olefins. Of the total deuterium incorporated, GC–MS indicates that ~90% is present in the 2-

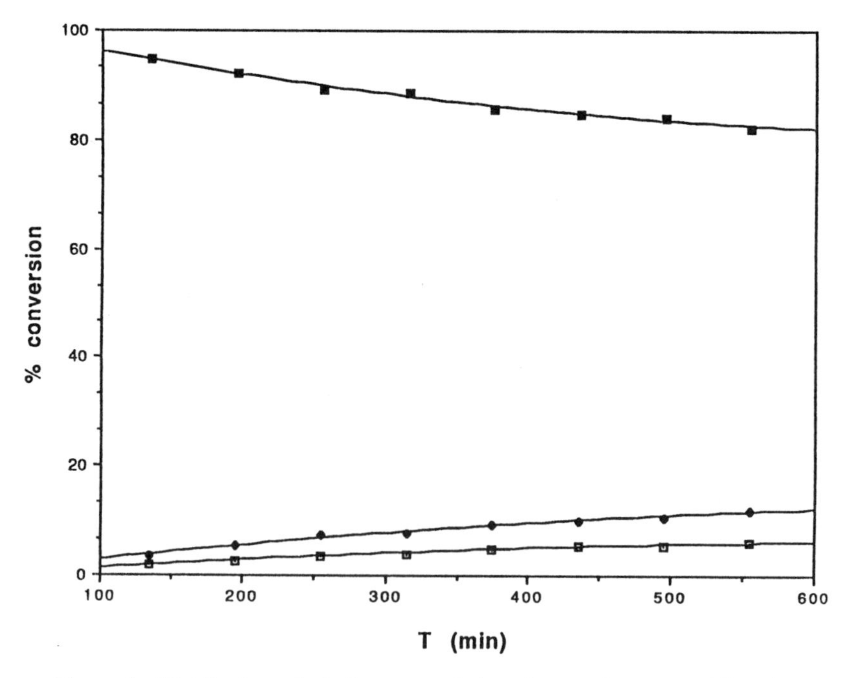

Figure 2. Distribution of cis-*2-pentene (□) and* trans-*2-pentene (◆) during 300-nm photolysis of (μ-H)$Ru_3(CO)_{10}$(μ-OSi) under 400 torr (53.32 kPa) of 1-pentene (■).*

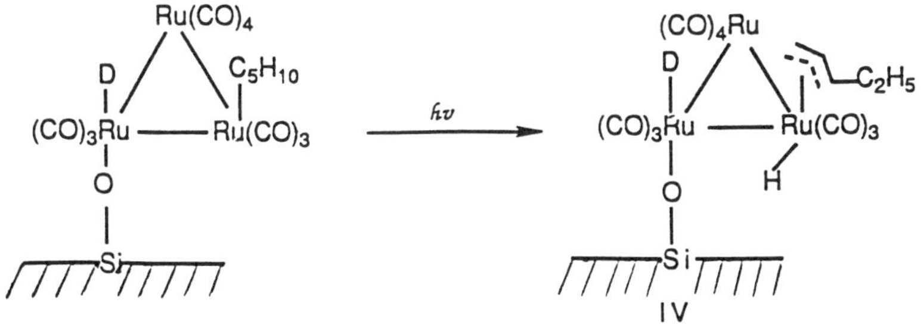

Scheme I.

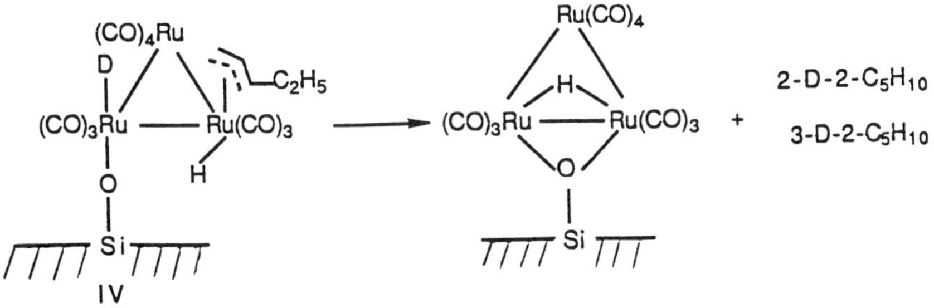

Scheme II.

pentenes, and the remainder is present principally as 1-D-1-pentene with trace amounts of 1-D-2-pentene.

Of the 2-pentenes formed, 2-D-2- and 3-D-2-pentenes are the dominant products, and the relative peak heights yield a 2-D-2-pentene:3-D-2-pentene ratio of $4.8 \pm 0.1{:}1$.

UV photolysis of 1-pentene physisorbed onto PVG does not result in olefin isomerization. The metal complex is essential to the conversion, but in this case, the reaction appears to be a photoassisted catalytic process in which excitation of **III** generates an excited state that promotes 1-pentene isomerization.

The formation of 2-D-2- and 3-D-2-pentenes suggests that the excited state may be similar to a pi-allyl complex, as is shown in Scheme I. In this configuration, the carbons in the 2- and 3-positions are susceptible to deuterium substitution. The displaced olefin hydrogen can be transferred to the metal complex to regenerate the surface grafted cluster, which then reacts with 1-pentene to re-form **III** (Scheme II).

Further experiments are necessary to determine the distribution of deuterium in the *cis*- and *trans*-2-pentenes. Nevertheless, the *trans–cis* ratio observed, 2.0 ± 0.1, is considerably smaller than the thermodynamic ratio, 4.82. One possible explanation, of course, is that the topology of the glass surface to which the complex is bound biases the isomer ratio.

However, if this were the sole determinant of the trans–cis ratio, we would expect a ratio similar to that obtained in the photocatalyzed isomerization of 1-pentene by $Fe(CO)_5$ on PVG, that is, 3.7 ± 0.2. In this case, the photochemical reaction generates a thermally activated ground-state catalyst, whereas in the Ru–PVG system, photoactivation generates an excited state that promotes olefin isomerization. Consequently, the smaller ratio found with this system may reflect not only the topology of the support, but also the specific catalytic species present on the glass surface.

Acknowledgments

Support of this research by the Research Foundation of the City University of New York and the National Science Foundation (Grant CHE–8913496) is gratefully acknowledged. H. D. Gafney thanks Corning Inc. for samples of porous Vycor glass.

References

1. Anpo, M.; Matsuura, T. *Photochemistry on Solid Surfaces;* Elsevier: New York, 1989.
2. Bailey, D. C.; Langer, S. H. *Chem. Rev.* **1981,** *81,* 109.
3. Avnir, D.; Ottolenghi, M. In *Photochemistry in Organized and Constrained Media;* Ramamurthy, V., Ed.; VCH: New York, 1991; p 535.
4. Zhang, Z.; Turro, N. J. In *Photochemistry on Solid Surfaces;* Anpo, M.; Matsuura, T., Eds.; Elsevier: New York, 1989; p 197.
5. Brinker, C. J.; Scherer, G. W. *Sol–Gel Science: The Physics and Chemistry of Sol–Gel Processing;* Academic: San Diego, CA, 1990; p 519.
6. Dieter, T.; Gafney, H. D. *Inorg. Chem.* **1988,** *27,* 1730.
7. Gafney, H. D. In *Photochemistry on Solid Surfaces;* Anpo, M.; Matsuura, T., Eds.; Elsevier: New York, 1989; p 272.
8. Iler, R. K. *The Chemistry of Silica,* Wiley-Interscience: New York, 1979; p 551.
9. Elmer, T. H. *J. Am. Ceram. Soc.* **1970,** *53,* 171.
10. Janowski, V. F.; Heyer, E. *Z. Chem.* **1979,** *19,* 1.
11. Hair, M. L.; Chapman, I. D. *J. Am. Ceram. Soc.* **1966,** *49,* 651; *Trans. Faraday Soc.* **1965,** *61,* 1507.
12. Mendoza, E. A.; Wolkow, E.; Sunil, D.; Wong, P.; Sokolov, J.; Rafailovich, M. H.; den Boer, M.; Gafney, H. D. *Langmuir,* **1991,** *7,* 3046.
13. Mendoza, E. A.; Wolkow, E.; Sunil, D.; Sokolov, J.; Rafailovich, M. H.; Gafney, H. D.; Long, G. G.; Jemian, P. R. *Chem. Phys. Lett.* **1990,** *57,* 209.
14. Wiltzius, P.; Bates, F. S.; Dierker, S. B.; Wignall, G. D. *Phys. Rev.* **1987,** *A36,* 2991.

15. Snyder, L. R.; Ward, J. W. *J. Phys. Chem.* **1966,** *70,* 3941.
16. Huber, T. E.; Huber, C. A. *J. Phys. Chem.* **1990,** *94,* 2505.
17. Simon, R. C.; Mendoza, E. A.; Gafney, H. D. *Inorg. Chem.* **1988,** *27,* 2733.
18. Darsillo, M. S.; Gafney, H. D.; Paquette, M. S. *J. Am. Chem. Soc.* **1987,** *109,* 3275.
19. Brenner, A.; Hucul, D. A. Hardwick, S. J. *Inorg. Chem.* **1979,** *18,* 1478.
20. Brenner, A.; Hucul, D. A. *J. Am. Chem. Soc.* **1980,** *102,* 2484.
21. Salvati, L.; Makovsky, L. E.; Stencel, J. M.; Brown, F. R.; Hercules, D. M. *J. Phys. Chem.* **1981,** *85,* 3700.
22. Murrell, L. L.; Grenoble, D. C.; Baker, R. T. K.; Prestridge, E. B.; Fung, S. C.; Chianelli, R. R.; Cramer, S. P. *J. Catal.* **1983,** *79,* 203.
23. Yan, Q.; Liu, H.; *Fenzi Cuihua,* **1988,** *2*(2), 87.
24. Reddy, B. M.; Rao, K. S. P.; Mastikhin, V. M. *J. Catal.* **1988,** *113,* 556.
25. Ward, J. W.; Habgood, H. W. *J. Phys. Chem.* **1966,** *70,* 1178.
26. Rethwisch, D. G.; Dumesic, J. A. *Langmuir,* **1986,** *2,* 73.
27. Cant, N. W.; Little, L. H. *Can. J. Chem.* **1964,** *42,* 802.
28. Darsillo, M. S.; Gafney, H. D.; Paquette, M. S. *Inorg. Chem.* **1988,** *27,* 2815.
29. Whetten, R. L.; Fu, K.-J.; Grant, E. R. *J. Am. Chem. Soc.* **1982,** *104,* 4270.
30. Wuu, Y.-M.; Bentsen, J. G.; Brinkley, C. G.; Wrighton, M. S. *Inorg. Chem.* **1987,** *26,* 530.
31. Schroeder, M. A.; Wrighton, M. S. *J. Am. Chem. Soc.* **1976,** *98,* 551.
32. Suib, S. L.; Kostapapas, A.; McHahon, K. C.; Baxter, J. C.; Winiecki, A. M. *Inorg. Chem.* **1985,** *24,* 858.
33. Liu, D. K.; Wrighton, M. S. *J. Am. Chem. Soc.* **1982,** *104,* 898.
34. Desrosiers, M. F.; Wink, D. A.; Trautman, R.; Friedman, A. E.; Ford, P. C. *J. Am. Chem. Soc.* **1986,** *108,* 1917.
35. Doi, Y.; Yano, K. *Inorg. Chem. Acta,* **1976,** *76,* L71.
36. Besson, B.; Choplin, A.; D'Ornelas, L.; Basset, J. M. *Chem. Commun.* **1982,** 842.
37. Nakamoto, K.; *Infrared and Raman Spectra of Inorganic and Coordination Compounds;* 3rd ed.; Wiley-Interscience: New York, 1980; p 290.

RECEIVED for review January 30, 1992. ACCEPTED revised manuscript June 29, 1992.

5

Metal–Organic Photochemistry in the Millisecond-to-Picosecond Time Domain

Formation and Dissociation of Cu–C Bonds

G. Ferraudi

Radiation Laboratory, University of Notre Dame, Notre Dame, IN 46556

Time-resolved techniques have been successfully applied to mechanistic studies of photosensitive metal–organic systems. New techniques have increasingly been used recently. Some of them have shown improvement in time resolution (i.e., from microsecond to femtosecond), and others represent the incorporation of various spectroscopies to the detection of reaction intermediates. Some applications of these new techniques are reviewed, and some of their intrinsic limitations are highlighted. Examples related to the formation and dissociation of carbon–copper bonds in the picosecond-to-millisecond time domain illustrate the application of flash photolysis to mechanistic studies.

Flash Photochemical Techniques

Flash photolysis has been widely used for the investigation of transient species (i.e., reaction intermediates and excited states) in the photochemistry of coordination compounds. Although considerable progress has been made since the introduction of the technique for the detection of electronic excited states of organic molecules, the basic idea behind these techniques remains the same (*1–4*). The method can be illustrated by considering a general photochemical reaction:

0065–2393/93/0238–0083$07.00/0

$$A \xrightarrow[\phi]{h\nu} T \xrightarrow{k} \text{products} \quad (1)$$

where a short-lived intermediate, T, is generated with a quantum yield, Φ, by flash irradiation of a photolyte, A. Assume that a mathematical function, $I_{ab}(t)$, of the time, t, gives the intensity of the flash absorbed by the sample at any instant of the irradiation. If the photogeneration of T in equation 1 is fast enough to follow the absorption of light, and the decay with a rate constant, k, is somehow a slower process, the integrated rate law for the concentration of T, [T], takes the form of equation 2:

$$[T] = C \exp -kt + \exp -kt \int I_{ab}(t) \exp kt \, dt \quad (2)$$

where C is an integration constant. Because the second term in equation 2 couples the dependence on time of the absorbed light, $I_{ab}(t)$, to the decay of T, mathematical deconvolution must be used in order to learn about the spectroscopic properties or the reaction kinetics of such an intermediate. The need for cumbersome calculations can be circumvented if the sample is irradiated with a short and intense flash that is, for practical purposes, extinguished at $t = \tau_{flash}$ (i.e., before the factor exp kt has undergone any significant change from unity). In these experimental conditions, equation 3 expresses the concentration of intermediate at any time, t > τ_{flash}, longer than the brief flash irradiation.

$$[T] = \phi n_{h\nu} \exp -kt \qquad t > \tau_{flash} \quad (3)$$

where $n_{h\nu}$ is the number of photons per unit volume absorbed from the flash. The product, $\phi n_{h\nu}$, gives the concentration of T left in solution by the flash irradiation, and the exponential term can be associated with the integrated rate law of a first-order reaction. Any chemical or physical property proportional to the concentration can be used to detect T and investigate the kinetics of the chemical reaction. A flash fall time, τ_{flash}, must be, therefore, shorter than the transient T lifetime, $1/k$, for its unobstructed observation in flash photolysis. In this regard, $\tau_{flash} < 0.1/k$ can be considered as a good experimental criterion (*5*). The direct conclusion from these arguments has fueled chemists' interest in the generation of flashes with very short rise times and fall times (i.e., with orders of femtoseconds) and correspondingly large powers (i.e., in the gigawatt range).

Equally important to the flash photolysis experiment is the procedure followed for the inspection of T as it undergoes chemical reactions. The

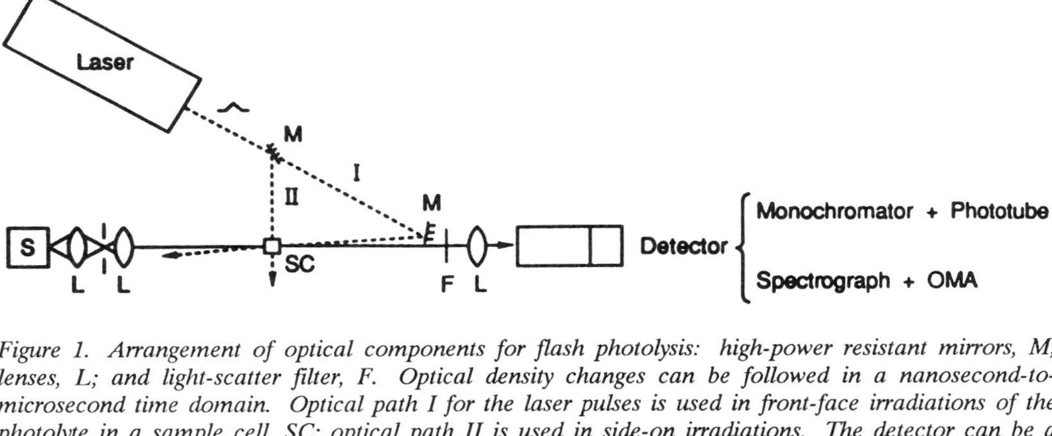

Figure 1. Arrangement of optical components for flash photolysis: high-power resistant mirrors, M; lenses, L; and light-scatter filter, F. Optical density changes can be followed in a nanosecond-to-microsecond time domain. Optical path I for the laser pulses is used in front-face irradiations of the photolyte in a sample cell, SC; optical path II is used in side-on irradiations. The detector can be a monochromator–phototube or a spectrograph–OMA. A gated OMA is required when a steady source of light, S, is used for probing the optical changes. Similar optics can be adapted to the study of time-resolved flash fluorescence.

optical arrangements in Figure 1 are commonly used for the measurement of transient spectra and reaction kinetics in a nanosecond-to-second time domain. Probing light from a lamp with a constant intensity, a Xe or quartz–halogen lamp, is used for experiments with microsecond-to-second time resolutions, and a Xe lamp, pulsed several orders of magnitude above the steady-state running conditions, is used for the nanosecond-to-microsecond time domain. Different electric wirings (divider circuitry) are required for the photomultiplier in order to get a response for the detection system that is compatible with the time domain of the chemical transformation (*6–8*). In some flash photolysis experiments in which a transient spectrum must be measured over a wide range of wavelengths with larger spectral than temporal resolution, a gated optical multichannel analyzer, OMA, may prove to be a more suitable detector than the photomultiplier (*9*).

Detection of optical transient events from a picosecond to several nanoseconds requires a different type of optical train; one of several in literature reports is shown in Figure 2 (*10–13*). Such optical arrangements are designed on the idea of delaying a pulse of white light for a well-established period with respect to a laser flash used for the irradiation of the photolyte (pump-probe method). For example, a flash with first harmonic light from a mode-locked Nd:YAG laser, λ = 1062 nm, is delayed with respect to another flash with light of a higher harmonic, λ = 531, 354, or 266 nm, when the probe light is made to travel a comparatively longer distance. When the delayed pulse is focused on a cell with some appropriate material, for example, a mixture of D_2O and H_2O or CS_2, Raman scatter or dielectric breakdown results in the generation of an intense, picosecond-lived, flash of white light, $\lambda \geq 420$ nm. Differences between the absorptions of the white light by a solution irradiated by the actinic (pump) pulse and unirradiated blanks are calculated into a differential spectrum that may signal the presence of transient species at an instant equal to the delay between laser pulses.

The mechanism of the reactions undergone by such species can be investigated by means of spectra measured with delays spanning several picoseconds to nanoseconds. A variety of flash photolysis apparatuses with picosecond time resolutions can be constructed by incorporating other optical elements for the delay of one pulse with respect to the other and by using detectors of various types.

Although the most popular flash photolysis apparatuses are based on the UV–visible emission or absorption of natural light as probes of chemical transformations, polarized light has been applied to the determination of circular dichroism (CD) (*14, 15*) and magnetic circular dichroism (MCD) (*9*) spectra of excited states and transient species. Other flash

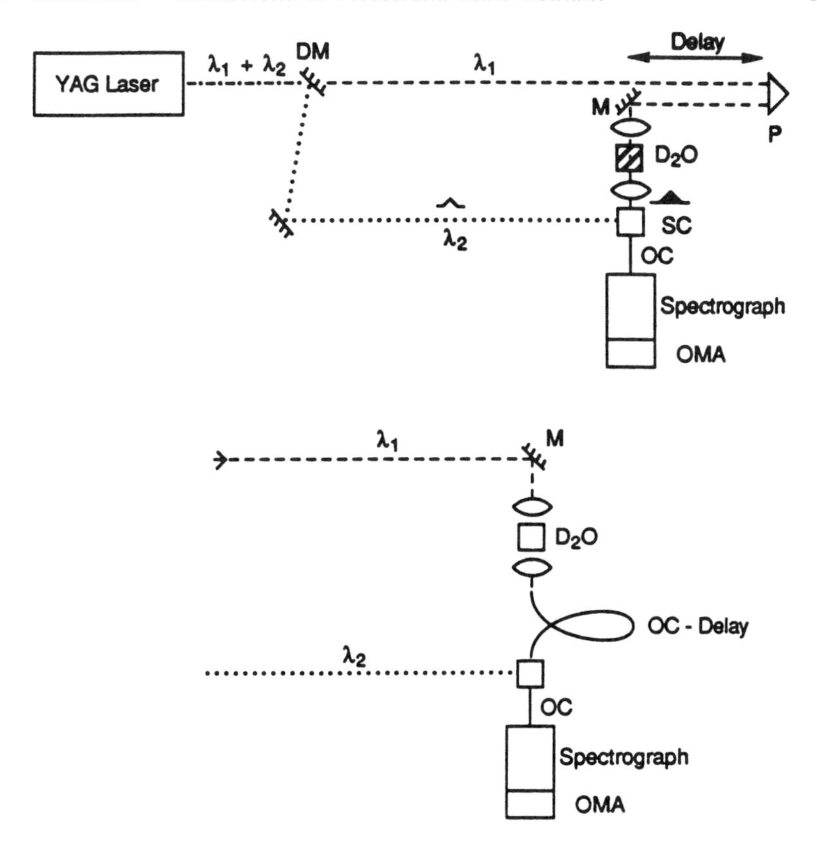

Figure 2. Typical optical arrangements for flash photolysis with picosecond time resolution. A pulse from the mode-locked YAG laser (top figure) composed of the fundamental, λ_1, and a higher harmonic, λ_2, are optically separated in a high-power resistant dielectric mirror, DM. A prism, P, and a high-power resistant mirror, M, are combined in an optical delay line. A cell with a dielectric medium, D_2O, is telescopically focused in the sample cell of the photolyte, SC. A delay line (bottom figure) can be constructed by using fiber optics, optical cable (OC) delay, of various lengths to carry the probing flash of white light.

photochemical techniques less widely used but extremely powerful for the study of reaction mechanisms make use of flash-induced changes of the electric conductance (*16, 17*), or dielectric constant (*18*). Time-resolved photoacoustic microcalorimetry (*19–23*) and time-resolved resonance Raman spectra, TRRR or TR^3 (*24*), have already proven to be extremely useful for the respective reaction heat measurement and structural characterization of transient species.

Picosecond-through-Millisecond Formation and Dissociation of Cu–C Bonds

Some of the technical points considered in the previous section will be illustrated next in a few examples connected to the dynamics of formation and dissociation of Cu–C bonds.

The mechanism for the oxidation of carbon-centered radicals by copper complexes, investigated by Jenkins and Kochi (*25*) with steady-state methods, concerns the oxidation of the radicals to the formation of metastable alkyl–copper intermediates. Such species were later detected in flash photolysis by following a rather common procedure in which one compound, the flash-irradiated photolyte, is used for the generation of one reactant (i.e., the radical) (*26*). The other reactant in the solution of the photolyte (i.e., the copper complex) does not absorb light from the flash and functions as a radical scavenger in these reactions. A number of Co(III)-carboxylato (*27, 28*) and alkyl–Co(III) complexes (*29*) have been used as sources of carbon-centered radicals in photochemical reactions.

In the experimental observations shown in Figure 3, ultraviolet irradiations of $Co(NH_3)_5OCOCH_3^{2+}$ were used for the generation of $CH_3\cdot$ radicals with $Cu^{2+}(aq)$ as a radical scavenger (*26*). The inset to Figure 3 shows a trace with time-resolved optical changes related to the formation and decay of $CuCH_3^{2+}$ according to the mechanism in equations 4–8:

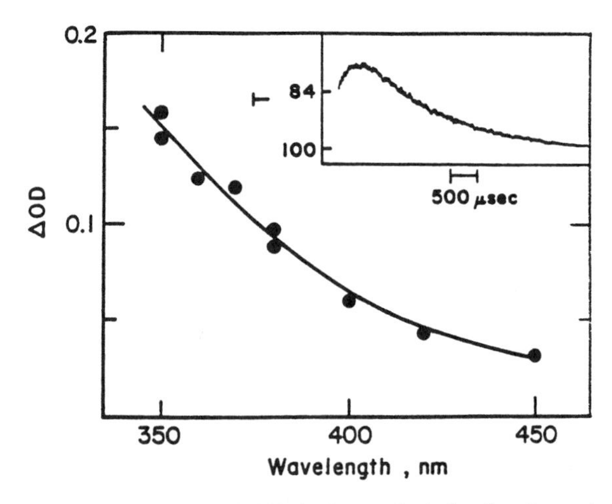

Figure 3. Transient spectra (ΔOD is the optical density change) observed in reactions of methyl radicals and $Cu^{2+}(aq)$ ions. The spectrum was recorded 500 ms after the flash irradiation, λ_{exc} = 240 nm, of 2 × 10^{-4} M $Co(NH_3)_5OCOCH_3^{2+}$ and 0.1 M $Cu^{2+}(aq)$ in 10^{-2} M $HClO_4$ (26). The inset shows time-resolved optical changes at λ_{ab} 370 nm; T *is transmittance.*

$$Co(NH_3)_5OCOCH_3^{2+} \xrightarrow[\phi]{h\nu,\, H^+} Co^{2+} + 5NH_4^+ + CO_2 + CH_3\cdot \quad (4)$$

$$CH_3\cdot + CH_3\cdot \longrightarrow C_2H_6 \quad (5)$$

$$Co(NH_3)_5OCOCH_3^{2+} + CH_3\cdot \xrightarrow{H^+} Co^{2+} + 4NH_4^+ + CH_3CO_2H + CH_4 + NH_3^{\cdot +} \quad (6)$$

$$CH_3\cdot + Cu^{2+}(aq) \longrightarrow CuCH_3^{2+} \quad (7)$$

$$CuCH_3^{2+} \xrightarrow{H_2O} Cu^+(aq) + CH_3OH + H^+ \quad (8)$$

The spectroscopic detection of $CuCH_3^{2+}$ was successful in flash photolysis because two experimental conditions were fulfilled. Optical absorptions of $Co(NH_3)_5OCOCH_3^{2+}$ and $Cu^{2+}(aq)$ were too weak to hinder the observation of the optical transient over a wide range of wavelengths, and the instrument's response was suitable to the span of time covered by the $CuCH_3^{2+}$ reactions. These conditions are not repeated in the formation and decomposition of $CuCH_3^+$ with a similar time resolution. The $CuCH_3^+$ species can be prepared by using $Cu^+(aq)$ as a scavenger of $CH_3\cdot$ radicals from the $Co(NH_3)_5OCOCH_3^{2+}$ photolysis:

$$CH_3\cdot + Cu^+(aq) \longrightarrow CuCH_3^+ \quad (9)$$

or in flash irradiations of $CuOCOCH_3^+$ at wavelengths of the charge-transfer absorption band (*26*). In the photodecarboxylation of $CuOCOCH_3^+$, equations 5 and 10–16 summarize the events leading to the formation and decay of $CuCH_3^+$.

$$CuOCOCH_3^+ \xrightarrow[\phi]{h\nu} [Cu^+, CH_3\cdot] + CO_2 \quad (10)$$

$$CuOCOCH_3^+ \xrightarrow[\phi]{h\nu} [Cu^+, \dot{O}COCH_3] \quad (11)$$

$$[Cu^+, \dot{O}COCH_3] \longrightarrow CuOCOCH_3^+ \quad (12)$$

$$[Cu^+, \dot{O}COCH_3] \longrightarrow CuCH_3^+ + CO_2 \quad (13)$$

$$[Cu^+, CH_3\cdot] \longrightarrow CuCH_3^+ \quad (14)$$

$$CuCH_3^+ \longrightarrow Cu^{2+} + CH_4 \quad (15)$$

$$CH_3\cdot + CuOCOCH_3^+ \longrightarrow Cu^+ + \begin{cases} CH_3OH + CH_3CO_2H \\ \text{and/or} \\ CH_3OCOCH_3 \end{cases} \quad (16)$$

Flash photochemical results in Figure 4 show the spectrum of the $CuCH_3^+$ measured before decomposition of the alkyl–copper complex into methane and Cu(II) aqua ions (eq 15). Time-resolved measurements (inset to Figure 4) reveal that such a decomposition process takes place in a microsecond-to-millisecond time domain. Reactions of products trapped

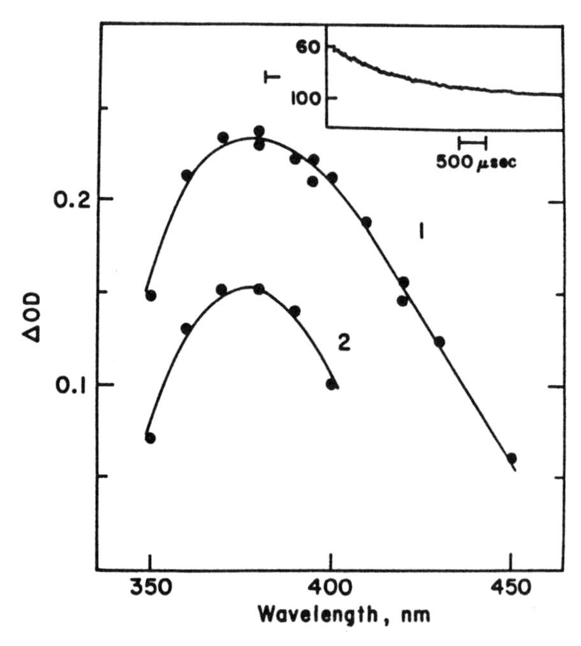

Figure 4. Absorptions (ΔOD is the optical density change) of transients generated in (1) flash photolysis of $CuOCOCH_3$ *(i.e.,* $[Cu_2] = 10^{-3}$ *M,* $[CH_3CO_2] = 5.0 \times 10^{-3}$ *M, and* $[CH_3CO_2H] = 5.0 \times 10^{-3}$ *M) and (2)* $Co(NH_3)_5OCOCH_3^{2+}$ *in the presence of* $Cu^+(aq)$ *(26). The inset shows a typical trace recorded in under the conditions indicated for curve 1 with* $\lambda_{exc} = 240$ *nm;* T *is transmittance.*

in the solvent cage (i.e., species shown inside square brackets in equations 10–14) lead to the formation of $CuCH_3^+$ (eqs 13 and 14) in times too short for flash photolysis with nanosecond or microsecond time resolutions. In the photohomolysis of alkylcobalamines, for example, recombination and diffusion out of the solvent cage occur over several hundred picoseconds (*30*). When Cu^{2+}(aq) is coordinated to polyacrylate the resulting complex undergoes a photodecarboxylation (eqs 17–21) similar to that just described for $CuOCOCH_3^+$ (*31*).

$$\cdots\left[\begin{array}{c}CO_2Cu^+\\ |\\ -CH-CH_2-\end{array}\right]\cdots \overset{h\nu}{\rightleftarrows} -\left[\begin{array}{c}(CT)CO_2Cu^+\\ |\\ \cdots CH-CH_2-\end{array}\right]\cdots \tag{17}$$

$$-\left[\begin{array}{c}(CT)CO_2Cu^+\\ |\\ \cdots CH-CH_2-\end{array}\right]\cdots \longrightarrow -\left[\begin{array}{c}CO_2{\cdot}Cu^+\\ |\\ \cdots CH-CH_2-\end{array}\right]\cdots \tag{18}$$

$$-\left[\begin{array}{c}CO_2{\cdot}Cu^+\\ |\\ \cdots CH-CH_2-\end{array}\right]\cdots \longrightarrow -\left[\begin{array}{c}CO_2Cu^+\\ |\\ \cdots CH-CH_2-\end{array}\right] \tag{19}$$

$$-\left[\begin{array}{c}CO_2{\cdot}Cu^+\\ |\\ \cdots CH-CH_2-\end{array}\right]\cdots \xrightarrow[-CO_2]{} -\left[\begin{array}{c}Cu^+\\ \\ \cdots\dot{C}H-CH_2-\end{array}\right]\cdots \tag{20}$$

$$\cdots\left[\begin{array}{c}Cu^+\\ \\ -\dot{C}H-CH_2-\end{array}\right]\cdots \longrightarrow \cdots\left[\begin{array}{c}Cu^+\\ |\\ -CH-CH_2-\end{array}\right]\cdots \tag{21}$$

Investigation of the processes shown in equations 17–21 by flash photolysis reveals that the spectrum of the alkyl–copper product, $\cdots CH(Cu)^+-CH_2\cdots$, is generated within the 15-ps flash (Figure 5) (*32*). These experiments indicate that decarboxylation (eq 20) and recombination processes (eqs 19 and 21) are faster than expected (i.e., with lifetimes $\tau < 15$ ps) by comparison to the decarboxylation of the $CH_3CO_2{\cdot}$ radical, $\tau \sim 300$ ps. The nature of the primary copper product dictates what events will lead to products. When $Cu(TIM)^{2+}$, where TIM is the macrocycle 2,3,9,10-$(CH_3)_4$-[14]1,3,8,10-tetraene-N_4 in Figure 5, is coordinated to the polyacrylate, the spectral changes recorded after the 15-ps irradiation corresponded to the formation of the primary products $Cu(TIM)^+$ and $\cdots(C(Cu^{III}(TIM)^{2+})\cdots$ (Figure 5) (*32*). Therefore, excess Cu(II) complex reacted with carbon-centered radicals within the flash, (eqs 22–24).

$$\cdots\left[\begin{array}{c}CO_2Cu^{II}(TIM)^+\\ |\\ -CH-CH_2\cdots\end{array}\right]^+_- \overset{h\nu}{\rightleftarrows} -\left[\begin{array}{c}(CT)CO_2Cu(TIM)^+\\ |\\ \cdots CH-CH_2\cdots\end{array}\right]^+_- \tag{22}$$

$$-\left[\begin{array}{c}(CT)CO_2Cu(TIM)^+\\ |\\ \cdots CH-CH_2\cdots\end{array}\right]^+_- \xrightarrow[-CO_2]{} \cdots\left[\begin{array}{c}Cu^{I}(TIM)^+\\ \\ -\dot{C}H-CH_2-\end{array}\right]^+\cdots \tag{23}$$

$$\cdots\left[\begin{array}{c}Cu^{I}(TIM)^+\\ \\ -\dot{C}H-CH_2-\end{array}\right]\cdots \xrightarrow[-Cu^{I}(TIM)^+]{-Cu^{II}(TIM)^{2+}} \cdots\left[\begin{array}{c}Cu(TIM)^{2+}\\ |\\ -CH-CH_2-\end{array}\right]\cdots \tag{24}$$

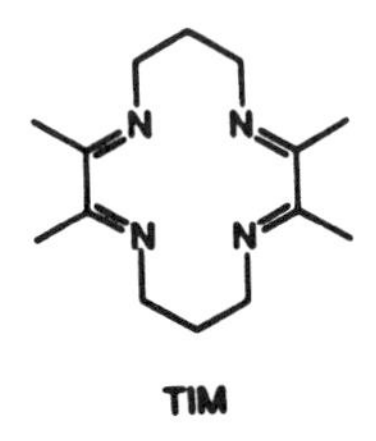

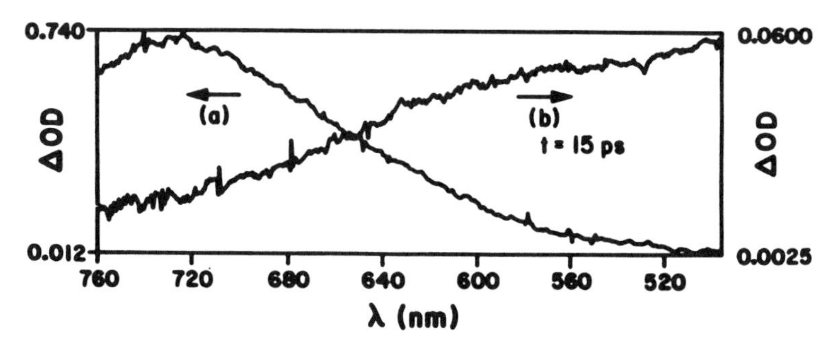

Figure 5. Spectra recorded 15 ps after the 266-nm irradiations of (a) 10^{-3} M $Cu(TIM)^{2+}$ and (b) 10^{-3} M $Cu^{2+}(aq)$; each of them in 6.0×10^{-3} M of a 2.0×10^{6} average-formula-weight polyacrylate.

Spectral transformations followed from the nanosecond to the microsecond (Figure 6) are in accord with the decomposition of the alkyl–copper intermediate into $Cu(TIM)^{+}$ (eq 25).

$$\cdots\left[-\underset{}{\overset{\displaystyle Cu(TIM)^{2+}}{\overset{|}{C}}}H-CH_2-\right]\cdots \xrightarrow[-H^+]{H_2O} \cdots\left[-\overset{\displaystyle Cu^{I}(TIM)^{+}}{\overset{|}{C}}H-CH_2-\right]\cdots + e_{aq}^{-} \qquad (25)$$

In flash photolysis experiments in which laser pulses with 10- to 10^2-ns widths are used for the irradiations and optical absorption for the detection of transient species, the average power of the flash is usually between one and several tens of megawatts for UV–visible light. Although powerful flashes can generate larger concentrations of reaction intermediates, it must be carefully considered what those experimental conditions mean in terms of the reaction mechanism. If a reaction intermediate or excited state undergoes transformation, for example, via two competitive parallel reactions of a first and a second order, the low concentrations of transient species generated with flash lamps in a long optical path cell (i.e., longer than 10 cm) can be conveniently used for the observation of a first-order kinetics (i.e., under conditions where the rate of the second-order reaction

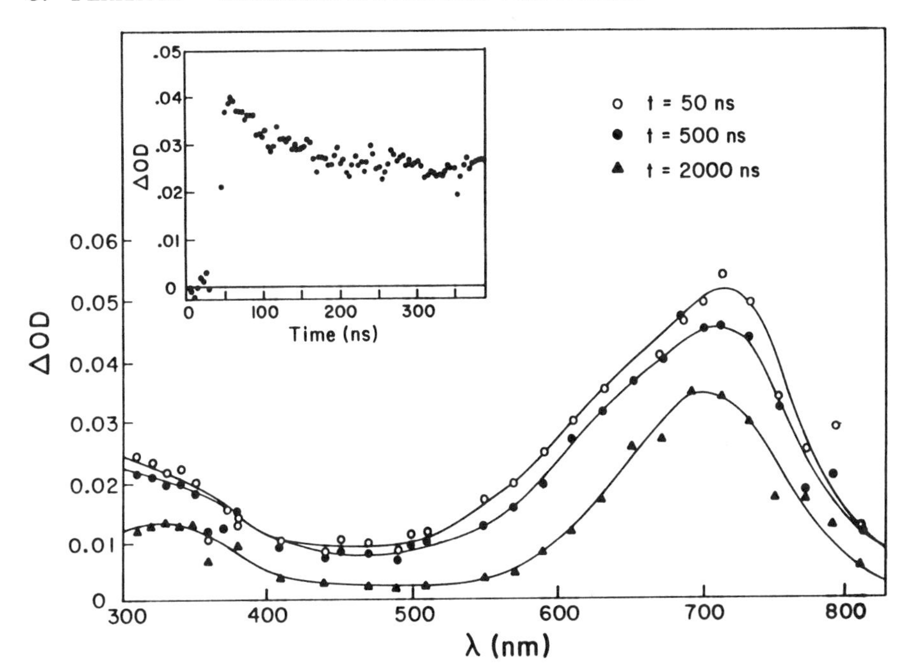

Figure 6. Transient spectra recorded in flash irradiations of 10^{-3} M $Cu(TIM)^{2+}$ in 6.0 × 10^{-3} M polyacrylate solutions from reference 32. The inset shows 650-nm time-resolved optical density changes following the conversion of a alkyl–copper into $Cu(TIM)^{+}$. Other conditions are as in Figure 5.

is very small). The observed decay kinetics will be more likely second-order under the conditions of laser flash photolysis, which must provide large concentrations of intermediates for the detection of optical changes with optical paths less than or equal to 1 cm. Comparisons of these photochemical observations with results of continuous photolyses with small light intensities may lead to quandaries about the reaction mechanism. There is also a limit to the power per unit volume that is convenient to deliver to the solution; if the power is too high, species photogenerated within the flash may absorb some of the excess light and undergo various phototransformations. An example of this photobehavior is found in the generation of solvated electrons by secondary photolysis of the alkyl–copper intermediates in polyacrylates, (eqs 17 and 26) (*31*).

$$\cdots\left[-\underset{}{\overset{Cu^{+}}{\overset{|}{C}}}H-CH_2-\right]\cdots \xrightarrow{h\nu} \cdots\left[-\overset{Cu^{2+}}{\overset{|}{C}}H-CH_2-\right]\cdots + e_{aq}^{-} \qquad (26)$$

In these regards, some flash photochemical (biphotonic) techniques allow one to generate an excited state or reaction intermediate with one laser pulse and irradiate again such species with a second laser pulse conveniently delayed with respect to the first. Some of these experiments are carried out with two lasers that are fired one after another with a fixed delay (i.e., one that can be adjusted to a convenient value between nanoseconds and milliseconds) (Figure 7) (*33, 34*). For example, the photochemical properties of $Cu(TIM)^+$ in polyacrylate and in methanolic

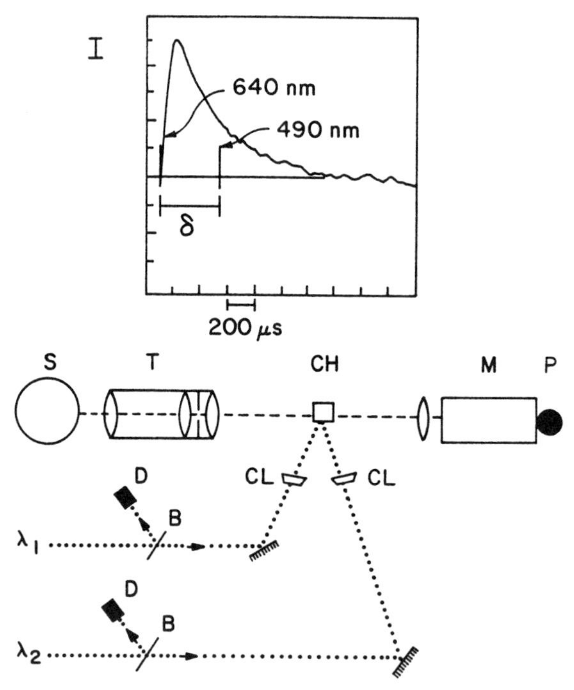

Figure 7. Diagram of an apparatus (bottom figure) for sequential, two-color, double-pulse flash photolysis experiments. Two synchronously fired lasers generate pulses of light with wavelengths λ_1 and λ_2. Beam splitters, B, direct a fraction of the light to photodiodes, D, and through the cylindrical lenses, CL, to the cell holder, CH. Optical changes induced by irradiation of the photolyte are probed with polychromatic light from a source, S, collimating optics, T, monochromator, M, and photomultiplier, P. A trace (top figure) shows the response of the phototube to changes of the probing light intensity when a transient species is photogenerated in a 640-nm flash irradiation. Superimposed on the trace are the respective photodiode signals; a 640-nm pulse (used for the photogeneration of a transient) and a 490-nm pulse (used for the photolysis of the transient). (Reproduced from references 33–35. Copyright 1984, 1987, 1988 American Chemical Society.)

solutions were investigated (*32*) by such a double-pulse–two-color experiment (Figure 8). The first laser pulse of 308 nm prepared $Cu(TIM)^+$ according to equations 22–25, and this species remains stable for several milliseconds. No photoreactions of the $Cu(TIM)^+$ complex were detected in methanol (Figure 8b). Only in polyacrylate are photochemical reactions observed when such a Cu(I) complex is irradiated with a second 760-nm laser pulse (Figure 8a). The optical transformations, a prompt bleach followed by a growth and a much slower decay of the near-IR (NIR) optical density to its original value, can be related to a photoreduction of the $Cu(TIM)^+$–carboxylate complex and to the slower formation and disappearance of a alkyl–copper species:

$$\left\{ \mathrm{H\overset{|}{\underset{|}{C}}{-}CO_2^{-}Cu^I(TIM)} \right\} \xrightarrow[-CO_2]{h\nu} \{\mathrm{HC^{\bullet}Cu^I(L^{\bullet})}\} \longrightarrow \left\{ \mathrm{H\overset{|}{\underset{|}{C}}{-}Cu^I(TIM)} \right\} \longrightarrow \text{products} \qquad (27)$$

The biphotonic technique has also been applied to the study of photoreactions initiated when a complex in a long-lived excited state, photo-

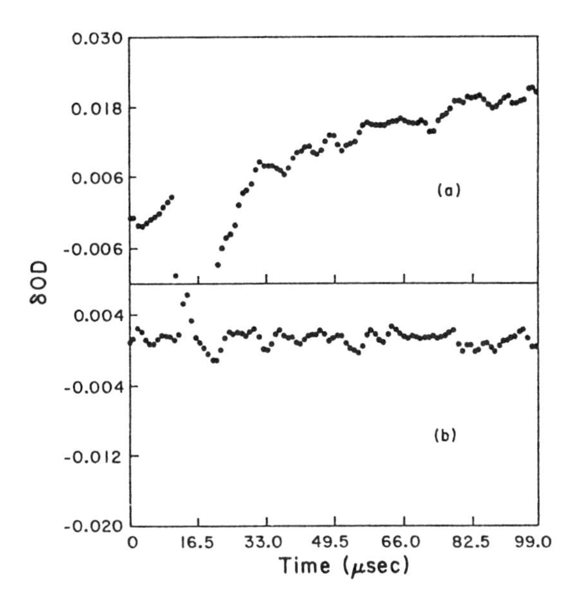

Figure 8. Transient optical changes generated by a 760-nm flash irradiation of $Cu(TIM)^+$ in (a) 6.0×10^{-3} M polyacrylate aqueous solutions and (b) methanolic solutions. In these solutions, the Cu(I) complex was generated by the 308-nm flash irradiation of $Cu(TIM)^{2+}$. (Reproduced from reference 32. Copyright 1990 American Chemical Society.)

generated with one laser pulse, is reexcited with a second laser pulse of another wavelength (*35, 36*). In Figure 9, $ClRe(CO)_3$(4-phenylpyridine)$_2$ is promoted to the metal-to-ligand charge-transfer (MLCT) state by irradiation with a 308-nm pulse (*36*). When light of a longer wavelength laser pulse is absorbed by such an excited state, the induced quenching of the emission and bleach of the excited-state spectrum can be correlated with photodecomposition of the complex. A number of complexes, unreactive in the lowest lying excited states, can be made photoreactive in biphotonic excitations. The resulting excited-state photoprocesses can be time-resolved for the determination of the spectra or the study of the reaction mechanism by using the technique based on two synchronously fired lasers.

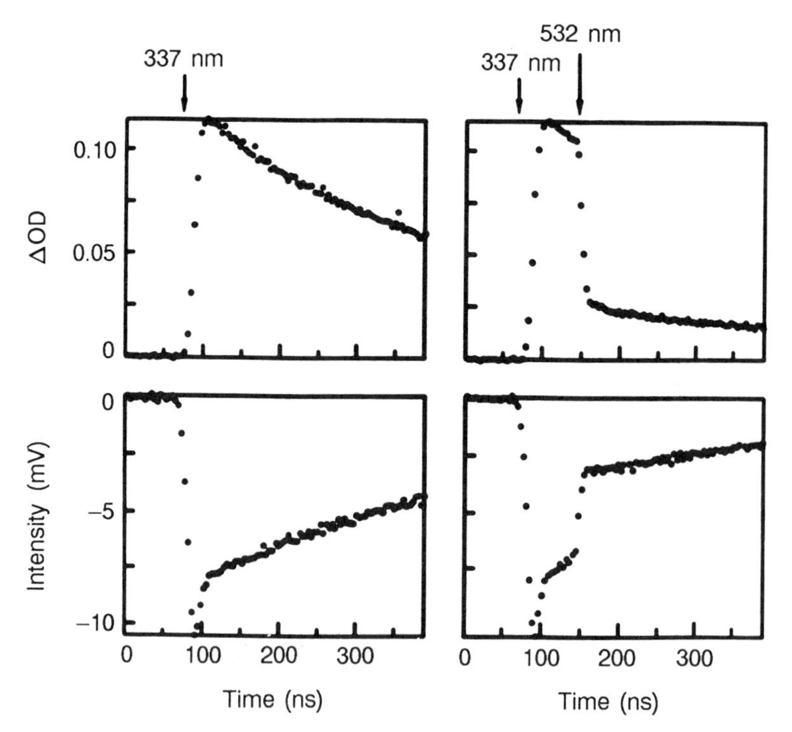

Figure 9. Traces for the time-resolved 560-nm absorption (top left) and 550-nm emission (bottom left) from the MLCT state generated with 337-nm flash irradiations of fac-*$ClRe(CO)_3$(4-phenylpyridine)$_2$. The absorption of a 532-nm pulse of laser light by the MLCT state (traces in the right side) causes a rapid bleach of the absorption (top) and emission quenching (bottom); both processes are associated with a photodecomposition of the Re complex. (Reproduced from reference 36. Copyright 1992 American Chemical Society.)*

A considerable number of reactions between carbon-centered radicals and coordination complexes have small rate constants; they must be investigated over an interval longer than several hundred microseconds. When these reactions are part of the photochemical mechanism, logical correlations between the results of flash and continuous photolysis can be established only if experimental observations are made over such an extended time range. The photochemistry of the Cu–olefin complexes provides interesting examples (*37*). The 254-nm irradiation of the ethylene complex, $Cu(H_2C{=}CH_2)^+$, in methanolic solutions saturated with ethylene catalyzes the formation of hexane and formaldehyde (eq 28 and Figure 10) (*38*).

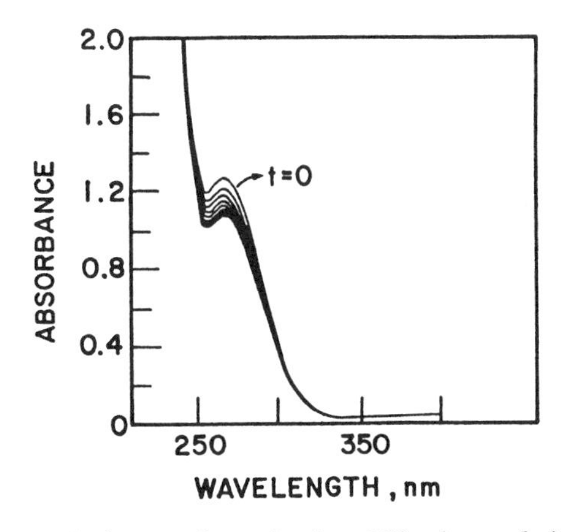

Figure 10. Spectral changes determined at 180-s intervals in 254-nm continuous photolysis of an argon-saturated methanolic solution of $Cu(C_2H_4)^+$. (Reproduced with permission from reference 38. Copyright 1985.)

$$3H_2C{=}CH_2 + CH_3OH \xrightarrow{h\nu} C_6H_{14} + CH_2O \qquad (28)$$

Because the polymerization of ethylene (eq 28) involves the oxidation of the solvent, CH_3OH, several intermediates, detected in flash photolysis, can be assigned to radical-like species. Spectra recorded at various instants of the reaction (Figure 11) suggest that they are species with alkyl–copper bonds. Moreover, the kinetics of the processes associated with the appearance and disappearance of the spectral features, investigated as a function of the concentration of ethylene and various intermediates concentrations, suggests the mechanism embodied in equations 29–38; it involves the primary photogeneration of an alkyl–copper diradical as an initiator of the polymerization.

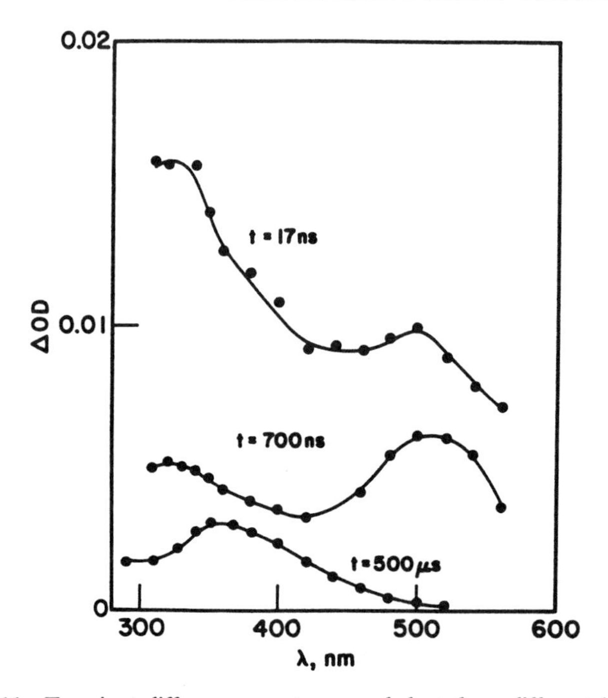

Figure 11. Transient difference spectra recorded at three different instants following the flash irradiation of $Cu(C_2H_4)^+$ in ethylene-saturated methanol. (Reproduced with permission from reference 38. Copyright 1985.)

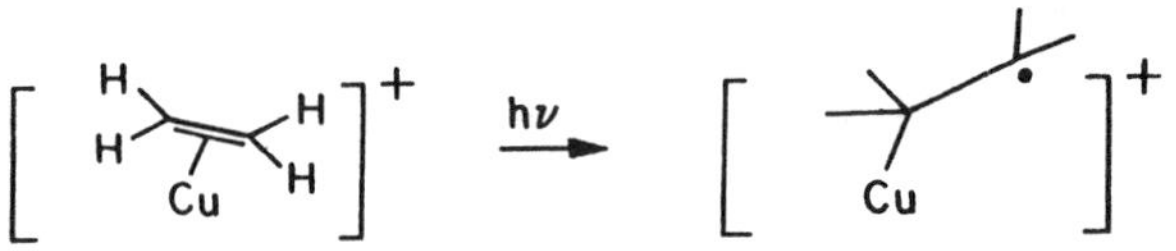

$$[Cu(C_2H_4)]^+ \xrightarrow{h\nu} [\text{IV}]^+ \qquad (29)$$

(IV)

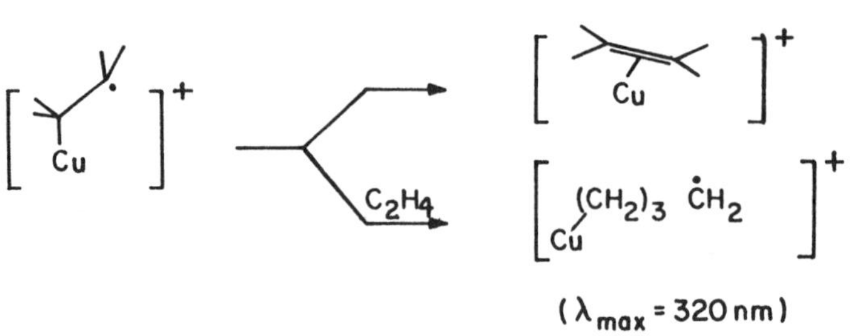

$$\xrightarrow{} [Cu(\text{alkene})]^+ \qquad (30)$$

$$\xrightarrow{C_2H_4} [Cu(CH_2)_3\dot{C}H_2]^+ \qquad (31)$$

$(\lambda_{max} = 320\ nm)$

$$\left[\underset{\text{Cu}}{\diagup}(CH_2)_3\dot{C}H_2\right]^+ \rightleftarrows \left[\text{Cu-cyclopentane}\right]^+ \quad (\lambda_{max} = 500 nm) \tag{32}$$

$$\left[\underset{\text{Cu}}{\diagup}(CH_2)_3\dot{C}H_2\right]^+ \xrightarrow{C_2H_4} \left[\underset{\text{Cu}}{\diagup}(CH_2)_5\dot{C}H_2\right]^+ \tag{33}$$

$$[\underset{\text{Cu}}{|}(CH_2)_5\dot{C}H_2]^+ + CH_3OH \xrightarrow{H^+} Cu^{2+} + C_6H_{14} + \dot{C}H_2OH \tag{34}$$

$$Cu^{2+} + \dot{C}H_2OH \longrightarrow Cu^+ + CH_2O + H^+ \tag{35}$$

$$2\dot{C}H_2OH \longrightarrow (CH_2OH)_2 \tag{36}$$

$$2\dot{C}H_2OH \longrightarrow CH_3OH + CH_2O \tag{37}$$

$$Cu^+ + CH_2{=}CH_2 \rightleftarrows Cu(CH_2{=}CH_2)^+ \tag{38}$$

The criteria followed for the assignment of the intermediates in equations 29–38 is based on previous knowledge of the chemical and spectroscopic properties of related species. Metal–alkyl diradicals have been proposed as intermediates in the photoisomerization of olefins coordinated to carbonyl complexes (*39*). The rapid reaction of such species with excess ethylene (eq 31) is in accordance with expectations for the addition of carbon-centered radicals to an olefinic double bond. The product of such an addition in equation 31, λ_{max} = 320 nm, has a spectrum and a lifetime similar to those of the Cu^{II}–alkyl intermediates in reactions between radicals and Cu(I) complexes. Equilibration between this diradical and the cuprocyclopentane, λ_{max} = 500 nm, in equation 32, was proposed to be a process parallel to the generation of a solvent-scavengeable species shown in equation 31, λ_{max} = 360 nm, on the basis of the kinetics of the spectral transformations and on the known chemical properties of metallocyclopentanes.

The decay of the 360-nm optical density with an 80-ms lifetime is slightly slower than expected for the reaction of carbon-centered radicals with methanol and leads to the regeneration of the $Cu(CH_2{=}CH_2)^+$ spectrum. This experimental observation, indicative of the catalytic recycling of Cu(I), was associated with the known diffusion-controlled reaction of solvent radicals with Cu(II) ions (eq 35). The catalyzed polymerization can be successful, according to the mechanism in equations 29–38, if the diradical species can be trapped by excess monomer with a rate much faster than the rate of back-electron-transfer reactions. The opposite conditions may lead to products other than those from polymerization or to no products at all. One example is the *cis–trans* photoisomerization of the *cis,cis*-cyclooctadiene (c,c-COD) in a Cu(I) complex (Figure 12) (*40, 41*). Flash photolysis experiments (Figure 13) reveal transient spectra corresponding to two kinetically significant species (eqs 39–41) whose lifetimes are too long for those involved in an intramolecular valence isomerizations (*42*). The spectral features of those species are those expected for a cuprodiradical, λ_{max} = 280 nm, and a copper–olefin complex, λ_{max} = 320 nm.

Cu $\xrightarrow[\phi]{h\nu}$ Cu (λ_{max} = 280nm) (39)

Cu S → $Cu(c,c\text{-}COD)_2^+$ (40)

Cu S → $Cu(c,c\text{-}COD)(c,t\text{-}COD)^+$ (41)

The lifetime of the cuprodiradical, λ_{max} = 280 nm, is too short for any appreciable reaction with solvent methanol or trace concentrations of the free c,c-COD ligand; the only reaction path is, therefore, isomerization into species, λ_{max} = 320 nm, that eventually form the terminal products. An application of flash photolysis with microsecond to millisecond irradiation has been, in this final example, to distinguish between a norbornadiene–quadricyclane valence isomerization type mechanism (*40*) and one based on the photogeneration of long-lived cuprodiradicals.

Limitations of Flash Photolysis to the Study of Photosensitive Metal–Organic Systems

The limited number of examples presented here hardly covers a minimal fraction of current applications of flash photolysis methods to the investi-

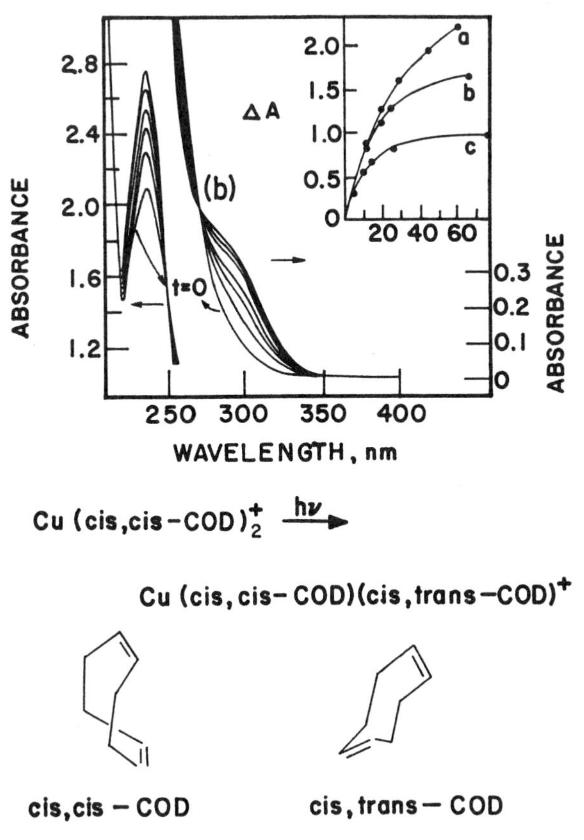

Figure 12. Spectral changes, determined at 180-s intervals in 254-nm continuous photolysis of Cu(c,c-COD)$^{2+}$ in argon-saturated methanol, were associated with a photoisomerization process shown on the bottom of the figure (38). The inset shows changes in the 290-nm optical density as a function of the irradiation time for various concentrations of the copper complex: 2.0 × 10^{-2} M (a), 5.0 × 10^{-3} M (b), and 10^{-3} M (c).

gation of reaction mechanisms in photosensitive metal–organic systems. They illustrate, however, the strong and weak points of the technique. A major problem, probably the weakest point of flash photolysis, is the correlation of transient optical spectra to chemical species (i.e., to characterize the reaction intermediates) when their detection is done by following changes in the absorption of UV–visible natural (versus polarized) light from picoseconds to seconds. In this task, there is always the need for additional experimental information from other techniques; often, the time researchers must use judicious chemical intuition educated by comparisons with chemically related systems.

In this last regard, it is sometimes possible to prepare and investigate a reaction intermediate of photochemical reactions by other techniques

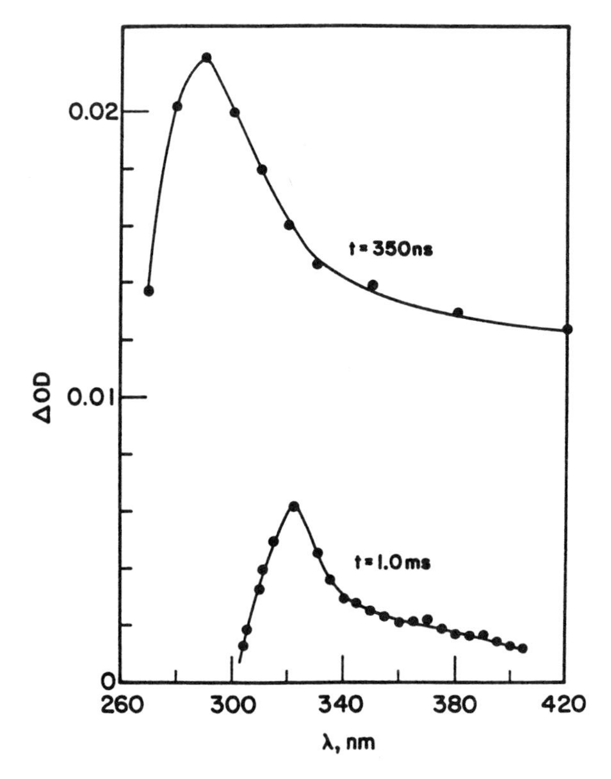

Figure 13. Transient difference spectra of the two intermediates detected in flash irradiations of $Cu(c,c\text{-}COD)^{2+}$ in Ar-saturated methanol. (Reproduced with permission from reference 38. Copyright 1985.)

(i.e., pulse radiolysis or stop-flow). For example, a number of early pulse radiolytic studies of reactions of carbon-centered radicals with copper complexes provided substantial spectroscopic information for the characterization of alkyl–copper complexes (*26, 43, 44*). Rates of reaction and spectra measured with these other techniques must be similar to data collected by flash photolysis to conclude that the same species has been generated by different routes. The type of reactions (e.g., redox and acid-base) that reaction intermediates undergo with purposely introduced organic or inorganic molecules (i.e., scavengers) can also provide information about the nature of the transient species (*45, 46*). Such reactions are particularly important when they lead to products that are already well-characterized and easy to detect. Indeed, a number of chemical methods can then be used for the measurement of the extinction coefficients and the formation of quantum yields of the intermediates (*47*).

Room-temperature electron spin resonance (ESR) spectra of transient species (i.e, metal–ligand radicals or coordination complexes with metals in unstable oxidation states) can be collected with solutions subjected to continuous or flash irradiations. The experimental information collected by ESR spectroscopy may help to characterize the species whose optical absorption spectrum is recorded by flash photolysis if an unequivocal link is established between these two sets of measurements. The same can be said about TRRR flash photochemical experiments.

Another promising flash photochemical technique for the characterization of reaction intermediates, not yet fully developed, is based on the time-resolved changes of the solution's IR spectrum (*48, 49*). However, some technical difficulties, such as weak sources of polychromatic IR radiation, low sensitivity of the detectors, and intense background absorptions of IR light, limit the extent of its application to the structural characterization of an intermediate. Sometimes, structural characterization can be more expeditiously achieved by taking ESR, IR, and the Raman spectra of products trapped when photolytes are irradiated in glassy solutions or matrices at low temperatures, for example, as was done with some Cr carbonyl complexes (*50*).

If structural characterization of intermediates constitutes the technique's major problem, its most appealing aspect is related to the kinetic investigation of reaction mechanisms. Studies based on the measurement of the products' quantum yields in steady-state photolyses are still necessary, but the flash photochemical observations provide a clear view of the intricacies of the photochemical processes leading to such products. The measurement of rate constants for fast reactions in simple photochemical systems, for example, electron-transfer quenching of excited states, has demonstrated once more the value of flash photolysis when it is applied to problems in the area of inorganic physical chemistry.

Acknowledgment

The work described herein was supported by the Office of Basic Energy Sciences of the Department of Energy. This is Contribution No. SR–147 from the Notre Dame Radiation Laboratory.

References

1. Porter, G. *Flash Photolysis; Technique of Organic Photochemistry;* Vol. VIII, 2nd ed.; Interscience Publishers: New York, 1963.
2. Claeson, S.; Lindquists, L. *Arkiv. Kemi* **1957,** *11,* 535.
3. Claeson, S.; Lindquists, L. *Arkiv. Kemi* **1958,** *12,* 1.
4. Ferraudi, G. *Elements of Inorganic Photochemistry;* Wiley Interscience: New York, 1988; pp 38–45.

5. Demas, J. N. *Excited State Lifetime Measurements;* Academic Press: New York, 1983; Chapters 5, 7, 8.
6. *RCA Photomultiplier Manual;* Technical Series PT–61; RCA Corporation: Harrison, NJ, 1970.
7. Hunt, J. W.; Greenstock, C. I.; Bronskill, M. J. *Int. J. Radiat. Phys. Chem.* **1972,** *4,* 87.
8. Hunt, J. W.; Thomas, J. K. *Radiat. Res.* **1967,** *32,* 149.
9. Perkovic, M.; Ferraudi, G., work in progress.
10. Ebbesen, T. W. *Rev. Sci. Instr.* **1988,** *59,* 1307 and references therein.
11. Netzel, T. L.; Rentzepis, P. M.; Leigh, J. S. *Science* **1973,** *182,* 238.
12. Rentzepis, P. M.; Jones, R. P.; Jortner, J. *J. Chem. Phys.* **1973,** *59,* 766.
13. Kenney-Wallace, G. A.; Walker, D. C. *J. Chem. Phys.* **1971,** *55,* 447.
14. Lewis, J. W.; Tilton, R. F.; Einterz, C. M.; Milder, S. J.; Kuntz, I. D.; Kliger, D. S. *J. Phys. Chem.* **1985,** *89,* 289.
15. Einterz, C. M.; Lewis, J. W.; Milder, S. J.; Kliger, D. S. *J. Phys. Chem.* **1985,** *89,* 3845.
16. Waltz, W. L.; Lilie, J.; Lee S. H. *Inorg. Chem.* **1984,** *23,* 1768.
17. Lilie, J.; Fessenden, F. W. *J. Phys. Chem.* **1973,** *77,* 674.
18. Toublanc, D. B.; Fessenden, R. W.; Hitachi, A. *J. Phys. Chem.* **1989,** *93,* 2893.
19. Linch, D.; Endicott, J. F. *Appl. Spectrosc.* **1989,** *43,* 826.
20. Song, X.; Endicott, J. F. *Inorg. Chem.* **1991,** *30,* 2214.
21. Tam, A. C. *Rev. Mod. Phys.* **1986,** *58,* 381.
22. Blimes, G. M.; Tocho, J. O.; Braslavsky, S. E. *J. Phys. Chem.* **1989,** *93,* 6696.
23. Rudzki, J. E.; Goodman, J. L.; Peters, K. S. *J. Am. Chem. Soc.* **1985,** *107,* 7849.
24. Schoonover, J. R.; Dallinger, R. F.; Killough P. M.; Sattelberger, A. P.; Woodruff, W. H. *Inorg. Chem.* **1991,** *30,* 1093 and references therein.
25. Jenkins, C. L.; Kochi, J. K. *J. Am. Chem. Soc.* **1972,** *94,* 843.
26. Ferraudi, G. *Inorg. Chem.* **1978,** *17,* 2506.
27. Roche, T.; Endicott, J. F. *Inorg. Chem.* **1974,** *13,* 1575.
28. Scaiano, J. C.; Leigh, W. J.; Ferraudi, G. *Can. J. Chem.* **1984,** *62,* 2355.
29. Bakac, A.; Espenson, J. H. *Inorg. Chem.* **1987,** *26,* 4353.
30. Netzel, T. L.; Endicott, J. F. *J. Am. Chem. Soc.* **1979,** *101,* 4000.
31. Das, S.; Ferraudi, G. *Inorg. Chem.* **1986,** *25,* 1066.
32. Baugartner, E.; Ronco, S.; Ferraudi, G. *Inorg. Chem.* **1990,** *29,* 4747.
33. Van Vlierberge, B.; Ferraudi, G. *Inorg. Chem.* **1987,** *26,* 337.
34. Ferraudi, G. *J. Phys. Chem.* **1984,** *88,* 3938.
35. Van Vlierberge, B.; Ferraudi, G. *Inorg. Chem.* **1988,** *27,* 1386.
36. Feliz, M.; Ferraudi, G. *J. Phys. Chem.* **1992,** *96,* 3059.
37. Kutal, C. *Coord. Chem. Revs.* **1990,** *99,* 213.
38. Geiger, D.; Ferraudi, G. *Inorg. Chim. Acta* **1985,** *101,* 197.
39. Geoffroy, G. L.; Wrighton, M. S. *Organometallic Photochemistry;* Academic Press: Orlando, FL, 1979; Chapter V.
40. Grobbelar, E.; Kutal, C.; Orchard, W. *Inorg. Chem.* **1982,** *21,* 414.
41. Whitesides, G. M.; Coe, G. L.; Cope, A. *J. Am. Chem. Soc.* **1969,** *91,* 2608.
42. Liaw, B.; Orchard, S. W.; Kutal, C. *Inorg. Chem.* **1988,** *27,* 1311.
43. Freiberg, M.; Meyerstein, D. *J. Chem. Soc. D* **1977,** 127.

44. Masarwa, M.; Cohen, H.; Glaser, R.; Meyerstein, D. *Inorg. Chem.* **1990,** *29,* 5031.
45. Hug, G. L. *Optical Spectra of Nonmetallic Inorganic Transient Species in Aqueous Solutions;* National Institute of Science and Technology: Washington, DC, 1981; Report Number NSRDS–NBS 69.
46. Ferraudi, G. *Elements of Inorganic Photochemistry;* Wiley Interscience: New York, 1988; pp 31-38.
47. Bonneau, R.; Carmichael, I.; Hug, G. L. *Pure Appl. Chem.* **1991,** *63,* 289.
48. Gordon, C. M.; Feltham, P. D.; Turner, J. J. *J. Phys. Chem.* **1991,** *95,* 2889.
49. Sedlacek, A. J.; Weston, R. E., Jr.; Flynn, G. W. *J. Phys. Chem.* **1991,** *95,* 6483.
50. Perutz, R. N.; Belt, S. T.; McCamley, A.; Wittlesey, M. K. *Pure Appl. Chem.* **1990,** *62,* 1539 and references therein.

RECEIVED for review November 7, 1991. ACCEPTED revised manuscript May 20, 1992.

6

Photosensitized Reduction of Alkyl and Aryl Halides Using Ru(II) Diimine Complexes

Inner- and Outer-Sphere Approaches

William F. Wacholtz[1], John R. Shaw, Staci A. Fischer, Melissa R. Arnold, Roy A. Auerbach, and Russell H. Schmehl*

Department of Chemistry, Tulane University, New Orleans, LA 70118

General schemes are presented for reduction of alkyl and aryl halides using Ru(II) diimine complexes as both inner-sphere and outer-sphere reductants and photosensitizers. The efficiency of outer-sphere reduction by a series of trisdiimineruthenium(II) complexes as a function of the one-electron reduction potential and charge of the complex is examined. The observed quantum yields for substrate reduction are analyzed in terms of the efficiencies for population of the reactive excited state, quenching, charge separation, and substrate reduction. Photoinduced inner-sphere reduction of halopyridines coordinated to Ru(II) sensitizers is also discussed. In these systems the excited complex undergoes either reductive quenching or ligand loss (of pyridine). The predominant photoreaction path can be controlled by regulating the temperature of the system.

MANY TRANSITION METAL COMPLEXES having metal-to-ligand charge-transfer (MLCT) excited states have been extensively used as sensitizers in energy- and electron-transfer reactions (*1–18*). In particular,

[1]Current address: Department of Chemistry, University of Wisconsin at Osh Kosh, Osh Kosh, WI 54901

*Corresponding author

0065–2393/93/0238–0107$07.00/0

M(II) (M is Ru or Os) diimine complexes have been widely applied (*2–6, 10, 12–14, 16, 19–23*). The complexes have excited states ranging in energy from 1.6 to 2.2 eV and excited-state lifetimes between 20 ns and 2 μs, and they are relatively stable upon one-electron oxidation and reduction (*1, 11, 24–28*). These characteristics make this class of complexes well suited as sensitizers in reaction schemes involving photoinduced electron transfer.

In devising catalytic schemes for reduction of substrates using sensitizers having MLCT excited states (*1*), the complex can be used either as an electron mediator (outer-sphere reductant) or it can be involved directly in substrate reduction via coordination of the substrate (inner-sphere reductant). This chapter discusses the use of Ru(II) diimine complex sensitizers as both inner- and outer-sphere reductants of alkyl and aryl halides (*19–23, 29, 30*). Factors defining the efficiencies of each type of process are discussed in terms of the development of homogeneous photocatalytic reduction processes.

MLCT Complexes as Sensitizers and Mediators in Substrate Reduction

Mechanistic Considerations. A simple mechanism for photosensitized substrate reduction via an outer-sphere process is shown in Scheme I. In the scheme, a Ru(II) complex sensitizer is reduced by a reversible electron donor (for example, *N,N,N′,N′*-tetramethylphenylenediamine, TMPD) following excitation of the complex. The reduced complex, $[Ru]^+$, can then either react with the $TMPD^+$ to regenerate starting materials (with rate constant k_b) or reduce the substrate S (k_{red}).

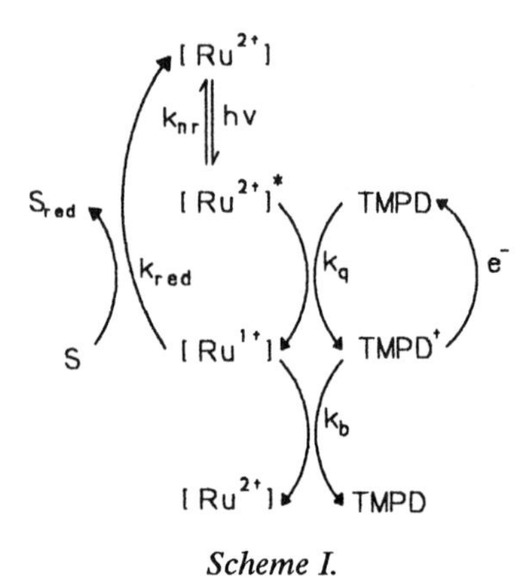

Scheme I.

In the presence of an external source of electrons (e.g., an electrode), the $TMPD^+$ can be reduced, and the scheme will be catalytic in $[Ru]^{2+}$ and TMPD, otherwise $TMPD^+$ will accumulate. Such a scheme requires only a reversible electron donor, a sensitizer, and the substrate.

The dynamics of each of the reactions involved in substrate reduction depend on the free energies of the reactions. The free energy of the photoinduced electron transfer (E_{pet}) is determined by the excited-state energy of the sensitizer (E_{oo}), the one-electron potentials of the sensitizer ($E_{Ru(2+/+)}$) and electron donor ($E_{TMPD(+/0)}$) and the work required to bring the two reactants together in solution ($w_{Ru(2+)/TMPD}$) (*31, 32*):

$$E_{pet} = E_{oo} - E_{TMPD(+/0)} + E_{Ru(2+/+)} + w_{Ru(2+)/TMPD} \quad (1)$$

The substrate reduction (E_{red}) and back-electron-transfer (E_b) free energies are defined by the one-electron reduction potentials of the species involved in the reaction and the work terms for formation of the reduced sensitizer–substrate complexes (eqs 2 and 3).

$$E_{red} = E_{S/S_{red}} - E_{Ru(2+/+)} + w_{Ru(+)/S} \quad (2)$$

$$E_b = E_{TMPD(+/0)} - E_{Ru(2+/+)} + w_{Ru(+)/TMPD(+)} \quad (3)$$

The overall free energy of substrate reduction can be evaluated if these ground-state potentials are known; the principal factor limiting the range of substrates capable of being reduced is the reduction potential of the sensitizer complex. For Ru(II) trisdiimine complexes, one-electron reduction potentials are between –0.8 and –1.6 V versus the sodium saturated calomel electrode (SSCE) (*26*), and thus only a small number of alkyl and aryl halides can be reduced by this approach (*17–23, 29, 32*).

The overall efficiency of substrate reduction can be experimentally evaluated in terms of the efficiency of each of the steps shown in Scheme I: formation of the reactive excited state, excited-state quenching by the donor, and substrate reduction. An additional important step not represented in Scheme I is separation of the products from the geminate ion pair formed in the photoreaction (k_{sep}):

$$[Ru]^{2+} \xrightarrow{h\nu} [Ru]^{2+*} \xrightarrow{TMPD} \{[Ru]^{2+*}, TMPD\} \quad (4a)$$

$$\{[Ru]^{2+*}, TMPD\} \xrightarrow{k_{et}} \{[Ru]^{+}, TMPD^{+\bullet}\} \quad (4b)$$

$$\{[Ru]^{+}, TMPD^{+}\} \underset{k_{-sep}}{\overset{k_{sep}}{\rightleftharpoons}} [Ru]^{+} + TMPD^{+} \quad (4c)$$

$$\{[Ru]^{+}, TMPD^{+}\} \xrightarrow{k_{gem}} [Ru]^{2+} + TMPD \quad (4d)$$

The reactive MLCT state of Ru(II) diimine complexes has triplet spin multiplicity (*24–28*), and the efficiency of forming this state is the intersystem crossing efficiency, η_{isc}. Figure 1 shows a state diagram typical of Ru(II) diimine sensitizers; in addition to the ^{3}MLCT state, many complexes have metal-centered (^{3}MC) excited states that are close in energy to the ^{3}MLCT state. For Ru(II) diimine complexes, η_{isc} can be estimated from limiting quantum yields for $S_2O_8^{2-}$ reduction in aqueous acid (*30, 33*). In the limit in which each excited complex reacts with $S_2O_8^{2-}$, 2 mol of oxidized sensitizer is produced (the intermediate SO_4^- radical is a powerful oxidant). Provided only the triplet excited state reacts with $S_2O_8^{2-}$, the limiting quantum yield is a direct measure of the intersystem crossing efficiency ($2\eta_{isc}$). This efficiency has been shown to be close to unity for several Ru(II) complexes (*26, 30 33–35*).

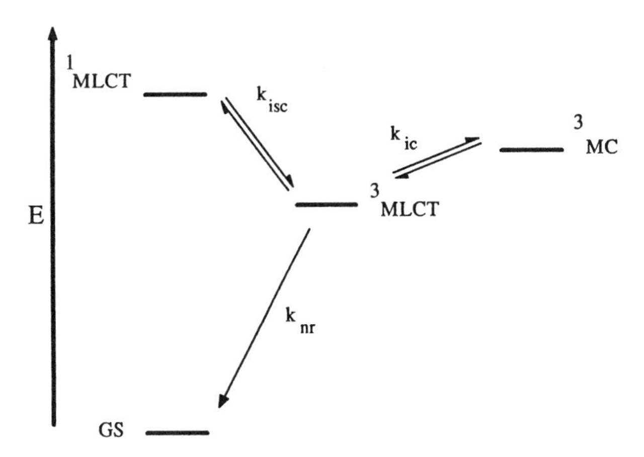

Figure 1. General state diagram for $[(diimine)_3Ru(II)]^{2+}$ complexes. Rate constant abbreviations are isc, intersystem crossing; ic, internal conversion; and nr, nonradiative–radiative. GS is ground state.

The quenching efficiency, η_q, is a function of the relative values of k_{nr} (representing the sum of the radiative and nonradiative decay rates of the excited state), k_q, and the concentration of quencher (TMPD in Scheme I) (*24, 25, 31*):

$$\eta_q = \frac{k_q[\mathrm{TMPD}]}{k_{nr} + k_q[\mathrm{TMPD}]} \qquad (5)$$

In practice, the efficiency of quenching can usually be made to be nearly unity by selecting a quencher having favorable thermodynamics for quenching the sensitizer ($E_{pet} > 0$) and adjusting the concentration of quencher (vide infra) (*14*).

The efficiency of charge separation, η_{cs}, of the initial photoproducts from the geminate ion pair (eq 6) depends on a variety of factors, including the charges of the ions, the spin multiplicity of the radical ions formed, the free energy of the geminate recombination, and the solvent (*36–45*).

$$\eta_{cs} = \frac{k_{sep}}{k_{sep} + k_{gem}} \tag{6}$$

High charge-separation yields are frequently observed when the geminate recombination free energy is much greater than the reorganizational energy associated with the process (vide infra). The effect of back-electron-transfer free energies on charge-separation efficiencies has been elegantly demonstrated by several techniques, including laser flash photolysis and time-resolved photoacoustic calorimetry (*46–48*).

The reaction steps in equations 4 and 5 define the efficiency for producing a strong reducing agent in solution. The remaining factor in determining the overall quantum yield of substrate reduction is competition between substrate reduction and back electron transfer of the reduced complex with the oxidized donor created in the photoreaction ($TMPD^+$):

$$\eta_{red} = \frac{k_{red}[S]}{k_{red}[S] + k_b[TMPD^+]} \tag{7}$$

In many systems the back-electron-transfer reaction occurs at rates approaching the diffusion limit (*12, 14, 31, 41, 42*), and substrate reduction must be relatively facile to compete, even though the concentration of the oxidized donor produced by excited-state quenching is typically much smaller than the steady-state substrate concentration (*16–18*). In principle, equation 7 should include a term for reduction of impurities in solution; impurity reduction becomes more important as the reducing agent produced ($[Ru]^+$) becomes more potent.

The overall photoreduction quantum yield for a one-electron substrate reduction is given in equation 8:

$$\Phi_{red} = \Phi_{abs}\, \eta_{isc}\, \eta_q\, \eta_{cs}\, \eta_{red} \tag{8}$$

where Φ_{abs} is the absolute yield reflecting the fraction of incident light absorbed by the sample. The η_{isc}, η_q, η_{cs} terms of equation 8 are associated with generation of a strong reducing agent in solution, and the η_{red} term is associated with the fate of the photoproduced reducing agent. The limiting yield for Scheme I is unity; however, if reduction of the substrate results in formation of radical intermediates or products that are themselves reductants, observed quantum yields can be much higher than unity (*1, 7, 8, 15*).

Electron-Transfer Dynamics. Scheme I includes three outer-sphere electron-transfer reactions: excited-state quenching (k_q), back reaction (k_b), and substrate reduction (k_{red}). The latter two reactions are in competition (eq 7), and high quantum yields for product formation are possible only when $k_{red}[S] >> k_b[TMPD^+]$. Each of these reactions involves association of the reduced complex with the reactant followed by electron transfer (eqs 9 and 10).

$$[Ru]^+ + S \rightleftarrows \{[Ru]^+, S\} \qquad K_{assoc} \tag{9}$$

$$\{[Ru]^+, S\} \longrightarrow \{[Ru]^{2+}, S^-\} \qquad k_{et} \tag{10}$$

The rate constant for the reaction, k_{obs} (either k_{red} or k_b), is given by

$$\frac{1}{k_{obs}} = \frac{1}{k_{diff}} + \frac{1}{K_{assoc}k_{et}} \tag{11}$$

where k_{diff} is the rate constant for the diffusion-limited reaction (*31, 32*). The association equilibrium constant, K_{assoc}, can be approximated by using the Fuoss equation:

$$K_{assoc} = \frac{4\pi N\sigma^3}{3000} \exp\left(-\frac{w(\sigma)}{RT}\right) \tag{12}$$

where σ is the sum of the radii of the reactants and $w(\sigma)$ is the work required to bring the reactants together (*32*), R is the gas constant, T is absolute temperature, and N is Avogadro's number. When one of the reactants is uncharged $w(\sigma)$ is zero; thus for association of $[Ru]^+$ (radius $r \approx 7$ Å) and an alkyl halide of radius 5 Å, $K_{assoc} \approx 5\ M^{-1}$. The association equilibrium constant and k_{diff} will be smaller when both reactants have like charges, and some measure of control over the photoreduction quantum yield results by considering charge in choosing the sensitizer, quencher, and substrate (vide infra). The electron-transfer rate constant, k_{et} (eq 11), can be evaluated by using classical electron-transfer theory (*2–5, 31, 32*). Equation 13 gives the classical expression

$$k_{et} = k_0 \exp\left[-\frac{(\lambda - \Delta G)^2}{4\lambda RT}\right] \qquad (13)$$

where k_0 is the nuclear frequency factor, λ is the reorganizational energy required to bring the reactants to the crossing point of the reaction coordinate (in volts), and ΔG is the free energy of the process (in volts). The electron transfer will be activationless when $\lambda = \Delta G$; Figure 2 illustrates the free energy dependence of k_{et} given by equation 13. The so-called Marcus inverted region occurs when ΔG is greater than λ and the rate constant becomes smaller than the activationless rate constant. Several excellent discussions of electron-transfer theories (*31, 32, 49, 50*) include evaluation of electron-transfer rate constants for bimolecular reactions in solution. Inverted region behavior is not commonly observed for bimolecular reactions because few reactions exist for which $K_{assoc}k_{et} < k_{diff}$ when $\lambda << -\Delta G$ and the observed rate constant equals the diffusion limited value (k_{diff}) (*1, 14*).

For the competing reactions of the photoreduction (k_b and k_{red}, eq 7), the parameters describing each of the electron transfers are not related; the only common factor is the reduction potential of the Ru(II) complex, $E_{Ru(2+/+)}$, which is required for determination of ΔG for both reactions. Thermal reduction of the substrate by the quencher (TMPD) is

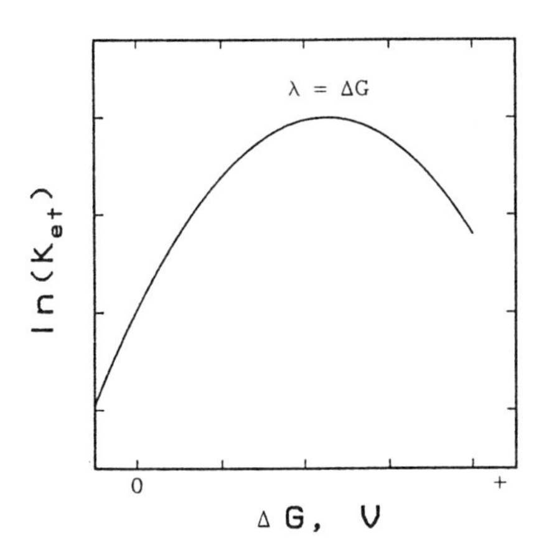

Figure 2. Free energy dependence of electron-transfer rate constants as described by eq 13. Activationless electron transfer occurs at the point where $\lambda = \Delta G$.

endoergic ($E_{S(0/-)} < E_{TMPD(+/0)}$), and therefore the free energy for substrate reduction by $[Ru]^+$ is necessarily less exoergic than that of back electron transfer. Thus, if the nuclear frequency factors and reorganizational energies for back electron transfer and substrate reduction are similar, k_b will necessarily be greater than or equal to k_{red}. The photoreduction should be most efficient when $k_b << k_{red}$; hence, reactants should be chosen for which the reorganizational barrier for back electron transfer is very large ($\lambda_b >> \lambda_{red}$) or the frequency factor for back electron transfer is much smaller than that for substrate reduction.

Photoreduction of Vicinal Dibromides

To examine the effect of variations in the ground-state reduction potential of the sensitizer on the overall quantum yield of a photoreduction reaction, the reduction of 1,2-dibromo-1,2-diphenylethane (DBDPE) and α,β-dibromoethylbenzene (DBEB) were studied by using a series of Ru(II) diimine complex sensitizers (*see* ligand structures on page 115) and TMPD as the electron-donating quencher. The net photoreaction involves reduction of the dibromoalkane ($OlBr_2$) to the olefin (Ol, eq 14) and is thus a two-electron process *(19–23, 30)*:

$$2TMPD + OlBr_2 \xrightarrow[h\nu]{[Ru]^{2+}} 2TMPD^+ + 2Br^- + Ol \qquad (14)$$

Quantum yields (Φ_{T+}) were measured for formation of $TMPD^+$, a deep blue radical ion that accumulates during the photolysis. Figure 3 shows spectrophotometric changes observed upon photolysis of $[(dmb)_3Ru]^{2+}$ (dmb is 4,4′-dimethyl-2,2′-bipyridine), DBDPE (0.005 M), and TMPD (0.05 M) in CH_3CN. Quantum yields for $TMPD^+$ formation at low conversions were measured for each of the complexes under these conditions and are listed in Table I. The limiting quantum yield for $TMPD^+$ formation (as the concentration of substrate was increased) was determined for two of the complexes to be 1.0 ± 0.05; thus, formation of $TMPD^+$ results from excited-state electron transfer alone and not from oxidation of TMPD by intermediate species produced in the dehalogenation of the vicinal dihalide. Table I also lists photophysical properties, one-electron reduction potentials of the complexes in CH_3CN, and rate constants for back electron transfer, k_b.

In evaluating the photoreaction it is assumed that the intersystem crossing efficiency is unity for each of the complexes; this fact has been established only for $[(bpy)_3Ru]^{2+}$ and $[(dmb)_3Ru]^{2+}$ *(30, 33–35)*. Approximate values of the excited-state energies of the complexes, E_{00},

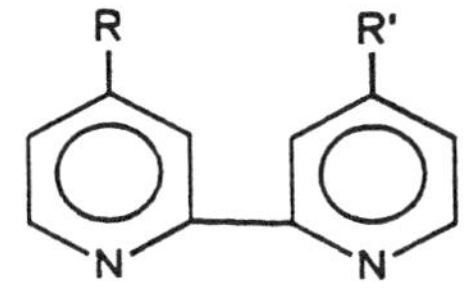

bpy R=R'=H

dmb R=R'=CH_3

decb R=R'=CO_2Et

dcab R=R'=$CON(Et)_2$

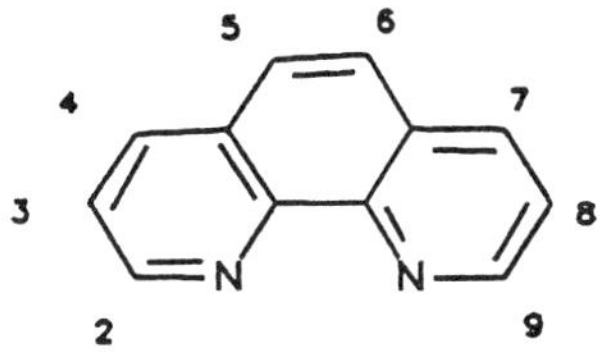

phen all H

tmphen 3,4,7,8 – tetramethyl

5–Cl–phen 5–chloro

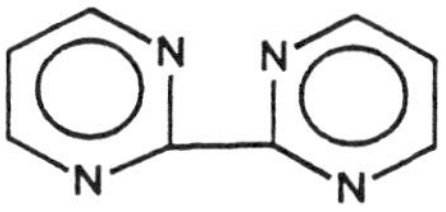

bpym

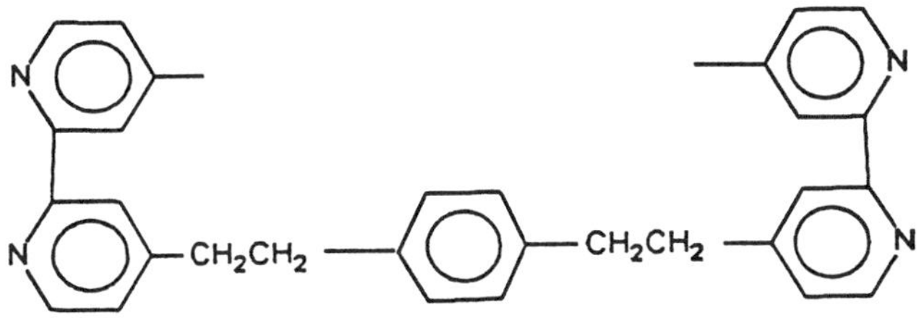

b–b

are obtained from emission spectral fitting techniques (*27, 51*). Values for the complexes of Table I range from 1.8 to 2.1 eV. Given the excited-state energies, one-electron reduction potentials of the complexes, $E_{Ru(2+/+)}$, and $E_{TMPD(+/0)}$ = 0.15 V versus SSCE (*52*), the quenching process is found to be exoergic (E_{pet} > 0) for all the complexes of the series (eq 1). Luminescence quenching rate constants, k_q, not reported in Table I, are greater than 10^9 M^{-1} s^{-1} in each case; for solutions containing 0.05 M TMPD observed quenching efficiencies (η_q) were greater than 0.95 for all the complexes.

The efficiency of charge separation from the geminate pair formed in the excited-state electron transfer (eqs 4 and 6) was determined by flash photolysis (*30, 33*). The measured charge-separation efficiency, η_{cs}^{obs},

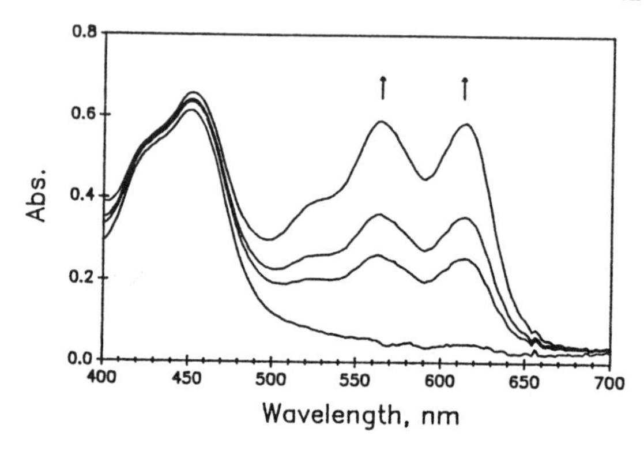

Figure 3. Absorption spectral changes observed at different times during photolysis of $[(dmb)_3Ru]^{2+}$ *(2 × 10^{-5} M), TMPD (0.05 M), and DBDPE (0.005 M) in deaerated* CH_3CN.

includes the efficiency for populating the reactive (^{3}MLCT) excited state, η_{isc}, and the fraction of excited states quenched, η_q:

$$\eta_{cs}{}^{obs} = \eta_{isc}\,\eta_q \left(\frac{k_{sep}}{k_{sep} + k_{gem}} \right) \tag{15}$$

In measurements made for the complexes of this series, η_{isc} is assumed to be unity and η_q is fixed to be between 0.3 and 0.6 by using appropriate TMPD concentrations. Thus, charge-separation yields determined by flash photolysis are simply the ratio of ions formed in solution immediately after the flash to the number of quenched excited states formed ($\eta_{isc}\eta_q$). The excited-state concentration and solvent-separated ion concentration were determined from the transient absorbance at 360 nm and the molar absorptivities of the Ru(II) ground state, ^{3}MLCT state (*53*) and the radical ions (determined by spectroelectrochemistry). Values of η_{cs} obtained by this approach were 0.66 for $[(bpy)_3Ru]^{2+}$ and 0.72 for $[(dmb)_3Ru]^{2+}$; these values represent upper limiting values for the yield of reducing ions in solution per photon absorbed.

Given that the observed yield of charge-separated ions can be determined experimentally, the overall quantum yield for the photoreduction, corrected for losses due to inefficiencies in producing solvent-separated ions, is simply the efficiency for the reduction step, η_{red}:

$$\eta_{red} = \frac{\Phi_{red}}{\Phi_{abs}\eta_{cs}{}^{obs}} = \frac{k_{red}[S]}{k_{red}[S] + k_b[\mathrm{TMPD}^+]} \tag{16}$$

Table I. Absorption and Luminescence Maxima, One-Electron Reduction Potentials, and Quantum Yields for Photoreduction of DBDPE and DBEB in CH_3CN at 298 K

Complex[a]	λ_{ab} *(nm)*	λ_{em} *(nm)*	τ_{em} *(ns)*[b]	$-E_{red}$ *(V)*	Φ_{T+} *for DBDPE*	Φ_{T+} *for DBEB*	k_b $(M^{-1}\,s^{-1})$
$[Ru(tmphen)_3](PF_6)_2$	440	595	290	1.60	—	0.82	—
$[Ru(tmphen)_2(dmb)](PF_6)_2$	438	630	967	1.50	0.88	0.78	—
$[Ru(dmb)_3](PF_6)_2$	458	635	576	1.46	0.63	0.58	1.0×10^{10}
$[Ru(phen)_3(PF_6)_2$	444	600	220	1.38	0.49	0.47	2.6×10^{10}
$[Ru(bpy)_3](PF_6)_2$	452	620	860	1.32	0.47	0.23	4.0×10^{10}
$[Ru(5\text{-}Cl\text{-}phen)_3](PF_6)_2$	448	605	320	1.24	0.24	0.21	—
$[Ru(dcab)_2(dmb)](PF_6)_2$	472	641	990	1.16	0.096	0.08	—
$[Ru(dmb)_2(decb)](PF_6)_2$	492	694	600	1.04	0.023	—	1.6×10^{10}
$[Ru(bpy)_2(bpym)](PF_6)_2$	480	720	88	1.025	0.006	0.002	1.5×10^{10}

[a]Ligand abbreviations are tmphen, 3,4,7,8-tetramethylphenanthroline; dmb, 4,4′-dimethyl-2,2′-bipyridine; phen, 1,10-phenanthroline; bpy, 2,2′-bipyridine; 5-Cl-phen, 5-chloro-1,10-phenanthroline; dcab, 4,4′-(*N,N*-diethylcarboxamido)-2,2′-bipyridine; decb, 4,4′-diethylcarboxy-2,2′-bipyridine; and bpym, 2,2′-bipyrimidine.

[b]τ_{em} is the luminescence lifetime of the unquenched complex in deaerated CH_3CN.

Evaluation of η_{red} requires knowledge of the rate constants for back electron transfer and substrate reduction and the concentrations of $TMPD^+$ and substrate in solution. In any steady-state photolysis the concentrations of both $TMPD^+$ and substrate will change during the photolysis, and the maximum overall quantum yield will be observed at the beginning of the photolysis. The back-electron-transfer rate constant can be measured directly via flash photolysis. Figure 4 shows a transient difference spectrum observed following 532-nm laser flash photolysis (Nd:YAG) of a solution of $[(dmb)_3Ru]^{2+}$ (10^{-4} M) and TMPD (0.05 M) in CH_3CN. The spectra of $TMPD^+$ and $[(dmb)_3Ru]^+$ can be obtained independently by spectroelectrochemistry, so the transient spectrum can be calculated (solid line of Figure 4). Decays of the absorbance change at wavelengths where the transient ions absorb follow equal concentration second-order kinetics, and the rate constants can be determined as long as the change in the molar absorptivity of the solution, $\Delta\epsilon$, is known at the wavelength used for kinetic analysis (Table I). Although only five of the complexes were examined in detail, k_b is greater than 10^{10} M^{-1} s^{-1} in each case.

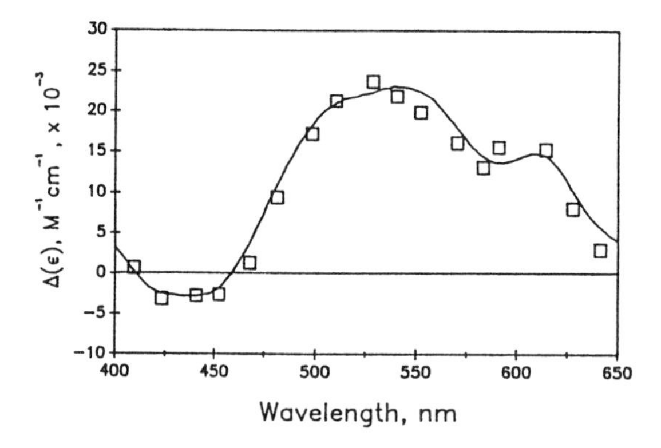

Figure 4. Transient difference spectrum obtained 1 μs after pulsed laser (532 nm, Nd:YAG, <10 mJ/pulse) photolysis of $[(dmb)_3Ru]^{2+}$ and TMPD (0.05 M) in deaerated CH_3CN. The solid line represents the difference spectrum calculated from the spectra of $[(dmb)_3Ru]^{2+}$, $[(dmb)_3Ru]^+$, and $TMPD^+$ ($\Delta\epsilon = \epsilon_{TMPD(+)} + \epsilon_{Ru(+)} - \epsilon_{Ru(2+)}$). ($\epsilon$ is the molar absorptivity of each species.)

These results suggest that values for η_{cs}, η_q, and k_b do not vary significantly for the series of complexes examined, yet initial quantum yields for formation of $TMPD^+$ vary by more than a factor of 100 (Table I) for reduction of each of the vicinal dihalides studied. The explanation for this variation is that although back electron transfer is diffusion-limited for each of the sensitizers studied, the rate constant for substrate reduction depends on the free energy of the reaction. Equation 16 may be restated as follows:

$$\ln\left(\frac{1}{\eta_{red}} - 1\right) = \ln\left(\frac{k_b[\mathrm{TMPD}^+]}{k_{red}[S]}\right) \tag{17a}$$

$$\ln\left(\frac{k_b[\mathrm{TMPD}^+]}{k_{red}[S]}\right) = \ln\left(\frac{k_b[\mathrm{TMPD}^+]}{k_0[S]}\right) + \frac{(\lambda - E_{red})^2}{4\lambda RT} \tag{17b}$$

By assuming that $\eta_{isc} = \eta_q = 1$ and that $\eta_{cs} = 0.75$ for all the complexes of the series, values of η_{red} can be approximated. Figure 5 shows $\ln(1/\eta_{red} - 1)$ for reduction of DBEB versus $E_{Ru(2+/+)}$ for the series of complexes. The solid line represents a fit to the data using equation 17b with $\lambda = 0.6$ V and $E_{S(0/-)} = -1.1$ V (vs. SSCE). The logarithmic term on the right side of equation 17b represents a minimum in the parabolic plot (when $\lambda = E_{red}$).

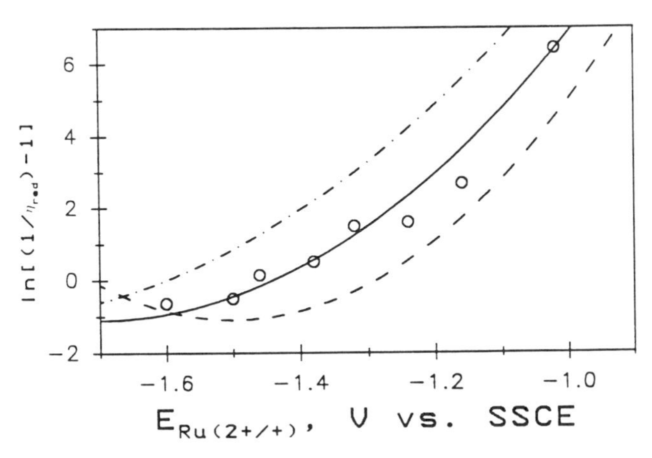

Figure 5. Variation of $\ln(1/\eta_{red} - 1)$ with mediator reduction potential, $E_{Ru(2+/+)}$, for DBEB reduction in CH_3CN. Fits to the data using eq 17b are shown for 0.4 V (– – –), 0.6 V (—), and 0.8 V (– • – •). $E_{S(0/-)} = -1.1$ V, and $\ln(k_b[TMPD^+]/k_0[S]) = 0.3$.

The preceding example illustrates some of the difficulties in developing efficient photocatalytic schemes for outer-sphere reduction of substrates. Even though high yields of charge-separated ions are formed in the photoreaction, yields for substrate reduction are limited by the fact that back electron transfer competes very effectively with substrate reduction, even when substrate concentrations are at least a factor of 100 larger than concentrations of $TMPD^+$ formed at early photolysis times. The fit of the data to equation 17 indicates that the reorganizational barrier for vicinal dihalide reduction is modest (0.6–0.8 V). Even so, the relatively small free energies associated with substrate reduction result in low reduction yields for all but the most strongly reducing complexes.

One approach to improving photoreduction yields in the systems discussed is to retard the back-electron-transfer reaction. This retardation can be accomplished by choosing an electron-donating quencher that has either a larger barrier for back electron transfer or a lower diffusion-limited rate (i.e., k_{diff} is smaller). When the positive charge on one or both of the ions of the geminate pair (Ru+ and $TMPD^+$) is increased by synthetic modification, the magnitude of K_{assoc} and k_{diff} are decreased (larger work term for association of the ions). This approach was tested by preparing a series of Ru complexes having identical spectroscopic and redox properties but differing in the overall charge of the complex (*54*). By using the bridging ligand b–b (*see* structures on page 115), the dinuclear and tetranuclear complexes $\{[(dmb)_2Ru](b\text{–}b)\}^{4+}$ and $\{[(dmb)_2\text{-}Ru(b\text{–}b)]_3\}Ru^{8+}$ were prepared and used in the scheme for photoreduction of DBDPE with TMPD as electron donor. Table II shows rate constants for quenching of the MLCT luminescence of each of the complexes by TMPD in CH_3CN, charge-separation efficiencies, and back-electron-transfer rate constants in the absence and presence of added electrolyte. In each case the quenching reaction is unaffected by the increased charge and size of the sensitizer complex. Charge-separation efficiencies measured for $[(dmb)_2Ru(b\text{–}b)]^{2+}$ and $\{[(dmb)_2Ru](b\text{–}b)\}^{4+}$ are nearly the same; this result is not surprising because η_{cs} is high even for the mononuclear complexes. Back-electron-transfer rate constants measured in the absence of added electrolyte decrease by a factor of 10 as the charge on the reduced metal complex increases from +1 to +7.

Figure 6 shows absorbance changes associated with $TMPD^+$ formation when absorbance-matched solutions of $[(dmb)_2Ru(b\text{–}b)]^{2+}$, $\{[(dmb)_2Ru](b\text{–}b)\}^{4+}$, and $\{[(dmb)_2Ru(b\text{–}b)]_3\}Ru^{8+}$ are irradiated (λ_{ir} > 400 nm) in the presence of TMPD (0.05 M) and DBDPE (0.0017 M). Initial quantum yields under these conditions are 0.25, 0.54, and 0.57 for the three complexes $[(dmb)_2Ru(b\text{–}b)]^{2+}$, $\{[(dmb)_2Ru](b\text{–}b)\}^{4+}$, and $\{[(dmb)_2Ru(b\text{–}b)]_3\}Ru^{8+}$, respectively.

The quantum yield results can be rationalized by restating equation 8 as follows:

$$\frac{1}{\Phi_{red}} = \frac{1}{\Phi_{abs}\eta_{cs}} + \frac{k_b[TMPD^+]}{\Phi_{abs}\eta_{cs}k_{red}[S]} \quad (18)$$

When k_b becomes sufficiently small, the observed quantum yield equals the charge-separation efficiency (the limiting quantum yield for the process). Increasing the charge of the reducing complex from +1 to +3 in this system causes the back-electron-transfer rate constant to decrease by a factor of 2.9, while the quantum yield for substrate reduction increases by

Table II. Rate Constants for Quenching and Back Electron Transfer between Complexes Having Varying Charges and TMPD in CH_3CN

Complex	k_q $(M^{-1}s^{-1})$[a]	I = *0 M*, k_b $(M^{-1}s^{-1})$[b]	I = *0.5 M* k_b $(M^{-1}s^{-1})$[b]	η_{cs}
$[Ru(dmb)_2(b{-}b)](PF_6)_2$	7.7×10^9	4.4×10^9	6.1×10^9	0.75 ± 0.09
$\{[(dmb)_2Ru]_2(b{-}b)\}(PF_6)_4$	8.3×10^9	1.5×10^9	1.2×10^{10}	0.71 ± 0.15
$\{[(dmb)_2Ru(b{-}b)]_3Ru\}(PF_6)_8$	7.0×10^9	4.6×10^8	3.9×10^9	—[c]

[a]Rate constants are ±5%.

[b]Rate constants are ±10%.

[c]No data.

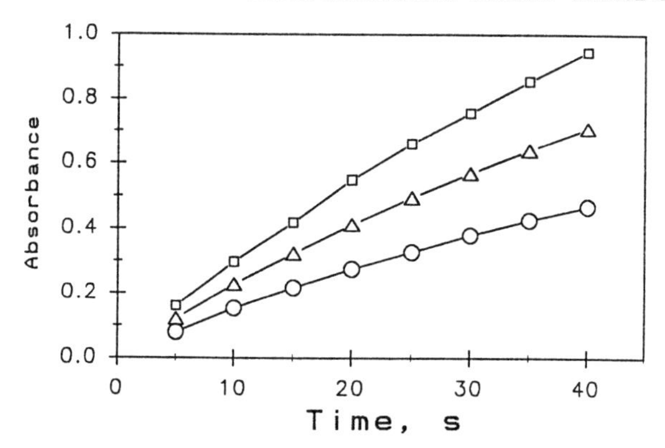

Figure 6. Absorbance changes observed at 612 nm upon photolysis of solutions containing TMPD (0.05 M), DBDPE (0.0017 M), and one of the following sensitizers: $[(dmb)_2Ru(b{-}b)]^{2+}$ (○), $\{[(dmb)_2Ru](b{-}b)\}^{4+}$ (Δ), *or* $\{[(dmb)_2Ru(b{-}b)]_3\}Ru^{8+}$ (□). *The absorbance of the sensitizers was matched at 1.0 optical density units (ODU).*

a factor of 2.2. This result suggests that most of the decrease in k_b is reflected in Φ_{red}. However, the further increase in charge of the reducing complex to +7 has little effect on the observed quantum yield, even though k_b decreases significantly relative to the +3 complex (factor of 3.2). Given that η_{cs} = 0.75 for the complexes, the difference between the dimer and tetramer is small because the measured quantum yield is nearing the limit of the charge-separation yield. The results thus show that exploiting charge effects to decrease k_b can have significant effects on the observed quantum yield as long as the yield is significantly below the limiting value. This feature may be particularly useful for cases in which only low substrate concentrations are possible because of solubility considerations.

MLCT Complexes as Sensitizers and Inner-Sphere Reductants

A significant problem with outer-sphere substrate reduction is that back electron transfer can compete very effectively with substrate reduction, and, in general, high substrate concentrations are required to obtain reasonable reduction quantum yields. An alternate approach is to devise systems in which the substrate is bound to the sensitizer or electron mediator so that substrate reduction becomes unimolecular. An example of such a scheme, which employs a MLCT complex mediator and a halopyridine substrate coordinated to the metal center (*29*), is shown in Scheme II.

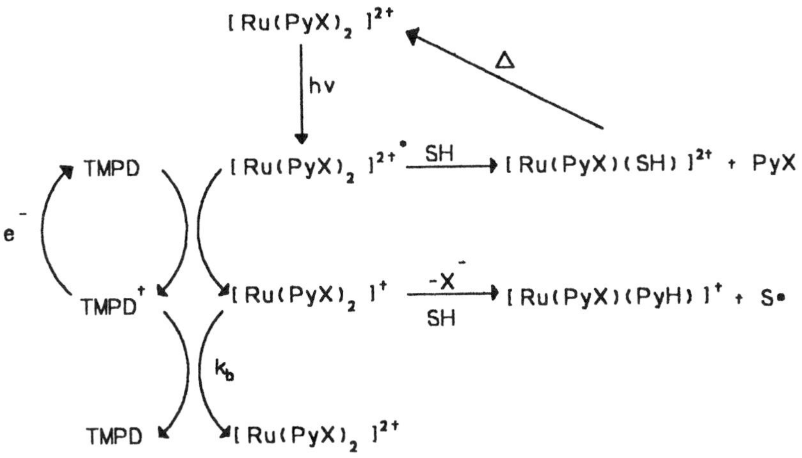

$[Ru(PyX)_2]^{2+}$ $\xrightarrow{h\nu}$ $[Ru(PyX)_2]^{2+*}$ $\xrightarrow{SH}$ $[Ru(PyX)(SH)]^{2+}$ + PyX $\xrightarrow{\Delta}$ $[Ru(PyX)_2]^{2+}$

TMPD + $[Ru(PyX)_2]^{2+*}$ $\rightarrow$ $TMPD^+$ + $[Ru(PyX)_2]^+$ (e^-)

$[Ru(PyX)_2]^+$ $\xrightarrow[SH]{-X^-}$ $[Ru(PyX)(PyH)]^+$ + S•

$TMPD^+$ + $[Ru(PyX)_2]^+$ $\xrightarrow{k_b}$ TMPD + $[Ru(PyX)_2]^{2+}$

Scheme II.

The excited complex, $[Ru(PyX)_2]^{2+*}$, has an additional decay pathway: substitution of the coordinated halopyridine (shown as a solvation reaction). Quenching by TMPD leads to formation of $TMPD^+$ and $[Ru(PyX)_2]^+$; in this case, however, back electron transfer competes with unimolecular C–X bond cleavage to yield a coordinated pyridyl radical. The radical formed will be very reactive and, in most organic solvents, hydrogen-atom abstraction from solvent will predominate (*55–57*). Such a scheme is considerably more complicated and very likely more limited in scope than outer-sphere reduction.

The additional excited-state decay path of Scheme II, ligand substitution, has been studied in detail for Ru(II) diimine complexes (*34, 58–63*). Evidence obtained for a wide range of complexes indicates that substitution results upon population of a metal-centered (MC) excited state that is produced by thermally activated internal conversion from the MLCT state, as shown in Figure 1 (*12, 24–28*). The implication for the scheme is that partitioning between photoreduction and substitution can be controlled by regulating population of the MC state (*18, 35*). This regulation is most easily accomplished by careful control of the temperature of the photolysis solution. Ideally a temperature is selected to allow both substitution and electron transfer to occur so that both reductive dehalogenation and replacement of product (PyH) with new substrate (PyX) can occur (via photosubstitution).

Such a scheme can be effective only if intramolecular reductive dehalogenation is rapid relative to the bimolecular back-electron-transfer reaction, k_b. One means of qualitatively assessing the rate of dehalogenation following reduction of the complex is cyclic voltammetry. Figure 7 shows reductive cyclic voltammograms obtained for three Ru(II) complexes having coordinated halopyridines: $[(dmb)_2Ru(3BrPy)_2]^{2+}$, $[(decb)_2Ru(3BrPy)_2]^{2+}$ (decb is 4,4′-diethylcarboxy-2,2′-bipyridine), and $[(dmb)_2Ru(3ClPy)_2]^{2+}$. Reduction of $[(dmb)_2Ru(3BrPy)_2]^{2+}$ at 200 mV/s is completely irreversible, and the shape of the wave indicates the reductive process may be catalytic. Exhaustive electrolysis of solutions of $[(dmb)_2Ru(3BrPy)_2]^{2+}$ at −1.5 V versus the SSCE in CH_3CN yields $[(dmb)_2Ru(Py)_2]^{2+}$ as the principal product. If, however, the spectator ligands of the complex have π^* levels significantly lower in energy than the halopyridine π^* levels, one-electron reduction of the complex will not result in facile dehalogenation.

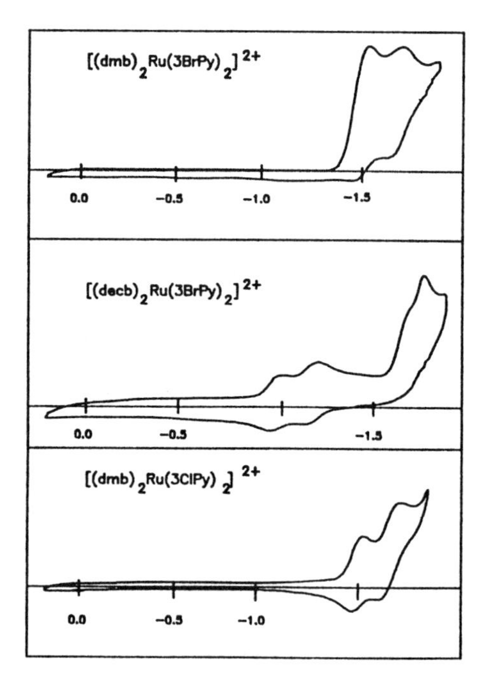

Figure 7. Reductive cyclic voltammograms of $[(dmb)_2Ru(3BrPy)_2]^{2+}$, $[(decb)_2Ru(3BrPy)_2]^{2+}$, and $[(dmb)_2Ru(3ClPy)_2]^{2+}$ in CH_3CN at 298 K. The supporting electrolyte was tetrabutylammonium perchlorate (TBAP); the sweep rate was 200 mV/s; the reference was the SSCE.

The cyclic voltammograms of $[(decb)_2Ru(3BrPy)_2]^{2+}$ and $[(dmb)_2$-$Ru(3ClPy)_2]^{2+}$ illustrate this effect. In $[(decb)_2Ru(3BrPy)_2]^{2+}$, the first two one-electron reductions are localized on the decb ligands, and both are reversible on the cyclic voltammetry time scale. The third reduction, at −1.8 V versus the SSCE, is irreversible and results in dehalogenation. Cyclic voltammograms of $[(dmb)_2Ru(3ClPy)_2]^{2+}$ in CH_3CN result in two quasi-reversible waves, a result suggesting that dehalogenation competes with reoxidation of the complex at the electrode. This result demonstrates that intramolecular reduction of a substrate is subject to thermodynamic constraints that are similar to those of outer-sphere substrate reduction.

A suitable temperature for photolysis, in which both substitution and electron transfer will occur, can be determined by examining the temperature dependence of the luminescence intensity and/or lifetime of the MLCT excited state. Figure 8 shows the temperature dependence of the luminescence lifetime of $[(dmb)_2Ru(3BrPy)_2]^{2+}$ in 4:1 ethanol:methanol. The decrease in the luminescence lifetime with increasing temperature can be attributed to thermally activated population of the MC excited state. The efficiency for internal conversion to the MC state can be approximated by eq 19 (*24–27*)

$$\eta_{ic} = 1 - \left(\frac{\tau_T}{\tau_{135}}\right) \qquad (19)$$

where τ_{135} is the lifetime of the complex in solution at the low-temperature limit (where $\eta_{ic} = 0$) and τ_T is the lifetime at a given temperature. The temperature at which 50% of the MLCT state decay results in population of a MC state is approximately 165 K for $[(dmb)_2$-$Ru(3BrPy)_2]^{2+}$. Ligand loss from the MC state is dependent upon a variety of factors including solvent and temperature, and it is possible that efficient substitutional photochemistry may be achieved only at temperatures at which η_{ic} is much higher than 0.5.

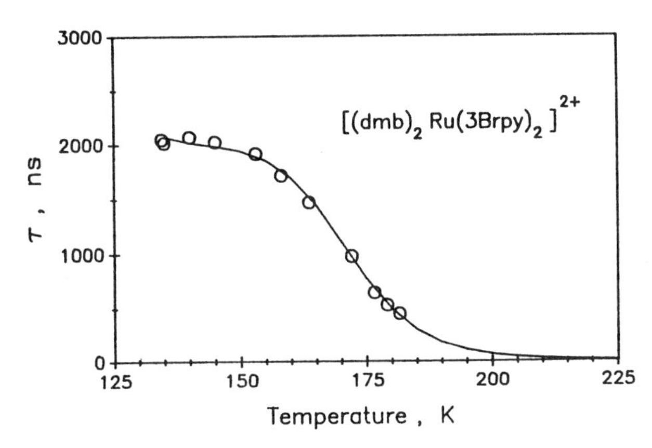

Figure 8. Temperature dependence of the luminescence lifetime of $[(dmb)_2Ru\ (3BrPy)_2]^{2+}$ in 4:1 ethanol:methanol.

Spectral changes observed upon photolysis of 4:1 ethanol:methanol solutions containing $[(dmb)_2Ru(3BrPy)_2]^{2+}$ and TMPD (0.05 M) at two different temperatures are shown in Figure 9. Room-temperature photolysis results in formation of $[(dmb)_2Ru(3BrPy)(ROH)]^{2+}$ based on the observed change of the MLCT maximum and the absorption spectrum of the monoalcohol complex prepared independently. At 150 K formation of $TMPD^+$ is observed, and only a slight change in the MLCT maximum of the complex occurs. This slight change is consistent with reductive dehalogenation followed by formation of the Py complex (Scheme II). Photolysis of solutions containing only TMPD at 150 K does not result in $TMPD^+$ formation.

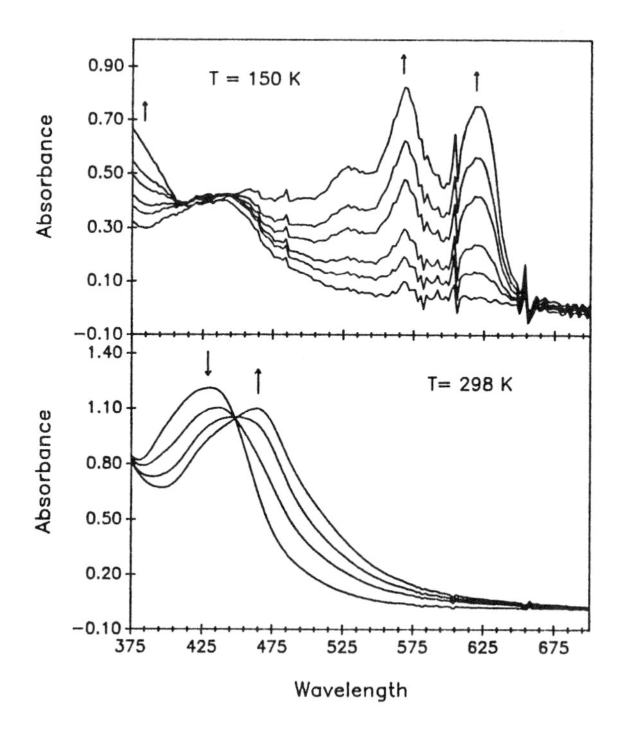

Figure 9. Spectral changes observed upon photolysis of mixtures of $[(dmb)_2Ru(3BrPy)_2]^{2+}$ *and TMPD (0.05 M) in 4:1 ethanol:methanol at 150 K and 298 K. Spectra were taken at 30-s intervals.*

These observations for $[(dmb)_2Ru(3BrPy)_2]^{2+}$ establish that both unimolecular reductive dehalogenation and photosubstitution can occur for a given complex. The efficiency of intramolecular dehalogenation (substrate reduction) is subject to thermodynamic constraints that are similar to those observed for outer-sphere substrate reduction; for Ru(II) diimine complexes, the first reduction of the complex must be partially localized on the substrate ligand. Partitioning of substitutional photochemistry and photoredox chemistry can be controlled by manipulation of the temperature.

Conclusion

Overall, use of Ru(II) diimine complexes as homogeneous photosensitizers for either inner-sphere or outer-sphere reduction of organic substrates appears to be limited to organic compounds that can be relatively easily reduced. The inner-sphere approach may be more useful in this regard because coordination can serve to lower the reduction potential of the substrate. In Schemes I and II, outer-sphere reduction of 3-bromopyridine is not possible using any of the trisdiimine complexes, yet the inner-sphere process occurs readily upon reduction of the halopyridine complex. We are presently working on the development of other systems in which the metal complex serves to coordinate the substrate and serve as the sensitizer for the photoinduced electron-transfer process.

Acknowledgments

The authors thank the Louisiana Educational Quality Support Fund, administered by the Louisiana State Board of Regents, for support of this work.

References

1. Chanon, F.; Chanon, M. In *Photocatalysis;* Serpone, N.; Pelizzetti, E., Eds.; Wiley-Interscience: New York, 1989; Chapter 15.
2. Sutin, N.; Creutz, C. *Pure Appl. Chem.* **1980,** *52,* 2717.
3. Keene, F. R.; Creutz, C.; Sutin, N. *Coord. Chem. Rev.* **1985,** *64,* 247.
4. Creutz, C.; Sutin, N. *Coord. Chem. Rev.* **1985,** *64,* 321.
5. Sutin, N.; Creutz, C. *J. Chem. Educ.* **1983,** *60,* 809.
6. Tazuke, S.; Kitamura, N. *Pure Appl. Chem.* **1984,** *56,* 1269.
7. Maverick, A. W.; Gray, H. B. *Pure Appl. Chem.* **1980,** *52,* 2339.
8. Gray, H. B.; Maverick, A. W. *Science* **1981,** *214,* 1201.
9. Hennig, H.; Rehorek, D.; Archer, R. D. *Coord. Chem. Rev.* **1985,** *61,* 1.
10. Prasad, D. R.; Mandal, K.; Hoffman, M. *Coord. Chem. Rev.* **1985,** *64,* 175.
11. Darwent, J. R.; Douglas, P; Harriman, A.; Porter, G.; Richoux, M.-C. *Coord. Chem. Rev.* **1982,** *44,* 83.
12. Kalyanasundaram, K. *Coord. Chem. Rev.* **1982,** *46,* 159.

13. Meyer, T. J. *Accts. Chem. Res.* **1989,** *22,* 163.
14. Bock, C. R.; Connor, J. A.; Gutierrez, A. R.; Meyer, T. J.; Whitten, D. G.; Sullivan, B. P.; Nagle, J. K. *J. Am. Chem. Soc.* **1979,** *101,* 4815.
15. Roundhill, D. M.; Gray, H. B. *Accts. Chem. Res.* **1989,** *22,* 55.
16. Whitten, D. G. *Accts. Chem. Res.* **1980,** *13,* 83.
17. Kavarnos, G. J.; Turro, N. J. *Chem. Rev.* **1986,** *86,* 401.
18. Julliard, M.; Chanon, M. *Chem. Rev.* **1983,** *83,* 425.
19. Willner, I.; Maidan, R.; Shapira, M. *J. Chem. Soc. Perkin Trans. 2* **1990,** 559.
20. Willner, I.; Steinberger-Willner, B. *Int. J. Hydrogen Energy* **1988,** *13,* 593.
21. Willner, I.; Mandler, D.; Maidan, R. *New J. Chem.* **1987,** *11,* 109.
22. Maidan, R.; Willner, I. *J. Am. Chem. Soc.* **1986,** *108,* 8100.
23. Willner, I.; Maidan, R.; Mandler, D.; Durr, H.; Dörr, G.; Zengerle, K. *J. Am. Chem. Soc.* **1987,** *109,* 6080.
24. Balzani, V.; Bolletta, F.; Moggi, L; Manfrin, M. F.; Laurence, G. S. *Coord Chem. Rev.* **1975,** *15,* 321.
25. Scandola, F.; Balzani, V. *J. Chem. Educ.* **1983,** *60,* 814.
26. Juris, A.; Balzani, V.; Barigelletti, F.; Campagna, S.; Belser, P.; von Zelewsky, A. *Coord. Chem. Rev.* **1988,** *52,* 85.
27. Meyer, T. J. *Pure Appl. Chem.* **1986,** *58,* 1193.
28. Sutin, N.; Creutz, C. J. *Chem. Educ.* **1983,** *60,* 809-814.
29. Wright, D. W.; Schmehl, R. H. *Inorg. Chem.* **1990,** *29,* 155.
30. Bensasson, R.; Salet, C.; Balzani, V. *J. Am. Chem. Soc.* **1976,** *98,* 3722.
31. Meyer, T. J. *Prog. Inorg. Chem.* **1983,** *30,* 389.
32. Sutin, N. *Prog. Inorg. Chem.* **1983,** *30,* 441.
33. Demas, J. N.; Taylor, D. G. *Inorg. Chem.* **1979,** *18,* 3177.
34. Wacholtz, W. F.; Auerbach, R. A.; Schmehl, R. H.; Ollino, M. A.; Cherry, W. R. *Inorg. Chem.* **1985,** *24,* 1758.
35. Wacholtz, W. F.; Auerbach, R. A.; Schmehl, R. H. *Inorg. Chem.* **1986,** *25,* 227.
36. Ohno, T.; Yoshimura, A.; Mataga, N.; Tazuke, S.; Kawanishi, Y.; Kitamura, N. *J. Phys. Chem.* **1989,** *93,* 3546.
37. Tazuke, S.; Kitamura, N.; Kim, H.-B. *Supramolecular Photochemistry;* North Atlantic Treaty Organization ASI Series, Series C, 1987, p 87.
38. Ohno, T.; Lichtin, N. *J. Am. Chem. Soc.* **1980,** *102,* 4636.
39. Ohno, T.; Kato, S.; Yamada, A.; Tanno, T. *J. Phys. Chem.* **1983,** *87,* 775.
40. Balzani, V.; Juris, A.; Scandola, F. In *Homogeneous and Heterogeneous Photocatalysis;* Pelizzetti, E.; Serpone, N., Eds.; D. Reidel: Dordrecht, Netherlands, 1986; p 1.
41. Olmsted, J.; Meyer, T. J. *J. Phys. Chem.* **1987,** *91,* 1649.
42. Mandal, K; Hoffman, M. Z. *J. Phys. Chem.* **1984,** *88,* 185.
43. Hoffman, M. Z. *J. Phys. Chem.* **1988,** *92,* 3458.
44. Chan, S.-F.; Chou, M.; Creutz, C.; Matsuhara, T.; Sutin, N. *J. Am. Chem. Soc.* **1981,** *103,* 369.
45. Jones, G.; Malbu, V. J. *Org. Chem.* **1985,** *50,* 5776.
46. Gould, I. R.; Ege, D.; Mattes, S. L.; Farid, S. *J. Am. Chem. Soc.* **1989,** *109,* 3794.

47. Gould, I. R.; Moser, J. E.; Armitage, B.; Farid, S.; Goodman, J. L.; Herman, M. S. *J. Am. Chem. Soc.* **1989,** *111,* 1917.
48. Ohno, T,; Yoshimura, A.; Mataga, N. *J. Phys. Chem.* **1986,** *90,* 3296.
49. Sutin, N.; Marcus, R. *Biochem. Biophys. Acta* **1985,** *811,* 265.
50. *Photoinduced Electron Transfer, Part A;* Fox, M. A.; Chanon, M., Eds.; Elsevier: Amsterdam, Netherlands, 1988.
51. Kober, E. M.; Caspar, J. V.; Lumpkin, R. S.; Meyer, T. J. *J. Phys. Chem.* **1986,** *90,* 3722.
52. Mann, C. K.; Barnes, K. K. *Electrochemical Reactions in Nonaqueous Systems;* Dekker: New York, 1970; p 278.
53. Atherton, S., Center for Fast Kinetics Research, Austin, TX, personal communication. The value for excited $[(bpy)_3Ru]^{2+}$ was used in the determination of values at 360 nm for the chromophores discussed in this chapter.
54. Wacholtz, W. F.; Auerbach, R. A.; Schmehl, R. H. *Inorg. Chem.* **1987,** *26,* 2989.
55. Saveant, J. M. *Accts. Chem. Res.* **1980,** *13,* 323.
56. Bunnett, J. F. *Accts. Chem. Res.* **1978,** *11,* 413.
57. Andrieux, C. P.; Blooman, C.; Dunas-Bouchiat, J-M.; Saveant, J.-M. *J. Am. Chem. Soc.* **1979,** *101,* 3431.
58. Durham, B.; Caspar, J. V.; Nagle, J. K.; Meyer, T. J. *J. Am. Chem. Soc.* **1982,** *104,* 4803.
59. Durham, B.; Walsh, J. L.; Carter, C. L.; Meyer, T. J. *Inorg. Chem.* **1980,** *19,* 860.
60. Caspar, J. V.; Meyer, T. J. *J. Am. Chem. Soc.* **1983,** *105,* 5583.
61. Van Houten, J.; Watts, R. J. *J. Am. Chem. Soc.* **1976,** *98,* 4853.
62. Van Houten, J.; Watts, R. J. *Inorg. Chem.* **1978,** *17,* 3381.
63. Hoggard, P. E.; Porter, G. B. *J. Am. Chem. Soc.* **1978,** *100,* 1457.

RECEIVED for review November 7, 1991. ACCEPTED revised manuscript May 27, 1992.

7

Photochemistry and Redox Catalysis Using Rhenium and Molybdenum Complexes

Andrew W. Maverick, Qin Yao, Abdul K. Mohammed, and Leslie J. Henderson, Jr.

Department of Chemistry, Louisiana State University, Baton Rouge, LA 70803–1804

Halide complexes of Mo(III), Re(IV), and Re(V) and complexes of Mo(V), Re(V), and Re(VI) containing one oxo ligand luminesce in the near-infrared region in room-temperature solutions. Many of them are photooxidized by electron acceptors; with $ReCl_6^{2-}$ *and several of the Mo(III) complexes, this is the first step in overall photoinitiated two- or three-electron transfers.* $ReCl_6^{2-}$ *catalyzes the oxidation and electrooxidation of* Cl^- *to* Cl_2 *and organic oxidations such as the conversion of toluene to benzaldehyde in aqueous HCl solution. The rhenium(V) complex* $ReCl_6^-$ *appears to be an important intermediate in the catalytic reactions. The fluorescences of the oxo-*d^1 *complexes* $Mo^VOX_4L^-$ *(X is Cl or Br and L is* H_2O *or* CH_3CN*) and* $Re^{VI}OCl_5^-$ *are surprisingly long-lived (lifetimes to ~100 ns in solution at room temperature), and these species are photoredox-active as well.*

EARLY TRANSITION METAL COMPLEXES are stable in oxidation states with the photophysically attractive d^3 electronic configuration, as well as in a number of adjacent oxidation states. Our goal has been to carry out an initial one-electron photooxidation reaction (eq 1; A is an electron acceptor) of a d^3 starting complex M followed by a second oxidation (eq 2, 3, or 4).

0065–2393/93/0238–0131$06.00/0

$$M + A \xrightarrow{h\nu} M^{+} + A^{-} \quad (1)$$

$$M^{+} + A \longrightarrow M^{2+} + A^{-} \quad (2)$$

$$2M^{+} \longrightarrow M^{2+} + M \quad (3)$$

$$M^{+} + A \xrightarrow{h\nu} M^{2+} + A^{-} \quad (4)$$

Previous transition metal photoredox experiments have concentrated on systems such as Ru(II) and Cr(III) polypyridine complexes. These often undergo efficient excited-state redox reactions (*1*), but they are largely limited to overall one-electron processes. The electron-transfer products, for example, $Ru(bpy)_3^{3+}$ (bpy is 2,2′-bipyridine), $Cr(phen)_3^{2+}$ (phen is 1,10-phenanthroline), reduced acceptors, and oxidized donors, react only very slowly with organic and inorganic substrates.

The value of multielectron systems in redox catalysis has been demonstrated elegantly for several Ru and Os complexes with labile coordination sites, which allow for multiple coupled proton and electron transfers. One such species, $[(bpy)_2(H_2O)Ru]_2(\mu\text{-}O)^{4+}$, catalyzes the oxidation of Cl^- to Cl_2 (*2, 3*) and H_2O to O_2 (*4*). In our work, we wished to combine photochemical and multielectron redox capabilities in the same complex. Early transition metal species appeared to offer the best opportunity for this combination.

The approaches we pursued for photochemical two-electron oxidations are outlined in equations 1–4. We have demonstrated that $V(phen)_3^{2+}$ is photooxidized to $[V^{III}(phen)_2]_2(\mu\text{-}O)^{2+}$ (*5*), which is oxidized spontaneously to $V^{IV}O(phen)_2^{2+}$ in basic solution (*6*). This scheme utilizes reactions 1 and 2, and requires that a single photon be capable of driving an overall two-electron process. A second scheme, represented by the combination of reactions 1 and 3, is illustrated by the disproportionation of $Mo^{IV}(NCS)_6^{2-}$ following initial photooxidation of $Mo^{III}(NCS)_6^{3-}$ (*7*). More recently, we also observed (*8*) disproportionation following photooxidation of the 1,4,7-trimethyl-1,4,7-triazacyclononane complexes $(Me_3[9]aneN_3)MoX_3$ ([9]ane is 1,4,7-triazacyclononane and X is Br or I). Finally, our interest in the photoredox reactions of $ReCl_6^{2-}$ was stimulated in part by the possibility that a photogenerated Re(V) complex might itself be photoactive (*see* eq 4). These latter schemes, in which reaction 1 is followed by reactions 3 and 4, both entail the absorption of two photons to effect a single net two-electron transfer.

Three groups of experiments are discussed in this chapter. First, our work using $ReCl_6^{2-}$ and its photoinitiated three-electron oxidation is reviewed. Exploration of the mechanism of this photooxidation led directly to the second topic, the discovery that $ReCl_6^{2-}$ has the surprising ability to catalyze the oxidation of Cl^- to Cl_2. We conclude with a discussion of the solution photophysics and photochemistry of metal–oxo complexes with the d^1 and d^2 configurations.

Closely related to the last of these three topics, and also to our own interest in the photochemistry of metal–oxo complexes, is the early work of Rillema and Brubaker (*9, 10*). They showed that the chloro–alkoxo species $M^VCl_5(OR)^-$ (M is Mo or W) thermally or photochemically eliminate the appropriate alkyl halides:

$$M^VCl_5(OR)^- \xrightarrow{\Delta \text{ or } h\nu} RCl\,(g) + M^VOCl_4^- \qquad (5)$$

This work is an excellent example of a photosensitive metal–organic system. The reaction also produces the oxo complexes $M^VOCl_4^-$, some of whose photoredox properties are reported herein.

Experimental Section

Materials and Procedures. Reagents and solvents were of the highest grade commercially available and were used as received. Samples for photochemical measurements were prepared by using dry-box, Schlenk, or high-vacuum techniques. General photochemical, spectroscopic, and electrochemical methods were reported previously (*11*).

The $ReCl_6^{-/2-}$ potential in 1 M HCl was estimated from cyclic voltammetry (half-wave potential $E_{1/2}$ at 20 V/s in order to minimize competing reactions of $ReCl_6^-$). Our estimate of the potential for the second oxidation of $ReCl_6^{2-}$ (to make $ReCl_6$) in aqueous HCl was derived from the work of Heath et al. in CH_2Cl_2 (*12*).

We used a chemical method to estimate the Ce^{IV}–Ce^{III} potential in 1 M HCl because the original value of 1.28 V versus the normal hydrogen electrode (NHE) (*13*) (based on electrochemical measurements), which is still quoted in reference works (*14*), was long ago called into question (*15*). (First, the original measurements were made at a Pt electrode, which is corroded at highly positive potentials in HCl(aq); Ce^{IV}–Ce^{III} electron-transfer reactions, on the other hand, are often quite slow. Second, solutions of Ce^{IV} in HCl(aq) are unstable with respect to Cl_2 evolution (*16, 17*). The 1.28-V value is likely to be associated with one or both of these competing reactions rather than the desired Ce^{IV}–Ce^{III} couple.) We estimated the value by using the known Cl_2–Cl^- potential and the equilibrium constant for the reaction $Ce^{IV} + Cl^- \rightleftarrows Ce^{III} + {}^1\!/_2Cl_2(g)$. The equilibrium constant was measured spectrophotometrically, by mixing solutions of Ce^{IV} and Ce^{III} separately with 1 M HCl under 1 atm (101.3 kPa) of Cl_2 and showing that both mixtures yielded the same final $[Ce^{IV}]$ to $[Ce^{III}]$ ratio (*18*).

Electrocatalysis and Spectroelectrochemistry. Our best results for electrocatalysis with $ReCl_6^{2-}$ were obtained at a freshly polished electrode surface. This approach contrasts with the systems of Meyer and co-workers (*2, 3*), which produce Cl_2 and O_2 most efficiently at highly oxidized glassy carbon electrodes (*19*).

Our spectroelectrochemical experiments utilized $(Bu_4N)_2ReCl_6$ in CH_3CN, in a sealed all-quartz cell (1-mm optical path length) fitted with a gold minigrid OTTLE (optically transparent thin-layer electrode) and Ag wire pseudoreference electrode; ferrocene was used as the reference redox couple. Electrolysis times with this apparatus were ~2 min. The reaction was studied in CH_3CN ($ReCl_6^{-/2-}$: $E_{1/2}$ 0.77 V vs. Fc^+/Fc (*11*); Fc is ferrocene) because the gold minigrid electrode is easily corroded under oxidizing conditions in aqueous HCl.

Photochemistry of $ReCl_6^{2-}$

We were originally interested in this area because an oxorhenium(V) complex, if it could be generated photochemically from $ReCl_6^{2-}$, might itself be photoactive. For example, dioxorhenium(V) complexes such as $ReO_2(py)_4^+$ (py is pyridine) are diamagnetic, with the $^1A_{1g}$ $((xy)_2)$ ground state (*see* Scheme I):

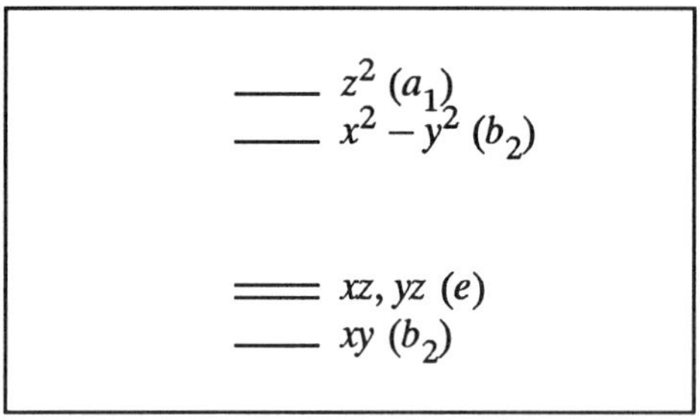

z^2 (a_1)
$x^2 - y^2$ (b_2)

xz, yz (e)
xy (b_2)

Scheme I. Energy diagram for metal–oxo and trans-*dioxo complexes.*

The lowest-lying ^3E_g $((xy)^1(xz, yz)^1)$ excited states of these complexes are long-lived (*20–22*). In contrast to the octahedral $ReCl_6^-$ ion, $ReOCl_5^{2-}$ is diamagnetic (*23*). Thus, it is likely to be electronically similar to $ReO_2(py)_4^+$. The present rhenium system therefore presents an unusual example of two adjacent oxidation states that are both potentially photoactive. A scheme taking advantage of this reactivity is illustrated in equations 6–8.

$$ReCl_6^{2-} + A \xrightarrow{h\nu} ReCl_6^- + A^- \qquad (6)$$

$$ReCl_6^- + H_2O \longrightarrow Re^{V}OCl_5^{2-} + 2H^+ + Cl^- \qquad (7)$$

$$Re^{V}OCl_5^{2-} + A \xrightarrow{h\nu} Re^{VI}OCl_5^- + A^- \qquad (8)$$

Thus, $ReCl_6^{2-}$ is a possible starting material for a two-electron sequence in which both steps are photochemical, that is, in which reaction 1 is followed by reaction 4. This sequence is similar to the one we observed for $Mo(NCS)_6^{3-}$ (*7*) and $(Me_3[9]aneN_3)MoX_3$ (*8*) (photoredox followed by disproportionation, or reactions 1 and 3) in that two photons must be absorbed for each net two-electron reaction. Both of these are potentially more versatile than the sequence we discovered for $V(phen)_3^{2+}$ (reactions 1 and 2), which requires that the entire two-electron transfer be driven by a single photon. Because two steps in the "two-photon" schemes can be endothermic, the doubly oxidized products can be more powerful oxidants than those formed in the "one-photon" mechanism.

Acceptors such as 2,3,5,6-tetrachloro- (chloranil) and 2,3-dichloro-5,6-dicyano-1,4-benzoquinone (DDQ) photooxidize $ReCl_6^{2-}$ reversibly. Tetranitromethane, on the other hand, functions as an irreversible oxidant. Irradiation of $ReCl_6^{2-}$ in the presence of $C(NO_2)_4$ yields ReO_4^- as the final product, along with reduction products from $C(NO_2)_4$ (*11*). This reaction represents an unusual example of a photoinitiated three-electron oxidation.

We have not yet determined the mechanism of the photooxidation of $ReCl_6^{2-}$ to ReO_4^- using $C(NO_2)_4$. In order to study possible rhenium-containing intermediates in the absence of the strongly absorbing $C(NO_2)_3^-$, we treated $ReCl_6^{2-}$ with the powerfully oxidizing cerium(IV) in aqueous HCl solution. However, this experiment is made more complicated by the fact that Ce^{IV} oxidizes chloride ion:

$$Ce^{IV}(aq) + Cl^-(aq) \longrightarrow Ce^{III}(aq) + \tfrac{1}{2}Cl_2(g) \qquad (9)$$

This reaction occurs slowly under ordinary conditions (*16, 17*), but we found that it is in fact *catalyzed by* $ReCl_6^{2-}$. (For example, a mixture that initially contains a 100-fold excess of Ce^{IV}, after evolution of chlorine is complete, still shows essentially quantitative recovery of $ReCl_6^{2-}$.) In separate experiments without cerium(IV) (*24*), $ReCl_6^{2-}$ also catalyzes the electrochemical oxidation of Cl^- to Cl_2 at a glassy-carbon electrode.

Mechanism of $ReCl_6^{2-}$-Catalyzed Cl_2 Evolution

These catalytic experiments are important because they indicate that an oxidized rhenium complex is a rapid oxidant. We first proposed the following mechanism for Cl_2 generation:

$$Re^{IV}Cl_6^{2-} \longrightarrow Re^{V}Cl_6^{-} + e^{-} \tag{10}$$

$$Re^{V}Cl_6^{-} + H_2O \longrightarrow Re^{V}OCl_5^{2-} + 2H^{+} + Cl^{-} \tag{11}$$

$$Re^{V}OCl_5^{2-} \longrightarrow Re^{VI}OCl_5^{-} + e^{-} \tag{12}$$

$$Re^{VI}OCl_5^{-} + Cl^{-} \longrightarrow [Cl_5Re(OCl)^{2-}] \tag{13a}$$

$$\xrightarrow{H^{+}, Cl^{-}} Re^{IV}Cl_6^{2-} + HOCl \tag{13b}$$

$$HOCl + H^{+} + Cl^{-} \longrightarrow Cl_2 + H_2O \tag{14}$$

This mechanism incorporates previously reported oxorhenium(V) and oxorhenium(VI) complexes. It is similar to the one proposed by Meyer and co-workers (*3*) for $[(bpy)_2(H_2O)Ru]_2(\mu\text{-}O)^{4+}$-catalyzed Cl_2 evolution. The latter mechanism also involved two one-electron oxidations followed by direct attack of Cl^- to form an intermediate containing coordinated hypochlorite (OCl^-) (*see* eqs 13a and 13b). The final step, the reaction of hypochlorous acid with HCl(aq) to form Cl_2, is rapid.

To test this mechanism, we prepared $K_2Re^{V}OCl_5$ (*23*) and $KRe^{VI}OCl_5$ (*25*) separately. We found that solutions of K_2ReOCl_5 in concentrated HCl are stable indefinitely (in agreement with the results of Casey and Murmann (*26*)); even in 1 M HCl, the ion persists for a short period of time. The d^1 complex $KReOCl_5$, on the other hand, is highly moisture-sensitive, reacting rapidly with aqueous HCl of any concentration to produce $Re^{V}OCl_5^{2-}$ and chlorine gas (*27*). None of the solutions prepared from K_5ReOCl_5 or $KReOCl_5$ showed measurable catalytic activity for Cl_2 formation. Thus, neither of these species is likely to be a part of the catalytic mechanism.

We next studied the direct oxidation of $ReCl_6^{2-}$ by spectroelectrochemistry in CH_3CN. The yellow-orange Re^V product in this case showed an absorption spectrum (Figure 1) very different from that of $Re^{V}OCl_5^{2-}$ (*26*); we assign the new bands to the non-oxo complex $Re^{V}Cl_6^{-}$. (Although previous reports on the synthesis of $Re^{V}Cl_6^{-}$ (*28, 29*) did not discuss its spectral properties, the bands are similar to those of the isoelectronic OsF_6 (*30*).) The addition of small amounts of water to these solutions has little effect on the electrolysis. Even when the electrolyzed solutions are added to HCl(aq), no hydrolysis to $Re^{V}OCl_5^{2-}$ occurs; instead, the spectral features of $Re^{V}Cl_6^{-}$ persist for several seconds, to be replaced by those of $ReCl_6^{2-}$. In contrast to the results obtained with oxorhenium complexes, the electrooxidized complex does show catalytic activity in 1 M HCl. Therefore, the singly oxidized species $Re^{V}Cl_6^{-}$ is likely to be an active member of the catalytic cycle.

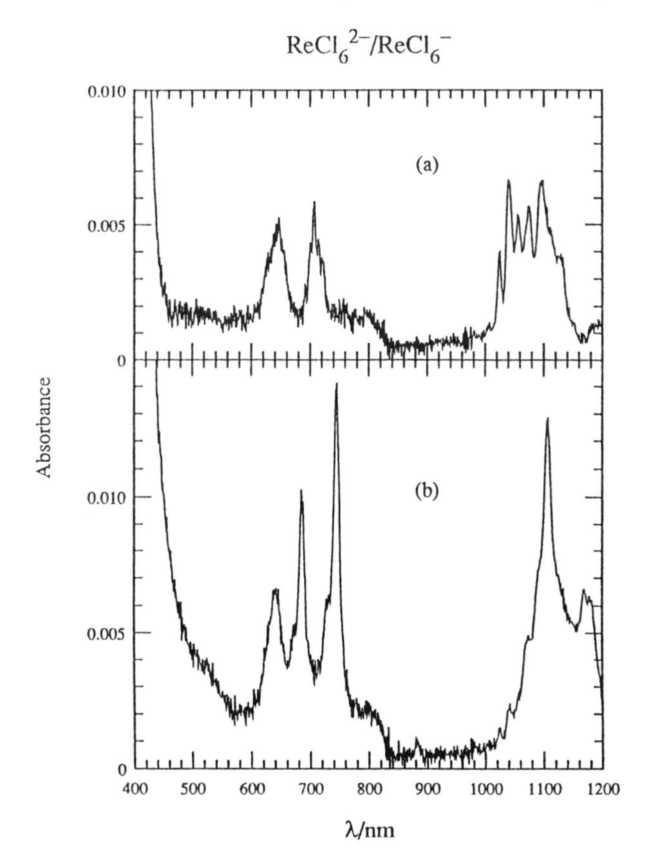

Figure 1. Portions of electronic absorption spectra recorded (a) before and (b) after one-electron electrooxidation (1.35 V vs. Fc^+/Fc; gold minigrid OTTLE) of $(Bu_4N)_2[ReCl_6]$ (5 mM) in CH_3CN.

Oxo complexes appear not to be involved in the catalysis, and hence the rhenium-centered redox reactions may all be outer-sphere. Two possible mechanisms, based on formation of $Re^VCl_6^-$ and $Re^{VI}Cl_6$ as the active oxidants for Cl^-, are outlined in Scheme II. Following initial one-electron oxidation (eq 15), $ReCl_6^-$ may react with Cl^- to form $ReCl_6^{2-}$ and a Cl_2^- radical anion (eq 16). Then, Cl_2^- could either disproportionate (eq 17) (*31*) or be oxidized directly (eq 18) either by the external oxidant or at the electrode, to produce Cl_2. An alternative route (reactions 19 and 20) involves further oxidation of $ReCl_6^-$ to $ReCl_6$, which could react with Cl^- to produce $ReCl_6^{2-}$ and Cl_2. Our results, however, indicate that the one-electron route of reactions 15–18 is the more reasonable.

$$ReCl_6^{2-} \longrightarrow ReCl_6^{-} + e^{-} \quad (15)$$

$$ReCl_6^{-} + 2Cl^{-} \longrightarrow ReCl_6^{2-} + Cl_2^{-} \quad (16)$$

$$2Cl_2^{-} \longrightarrow Cl_2 + 2Cl^{-} \quad (17)$$

$$Cl_2^{-} \longrightarrow Cl_2 + e^{-} \quad (18)$$

$$ReCl_6^{-} \longrightarrow ReCl_6 + e^{-} \quad (19)$$

$$ReCl_6 + 2Cl^{-} \longrightarrow ReCl_6^{2-} + Cl_2 \quad (20)$$

Scheme II. Possible mechanisms for $ReCl_6^{2-}$-catalyzed oxidation of Cl^- to Cl_2.

Electrode potentials for the $Ce^{IV/III}$ and $ReCl_6^{-/2-}$ couples in 1 M HCl are ~1.46 and 1.70 V vs. NHE, respectively (*see* Experimental Section); estimates are available for the Cl_2^- to Cl^- ratio (~2.2 V) (*32, 33*) and $ReCl_6^{0/-}$ (>2.5 V) (*12*). Generation of Cl_2^- (reaction 16) is considerably more favorable than oxidation of $ReCl_6^-$ by Ce^{IV} (reaction 19); reaction 15 might be expected to be faster as well. Also supporting the mechanism of reactions 15–18 are our cyclic voltammograms for $ReCl_6^{2-}$ in aqueous HCl. These show the most chemically reversible behavior (the ratio of cathodic to anodic peak current is closest to 1) at high scan rates and low [Cl^-]. The less reversible nature of the process at low scan rates and high [Cl^-] is consistent with a model involving scavenging of electrogenerated $ReCl_6^-$ by Cl^- (eq 16).

Surprisingly, we also observed weak luminescence (λ_{max} = 1500 nm; lifetime about 100 ns) in the CH_3CN solutions containing electrogenerated $ReCl_6^-$. This discovery suggests that even simple non-oxo d^2 complexes may be useful for photochemical reactions. In particular, because $ReCl_6^-$ is a powerful oxidant even in its ground state, we expect it to be a still more reactive species in its luminescent excited state.

Catalytic Oxidation of Other Substrates

Oxidation of Cl^- to Cl_2 in the chlor-alkali process is perhaps the best example of a catalyzed oxidation carried out in aqueous chloride-containing solutions. This oxidation was traditionally carried out at graphite electrodes, which are slowly consumed under the operating conditions of high potential and current. However, the so-called dimensionally stable anode (DSA), which contains RuO_2 as the active catalyst, has become increasingly popular (*34*).

Both because of the success of the DSA in chlorine production and because of forecasts that the demand for Cl_2 is expected to decrease over the next few decades (*35*), we were interested in using our system to oxidize substrates other than chloride ion. We therefore prepared a standard catalytic mixture containing Ce(IV) in 0.1 M HCl, with and without K_2ReCl_6. (The smaller acid concentration was chosen so as to decrease the *uncatalyzed* reaction rate and make catalytic effects easier to discern.) We then added a variety of organic compounds to these mixtures.

Nearly all compounds tested, including CH_2Cl_2 and CH_3CN (which are ordinarily considered to be resistant to oxidation), decolorized the Ce^{IV} test solution at least slowly, even in the absence of $ReCl_6^{2-}$. However, with several substrates, oxidation was markedly accelerated when $ReCl_6^{2-}$ was added. Toluene, for example, was oxidized ~60 times faster in the presence of $ReCl_6^{2-}$ (8×10^{-4} M); under similar conditions, isopropyl alcohol was oxidized about 5 times faster when $ReCl_6^{2-}$ was added. The oxidation products in these reactions (primarily benzaldehyde from toluene, with smaller amounts of benzoic acid also detected; exclusively acetone from isopropyl alcohol) were the same whether or not $ReCl_6^{2-}$ was added.

More detailed mechanistic studies of these systems, for example, via stopped-flow methods, may help to determine whether specific complexes among the species in our proposed Cl_2-evolution mechanism function as rapid oxidants in the oxidation of organic substrates as well. However, clearly, the simple complex $ReCl_6^{2-}$ participates in several unusual types of redox reactions. (Experiments with benzyl alcohol as the substrate give little rate enhancement with Re; however, this molecule is oxidized primarily to benzoic acid, and very little benzaldehyde is formed. This chemoselectivity, and that for oxidation of toluene to benzaldehyde mentioned previously, may also provide useful mechanistic information.)

Photophysics and Photochemistry of Oxo Complexes

The general features of the electronic structure of metal–oxo complexes were discussed extensively in the 1960s, based on absorption spectra of the "vanadyl" ($V^{IV}O^{2+}$) and "molybdenyl" ($Mo^{V}O^{3+}$) ions (*36–38*). Interest in these species has been rekindled in recent years by the discovery that closely related *trans*-dioxometal complexes with the d^2 configuration (such as $ReO_2(py)_4^+$ and $ReO_2(CN)_4^{3-}$) phosphoresce in solution and undergo efficient excited-state redox reactions (*20–22*). An electronic absorption and emission study of "molybdenyl" (d^1) complexes in the solid state (*39*) has also contributed to this new area of inorganic photochemistry.

The d^1 and d^2 species are both found with one oxo ligand, as MOL_4

or MOL_5; the d^2 species are also found as *trans*-MO_2L_4. All of these complexes can be treated with the qualitative energy diagram of Scheme I. The lowest-energy absorption band in both cases is assigned to the $b_2 \rightarrow e$ ($xy \rightarrow xz, yz$) transition. (The orbital designations used here are those for C_{4v} symmetry.) In the d^1 complexes there is only one such transition, and luminescence from the lowest-energy excited state is always fluorescence ($^2E \rightarrow {}^2B_2$). The d^2 configuration, on the other hand, leads to singlet and triplet excited states. In this case, the two lowest-energy bands are $^1A_1 \rightarrow {}^3E$ and, at higher energy, $^1A_1 \rightarrow {}^1E$; phosphorescence is observed from ^3E (*20–22*).

Oxo-d^2 Systems

Among the most promising d^2 systems reported so far is $ReO_2(py)_4{}^+$, whose phosphorescence occurs with a maximum at 640 nm and a lifetime of 10 μs in CH_3CN at room temperature. This complex can be photooxidized to the powerful oxidant $ReO_2(py)_4{}^{2+}$, which is in turn capable of attacking organic substrates, apparently by H-atom abstraction (*22*). More recent experiments have dealt with nitridorhenium(V) complexes, whose electronic spectral features are similar to those of the $ReO_2{}^+$ species (*40*).

We have now also observed luminescence and photoredox reactions for the monooxo complex $Re^VOCl_4(CH_3CN)^-$ in acetonitrile solution. (The absorption spectrum of this ion, shown in Figure 2, is similar to that reported for $Re^VOCl_4(H_2O)^-$ in concentrated HCl(aq) (*26*).) We assign the two lowest-energy absorption bands to transitions from $^1A_1$ to ^3E and ^1E. (The fact that the intensities of the two bands are similar is probably the result of substantial mixing via spin-orbit coupling.) The maxima in the luminescence (~1300 nm) and absorption spectra (1175 nm) suggest an excited-state energy of approximately 8000 cm^{-1}; the lifetime of the emitting state is about 100 ns.

Cyclic voltammetry performed on $Re^VOCl_4(CH_3CN)^-$ reveals a quasi-reversible oxidation wave, with half-wave potential $E_{1/2}$ ~1.0 V vs. Ag–AgCl. Accordingly, we have also carried out photochemical electron-transfer experiments with the complex. Irradiation (λ >650 nm) in the presence of DDQ (2,3-dichloro-5,6-dicyano-1,4-benzoquinone) leads to formation of the radical anion DDQ^-; this ion then disappears with second-order kinetics. These observations are consistent with reversible one-electron transfer:

$$ReOCl_4(CH_3CN)^- + DDQ \xrightarrow{h\nu} ReOCl_4(CH_3CN) + DDQ^- \qquad (21)$$

$$ReOCl_4(CH_3CN) + DDQ^- \xrightarrow{k_b} ReOCl_4(CH_3CN)^- + DDQ \qquad (22)$$

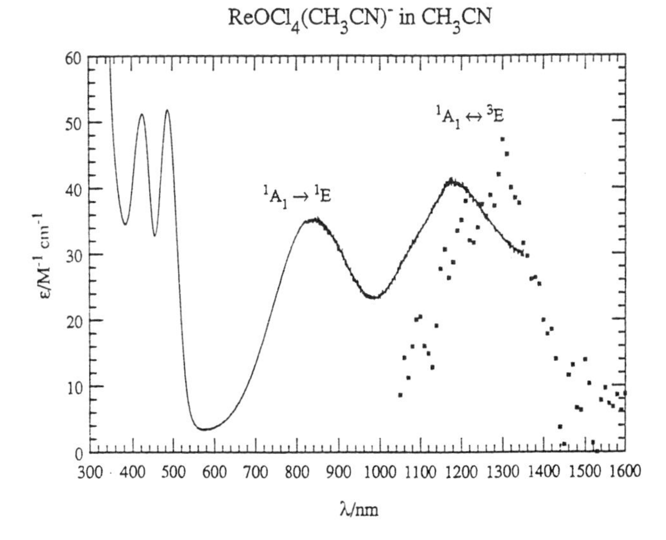

Figure 2. Electronic absorption (—) and luminescence (■) spectra for $(Bu_4N)[Re^VOCl_4(CH_3CN)]$ in CH_3CN at room temperature. The emission spectrum was recorded by using a 5 × 10^{-2} M solution, with 436-nm excitation and emission bandpass of ~30 nm.

Kinetic analysis indicates a value of 3 × 10^9 M^{-1} s^{-1} for k_b. We are now exploring the reactivity of the photogenerated Re(VI) complex in this system and the suitability of other Re(V) oxo complexes for photoredox reactions.

Oxo-d^1 Systems

As part of our study of the photochemical properties of octahedral Mo(III) complexes, we attempted to prepare $HB(Me_2pz)_3Mo^{III}Br_3^-$ (Me_2pzH = 3,5-dimethylpyrazole), for comparison with the previously reported $HB(Me_2pz)_3Mo^{III}Cl_3^-$ (*41*). Reaction of $HB(Me_2pz)_3Mo(CO)_3^-$ with concentrated HBr(aq) yields $MoOBr_4(H_2O)^-$ instead of the desired complex; however, solutions prepared from this Mo(V) product are luminescent!

Electronic absorption and emission spectra of $Mo^VOBr_4(CH_3CN)^-$, obtained by dissolving $(Bu_4N)[MoOBr_4(H_2O)]$ in CH_3CN, are illustrated in Figure 3. Absorption (*36–38*) and solid-state luminescence (*39*) studies of these oxo-d^1 species have established that the lowest-energy absorption and emission bands are associated with the $^2B_2 \leftrightarrow {}^2E$ transition. However, these species have not previously been reported to fluoresce in solution at

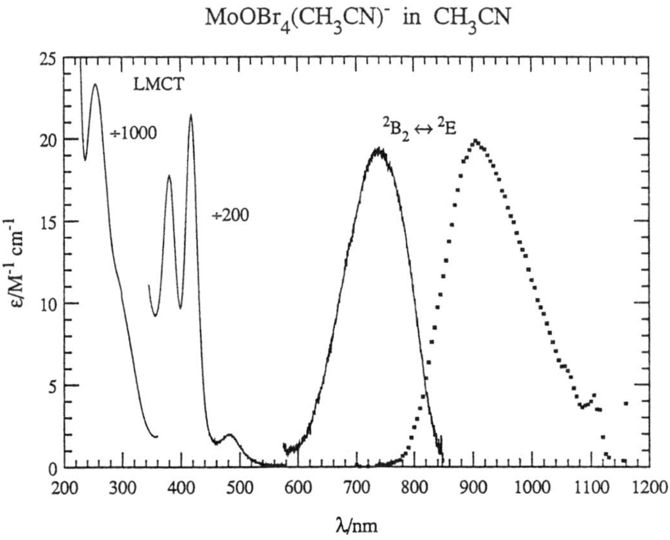

$MoOBr_4(CH_3CN)^-$ in CH_3CN

Figure 3. Electronic absorption (—) and corrected fluorescence (■) spectra for $(Bu_4N)[Mo^VOBr_4(CH_3CN)]$ in CH_3CN at room temperature. The emission spectrum was recorded by using a 1×10^{-3} M solution, with 436-nm excitation and emission bandpass of ~15 nm.

room temperature. (The fluorescence quantum yields for Mo^VOCl_4-$(CH_3CN)^-$ and $Mo^VOBr_4(CH_3CN)^-$ are ~4.4×10^{-4} and 1.4×10^{-4} mol/einstein, respectively.)

The photophysical properties of several Mo(V) and Re(VI) oxo complexes are presented in Table I. The most important features of these complexes are as follows. First, the longer-lived excited states are well suited to bimolecular reactions; in separate experiments, we showed that both $MoOCl_4(CH_3CN)^-$ and $MoOBr_4(CH_3CN)^-$ undergo photooxidation in the presence of acceptors such as tetracyanoethylene (*42*). Second, the aqua complexes $MoOX_4(H_2O)^-$ are also fluorescent in CH_2Cl_2. These are among the very few known aqua complexes that luminesce in room-temperature solution. Third, we were surprised to find relatively intense fluorescence with these species, since Winkler (*39*) found that $MoOCl_4^-$ (in which the axial ligand *trans* to the oxo group is absent) did not fluoresce measurably in solution. We repeated Winkler's work, and we found that the fluorescence of $MoOCl_4^-$ in CH_2Cl_2 solution, although present, is indeed much weaker than that of the solvated species $MoOCl_4(L)^-$. Fourth, and again unexpectedly, complexes with chelating ligands such as $Me_3[9]aneN_3$ (*see* Table I) are also poorer emitters than $MoOCl_4(L)^-$. And finally, other d^1 oxo complexes, such as those of Re(VI) (data for $ReOCl_5^-$ are included in Table I and Figure 4) and W(V) (*43*), also fluoresce. These results make it likely that a large number of other d^1 species will prove to be effective photoredox agents.

Table I. Photophysical Properties of Mo(V) and Re(VI) Oxo Complexes

		λ_{max} (nm), $^2B_2 \leftrightarrow {}^2E$			
Complex	*Solvent*	*Abs.*	*Fluor.*	$E\ (^2E)\ (cm^{-1})^a$	τ *(ns)*
$MoOCl_4(CH_3CN)^-$	CH_3CN	750	950	11,700	100
$MoOBr_4(CH_3CN)^-$	CH_3CN	750	900	12,100	40
$MoOCl_4(H_2O)^-$	CH_2Cl_2	720	950	12,000	60
$[(Me_3[9]aneN_3)MoOBr_2]^+$	CH_3CN	700	1100	11,100	<10
$MoOCl_4^{-b}$	crystal	630	890	13,100	160
$ReOCl_5^-$	$HOAc–Ac_2O$	920	ca. 1110[c]	ca. 10,000	—[d]

NOTE: All values are given for solutions at room temperature, unless otherwise noted.

[a]Estimated from average of energies of absorption and emission maxima.

[b]Data are taken from ref 39.

[c]Uncorrected.

[d]Not measured.

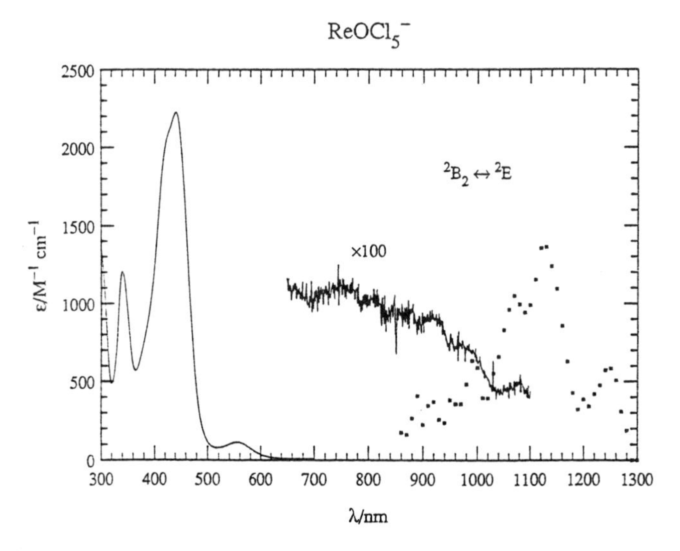

Figure 4. Electronic absorption (—) and fluorescence (■) spectra for $KRe^{VI}OCl_5$ in $HOAc–Ac_2O$ at room temperature (300–830 nm, 5×10^{-4} M; 830–1000 nm, 4×10^{-3} M, expanded ×50). The emission spectrum was recorded with a 0.023 M solution, 436-nm excitation, and emission bandpass of ~30 nm.

Summary

We began our work with an exploration of the properties of d^3 complexes, using V(II) at first, followed by Mo(III) and Re(IV). We developed photoinitiated two- and three-electron-transfer processes based on several of these starting complexes. Two important new developments are reported here. First, $ReCl_6^{2-}$ is an unusual example of a simple complex, without specially designed ligands, that shows high catalytic activity for oxidation of Cl^- and organic substrates. Second, oxo-d^1 complexes, as well as oxo and non-oxo d^2 species, closely connected with these photochemical and redox processes are themselves photoactive. Work now in progress includes combining the catalytic and photoredox properties of these complexes into photochemically driven catalytic cycles, and further exploration of excited-state electron- and atom-transfer reactions of the d^1 and d^2 oxo complexes.

Acknowledgments

We are grateful to Professor Russell H. Schmehl (Tulane University) for numerous helpful discussions and for assistance with excited-state lifetime measurements. This research was supported by grants from the National Science Foundation (CHE–8601008) and the Louisiana Educational Quality Support Fund (LEQSF(1990–92)–RD–A–06) administered by the Louisiana Board of Regents. Some of the experiments and data analyses were performed at the Center for Fast Kinetics Research, which is supported jointly by the Biomedical Research Technology Program of the Division of Research Resources of the National Institutes of Health (RR00886) and by The University of Texas at Austin.

References

1. Hoffman, M. Z.; Bolletta, F.; Moggi, L.; Hug, G. L. *J. Phys. Chem. Ref. Data* **1989,** *18,* 219–543; and references therein.
2. Ellis, C. D.; Gilbert, J. A.; Murphy W. R., Jr.; Meyer, T. J. *J. Am. Chem. Soc.* **1983,** *105,* 4842–4843.
3. Vining, W. J.; Meyer, T. J. *Inorg. Chem.* **1986,** *25,* 2023–2033.
4. Gilbert, J. A.; Eggleston, D. S.; Murphy, W. R., Jr.; Geselowitz, D A.; Gersten, S. W.; Hodgson, D. J.; Meyer, T. J. *J. Am. Chem. Soc.* **1985,** *107,* 3855–3864.
5. Shah, S. S.; Maverick, A. W. *Inorg. Chem.* **1986,** *25,* 1867–1871.
6. Shah, S. S.; Maverick, A. W. *Inorg. Chem.* **1987,** *26,* 1559–1562.
7. Yao, Q.; Maverick, A. W. *J. Am. Chem. Soc.* **1986,** *108,* 5364–5365.
8. Mohammed, A. K.; Maverick, A. W., unpublished results.
9. Rillema, D. P.; Brubaker, G. R. *Inorg. Chem.* **1969,** *5,* 587–590.
10. Rillema, D. P.; Brubaker, G. R. *Inorg. Chem.* **1969,** *5,* 1645–1649.
11. Maverick, A. W.; Lord, M. D.; Yao, Q.; Henderson, L. J., Jr. *Inorg. Chem.* **1991,** *30,* 553–558.
12. Heath, G. A.; Moock, K. A.; Sharp, D. W. A.; Yellowlees, L. J. *J. Chem. Soc. Chem. Commun.* **1985,** 1503–1505.
13. Smith, G. F.; Goetz, C. A. *Ind. Eng. Chem., Anal. Ed.* **1938,** *10,* 191–195.
14. See, for example: Morss, L. R. In *Standard Potentials in Aqueous Solution;* Bard, A. J.: Parsons, R.; Jordan, J., Eds.; Marcel Dekker: New York, 1985; Chapter 20.
15. Wadsworth, E.; Duke, F. R.; Goetz, C. A. *Anal. Chem.* **1957,** *29,* 1824–1825.
16. Koppel, J. *Z. Anorg. Chem.* **1898,** *18,* 305–311.
17. Duke, F. R.; Borchers, C. A. *J. Am. Chem. Soc.* **1953,** *75,* 5186–5188.
18. Maverick, A. W.; Yao, Q., unpublished results.
19. Cabaniss, G. E.; Diamantis, A. A.; Murphy, W. R., Jr.; Linton, R. W.; Meyer, T. J. *J. Am. Chem. Soc.* **1985,** *107,* 1845–1853.
20. Winkler, J. R.; Gray, H. B. *J. Am. Chem. Soc.* **1983,** *105,* 1373–1374.
21. Winkler, J. R.; Gray, H. B. *Inorg. Chem.* **1985,** *24,* 346–355.
22. Thorp, H. H.; Van Houten, J.; Gray, H. B. *Inorg. Chem.* **1989,** *28,* 889–892.
23. Fergusson, J. E.; Love, J. L. *Aust. J. Chem.* **1971,** *24,* 2689–2693.
24. Maverick, A. W.; Henderson, L. J., Jr.; Yao, Q., unpublished results.
25. Yatirajam, V.; Singh, H. *J. Inorg. Nucl. Chem.* **1975,** *37,* 2006–2008.
26. Casey, J. A.; Murmann, R. K. *J. Am. Chem. Soc.* **1970,** *95,* 78–84.

27. Brisdon, B. J.; Edwards, D. A. *Inorg. Chem.* **1968,** *7,* 1898–1903.
28. Lock, C. J. L.; Frais, P. W.; Guest, A. *J. Chem. Soc., D* **1970**, 1612–1613.
29. Virovets, A. V.; Sokolov, M. N.; Fedin, V. P.; Fedorov, V. E. *Bull. Acad. Sci. USSR, Div. Chem. Sci.* **1985,** *34,* 2236.
30. Eisenstein, J. C. *J. Chem. Phys.* **1961,** *34,* 310–318.
31. Neta, P.; Huie, R. E.; Ross, A. B. *J. Phys. Chem. Ref. Data* **1988,** *57,* 1027–1284, and references therein.
32. Wardman, P. *J. Phys. Chem. Ref. Data* **1989,** *55,* 1637–1755.
33. Stanbury, D. M. *Adv. Inorg. Chem.* **1989,** *33,* 69–138.
34. Leddy, J. J.; Jones, I. C., Jr.; Lowry, B. S.; Spillers, F. W.; Wing, R. E.; Binger, C. D. In *Kirk–Othmer Encyclopedia of Chemical Technology;* 3rd Ed.; Vol. 1; Wiley: New York, 1978; pp 799–865.
35. *Chem. Eng. News* **1990,** *65(41),* 18–19.
36. Ballhausen, C. J.; Gray, H. B. *Inorg. Chem.* **1962,** *1,* 111–122.
37. Gray, H. B.; Hare, C. R. *Inorg. Chem.* **1962,** *1,* 363–368.
38. Winkler, J. R.; Gray, H. B. *Comments Inorg. Chem.* **1981,** *1,* 257–263.
39. Winkler, J. R. Ph. D. Thesis, California Institute of Technology, Pasadena, CA, 1984.
40. Neyhart, G. A.: Seward, K. J.; Boaz, J., II; Sullivan, B. P. *Abstracts of Papers* 201st National Meeting of the American Chemical Society, Atlanta, GA; American Chemical Society: Washington, DC, April 1991; INOR 455.
41. Millar, M.; Lincoln, S.; Koch, S. A. *J. Am. Chem. Soc.* **1982,** *104,* 288–289.
42. Mohammed, A. K.; Maverick, A. W. *Inorg. Chem.* **1992,** *31,* 4441–4443.
43. Mohammed, A. K. Ph. D. Thesis, Louisiana State University, Baton Rouge, LA, 1992.

RECEIVED for review November 7, 1991. ACCEPTED revised manuscript June 19, 1992.

8

Photoredox Chemistry of d^4 Bimetallic Systems

Colleen M. Partigianoni, Claudia Turró, Carolyn Hsu, I-Jy Chang, and Daniel G. Nocera*

Department of Chemistry, Michigan State University, East Lansing, MI 48824

Excited states of quadruply bonded metal–metal complexes exhibit a rich oxidation–reduction chemistry with organic substrates. Irradiation of the $Mo_2(II,II)$ diarylphosphate, $Mo_2[O_2P(OC_6H_5)_2]_4$, in the presence of dihalocarbons yields olefin and the one-electron oxidized mixed-valence complex, $Mo_2[O_2P(OC_6H_5)_2]_4^+$. Photodehalogenation proceeds from the $^1(\delta\delta^*)$ excited state; the primary photoevent involves the one-electron reduction of substrate. Discrete two-electron reduction of alkyl halides is observed when the photoreagent is $W_2Cl_4(dppm)_2$ (dppm is diphenylphosphinomethane). The ability of the photogenerated intermediates and products to assume a bioctahedral geometry is an important factor in determining the two-electron photochemistry of this complex. The photoreactivity of $W_2Cl_4(dppm)_2$ provides an excited-state complement to the oxidative-addition chemistry of Vaska's complex.

BINUCLEAR METAL COMPLEXES continue to assume a prominent role in the light-initiated activation of organic substrates. The capacity of a bimetallic core to coordinate organic molecules at multiple metal sites, which feature complementary redox function upon light excitation, provides the opportunity to photochemically promote the multielectron activation of a variety of substrates. The most popular and successful approaches to the photochemical activation of organic substrates by binuclear complexes have been predicated on the chemistry of d^7–d^7 and

*Corresponding author

0065–2393/93/0238–0147$06.00/0

$d^8 \cdots d^8$ complexes (d^7 binuclear complexes feature a formal single metal–metal bond in the ground state (*1*), which is indicated by a line; the d^8 binuclear complexes are predicted to a first approximation to have no metal–metal bond in the ground state, but spectroscopic studies show a weak interaction (*2, 3*) that is indicated by the dotted line). Oxidation of organic substrates by net atom abstraction is the prominent reaction pathway for both classes of bimetallic compounds (*4, 5*).

Qualitatively, the primary photochemical event of d^7 and d^8 binuclear complexes produces similar reactive intermediates. For the d^7 complexes, irradiation of metal-localized $\sigma \rightarrow \sigma^*$ and $\pi \rightarrow \sigma^*$ transitions typically results in the cleavage of the metal–metal bond to produce a "dissociative diradical" pair (•M, M•, Figure 1a) (*5–9*). Conversely, excitation of the lowest energy $d\sigma^* \rightarrow \pi\sigma$ transition of $d^8 \cdots d^8$ complexes (*10–12*) yields an "associative diradical" pair (•M–M•, Figure 1b) wherein the electrons of this triplet-configured lowest-energy excited state are localized formally on the metal atoms (*11–14*). Thus the excited state of binuclear d^8 compounds may be described chemically as a diradical tethered by a metal–metal bond, as compared to the dissociated diradical formed upon d^7–d^7 photoexcitation.

Inasmuch as the d^7–d^7 and $d^8 \cdots d^8$ systems produce singly occupied metal-centered orbitals upon excitation, their activation chemistry is simi-

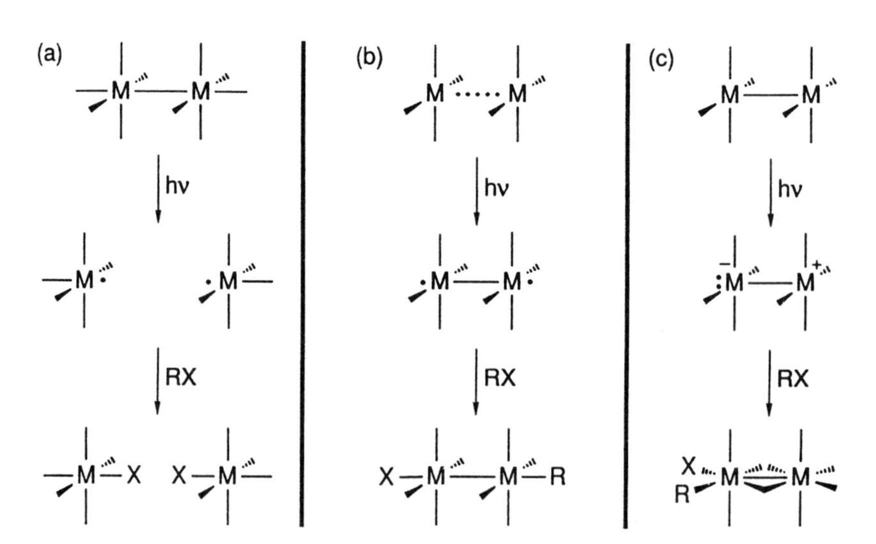

Figure 1. Photooxidation mechanisms for the reaction of an organic substrate (RX) with the photogenerated (a) "dissociative diradical" of d^7 and (b) "associative diradical" of d^8 binuclear complexes. Part c: The possible reaction of RX with a photogenerated charge-separated state of a binuclear complex.

lar. Namely, the occurrence of a single electron at a coordination vacancy of an individual metal center provides a site for substrate activation by atom transfer (*4, 5, 7, 14–16*). Yet closer inspection reveals distinct differences between the d^7 and d^8 bimetallic systems. The electronic energy of d^7–d^7 excited states is nonradiatively dissipated by metal–metal bond cleavage to produce reactive primary photoproducts. The facility of the dissociative decay channel renders the d^7–d^7 excited state too short-lived (*17*) to permit its direct participation in substrate activation. On the other hand, the photogenerated associative diradical of $d^8 \cdots d^8$ complexes represents a discrete excited state that is long-lived and capable of direct bimolecular reaction with substrate (*18*). This diradical is attractive because the thermodynamic driving force for the oxidation–reduction chemistry of a molecule in an electronic excited state is greater than that from its corresponding ground state. For this reason, substrate activation processes of binuclear d^8 molecules in electronic excited states offer the opportunity to design light-to-chemical-energy conversion schemes (*19*).

Owing to the diradical nature of the binuclear d^7 and d^8 intermediates, multielectron phototransformations of organic reactants by these system are confined to coupling sequential one-electron reactions. Because the diradicals photogenerated from d^7 binuclear complexes can quickly recombine (*20, 21*) and are uncoupled, selective multielectron activation of substrates by these complexes is difficult to control. To this end, the d^8 systems are attractive because the oxidation–reduction processes of the individual metals are confined to a single metal–metal core. The associative diradical permits the individual metal centers of the bimetallic core to cooperatively interact such that atom abstraction reactions can be coupled to effect the selective multielectron transformation of substrates. For instance, the excited states of the $d^8 \cdots d^8$ dimers $Pt_2(P_2O_5H_2)_4^{4-}$ (*14, 22–24*), Ir_2(2,5-diisocyano-2,5-dimethylhexane)$_4^{2+}$ (*25*), and [Ir(pyrazole)-1,5-cyclooctadiene]$_2$ (*26*) promote the two-electron photoreductions of a variety of organic and organometallic substrates. The primary photoprocess involves hydrogen or halogen abstraction by the $^3(d\sigma^* p\sigma)$ excited state to give a mixed valence d^7–d^8 intermediate and the corresponding organic radical. Subsequent trapping of the radical by the d^7–d^8 intermediate effects the overall two-electron reduction of the substrate. This mechanism successfully explains the photocatalytic conversion of 2-propanol to acetone by $Pt_2(P_2O_5H_2)_4^{4-}$ (*14*) and the dehydrogenation of selected hydrocarbon substrates by Ir_2(2,5-diisocyano-2,5-dimethylhexane)$_4^{2+}$ (*25*).

In contrast to a diradical approach, we have become interested in exploring the chemistry of excited states in which a redox pair of electrons of a binuclear core are singlet coupled ($:M^-$–M^+, Figure 1c). New photoreactivity of electronically excited binuclear cores of this type may be expected because the system is ideally suited to the multielectron activation of substrates at a single metal center of the bimetallic core. Two-electron reductions of substrate may be promoted at the $:M^-$ site, whereas substrates susceptible to two-electron oxidation may react at the M^+ site (*27–29*). A $:M^-–M^+$ excited state may be generated from a symmetric binuclear complex when charge is transferred from one metal to the other upon excitation (Figure 1c). To this end, the metal-to-metal charge-transfer (MMCT) character of the lowest-energy excited states of quadruply bonded metal–metal (M≡M) dimers (*30*) suggested to us that these species were logical candidates as $:M^-–M^+$ photoreagents.

Electronic Structure of d^4 Bimetallic Systems

The d^4 electron count of the metals comprising the quadruply bonded bimetallic complexes (D_{4h} symmetry) renders a $\sigma^2\pi^4\delta^2$ ground-state electronic configuration with a δ-HOMO (highest occupied molecular orbital) and a δ^*-LUMO (lowest unoccupied molecular orbital) formed from the interaction of the d_{xy} orbitals of the individual metal centers. Transitions involving the promotion of electrons to and from the HOMO and LUMO levels such as $\delta \rightarrow \delta^*$, $\pi \rightarrow \delta^*$, and $\delta \rightarrow \pi^*$ are predicted to exhibit MMCT character (*31*). Experimental studies (*30*) have shown this to be the case for the spin and dipole-allowed $\delta \rightarrow \delta^*$ ($^1A_{2u} \leftarrow {}^1A_{1g}$) transition. Winkler and co-workers (*32*) determined a dipole moment of 4.0 D for the $\delta\delta^*$ excited state of $Mo_2Cl_4(PMe_3)_4$ (relative to the nonpolar ground state), which corresponds to a partial charge transfer of 0.4 electron between the metal centers. Consistent with this prediction is their observation that the temporal evolution of the emission spectra occurs on the time scale of the microscopic solvent relaxation time. This result shows that solvent is an important controlling factor in the dynamics of $\delta\delta^*$ luminescence and is indicative of dielectric coupling between the solvent dipoles and the developing charge distribution of the M≡M excited state. Within the context of this spectroscopic framework, we have undertaken a comprehensive study of the photochemistry of binuclear quadruply bonded metal–metal complexes.

Metal–Organic Photochemistry of d^4 Bimetallic Phosphates

The photoactivation chemistry of organic reactants by Mo≡Mo dialkyl and diaryl phosphates, $Mo_2[O_2P(OR)_2]_4$ (R = C_6H_5, C_2H_5, or C_4H_9), occurs from low-energy excitation of the binuclear core. This condition is in contrast to previous reports of M≡M photochemistry, which has been restricted exclusively to ultraviolet excitation (*33*). Most of our work has focused on the diphenylphosphate complex $Mo_2[O_2P(OC_6H_5)_2]_4$, which displays weak luminescence (ϕ_{em} ~5×10^{-4}) in nonaqueous solutions and a sufficiently long lifetime (τ_0 = 68 ns) to permit excited-state reaction upon $\delta \rightarrow \delta^*$ excitation (*34, 35*). For example, the two-electron conversion of 1,2-dichloroethane (DCE) to ethylene and the mixed-valence Mo_2(II,III) complex is promoted by visible light according to the reaction

$$2Mo_2[O_2P(OC_6H_5)_2]_4 + ClCH_2CH_2Cl \xrightarrow{h\nu} 2Mo_2[O_2P(OC_6H_5)_2]_4Cl + CH_2CH_2 \quad (1)$$

The production of 0.5 equivalents of ethylene in conjunction with the quantitative formation of the mixed-valence complex has lead us to propose the photochemical reaction mechanism shown in Scheme I. An estimated electronic origin $E_{0,0}$ ($Mo_2[O_2P(OC_6H_5)_2]_4$*) = 2.3 eV and a measured half-wave potential $E_{1/2}$ [Mo_2(II,II)/Mo_2(II,III)] = –0.1 V versus the saturated calomel electrode (SCE) reveals that the electronically excited $Mo_2[O_2P(OC_6H_5)_2]_4$ is sufficiently energetic to directly react with DCE to yield $\cdot CH_2CH_2Cl$ and $Mo_2[O_2P(OC_6H_5)_2]_4Cl$. Subsequent reaction of the radical with another equivalent of Mo_2(II,II) starting complex directly yields the observed photoproducts. Alternatively, if the organic radical intermediate is scavenged by the mixed-valence complex within the solvent cage, then the fully oxidized Mo_2(III,III) complex is produced. The energetics of the Mo_2(II,II/II,III) and Mo_2(II,III/III,III) couples are such that comproportionation of Mo_2(III,III) and unreacted Mo_2(II,II)

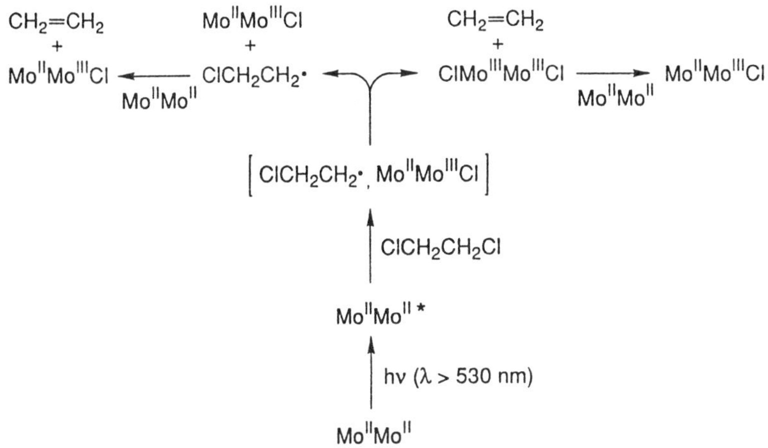

$CH_2{=}CH_2$ + $Mo^{II}Mo^{III}Cl$ ← ($Mo^{II}Mo^{II}$) — $ClCH_2CH_2\cdot$ + $Mo^{II}Mo^{III}Cl$ ← [$ClCH_2CH_2\cdot$, $Mo^{II}Mo^{III}Cl$] → $CH_2{=}CH_2$ + $ClMo^{III}Mo^{III}Cl$ — ($Mo^{II}Mo^{II}$) → $Mo^{II}Mo^{III}Cl$

$Mo^{II}Mo^{II}$ —hν (λ > 530 nm)→ $Mo^{II}Mo^{II}$* —$ClCH_2CH_2Cl$→ [$ClCH_2CH_2\cdot$, $Mo^{II}Mo^{III}Cl$]

Scheme I. Proposed mechanism for the photochemical reduction of 1,2-dichloroethane by $Mo_2[O_2P(OC_6H_5)_2]_4$.

diphenylphosphate will yield the mixed-valence species. Thus, the stoichiometry of the photoreaction is independent of whether the radical reacts inside or outside the solvent cage of the primary photoproducts.

The $Mo_2[O_2P(OC_6H_5)_2]_4$ complex also reacts with dihalocarbons other than DCE. Table I shows the quantum yields for reaction of selected substrates as determined by monitoring the disappearance of the $\delta \rightarrow \delta^*$ transition of $Mo_2[O_2P(OC_6H_5)_2]_4$ at 515 nm and the concomitant appearance of the $\delta \rightarrow \delta^*$ absorption band for $Mo_2[O_2P(OC_6H_5)_2]_4Cl$ at 1492 nm. The reduction of alkyl halides by transition metal donors can occur by outer-sphere or inner-sphere electron-transfer pathways (*36, 37*), the inner-sphere pathway being especially important for transition metal reductants featuring open coordination sites (*38–40*). In view of the vacant axial coordination sites of the metal–metal core, an inner-sphere reaction pathway is likely to play a significant role in the photoreduction of alkyl halides by electronically excited Mo_2(II,II) diphenylphosphate complex. Computer-generated space-filling models show that the metal–metal core is readily accessible to axially approaching substrates.

Quantum yield data for the photoreduction of DCE by $Mo_2[O_2P(OC_6H_5)_2]_4$ in a variety of solvents suggest that reaction is indeed confined to the axial coordination site of the metal–metal core. As shown in Table I, the quantum yield for photoreaction decreases dramatically with the increasing ability of solvent to ligate the metal core at the axial coordination site. For tetrahydrofuran (THF), the crystal structure of $Mo_2[O_2P$-

Table I. Quantum-Yield Data for the Reaction of $Mo_2[O_2P(OC_6H_5)_2]_4$ and Dichlorocarbon in Various Nonaqueous Solvents

Dichlorocarbon	*Solvent*	Φ_p[a]
1,2-Dichloroethane	C_6H_6[b]	0.031
1,2-Dichloroethane	THF[b]	0.014
1,2-Dichloroethane	CH_3CN[b]	0.0012
1,2-Dichlorocyclohexane	C_6H_6[c]	0.040
1,2-Dichloroethylene	C_6H_6[c]	$<5 \times 10^{-6}$
o-Dichlorobenzene	C_6H_6[c]	NR

[a] Quantum yield for the reaction of $Mo_2[O_2P(OC_6H_5)_2]_4$ with dichlorocarbon as determined by using a ferrioxalate actinometer.
[b] Concentration of chlorocarbon is 9 M.
[c] Concentration of chlorocarbon is 7 M.

$(OC_6H_5)_2]_4 \cdot 2THF$ shows the solvent to completely block the metal–metal core from substrate. Not surprisingly, the photoreduction of DCE by $Mo_2[O_2P(OC_6H_5)_2]_4$ in this solvent is severely impeded.

Metal–Organic Photochemistry of M_2Cl_4(diphenylphosphinomethane)$_2$ Systems

Although the net photoreduction of dihalocarbons is promoted by $Mo_2[O_2P(OC_6H_5)_2]_4$, two-electron reaction of the binuclear core is hindered by the fact that oxidation to Mo_2(III,III) diphenylphosphate is a thermodynamically unfavorable process ($E_{1/2}$ [Mo_2(III,III/II,III)] = 1.00 V vs. SCE). A recurrent theme in M≡M chemistry is that oxidation of the binuclear metal core is accompanied by rearrangement of the ligating sphere to confacial, **1** (*41–47*), or edge-sharing bioctahedral, **2** (*48–52*), geometries, which stabilize the binuclear metal core by ensuring an octahedral coordination geometry about the oxidized metal centers. With the D_{4h} "lantern" structure, **3**, the rigid tetrakis ligation geometry of the bidentate phosphate ligands cannot accommodate such a bioctahedral rearrangement. For these reasons we turned our attention to $M_2X_4(\widehat{P\ P})_2$ complexes ($\widehat{P\ P}$ is bridging phosphine and X is halide), **4**, where the terminal halide ligands can rearrange to form an edge-sharing bioctahedral coordination (*53–56*) about photo-oxidized binuclear cores.

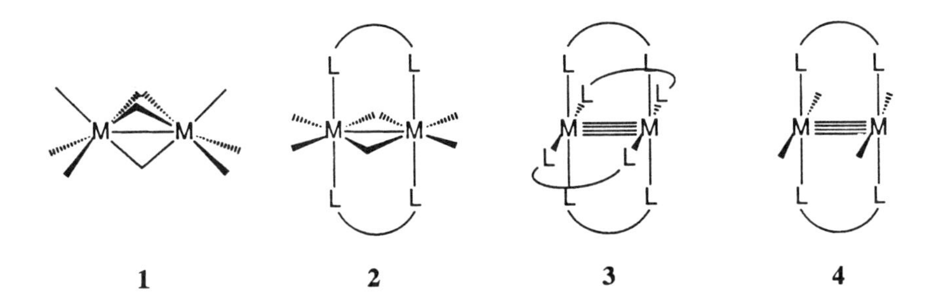

Transient Absorption Spectroscopy. The importance of bioctahedral geometries in the photochemistry of $M_2X_4(\widehat{P\ P})_2$ complexes is evident from their transient absorption spectroscopy. Figure 2 displays the time-resolved absorption spectrum of a deoxygenated C_6H_6 solution of $W_2Cl_4(dppm)_2$ (dppm is diphenylphosphinomethane) collected 100 ns after a 532-nm excitation pulse of a Nd:YAG laser. The transient absorption spectrum features prominent maxima at 390 and 470 nm and decays to ground state within 46 μs. Although not identified, long-lived absorp-

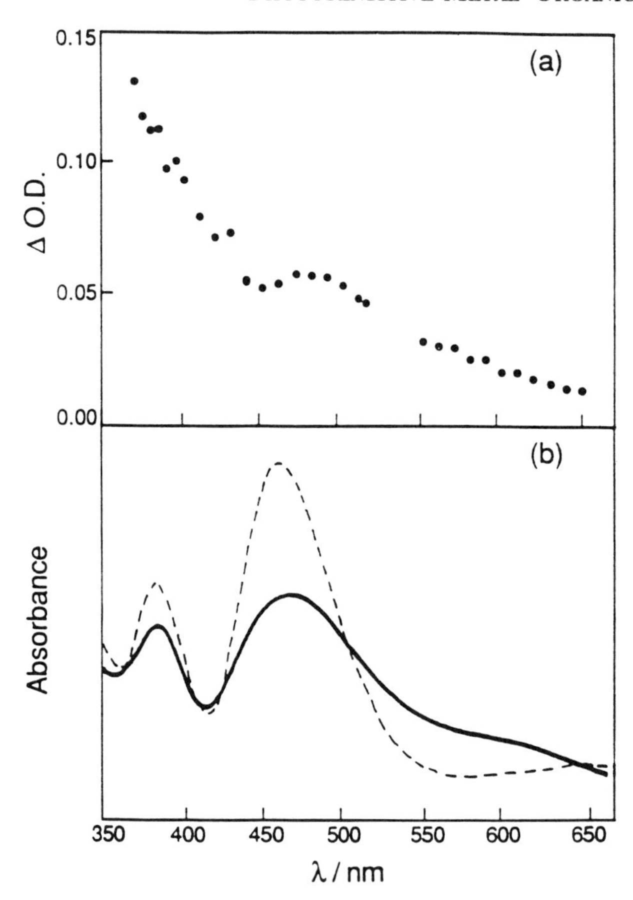

Figure 2. Part a: Transient difference spectrum of a deoxygenated benzene solution of $W_2Cl_4(dppm)_2$ (2×10^{-3} M) collected 100 ns after the excitation pulse of the second harmonic of a Nd:YAG laser. Part b: Electronic absorption spectrum of $W_2Cl_6(PEt_3)_4$) (—) and $W_2Cl_6(dppm)_2$ (- - -) in toluene and CH_2Cl_2, respectively, at room temperature.

tion profiles of nonluminescent transients of this type for M≡M complexes have been suggested to arise from chemically distorted intermediates (*57*). Indeed, near-UV absorptions with additional prominent features in the 450- to 500-nm spectral range are ubiquitous to the spectra of dimolybdenum and ditungsten edge-sharing bioctahedral complexes. As shown by Figure 2, the transient absorption features are energetically comparable to those observed in the spectra of the edge-sharing bioctahedral species ($W_2Cl_6(dppm)_2$: $\lambda_{abs,max}$ = 387 and 486 nm (*55*); W_2Cl_6(pyridine)$_4$: $\lambda_{abs,max}$ = 379 and 481 nm (*58*); $W_2Cl_6(PEt_3)_4$: 380 and 470 nm).

The spectral profiles of the $W_2Cl_4(dppm)_2$ transient and edge-sharing bioctahedral complexes exhibit common features despite the differences in d-electron counts for the two systems. Such similarities may be accounted for by the prediction of a small gap in the HOMO–LUMO manifold (*49, 59–63*). Inasmuch as the lowest energy transitions would be predicted to lie in the near-IR region, higher energy visible absorptions for these systems may have similar transitions that are not of HOMO–LUMO parentage. Notwithstanding, the similarities between $W_2Cl_4(P\ P)_2$ transients and edge-sharing bioctahedral complexes are striking and have led us to suggest that the chemically distorted intermediate responsible for the long-lived nonluminescent transient spectra of $M_2X_4(P\ P)_2$ species is an edge-sharing bioctahedron (eq 2).

$$\xrightarrow{h\nu} \qquad (2)$$

An edge-sharing ligand rearrangement provides cooperative stabilization of a charge-separated core by achieving an octahedral geometry about the oxidized metal center and by diminishing the donation of electron density from the halides about the reduced metal center. Charge transfer within the homobimetallic core does not appear to be solely sufficient to promote rearrangement because the structurally distorted intermediate is not observed to form upon $\delta\delta^*$ excitation, but rather is observed only when the higher energy $\pi\delta^*$ or $\delta\pi^*$ transitions are excited. These states not only possess charge-transfer character, but their population should also lead to a diminished metal–metal π-bonding interaction relative to that of the ground-state molecule. This feature is expected to enhance formation of a bioctahedral intermediate because interactions of the metal d_{yz} (or d_{xz}) orbitals with those of ligands in the equatorial plane of an edge-bridging bioctahedron occur at the expense of M–M π interactions.

Excited-State Reactivity. The photochemistry of the W_2Cl_4-$(dppm)_2$ binuclear complex is consistent with transient spectroscopic results. Although CH_3I (MeI) solutions of $W_2Cl_4(dppm)_2$ are indefinitely stable at room temperature in the absence of light, excitation with frequencies energetically coincident with the metal-localized $\delta\pi^*$ and $\pi\delta^*$ transitions leads to the quantitative production of a single product. Chemical and spectroscopic analyses have shown that the product corresponds to the addition of MeI to the tungsten–tungsten bond to yield $W_2Cl_4(dppm)_2(CH_3)(I)$ (*62*). Although the unequivocal determination of the coordination geometry about the tungsten–tungsten bond has been impeded by our inability to obtain single crystals for X-ray structural analysis, the electronic absorption spectrum of the photoproduct is similar to that of edge-sharing bioctahedral species, and ^{13}C NMR spectroscopy indicates that the methyl group is more likely to be in a terminal rather than a bridging position. Terminal coordination of the methyl group is consistent with addition of the substrate to an open coordination site of a photogenerated edge-sharing bioctahedral intermediate.

Simple addition of substrate to the binuclear core is not observed for photoactivation reactions involving free radical mechanisms. Photolysis of CH_3CH_2I (EtI) solutions of $W_2Cl_4(dppm)_2$ yields two products, one of which displays the spectral properties of independently prepared W_2Cl_4-$(dppm)_2I_2$. The parent ion cluster of this product appears at 1531 amu in the fast-atom bombardment mass spectrum (FABMS) of solids isolated from photolyzed solutions (Figure 3a). In addition the parent ion cluster of $W_2Cl_5(dppm)_2I$ at 1441 amu is also observed in the FABMS of the photoproduct. This pentachloro product is a signature of the free radical photoreactions of $M_2Cl_4(\overline{P\ P})_2$ complexes. We have accounted for the formation of $M_2Cl_5(\overline{P\ P})_2X$ and $M_2Cl_4(\overline{P\ P})_2X_2$ products with a radical mechanism wherein the $M_2Cl_4(\overline{P\ P})_2$ complex photoreacts with substrate to produce a mixed-valence $M_2Cl_4(\overline{P\ P})_2X$ primary photoproduct (eq 3) that disproportionates by either chlorine (eq 4) or X atom (eq 5) transfer (*35, 62*).

A similar disproportionation mechanism involving halogen atom transfer between mixed-valence intermediates for the photochemical reaction of diplatinum pyrophosphite with aryl halides has been proposed (*64*). By invoking the mechanism here, the observed photoproducts are accounted for as follows:

$$W_2Cl_4(dppm)_2 + EtI \xrightarrow{h\nu} W_2Cl_4(dppm)_2X + \cdot Et \quad (3)$$

$$2W_2Cl_4(dppm)_2X \longrightarrow \frac{1}{2}[W_2Cl_5(dppm)_2X + W_2Cl_3X(dppm)_2] \quad (4)$$

$$2W_2Cl_4(dppm)_2X \longrightarrow \frac{1}{2}[W_2Cl_5(dppm)_2X + W_2Cl_4(dppm)_2] \quad (5)$$

This reaction sequence is supported by the electrochemistry of $W_2Cl_4(dppm)_2$ in the presence of I^-, which permits the mixed-valence $W_2Cl_4(dppm)_2I$ intermediate to be accessed independently of a photochemical reaction pathway. Cyclic voltammograms of toluene–MeI solutions of $W_2Cl_4(dppm)_2$ show the reversible one-electron $W_2Cl_4(dppm)_2^{+/0}$ wave at +0.1 V versus a Ag wire quasi-reference electrode (Figure 4a). Upon the addition of I^-, the cathodic component of the one-electron oxi-

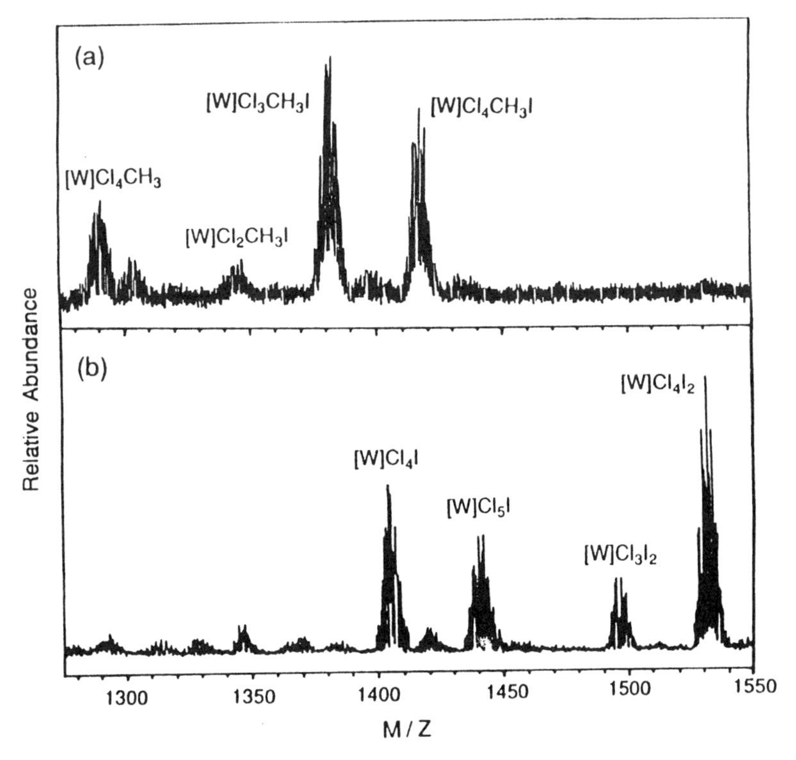

Figure 3. Fast-atom bombardment mass spectra of photolyzed (λ_{exc} > 436 nm) solutions of (a) $(dppm)_2$ and MeI and (b) $(dppm)_2$ and EtI in CH_2Cl_2. Selected peak assignments are shown with [W] = $W_2(dppm)_2$. Each ion cluster is flanked by a solvent adduct peak that is indicated by an asterisk.

dation wave disappears, and scanning cathodically gives rise to a reversible wave at –0.82 V. This reduction process coincides with the one-electron reduction (shown by the dashed line in Figure 4b) of independently prepared $W_2Cl_4(dppm)_2I_2$. (Unfortunately, we have not been able to prepare and cleanly isolate $W_2Cl_5(dppm)_2I$, and thus its cyclic voltammogram is not available for comparison. However the potential for $W_2Cl_6(dppm)_2$ is only 0.13 V greater than that of $W_2Cl_4(dppm)_2I_2$, and thus the wave at –0.82 V may very well reflect the reduction of both $W_2Cl_5(dppm)_2I$ and $W_2Cl_4(dppm)_2I_2$). These data are consistent with the disproportionation mechanism shown in reactions 3 and 4. The efficient trapping of $W_2Cl_4(dppm)_2^+$ by I^- is in accordance with the absence of the cathodic component of the $W_2Cl_4(dppm)_2^{+/0}$ redox wave when I^- is present, and the disproportionation of the $W_2Cl_4(dppm)_2I$ is suggested by the appearance of the reduction wave of $W_2Cl_4(dppm)_2I_2$.

The observations of simple photoaddition of MeI to terminal coordi-

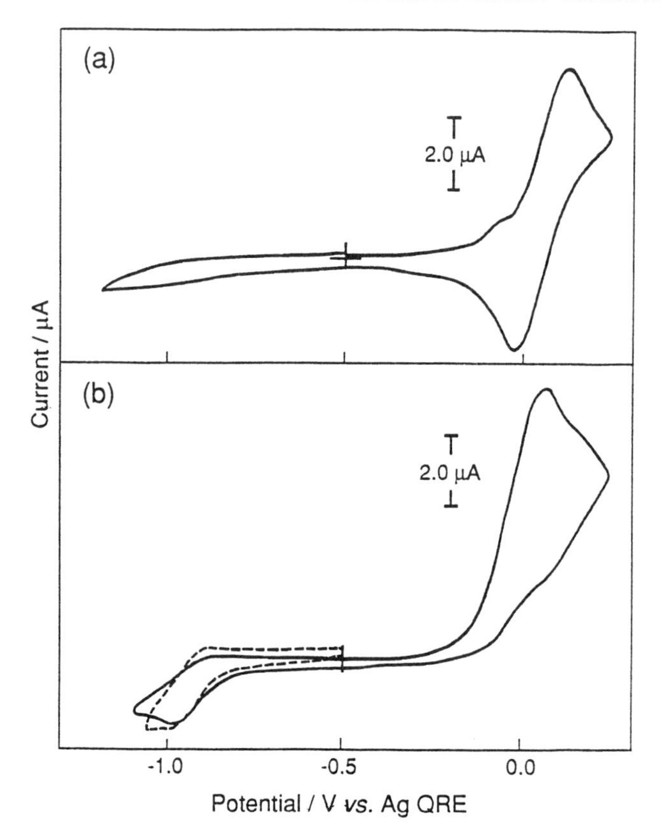

Figure 4. Cyclic voltammograms of toluene–MeI solutions containing (a) $W_2Cl_4(dppm)_2$ (1 mM) and (b) $W_2Cl_6(dppm)_2$ (1 mM) plus tetrahexylammonium (THA) iodide (5 mM) (—) and independently prepared $W_2Cl_4(dppm)_2I_2$ (- - -). The supporting electrolyte is 0.1 M $THAPF_6$.

nation sites of the ditungsten core, coupled with the absence of radical photoproducts including ethane, $W_2Cl_5(dppm)_2I$, and $W_2Cl_4(dppm)_2I_2$ (*see* Figure 3b) suggest a photoreaction pathway in which substrate is activated at a distinct tungsten atom of the photoexcited bimetallic core. The formation of the edge-sharing bioctahedral intermediate appears to be central to the photoreactivity of $W_2Cl_4(dppm)_2$. This distortion enhances the charge-separated character of the photoexcited binuclear core and simultaneously provides two open coordination sites at the reduced metal center. In this regard, the photoreactivity of the $W_2Cl_4(dppm)_2$ system parallels the oxidative–addition chemistry of mononuclear d^8 square-planar metal complexes (*65, 66*). In both cases addition of substrate is promoted at a low-valent, coordinatively unsaturated ML_4 center to yield a saturated octahedral geometry about the active site. Additionally, activation of substrate by photoexcited quadruply bonded complexes and ground-state d^8 square-planar metal complexes exhibit parallel reactivity trends.

The variation in photochemical reactivity of $W_2Cl_4(dppm)_2$ with MeI and EtI by simple addition and one-electron radical pathways, respectively, is also observed for the reaction of these substrates with Vaska's complex, *trans*-$IrCl(CO)(PR_3)_2$ *(67)*. A radical chain mechanism was proposed *(67)* for the oxidative–addition of EtI based on the attenuation of the rate in the presence of radical inhibitors. Conversely the rate is unaffected by radical inhibitors when MeI is the reactant. The solvent dependence and large negative activation entropy with MeI as the oxidative–addition substrate are consistent with a nucleophilic S_N2 mechanism that entails the addition of the electropositive $H_3C^{\delta+}$ to the metal center followed by rapid addition of the displaced iodide. However, a mechanism involving a very short-lived radical cage cannot be ruled out.

Similarly, this is the case for the photoreaction of MeI with $W_2Cl_4(dppm)_2$. The addition of MeI to the bimetallic core may proceed by sequential one-electron transfer within a solvent cage or via a concerted transition state. Our ability to open the temporal window of reactivity by photochemically promoting the addition of substrate to a bimetallic core should provide us with the opportunity to explore the most fundamental mechanistic issues of multielectron transformations of substrate by transient laser methodologies.

Conclusions

The photochemistry of the d^7–d^7 complexes with benzyl chloride ($PhCH_2Cl$), and $d^8 \cdots d^8$ and $d^4 \equiv d^4$ complexes with MeI is evocative of the primary photoprocesses of these bimetallic cores. Isolation of $Mn(CO)_5Cl$ from irradiated solutions of $Mn_2(CO)_{10}$ and $PhCH_2Cl$ *(68)* is in accordance with metal–metal-bond homolysis to produce a mononuclear fragment with a hole in a $d\sigma$ orbital, which is rendered inactive upon abstraction of an chloride atom from $PhCH_2Cl$ to produce a coordinatively saturated primary photoproduct. Consequently subsequent interaction of the metal center with the benzyl radical is obviated.

This is not the case for $d^8 \cdots d^8$ systems. The addition of substrate to the axial sites of the bimetallic core *(69)* is consistent with the diradical nature of $d^8 \cdots d^8$ electronic excited states. Initial abstraction of iodide with subsequent trapping of the incipient methyl radical by the neighboring radical center establishes a cooperative redox function between the individual metals comprising the bimetallic core. Similar cooperative redox function is observed for the $d^4 \equiv d^4$ photoactivation chemistry. In the

M≡M photochemistry, although one metal may not be directly involved in activation of substrate, it plays an important role in providing the reducing equivalents to its neighboring active site upon photoexcitation.

Charge transfer within the bimetallic core is a crucial primary step to the novel reactivity offered by d^4 binuclear complexes. Our current studies are centered on further enhancing the MMCT character of M≡M excited states by introducing chemical asymmetry in or about the bimetallic core. Certain ligands coordinate M≡M cores with an asymmetric disposition. For instance, the introduction of a fluorine at the 6-position of the 2-hydroxypyridine prompts one metal center to be coordinated entirely by nitrogen atoms and the other metal center entirely by oxygen atoms (*70*). Alternatively, heterobimetallic Mo≡W cores (*71*) present a straightforward means to further accentuate the charge-transfer character in metal-localized excited states. Studies are underway to elaborate the photochemistry of Mo≡W complexes, and to assess the influence of the asymmetric core on photoreactivity with comparative studies of the appropriate Mo_2 and W_2 complexes.

Finally, we continue to explore the scope of chemical reactivity of M≡M excited states. As inferred by Creutz and co-workers (*72*) from their work on the self-exchange kinetics of the $CpM(CO)_3^-$–$CpM(CO)_3X$ (M = Mo or W; X = Cl, Br, or I) systems, an electron pair situated on a coordinatively unsaturated metal center is well-suited to receive an X^+ atom (*72*). On this basis, the $:M^-–M^+$ character of M≡M excited states provides the underpinning for the design of photochemical activation schemes based not only on oxidative–addition but X^+ transfer as well.

The discrete multielectron photochemistry of $d^4≡d^4$ compounds further expands upon the coupled electron transfer and X• atom transfer multielectron activation chemistry offered by the $d^7–d^7$ and $d^8\cdots d^8$ complexes. In this sense bimetallic complexes assume a unique position as photosensitive metal–organic systems with their potential to offer a full complement of photoreactivity. With the continued discovery of new excited-state reaction pathways of bimetallic systems, it should soon be possible to tailor the diverse chemistry required for the activation of organic substrates to a specific family of bimetallic photoreagents.

Acknowledgments

Our studies of the photochemistry of d^4 binuclear complexes have been supported generously by the National Science Foundation (Grant No. CHE–9100532).

References

1. Albright, T. A.; Burdett, J. K.; Whangbo, M.-H. *Orbital Interactions in Chemistry;* Wiley-Interscience: New York, 1985; Chapter 17.
2. Mann, K. R.; Gray, H. B. In *Inorganic Compounds with Unusual Properties—II;* King, R. B., Ed.; Advances in Chemistry 173; American Chemical Society: Washington, DC, 1979; pp 225–235.
3. Rice, S. F.; Milder, S. J.; Goldbeck, R. A.; Kliger, D. S.; Gray, H. B. *Coord. Chem. Rev.* **1982,** *43,* 349.
4. Smith, D. C.; Gray, H. B. In *The Challenge of d and f Electrons;* Salahub, D. R.; Zerner, M. C., Eds.; ACS Symposium Series 394; American Chemical Society: Washington, DC, 1989; pp 356–365.
5. Meyer, T. J.; Caspar, J. V. *Chem. Rev.* **1985,** *85,* 187.
6. Wrighton, M. S.; Graff, J. L.; Luong, J. C.; Reichel, C. L.; Robbins, J. L. In *Reactivity of Metal–Metal Bonds;* Chisholm, M. H., Ed.; ACS Symposium Series 155; American Chemical Society: Washington, DC, 1981; pp 85–110.
7. Reinking, M. K.; Kullberg, M. L.; Cutler, A.R.; Kubiak, C. P. *J. Am. Chem. Soc.* **1985,** *107,* 3517.
8. Yasufuku, K.; Hiraga, N.; Ichimura, K.; Kobayashi, T. In *Photochemistry and Photophysics of Coordination Compounds;* Yersin, H.; Vogler, A., Ed.; Springer-Verlag: New York, 1987; p 271.
9. Geoffroy, G. L.; Wrighton, M. S. *Organometallic Photochemistry;* Academic Press: Orlando, FL, 1979.
10. Mann, K. R.; Thich, J. A.; Bell, R. A.; Coyle, C. L.; Gray, H. B. *Inorg. Chem.* **1980,** *19,* 2462.
11. Rice, S. F.; Miskowski, V. M.; Gray, H. B. *Inorg. Chem.* **1988,** *27,* 4704.
12. Rice, S. F.; Gray, H. B. *J. Am. Chem. Soc.* **1981,** *103,* 1593.
13. Dallinger, R. F.; Miskowski, V. M.; Gray, H. B.; Woodruff, W. H. *J. Am. Chem. Soc.* **1981,** *103,* 1595.
14. Roundhill, D. M.; Gray, H. B.; Che, C.-M. *Acc. Chem. Res.* **1989,** *22,* 55.
15. Lee, K.-W.; Brown, T. L. *J. Am. Chem. Soc.* **1987,** *109,* 3269.
16. Stiegman, A. E.; Tyler, D. R. *Comments Inorg. Chem.* **1986,** *5,* 215.
17. Rothberg, L. J.; Cooper, N. J.; Peters, K. S.; Vaida, V. *J. Am. Chem. Soc.* **1982,** *104,* 3536.
18. Marshall, J. L.; Stobart, S. R.; Gray, H. B. *J. Am. Chem. Soc.* **1984,** *106,* 3027.
19. Marshall, J. L.; Stiegman, A. E.; Gray, H. B. In *Excited States and Reactive Intermediates: Photochemistry, Photophysics, and Electrochemistry;* Lever, A. B. P., Ed.; ACS Symposium Series 307; American Chemical Society: Washington, DC, 1986; pp 166–176.
20. Lemke, F. R.; Granger, R. M.; Morgenstern, D. A.; Kubiak, C. P. *J. Am. Chem. Soc.* **1990,** *112,* 4052.
21. Lee, K.-W.; Hanckel, J. M.; Brown, T. L. *J. Am. Chem. Soc.* **1986,** *108,* 2266.
22. Harvey, E. L.; Stiegman, A. E.; Vlcek, A., Jr.; Gray, H. B. *J. Am. Chem. Soc.* **1987,** *109,* 5233.
23. Vlcek, A., Jr.; Gray, H. B. *Inorg. Chem.* **1987,** *26,* 1997.
24. Roundhill, D. M. *J. Am. Chem. Soc.* **1985,** *107,* 4354.
25. Smith, D. C.; Gray, H. B. *Coord. Chem. Rev.* **1990,** *100,* 169.
26. Caspar, J. V.; Gray, H. B. *J. Am. Chem. Soc.* **1984,** *106,* 3029.

27. Dulebohn, J. I.; Ward, D. L.; Nocera, D. G. *J. Am. Chem. Soc.* **1988,** *110,* 4054.
28. Dulebohn, J. I.; Ward, D. L.; Nocera, D. G. *J. Am. Chem. Soc.* **1990,** *112,* 2969.
29. Partigianoni, C. M.; Turro, C.; Shin, Y.-g. K.; Motry, D. H.; Kadis, J.; Dulebohn, J. I.; Nocera, D. G. In *Mixed Valency Systems: Applications in Chemistry, Physics, and Biology;* Prassides, K., Ed.; Kluwer Academic: Dordrecht, 1991, p 91.
30. Hopkins, M. D.; Gray, H. B.; Miskowski, V. M. *Polyhedron* **1987,** *6,* 705.
31. Hay, P. J. *J. Am. Chem. Soc.* **1982,** *104,* 7007.
32. Zhang, X.; Kozik, M.; Sutin, N.; Winkler J. R. In *Electron Transfer in Inorganic, Organic, and Biological Systems;* Bolton, J. R.; Mataga, N.; McLendon, G., Eds.; Advances in Chemistry Series No. 228; American Chemical Society: Washington, DC, 1991; pp 247–264.
33. Chang, I-J.; Nocera, D. G. *J. Am. Chem. Soc.* **1987,** *109,* 4901.
34. Chang, I-J.; Nocera, D. G. *Inorg. Chem.* **1989,** *28,* 4309.
35. Partigianoni, C. M.; Chang, I-J.; Nocera, D. G. *Coord. Chem. Rev.* **1990,** *97,* 105.
36. Bakac, A.; Espenson, J. H. *J. Am. Chem. Soc.* **1986,** *108,* 713.
37. Andrieux, C. P.; Gallardo, I.; Saveant, J.-M.; Su, K.-B. *J. Am. Chem. Soc.* **1986,** *108,* 638.
38. Saveant, J.-M. *J. Am. Chem. Soc.* **1987,** *109,* 6788.
39. Lexa, D.; Saveant, J.-M.; Su, K.-B.; Wang, D.-L. *J. Am. Chem. Soc.* **1988,** *110,* 7617.
40. Lexa, D.; Saveant, J.-M.; Schafer, H. J.; Su, K.-B.; Vering, B.; Wang, D. L. *J. Am. Chem. Soc.* **1990,** *112,* 6162.
41. Moynihan, K. J.; Gao, X.; Boorman, P. M.; Fait, J. F.; Freeman, G. K. W.; Thornton, P.; Ironmonger, D. *J. Inorg. Chem.* **1990,** *29,* 1648.
42. Bott, S. G.; Clark, D. L.; Green, M. L. H.; Mountford, P. *J. Chem. Soc., Chem. Commun.* **1989,** 418.
43. Cotton, F. A.; Luck, R. L. *Inorg. Chem.* **1989,** *28,* 182.
44. Cotton, F. A.; Poli, R. *Inorg. Chem.* **1987,** *26,* 3310.
45. Bergs, D. J.; Chisholm, M. H.; Folting, K.; Huffman, J. C.; Stahl, K. A. *Inorg. Chem.* **1988,** *27,* 2950.
46. Chisholm, M. H.; Eichhorn, B. W.; Folting, K.; Huffmann, J. C.; Ontiveros, C. D.; Streib, W. E.; Van Der Sluys, W. G. *Inorg. Chem.* **1987,** *26,* 3182.
47. Nocera, D. G.; Gray, H. B. *Inorg. Chem.* **1984,** *23,* 3686.
48. Cotton, F. A.; Eglin, J. L.; Luck, R. L.; Son, K. *Inorg. Chem.* **1990,** *29,* 1802.
49. Poli, R.; Mui, H. D. *Inorg. Chem.* **1990,** *30,* 2509.
50. Mui, H. D.; Poli, R. *Inorg. Chem.* **1989,** *28,* 3609.
51. Schrock, R. R.; Sturgeoff, L. G.; Sharp, P. R. *Inorg. Chem.* **1983,** *22,* 2801.
52. Jackson, R. B.; Streib, W. E. *Inorg. Chem.* **1971,** *10,* 1760.
53. Cotton, F. A. *Polyhedron* **1987,** *6,* 667.
54. Chisholm, M. H. *Acc. Chem. Res.* **1990,** *23,* 419.
55. Fanwick, P. E.; Harwood, W. S.; Walton, R. A. *Inorg. Chem.* **1987,** *26,* 242.
56. Cotton, F. A.; Daniels, L. M.; Dunbar, K. R.; Falvello, L. R.; O'Connor, C. J.; Price, A. C. *Inorg. Chem.* **1991,** *30,* 2509.

57. Winkler, J. R.; Nocera, D. G.; Netzel, T. L. *J. Am. Chem. Soc.* **1986,** *108,* 4451.
58. Saillant, R.; Hayden, J. L.; Wentworth, R. A. D. *Inorg Chem.* **1967,** *6,* 1497.
59. Shaik, S.; Hoffman, R.; Fisel, R.; Summerville, R. H. *J. Am. Chem. Soc.* **1980,** *102,* 4555.
60. Chakravarty, A. R.; Cotton, F. A.; Diebold, M. P.; Lewis, D. B.; Roth, W. J. *J. Am. Chem. Soc.* **1986,** *108,* 971.
61. Cotton, F. A.; Diebold, M. P.; O'Connor, C. J.; Powell, G. L. *J. Am. Chem. Soc.* **1985,** *107,* 7438.
62. Partigianoni, C. M.; Nocera, D. G. *Inorg. Chem.* **1990,** *29,* 2033.
63. Cotton, F. A.; Walton, R. A. *Multiple Bonds between Metals;* Wiley-Interscience: New York, 1982; p 221.
64. Roundhill, D. M. *Inorg. Chem.* **1986,** *25,* 4071.
65. Collman, J. P.; Hegedus, L. S.; Norton, J. R.; Finke, R. G. *Principles and Applications of Organotransition Metal Chemistry;* University Science: Mill Valley, CA, 1987; p 177.
66. Halpern, J. *Acc. Chem. Res.* **1970,** *3,* 386.
67. Labinger, J. A.; Osborn, J. A.; Coville, N. J. *Inorg. Chem.* **1980,** *19,* 3236.
68. Wrighton, M. S.; Ginley, D. S. *J. Am. Chem. Soc.* **1975,** *97,* 2065.
69. Che, C.-M.; Mak, T. C. W.; Gray, H. B. *Inorg. Chem.* **1984,** *23,* 4386.
70. Cotton, F. A.; Falvello, L. R.; Han, S.; Wang, W. *Inorg. Chem.* **1983,** *22,* 4106.
71. Luck, R. L.; Morris, R. H. *J. Am. Chem. Soc.* **1984,** *106,* 7978.
72. Schwarz, C. L.; Bullock, R. M.; Creutz, C. *J. Am. Chem. Soc.* **1991,** *113,* 1225.

RECEIVED for review November 7, 1991. ACCEPTED revised manuscript June 19, 1992.

9

Patterned Imaging of Palladium and Platinum Films

Electron Transfers to and from Photogenerated Organometallic Radicals

Clifford P. Kubiak, Gregory K. Broeker, Robert M. Granger, Frederick R. Lemke, and David A. Morgenstern

Department of Chemistry, Brown Laboratories, Purdue University, West Lafayette, IN 47907

Photogenerated $[M(CNR)_3]^{\cdot+}$ radicals, formed by σ,σ^* homolysis of the M–M bond of $M_2(CNR)_6(PF_6)_2$ (M = Pd or Pt; R = CH_3 or tert-*C_4H_9), were found to be significantly stronger reductants and oxidants compared to their parent ground-state dimers. The kinetics of electron transfers to and from photogenerated $[M(CNR)_3]^{\cdot+}$ radicals were examined by laser flash photolysis. Photogenerated $[M(CNR)_3]^{\cdot+}$ radicals were found to be single-electron reductants of various electron acceptors with electron-transfer rate constants of up to 2×10^8 M^{-1} s^{-1}. Photogenerated $[M(CNMe)_3]^{\cdot+}$ radicals were also found to be single-electron oxidants of a homologous series of redox-tuned ferrocene electron donors. The electron-transfer rate constants,* k_e, *exhibit Marcus–Agmon–Levine dependence on reduction potential* E^o *(ferricinium–ferrocene), modulated by preequilibrium loss of one isocyanide ligand. The radical species $[PdCNMe)_3]^{\cdot+}$ may also be prepared in the gas phase where electron-transfer reactivity also has been observed. Patterned metal films were prepared by exploiting the enhanced redox properties of photogenerated $[M(CNR)_3]^{\cdot+}$ radicals.*

LASER DIRECT WRITING of conducting metal features has received considerable attention as a means of defining and wiring a micrometer-

0065–2393/93/0238–0165$06.00/0

scale circuit in a single step (*1–4*). We have employed the significantly enhanced oxidation–reduction properties of organometallic radicals that are formed by single-photon excitation of metal–metal (M–M) bonded binuclear species to optically image metal films. The photochemistry of metal–metal σ-bonds has been an area of considerable interest over the past several years (*5*). The σ,σ^* or $d\pi,\sigma^*$ excited states of M–M bonded species generally are dissociative with respect to the formation of reactive organometallic free radicals. We first reported the σ,σ^* photochemistry of the d^9,d^9 hexakis(isocyanide) binuclear palladium and platinum complexes, $M_2(CNR)_6(PF_6)_2$ (*6*), in 1985 (*7*). Photogenerated $[M(CNR)_3]^{\bullet+}$ (M = Pd or Pt) radicals are reactive toward ligand substitution and atom abstraction (*7, 8*).

More recently, we studied the electron-transfer chemistry of photogenerated $[M(CNR)_3]^{\bullet+}$ (M = Pd or Pt) radicals (*9, 10*). The reactivity of photogenerated $[M(CNR)_3]^{\bullet+}$ radicals toward electron transfers provides the basis for photochemical metal-film deposition. Photogenerated metal radicals are both potentially stronger oxidants and reductants compared to their parent ground-state M–M bonded complexes. The enhancement of oxidation–reduction reactivity is the predicted result of a nonbonded electron of the radical that is more energetic than the corresponding M–M σ-bonding electrons in the ground configuration; and a hole in the radical's electronic configuration that is deeper than the corresponding antibonding holes in the ground configuration. An interaction diagram for a pair of $[M(CNR)_3]^{\bullet+}$ radicals, the resulting σ and σ^* energy levels of $[M_2(CNR)_6]^{2+}$, and the representative structure of $[Pd_2(CNMe)_6]^{2+}$ are shown in Figure 1.

The $[M(CNR)_3]^{\bullet+}$ (M = Pd or Pt) radical systems provide an unprecedented opportunity to study the energetics and dynamics of both the oxidative and reductive electron-transfer processes of an organometallic radical (eq 1).

$$[M(CNR)_3]^{2+}\ (d^8) \underset{-e^-}{\overset{e^-}{\rightleftharpoons}} [M(CNR)_3]^{+}\ (d^9) \underset{-e^-}{\overset{e^-}{\rightleftharpoons}} M(CNR)_3\ (d^{10}) \qquad (1)$$

The electron transfers are examined within the context of the Marcus–Agmon–Levine (*11–18*) theoretical framework for electron transfer in order to elucidate (1) the energetics of electron transfers, and

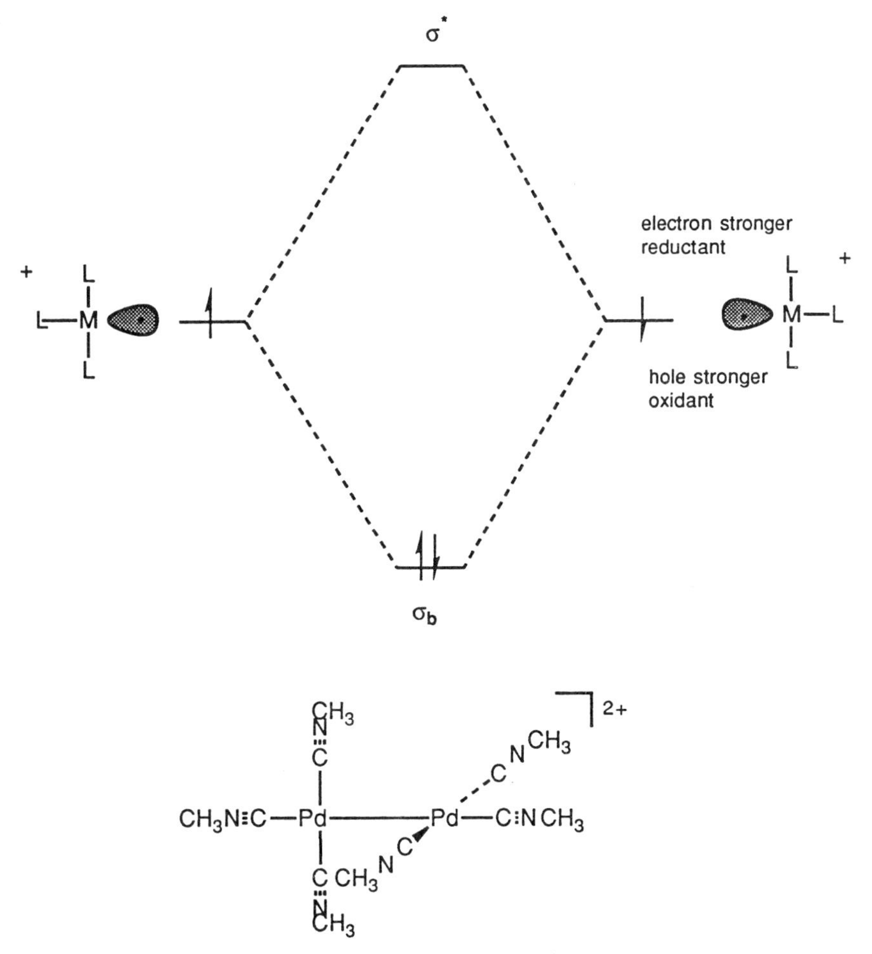

Figure 1. Interaction diagram for two d^9 $ML_3^{\bullet +}$ fragments and the structure of $[Pd_2(CNMe)_6]_{2+}$.

(2) the relative importance of structural and solvent reorganizational barriers for electron transfers. Furthermore, the same highly energetic $[Pd(CNMe)_3]^{\bullet +}$ radical ion that can be prepared by photochemical σ,σ^* excitation in solution can be prepared in the gas phase where its electron transfers can be studied by Fourier transform mass spectrometry (FTMS) in the absence of solvent (*19*). Both the reductive and oxidative electron-transfer reactivities can be exploited for the photochemical deposition of metal films. The properties of metal films that were prepared by direct chemical and electrochemical reduction of $[M(CNR)_3]^{\bullet +}$ radicals are discussed.

Electron Transfers to and from $[Pd(CNMe)_3]^{+\cdot}$: Kinetics and Mechanism

Laser transient absorbance (TA) studies have provided many of the essential kinetic and mechanistic details of the chemistry of photogenerated $[M(CNR)_3]^{\cdot+}$ radicals. Here we will consider both the reductive and oxidative electron-transfer behavior of the palladium system, $[Pd(CNMe)_3]^{\cdot+}$. Laser flash photolysis of $Pd_2(CNMe)_6(PF_6)_2$ produces an intense transient absorbance at λ_{max} = 405 nm in acetonitrile, a wavelength at which there is no interference from the parent dimer (λ_{max} = 307 nm). The transient absorbance decays by second-order kinetics, corresponding to the recombination of $[Pd(CNMe)_3]^{\cdot+}$ radicals (eq 2).

$$[(MeN{:}C)_3Pd{-}Pd(C{:}NMe)_3]^{2+} \underset{k_r}{\overset{h\nu}{\rightleftharpoons}} 2\,[Pd(C{:}NMe)_3]^{\cdot+} \qquad (2)$$

The transient absorbance instrumentation in use in our laboratories was described elsewhere (*10, 20*). An unusual characteristic of $[Pd(CNMe)_3]^{\cdot+}$ radicals is their intense molar absorptivity, ϵ = 50,000 M^{-1} cm^{-1} at 405 nm. The signature electronic transition of the radical has been assigned to an allowed d,p transition, which becomes accessible in a d^9 electronic configuration of a metal with a relatively small d,p energy gap. A similar assignment was reported by Caspar (*21*) for d^{10} Pd(0) diphosphine complexes. The recombination rate constant, k_r, was determined to be 1×10^9 M^{-1} s^{-1}, approaching the diffusion-controlled limit in acetonitrile (*9*).

The first direct observation of electron transfer from a photogenerated organometallic radical of the platinum group was reported in 1986 (*9*). Flash photolysis of an acetonitrile solution of $Pd_2(CNMe)_6(PF_6)_2$ in the presence of one of the electron acceptors, A [A = *N,N'*-propylene-1,10-phenanthrolinium (PPQ^{2+}), benzyl viologen (BV^{2+}), methyl viologen (MV^{2+}), and 2,5-dichlorobenzoquinone (DCBQ)] leads to rapid electron transfer from $[Pd(CNMe)_3]^{\cdot+}$ to the acceptor. The disappearance of $[Pd(CNMe)_3]^{\cdot+}$ absorbance at 405 nm is observed together with the synchronous appearance of the reduced form of the acceptor ($PPQ^{\cdot+}$, $BV^{\cdot+}$, $MV^{\cdot+}$, etc.) at a different wavelength. Electron-transfer rate constants from $[Pd(CNMe)_3]^{\cdot+}$ radicals to the different electron acceptors, A, are

Table I. Electron-Transfer Rate Constants for the Reduction of Electron Acceptors by Photogenerated $[Pd(CNMe)_3]^{\bullet+}$ Radicals in CH_3CN

Electron Acceptor (A)	E^o $(A/A^{\bullet-})$ *vs. SCE*	k_e $(M^{-1}\ s^{-1})$
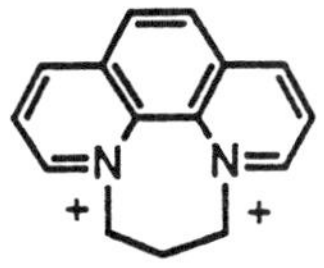 [diquat-type phenanthrolinium dication]	–0.13	2×10^8
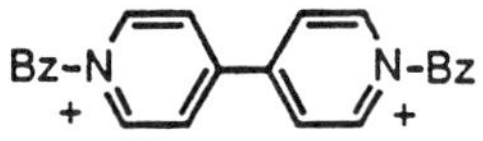 Bz-N+ ... N+-Bz (benzyl viologen)	–0.36	1×10^8
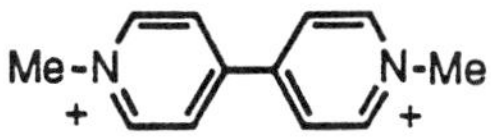 Me-N+ ... N+-Me (methyl viologen)	–0.45	3×10^7
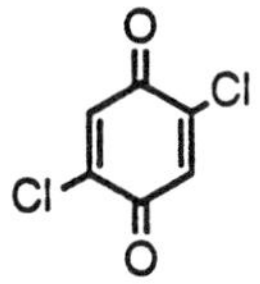 2,5-dichloro-1,4-benzoquinone (O, Cl, Cl, O)	–0.18	3×10^7

presented in Table I. These rate constants, ranging from 3×10^7 to 2×10^8 M^{-1} s^{-1}, show that electron transfers are rapid for electron acceptors as difficult to reduce as methyl viologen, E^o $(MV^{2+/+})$ = –0.45 versus the saturated calomel electrode (SCE). These results also indicate that photogenerated $[Pd(CNMe)_3]^{\bullet+}$ radicals are sufficiently potent reagents for the reduction of other metal ions, including Ag^+ and $Au(CN)_2^-$ to metal films. Results from our laboratory show that this is indeed the case. Palladium radicals can be used to mediate the photodeposition of both silver and gold. Well-established methods for depositing these metals already exist and do not involve sacrificial palladium, but very few methods exist for the direct photodeposition of the technologically important metals palladium and platinum. Accordingly, we have focused our studies on the reductive couple, $[M(CNR)_3]^{+/0}$, of photogenerated $[M(CNR)_3]^{\bullet+}$ (M = Pd or Pt) radicals.

Irradiation (λ = 313 nm) of $Pd_2(CNMe)_6(PF_6)_2$ in the presence of decamethylferrocene in acetonitrile leads to the disappearance of the dimer and the appearance of a new absorbance at 777 nm, corresponding to decamethylferricinium. Another new absorbance at 410 nm corresponds to $[Pd(CNMe)_2]_n$ (*22, 23*). The oligomeric Pd(0) species $[Pd(CNR)_2]_n$ (R = isopropyl, C_6H_{11}, Ph, *p*-MeC_6H_4, or *p*-$MeOC_6H_4$)

have been reported (*22, 23*) to be unstable in polar solvents, and indeed, within a few minutes the absorbance at 410 nm disappears and metallic palladium is deposited in the photolysis cell. Irradiation of an identical solution at 438 nm (within the decamethylferrocene absorption band) leads to no reaction. Refluxing the solution also has no effect.

For our studies of the electron-transfer kinetics for the reductive couple, $[Pd(CNMe)_3]^{+/0}$, we chose to use a homologous series of electron-donor ferrocenes. For a structurally similar, homologous series of electron donors, D, of differing E^o $(D^{+/0})$, the barrier to self-exchange (eq 3) can be considered to be comparable throughout the series,

$$D + D^+ \longrightarrow D^+ + D \tag{3}$$

and the rate constants that describe the formation and breakup of the activated complex, k_d, k_{-d}, and k_{-d}', are also comparable (eq 4).

$$Pd(CNMe)_3^{\bullet+} + D \underset{k_{-d}}{\overset{k_d}{\rightleftarrows}} [Pd(CNMe)_3^{\bullet+}, D] \underset{k_{-et}}{\overset{k_{et}}{\rightleftarrows}} [Pd(CNMe)_3^0, D^+] \xrightarrow{k_{-d}'} Pd(CNMe)_3^0 + D^+ \tag{4}$$

Under these conditions, the Marcus–Agmon–Levine (*11–18*) theory allows a quantitative description of the dependence of electron-transfer rate on driving force within the series. The kinetic scheme of equation 4 has been discussed by several authors (*24–26*) and leads to the key definitions summarized as follows. The overall electron-transfer rate constant, k_e, is given by equation 5 (*27*).

$$k_e = \frac{k_d}{1 + \dfrac{k_{-d}}{k_{et}} + \dfrac{k_{-d}}{k_{-d}'}\dfrac{k_{-et}}{k_{et}}} \tag{5}$$

where k_{et} and k_{-et} are the unimolecular forward and reverse electron-transfer rate constants, respectively, within the activated complex. For electron transfer between a singly charged ($[Pd(CNMe)_3]^{\bullet+}$) and neutral (ferrocene) species, no complications result from electrostatics, and thus from equations 6 and 7, the overall expression for the electron-transfer rate constant, equation 8, can be obtained.

$$\frac{k_{-et}}{k_{et}} = \exp\left(\frac{\Delta G}{RT}\right) \tag{6}$$

$$k_{et} = k_{et}{}^{0} \exp\left(\frac{-\Delta G^{*}}{RT}\right) \tag{7}$$

$$k_{et} = \frac{k_{d}}{1 + \left[\frac{k_{-d}}{k_{et}{}^{0}} \exp\left(\frac{\Delta G^{*}}{RT}\right)\right] + \left[\frac{k_{-d}}{k_{-d}{}'} \exp\left(\frac{\Delta G}{RT}\right)\right]} \tag{8}$$

where ΔG is the free energy change associated with the electron transfer, ΔG^{*} is the free energy of activation, R is the gas constant, and T is absolute temperature. Both Meyer and co-workers (*24*) and Balzani et al. (*25*) developed kinetic interpretations based upon these equations, and their results are essentially equivalent. Only one other consideration remains: the expression that relates ΔG^{*} to ΔG. Several conflicting equations have appeared in the literature (*28*). They are distinguishable only in the region of very large driving force. We use the model in equation 9, originally developed by Marcus (*18*).

$$\Delta G^{*} = \Delta G + \frac{\Delta G^{*}(0)}{\ln 2} \ln\left[1 + \exp\left(\frac{\Delta G \ln 2}{\Delta G^{*}(0)}\right)\right] \tag{9}$$

If the reaction is adiabatic and the driving force is large, the electron transfer should be substantially faster than the breakup of the activated complex ($k_{et} >> k_{-d}$). The model then predicts that at large driving force ($k_{et} >> k_{-et}$), the overall electron-transfer rate, k_{e}, will saturate at $k_{e} = k_{d}$. The rate constant for formation of the activated complex, k_{d}, is approximately equal to the diffusion-controlled rate constant. In acetonitrile at 25 °C, the diffusion-controlled rate constant is calculated to be $1.9 \times 10^{10}\ M^{-1}\ s^{-1}$ from the Stokes–Einstein equation. With this background, we will now consider the kinetics of electron transfers between photogenerated $[Pd(CNMe)_3]^{\bullet +}$ radicals and a homologous series of ferrocene electron donors.

The electron-transfer rate constants, determined by transient absorbance kinetics, for the reduction of photogenerated $[Pd(CNMe)_3]^{\bullet +}$ radicals by a homologous series of ferrocenes are shown in Table II. The ferrocenes span nearly 1 V in their oxidation potentials (*28, 29*). The electron-transfer rate for 1,1′-dichloroferrocene, the ferrocene most stable to oxidation (half-wave potential $E_{1/2} = +0.77$ V vs. SCE), was too slow to measure. Figure 2 presents the data as a plot of $\log(k_{e})$ vs. $E_{1/2}$ (ferrocene–ferricinium). A surprising result of these studies is that the measured electron-transfer rate constants, k_e, saturate 2 orders of magnitude below the diffusion-controlled rate in acetonitrile. The theoretical framework of Marcus–Agmon–Levine allows two possible explanations for the observed saturation in rate. First, it may be that the electron-transfer reaction is nonadiabatic. In this case, poor electronic coupling between the initial and final state make $k_{et} < k_{-d}$, and so, even with a large driving force, the activated complex is most likely to fall apart before the electron transfer occurs. A second explanation is that one of the reagents (most likely $[Pd(CNMe)_3]^{\bullet +}$) undergoes changes governed by a preequilibrium constant, P_1^* (*19*). In this case, the overall observed rate constant for electron transfer, k_e^{obs}, will consist of the actual electron transfer, k_e, modulated by the preequilibrium constant, P_1^*.

Table II. Electron-Transfer Rate Constants for the Oxidation of Electron Donors, D, by Photogenerated $[Pd(CNMe)_3]^{\bullet +}$ and $[Pd(CN\text{-}tert\text{-}Bu)_3]^{\bullet +}$ Radicals in CH_3CN

		k_e (M^{-1} s^{-1})	
D	E^o ($D/D^{\bullet +}$) *vs. SCE*	$Pd_2(CNMe)_6^{2+}$	$Pd_2(CN\text{-t-}Bu)_6^{2+}$
$FeCp_2^*$	−0.09	9×10^7	1.5×10^8
$FeCpCp^*$	0.13	6×10^7	–
$Fe(CpMe)_2$	0.31	3×10^7	2.5×10^7
$FeCp(CpCH_2OH)$	0.38	3×10^6	9.0×10^6
$FeCp_2$	0.42	2×10^6	6.4×10^6
FeCp(CpI)	0.54	5×10^4	1.5×10^4

The electron-transfer kinetic data in Table II are well accommodated by the model represented by equations 8 and 9. However, the value obtained for $\Delta G^*(0)$ was only 1.3 kcal/mol. The Marcus–Agmon–Levine theory predicts that $\Delta G^*(0)$ is the average of the two-component self-exchange free energies (eq 10).

$$\Delta G^*(0) = \frac{1}{2}\Delta G^*(PdL_3{}^{+/0}) + \frac{1}{2}\Delta G^*(Fc^{+/0}) \qquad (10)$$

where L is ligand and Fc is ferrocene. The self-exchange barrier for ferrocene is 5.3 kcal/mol (*29, 30*). Therefore, even if the self-exchange barrier for $[Pd(CNMe)_3]^{+/0}$ were zero, it would be impossible to satisfy equation 10. Hence, we reject the nonadiabaticity hypothesis and turn instead to the possibility of a preequilibrium.

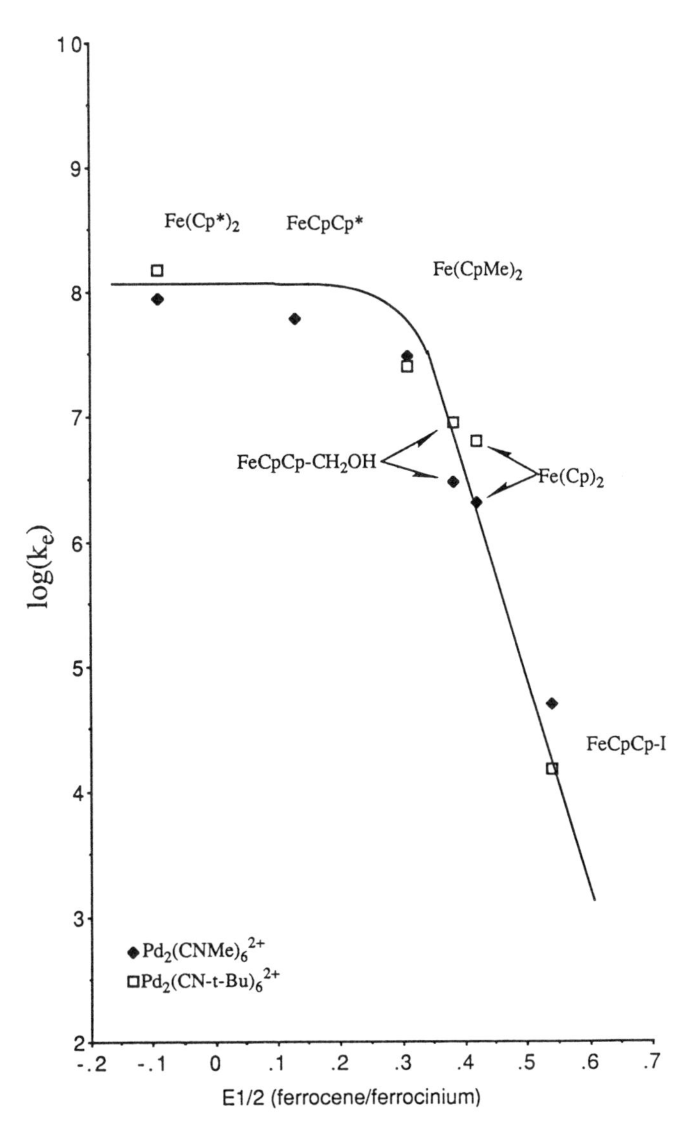

Figure 2. Plot of $log(k_e)$ *vs.* $E_{1/2}$ *(ferrocene–ferricinium) for electron transfers between photogenerated radicals derived from photolysis of* $Pd_2(CNMe)_6(PF_6)_2$ *and* $[Pd_2(CN\text{-tert-}Bu)_6][PF_6]_2$ *and variously substituted ferrocene electron donors.*

The kinetics of electron transfer between $[Pd(CNMe)_3]^{\bullet+}$ and the various ferrocenes were reexamined in the presence of different concentrations of CNMe. The electron-transfer rate constants, k_e^{obs}, display an inverse linear dependence on CNMe. These results lead to the surprising conclusion that preequilibrium dissociation of methyl isocyanide is required before electron transfer (eqs 11 and 12).

$$[Pd(CNMe)_3]^{+\bullet} \xrightleftharpoons{K_{diss}} [Pd(CNMe)_2]^{+\bullet} + CNMe \tag{11}$$

$$[Pd(CNMe)_2]^{+\bullet} + e^- \xrightarrow{k_e} Pd(CNMe)_2 \tag{12}$$

The average dissociation constant for the various ferrocenes examined was K_{diss}/CNMe = 0.005.

Several of the most surprising aspects of the chemistry of photogenerated $[Pd(CNMe)_3]^{\bullet+}$ radicals in solution gain support from an examination of their gas-phase chemistry. Gord and Freiser (*11*) prepared the ions $[Pd(CNMe)_n]^{\bullet+}$ (n = 1, 2, or 3) in the gas phase by laser ionization of palladium metal followed by an ion–molecule reaction with CNMe. The chemistry of the differently substituted ions $[Pd(CNMe)_n]^{\bullet+}$ (n = 1, 2, or 3) may then be followed by Fourier transform mass spectrometry. A significant finding is that none of the four-coordinate, 17e$^-$ species $[Pd(CNMe)_4]^{\bullet+}$ is produced in the gas phase. The technique of swept double resonance was employed in the Fourier transform mass spectrometer (Nicolet FTMS-2000) to isolate each of the ions $[Pd(CNMe)_3]^{\bullet+}$, $[Pd(CNMe)_2]^{\bullet+}$, and $PdCNMe^{\bullet+}$ such that the dynamics of the reactions of each ion with electron-donor ferrocenes in the gas phase could be examined (*11*). The ions $[Pd(CNMe)_3]^{\bullet+}$, $[Pd(CNMe)_2]^{\bullet+}$, and $PdCNMe^{\bullet+}$ exhibited straightforward first-order exponential decay in the presence of ferrocenes. The relative order of reactivity toward electron transfer from ferrocenes in the gas phase to the palladium ions was as follows: $PdCNMe^{\bullet+} > [Pd(CNMe)_2]^{\bullet+} > [Pd(CNMe)_3]^{\bullet+}$.

It is difficult to make inferences about condensed-phase chemical reactivity from gas-phase results; however, several of the gas-phase observations appear to be significant. The fact that no $[Pd(CNMe)_4]^{\bullet+}$ was observed suggests that the 15e$^-$ $[Pd(CNMe)_3]^{\bullet+}$ radicals may be rather stable to association or associative substitution by a fourth ligand. The relative ordering of reactivity, which showed that the less substituted ions are more reactive, suggests that the ion electron affinities increase with decreasing substitution. In our condensed-phase studies, this increase may be reflected by increased driving force for the reduction of $[Pd(CNMe)_2]^{\bullet+}$ and formation of $Pd(CNMe)_2$ (eq 12) because of the special stability of linear ML_2 complexes of d^{10} metals.

We attempted to determine whether steric effects arising from more bulky isocyanide ligands influence the dissociation of an isocyanide ligand from $[Pd(CNR)_3]^{\bullet +}$ in solution. The electron-transfer rate constants for the reduction of photogenerated $[Pd(CN\text{-}tert\text{-}Bu)_3]^{\bullet +}$ radicals by various ferrocenes are compared to the data for $[Pd(CNMe)_3]^{\bullet +}$ radicals in Table II. Figure 2 compares the data as a plot of $\log(k_e)$ vs. $E_{1/2}$ (ferricinium–ferrocene). In general, there is excellent agreement between the behavior of the two differently substituted palladium radicals. The values of $\log(k_e)$ for $[Pd(CN\text{-}tert\text{-}Bu)_3]^{\bullet +}$ show a linear dependence on $E_{1/2}$ (ferricinium–ferrocene) for the moderate electron-donor ferrocenes, then saturate well below the expected diffusion-controlled limit at high driving force. The small differences between the electron-transfer rates for $[Pd(CNMe)_3]^{\bullet +}$ and $[Pd(CN\text{-}tert\text{-}Bu)_3]^{\bullet +}$ suggest that the bulkier *tert*-BuNC ligand is not sufficiently more readily dissociated to accelerate electron transfer.

Photochemistry of $Pt_2(CNMe)_6(BF_4)_2$

Several details of the photochemistry of $Pt_2(CNMe)_6(BF_4)_2$ are distinctly different from the related palladium complex. The net photochemistry of the complexes of the two metals, however, shows many parallels. Laser (Nd:YAG, 266 nm ($4\nu_0$), 7 ns) excitation of $Pt_2(CNMe)_6(BF_4)_2$ in acetonitrile results in a transient absorbance that is significantly weaker than that of the palladium complex. The spectroscopic differences between the palladium and platinum $[M(CNMe)_3]^{\bullet +}$ transients may be understood in terms of the larger d,d energy splittings of third- versus second-row transition metal ions. The intense phototransient observed for the $[Pd(CNMe)_3]^{\bullet +}$ radical is assigned to a fully allowed d,p transition, a result of decreased $d_{x^2-y^2}$,p orbital splittings relative to d_{xy}, $d_{x^2-y^2}$ splittings. In the $[Pt(CNMe)_3]^{\bullet +}$ system, this transition is expected to occur at lower energy, perhaps in the near IR region and beyond the response of our photomultiplier, because of the increased d_{xy}, $d_{x^2-y^2}$ splittings for platinum. The observed transient absorbance of the radical $[Pt(CNMe)_3]^{\bullet +}$ has tentatively been assigned to a d,d transition.

Photogenerated $[Pt(CNMe)_3]^{\bullet +}$ radicals also appear to be better oxidants and reductants than the parent ground-state dimer. Reduction of $[Pt(CNMe)_3]^{\bullet +}$ has been demonstrated by photolysis of a 0.1 mM acetonitrile solution of decamethylferrocene and $Pt_2(CNMe)_6(BF_4)_2$. The ground-state platinum dimer does not react thermally with $FeCp_2^*$ (Cp* is $\eta^5\text{-}C_5Me_5$). Upon irradiation, the disappearance of $Pt_2(CNMe)_6(BF_4)_2$ is accompanied by the appearance of a new absorption band in the UV–visible spectrum corresponding to decamethylferricinium, $\lambda_{max} = 777$ nm (Figure 3). Oxidation of the $[Pt(CNMe)_3]^{\bullet +}$ radical has been demon-

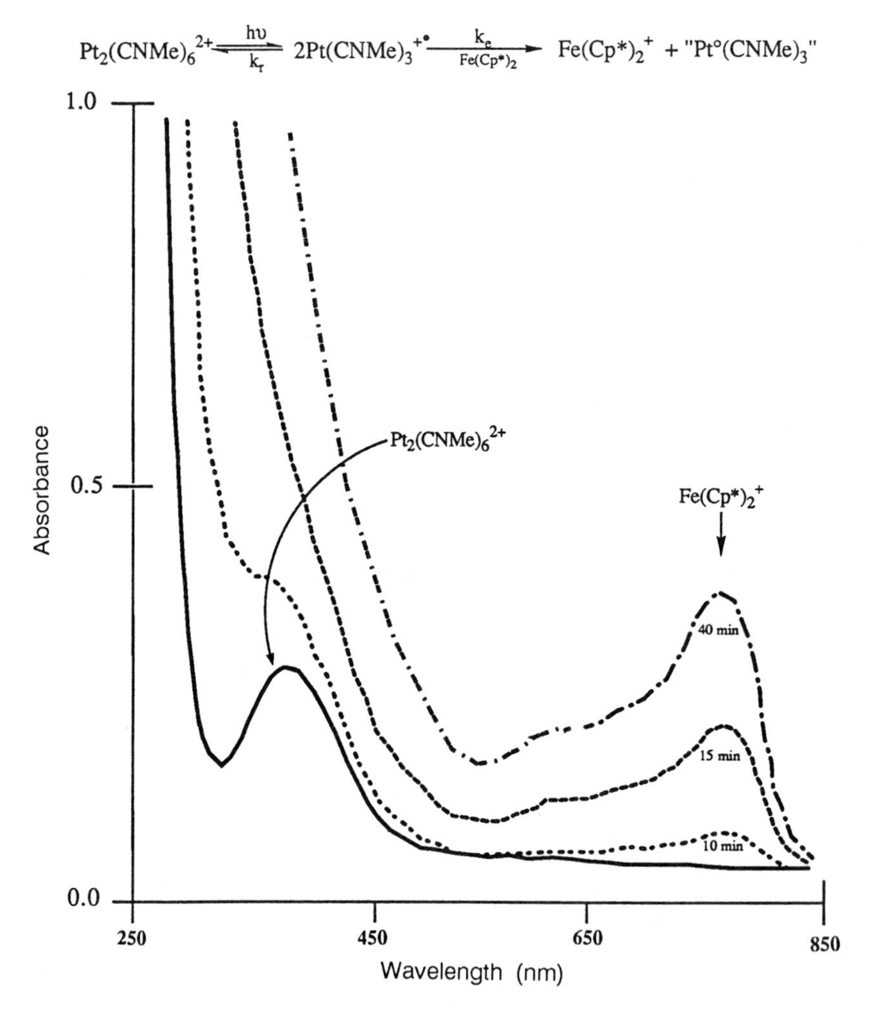

Figure 3. Changes in the electronic absorption spectrum of Pt_2-$(CNMe)_6(BF_4)_2$ (1×10^{-3} M) and FeCp (1×10^{-3} M) in acetonitrile upon irradiation (λ >290 nm).*

strated in the presence of benzyl viologen (BV^{2+}). The parent ground-state dimer shows no thermal reaction with BV^{2+}. However, when the solution is photolyzed, oxidation of the $[Pt(CNMe)_3]^{\bullet +}$ radicals is indicated by the appearance of the radical cation of benzyl viologen ($BV^{\bullet +}$) at 607 nm.

Transient absorbance (TA) kinetic studies of photogenerated $[Pt(CNMe)_3]^{\bullet +}$ radicals with ferrocenes has been hindered by very sluggish electron-transfer kinetics. Reinking et al. (7) showed that the substitutional lability of the methyl isocyanide ligands of the palladium and platinum dimers differ significantly. For the palladium dimer, it is not possible to distinguish the axial and equatorial ligands in MeCN, acetone, or methylene chloride at 25 °C by ^1H NMR spectroscopy, a result indicating rapid exchange on the NMR time scale. However, the 25 °C ^1H NMR spectrum of $Pt_2(CNMe)_6(BF_4)_2$, does show both axial and equatorial isocyanide methyl signals in the same solvents. The significantly slower ligand-exchange kinetics for $Pt_2(CNMe)_6(BF_4)_2$ may be manifested in the electron-transfer kinetics of $[Pt(CNMe)_3]^{\bullet +}$. If the mechanism of electron transfer from a ferrocene donor to a photogenerated $[Pt(CNMe)_3]^{\bullet +}$ radical requires preequilibrium dissociation of CNMe, similar to that of the palladium analog, then the electron-transfer kinetics for $[Pt(CNMe)_3]^{\bullet +}$ will probably be retarded by a significantly lower dissociation constant.

Photodeposition of Palladium and Platinum Films by Electron Transfer

An important goal of our research has been to develop new means of preparing conducting materials at solid–solid interfaces with submicrometer-scale resolution. Possible applications include the fabrication of metal contact lines, metal mask repair, and localized fine-scale etching. "Laser writing" of fine metal lines is an area of growing technological importance and is finding application in the fabrication of integrated circuits (*2–4*) and in packaging and interconnection of integrated circuits (*5*). Despite their cost, palladium and platinum are often the metals of choice for semiconductor metallization. In part, the reason for this choice is that these metals do not readily diffuse into the bulk of the semiconductor (*31*). In addition, platinum silicides (*32*) and palladium silicides (*33*) both form excellent ohmic contacts to silicon, and both have a close lattice match to Si (111) for good epitaxy (*34, 35*). PtSi can also be used to form Schottky barriers in polysilicon (*36*). Neither metal, by itself, is suitable for contacts to GaAs, because of the formation of a number of intermetallic phases at the interface (*37, 38*). However, Ti–Pd–Au and Ti–Pt–Au alloys are effective for both ohmic and Schottky contacts (*39, 40*).

Several advantages of $M_2(CNR)_6(PF_6)_2$ (M = Pd or Pt) as a metal-film precursor are apparent. The deposition can be achieved by using any of a number of available lasers below 360 nm, notably the third harmonic of the Nd:YAG laser at 355 nm. There are no known competing photoreactions or side reactions. Substrate heating is minimized both by the presence of the solvent and by the comparatively low powers required to effect the photolyses. Side reactions such as etching that are typically associated with gas-phase pyrolysis reactions are thus avoided (*41*). Most importantly, unreacted $[Pd(CNMe)_3]^{\bullet +}$ recombines efficiently and rapidly (typically in less than 1 ms) to regenerate the parent dimer, thus avoiding diffusion of the reactive species out of the illuminated area and eliminating loss of unreduced palladium.

Metallization by laser lithography is now a well-developed field, and laser deposition schemes have already been developed for many metals (*42*). However, given the vulnerability of III–V semiconductors to sample heating associated with pyrolytic deposition, no fully satisfactory method for direct optical writing of palladium and platinum can yet be said to exist.

Laser deposition schemes for most metals depend on gas-phase precursors. Although MOCVD (metal–organic chemical vapor deposition) methods for depositing unpatterned Pd and Pt films have been developed (*43*), only one report (*44*) of gas-phase optical writing for platinum (a pyrolytic technique) has appeared, and none for palladium. A small number of solution-phase techniques, mostly pyrolytic, for optical writing with these metals are listed in reference 31. An additional pyrolytic method was recently reported by Partridge and Chen of IBM (*45*).

Ion beams have also been used for direct writing of palladium and platinum; the deposition mechanism is, of course, pyrolytic (*46*).

Montgomery and Mantei (*47, 48*), in collaboration with our research group, succeeded in laser-writing palladium lines onto transparent substrates from methanol–acetonitrile solutions of $[Pd_2(CNMe)_6]^{2+}$. Light projected through the back side of a transparent substrate generated $[Pd(CNMe)_3]^{\bullet +}$ radicals in solution at the solution–substrate interface. The mechanism by which these radicals were reduced is unclear; the authors suggest the oxidation of methanol. A limitation of the system is that because the photochemical process occurs in bulk solution (yielding palladium colloid), the use of transparent substrates is mandatory. Clearly, this restriction is unacceptable on any practical scheme for metallization of semiconductors.

A method exists to ensure that reduction of metal occurs only at the semiconductor surface and not in bulk solution. The semiconductor is biased to a potential that will reduce the photoproduct but not its parent compound. This photoelectrochemical technique was used by Micheels et al. (*49*) in the only reported example of direct optical writing of palladium onto Si. The p-type silicon wafer, immersed in an aqueous palladium–cyanide solution, was biased between –1.0 and –2.0 V vs. a Pt counter electrode and illuminated with a halogen lamp to yield patterned palladium films.

Photodeposition of Metal Films onto Semiconductor Substrates

The enhanced redox properties of $[Pd(CNMe)_3]^{\bullet +}$ radicals compared to $Pd_2(CNMe)_6(PF_6)_2$ provide the opportunity for producing spatially well-resolved metallic films. The cyclic voltammogram of $Pd_2(CNMe)_6(PF_6)_2$ at a glassy carbon electrode in acetonitrile exhibits two irreversible reductions at $E = -0.9$ and $-1.1V$ vs. SCE. The electrochemical properties of photogenerated $[Pd(CNMe)_3]^{\bullet +}$ radicals can be conveniently monitored by transient photovoltammetry. Irradiation of the analyte–supporting electrolyte solution at the working electrode by a 1000-W Hg–Xe arc modulated at 13 Hz provided a convenient method to measure photocurrents derived from the formation of organometallic radicals. A block diagram of the instrumentation used to measure photocurrents and for the photodeposition of metal films is shown in Figure 4.

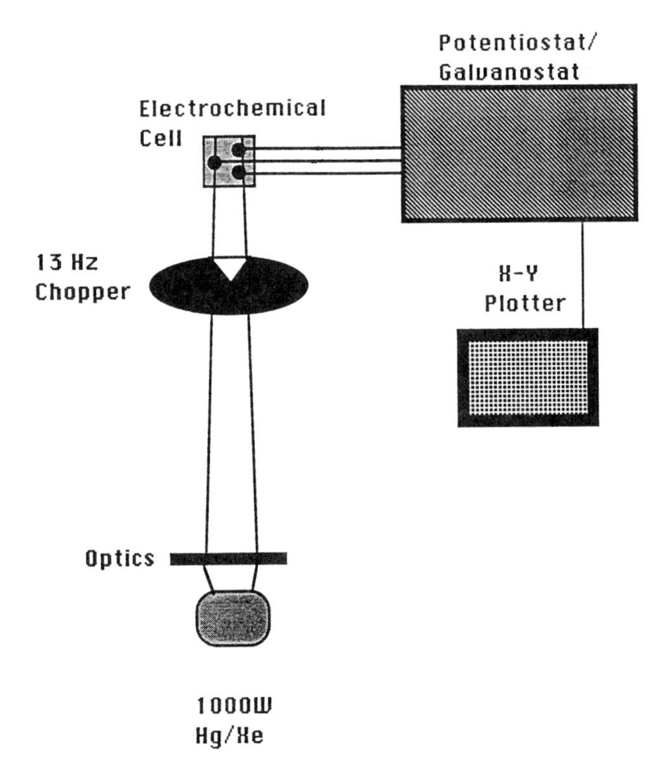

Figure 4. Block diagram for photovoltammetric instrumentation used in the photodeposition of palladium and platinum films on metallic and semiconducting electrode surfaces.

Chopped (13 Hz) UV irradiation of a glassy carbon electrode at –0.1 V vs. SCE in an electrochemical cell containing $Pd_2(CNMe)_6(PF_6)_2$ (1 mM) in 0.1 M NBu_4PF_6 led to cathodic transient photocurrents of 0.5 mA and the deposition of palladium films on the electrode surface. Maximum photocurrents in the absence of $Pd_2(CNMe)_6(PF_6)_2$ were less than 2% of those in the presence of the dimer. Analysis by X-ray photoelectron spectroscopy (XPS) of palladium films deposited onto glassy carbon indicated the presence of metallic palladium ($3d_{5/2}$, 335.8 eV and $3d_{3/2}$, 341.1 eV) of high purity (Figure 5).

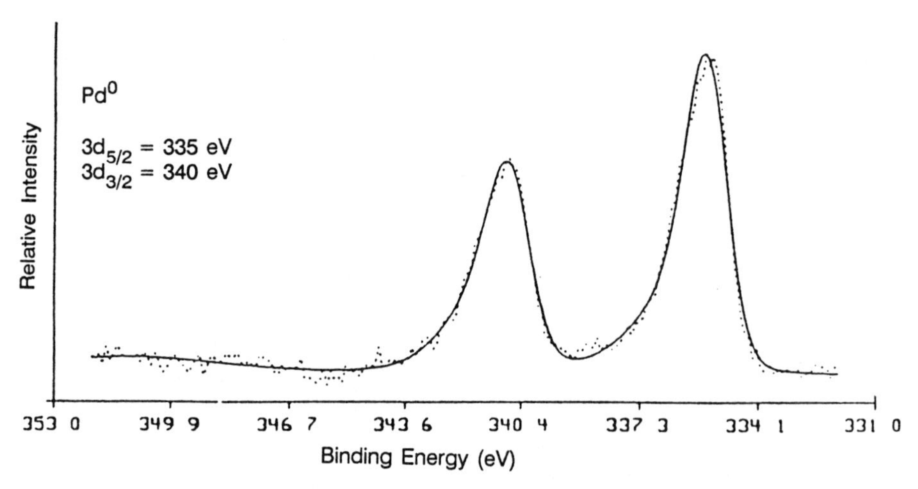

Figure 5. X-ray photoelectron spectrum of palladium metal film on glassy carbon.

Similar experiments with an n-Si (8.00–12.0 Ω-cm, (100)) working electrode at –0.5 V vs. SCE also resulted in palladium films. The palladium adhering to the electrode's surface is predominantly metallic ($3d_{5/2}$, 334.7 eV and $3d_{3/2}$, 340.1 eV) (Figure 6). When the experiment is repeated in the absence of light, no palladium was detectable on the surface of Si by energy-dispersive X-ray (EDX) spectrometry. Thus, photogenerated palladium radicals, not the dimer, are responsible for the films. A key result is that the photodeposition of metal films can be achieved at potentials that are far less reducing than those required to reduce the parent ground-state dimers because of the significant enhancement of the redox characteristics of the radicals.

Films containing platinum have been fabricated with $Pt_2(CNMe)_6$-$(BF_4)_2$. The cyclic voltammogram of the platinum dimer has two irrever-

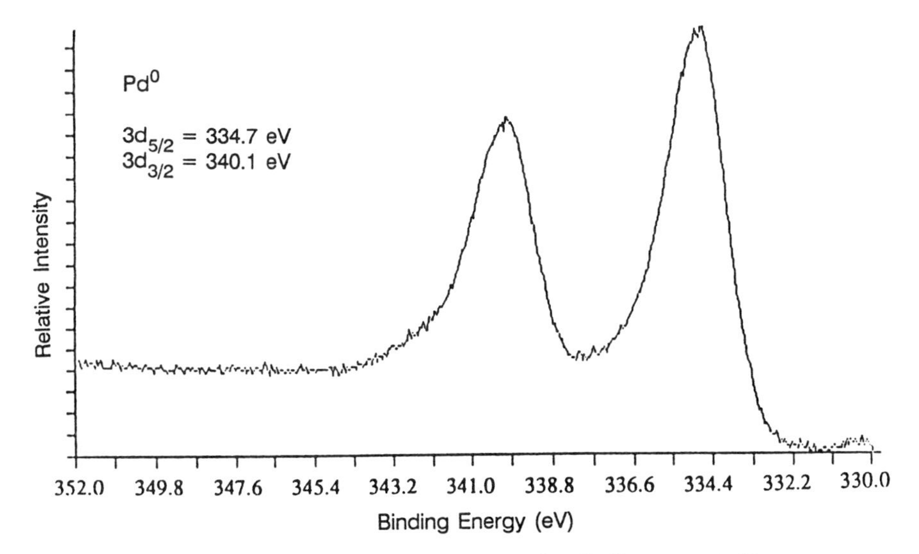

Figure 6. X-ray photoelectron spectrum of palladium metal film on n-silicon.

sible reductions at −1.00 and −1.45 V vs. a AgCl–Ag electrode. Irradiation of a glassy carbon electrode at −0.1 V vs. SCE in the presence of $Pt_2(CNMe)_6(BF_4)_2$ in 0.1 M $[NBu_4][BF_4]$–MeCN afforded transient photocurrents of 0.25 mA, greater than twice those in the absence of the platinum dimer. Films containing platinum resulted. Analysis by XPS of the films deposited onto glassy carbon indicated the presence of platinum oxide ($4f_{7/2}$, 73.4 eV and $4f_{5/2}$, 76.6 eV). Figure 7 shows the XPS spectrum indicating the presence of PtO in the films. The formation of the platinum oxide occurred from post-experiment exposure to air.

Conclusion

The photodeposition of palladium and platinum films by electron transfers to organometallic radicals derived from photolysis of $M_2(CNMe)_6(PF_6)_2$ (M = Pd or Pt) affords the opportunity to achieve photolytic metallization at relatively long wavelengths and high quantum efficiencies. The method produces high quality films of useful metals with minimal sample heating and has the potential for good spatial resolution. As fundamental understanding of the electron-transfer chemistry of photogenerated organometallic radicals continues to evolve, important new advances in optically patterned metal-film images can be expected to develop.

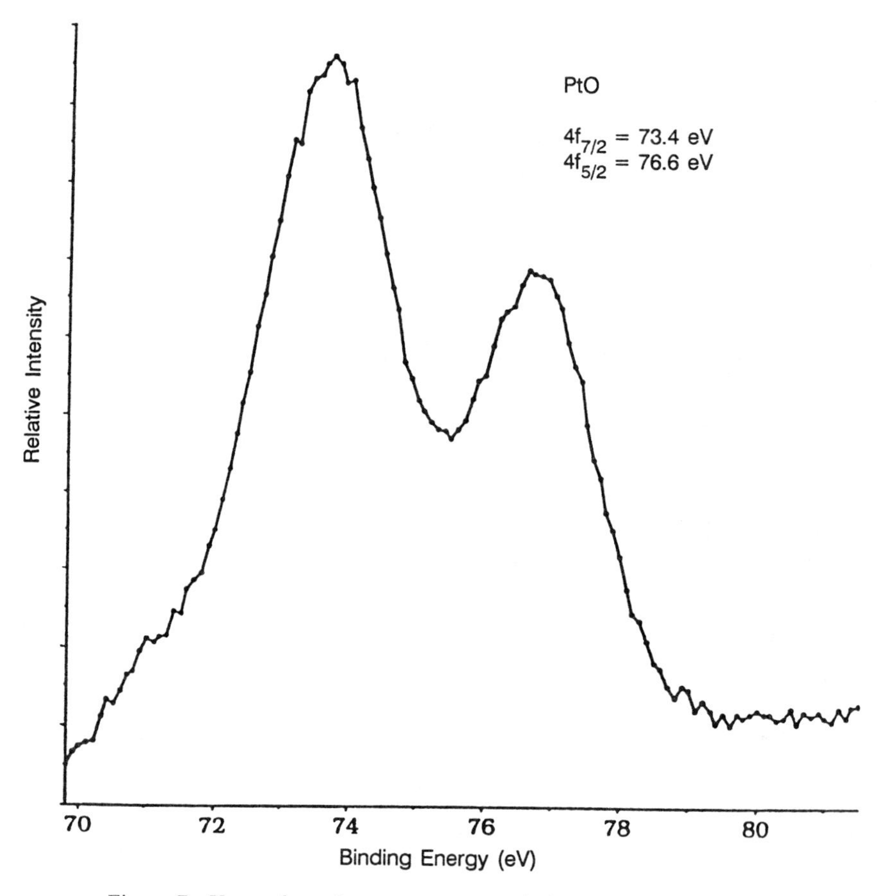

Figure 7. X-ray photoelectron spectrum of platinum oxide on glassy carbon.

Acknowledgements

This work was supported by the National Science Foundation (NSF) and the National Aeronautics and Space Administration (NASA). We acknowledge the important early contributions to the chemistry of photogenerated palladium and platinum radicals made by Marty St. Clair, Terry Miller, Marc Kullberg, Mark Reinking, and Patty Metcalf. We are grateful to Lynn Eshelman and Brett Cowans at Purdue for XPS analysis of metal films. R. M. Granger acknowledges a NASA graduate fellowship. C. P. Kubiak acknowledges a Research Fellowship from the Alfred P. Sloan Foundation, 1987–1991.

References

1. Murarka, S. P.; Peckerar, M. C. In *Electronic Materials: Science and Technology;* Academic Press: Orlando, FL, 1989; pp 566-572.
2. Ehrlich, D. J.; Tsao, J. Y. *J. Vac. Sci. Technol. B* **1983,** *1,* 969.
3. Gross, M. E.; Appelbaum, A.; Gallagher, P. K. *J. Appl. Phys.* **1987,** *61,* 1628.
4. Miracky, R. F. *Laser Focus World* **1991,** *27,* 85.
5. Meyer, T. J.; Caspar, J. V. *Chem. Rev.* **1985,** *85,* 187.
6. Balch, A. L.; Doonan, D. J.; R., B. J. *J. Am. Chem. Soc.* **1976,** *98,* 4845.
7. Reinking, M. K.; Kullberg, M. L.; Cutler, A. R.; Kubiak, C. P. *J. Am. Chem. Soc.* **1985,** *107,* 3517.
8. Lemke, F. R. Ph.D. Thesis, Purdue University, 1988.
9. Metcalf, P. A.; Kubiak, C. P. *J. Am. Chem. Soc.* **1986,** *108,* 4682.
10. Lemke, F. R.; Granger, R. M.; Morgenstern, D. A.; Kubiak, C. P. *J. Am. Chem. Soc.* **1990,** *112,* 4052.
11. Marcus, R. A. *Annu. Rev. Phys. Chem.* **1964,** *15,* 155.
12. Marcus, R. A. *J. Chem. Phys.* **1965,** *43,* 679.
13. Marcus, R. A. *J. Chem. Phys.* **1968,** *72,* 891.
14. Marcus, R. A.; Sutin, N. *Inorg. Chem.* **1975,** *14,* 213.
15. Agmon, N.; Levine, R. D. *Chem. Phys Lett.* **1977,** *52,* 197.
16. Agmon, N. *Int. J. Chem. Kinet.* **1981,** *13,* 333.
17. Levine, R. D. *J. Phys. Chem.* **1979,** *83,* 159.
18. Levine, R. D. *J. Phys. Chem.* **1979,** *83,* 159.
19. Gord, J.; Freiser, B. *Anal. Chim. Acta* **1989,** *225,* 11.
20. Bao, F.; Kubiak, C. P. *Photochem. Photobiol.* **1992,** *55(4),* 479.
21. Caspar, J. V. *J. Am. Chem. Soc.* **1985,** *107,* 6718.
22. Malatesta, L. *J. Chem. Soc.* **1955,** 3924.
23. Fischer, E. O.; Werner, H. *Chem. Ber.* **1962,** *95,* 703.
24. Bock, C. R.; Connor, J. A.; Gutierrez, A. R.; Meyer, T. J.; Whitten D. G.; Sullivan B. P.; Nagle, J. K. *J. Am. Chem. Soc.* **1979,** *101,* 4815.
25. Balzani, V.; Scandola, F.; Orlandi, G.; Sabbatini, N.; Irdelli, M. T. *J. Am. Chem. Soc.* **1981,** *103,* 3370.
26. Sutin, N. *Acc. Chem. Res.* **1982,** *15,* 275.
27. Rehm, D.; Weller, A. *Isr. J. Chem.* **1970,** *8,* 259.
28. Scandola, F.; Balzani, V. *J. Am. Chem. Soc.* **1979,** *101,* 6140.
29. McManis, G. E.; Nelson, R. M.; Gochev, A.; Weaver, M. J. *J. Am. Chem. Soc.* **1989,** *111,* 5533.
30. Yang, E. S.; Chan, M.-S.; Wahl, A. C. *J. Phys. Chem.* **1980,** *84,* 3094.
31. Haigh, J.; Aylett, M. R. In *Laser Microfabrication: Thin Film Processes and Lithography;* Ehrlich, D. J.; Tsao, J. Y., Eds.; Academic Press: Orlando, FL, 1988; pp 453.
32. Murarka, S. P.; Kinsbron E.; Fraser, D. B.; Andrews, J. M.; Lloyd E. J. *J. Appl. Phys.* **1983,** *54,* 6943.
33. Singh, R. N.; Skelly D. W.; Brown, D. M. *J. Electrochem. Soc.* **1986,** *133,* 2390.
34. Ho, P. S.; Tan, T. Y.; Lewis J. E.; Rubloff G. W. *J. Vac. Sci Technol.* **1979,** *16,* 1120.

35. Ottaviani, G. *J. Vac. Sci. Technol.* **1979,** *16,* 1112.
36. Sagara, K.; Tamaki, Y. *J. Electrochem. Soc.* **1991,** *138,* 616.
37. Lin, J.-C.; Schulz, K. J.; Hsieh, K.-C.; Chang Y. A. *J. Electrochem. Soc.* **1989,** *136,* 3006.
38. Kobayashi, A.; Sakurai, T.; Hashizumi, T.; Sakata, T. *J. Appl. Phys.* **1986,** *59,* 3448.
39. Sharma, B. L. In *Metal–Semiconductor Schottky Barrier Junctions and Their Applications;* Sharma, B. L., Ed.; Plenum: New York, 1984; p 139.
40. Welch, B. M.; Nelson, D. A.; Shen, Y. D.; R., V. In *VLSI Electronics Microstructure Science;* Einspruch, N. G.; Cohen, S. S.; Gildenblat, G. S., Eds.; Academic Press: Orlando, FL, 1987; Vol. 15; p 420.
41. Kowalczyk, S. P.; Miller D. L. *J. Appl. Phys.* **1986,** *59,* 287.
42. Rytz-Froidevaux, Y. R.; Salathé, R. P.; Gilgen, H. H. *Appl. Phys. A* **1985,** *37,* 121.
43. Gozum, J. E.; Pollina, D. M.; Jensen, J. A.; Girolami, G. S. *J. Am. Chem. Soc.* **1988,** *110,* 2688.
44. Mingxin, Q.; Monot, R.; Van den Bergh, H. *Scientia (Peking) (Engl. Transl.)* **1984,** *27,* 531.
45. Partridge, J. P.; Chen, C. J. *MRS Symp. Proc.* **1991,** *203,* 375.
46. Gross, M. E.; Brown, W. L.; Harriott, L. R.; Cummings K. D.; Linnros, J.; Funsten, H. *J. Appl. Phys.* **1989,** *66,* 1403.
47. Mantei, T. D.; Montgomery, R. K. *Proc. Electrochem. Soc.* **1986,** *86,* 478.
48. Montgomery, R. K.; Mantei T. D. *Appl. Phys. Lett.* **1986,** *48,* 493.
49. Micheels, R. H.; Darrow, A. D. I.; Rauh, R. D. *Appl. Phys. Lett.* **1981,** *39,* 418.

RECEIVED for review November 7, 1991. ACCEPTED revised manuscript May 4, 1992.

10

Photoinduced Electron-Transfer Reactions between Excited Transition Metal Complexes and Redox Sites in Enzymes

Itamar Willner and Noa Lapidot

Department of Organic Chemistry, The Hebrew University of Jerusalem, Jerusalem 91904, Israel

Photogenerated reduced relays act as electron mediators in biocatalyzed photosynthetic biotransformations. Photogenerated N,N′-*dimethyl-4,4′-bipyridinium radical cation acts as diffusional electron carrier for the enzymes nitrate reductase and nitrite reductase. In photochemical assemblies that include these biocatalysts, photosensitized reduction of nitrate to nitrite and of nitrite to ammonia proceeds with quantum yields of* $\Phi = 0.08$ *and* 0.06*, respectively. In a photochemical system that includes the two biocatalysts, photoreduction of nitrate to ammonia proceeds with a quantum yield of* $\Phi = 0.08$*. Electrical communication of protein active sites with photoexcited species is also accomplished in organized redox assemblies. Immobilization of nitrate reductase in a functionalized bipyridinium acrylamide–acrylamide copolymer results in an organized assembly exhibiting electrical communication that leads to photosensitized reduction of nitrate to nitrite. Chemical modification of the enzyme glutathione reductase by bipyridinium electron-relay components generates a functionalized protein that acts as an electron-transfer quencher of excited species and is electrically wired toward biocatalyzed reduction of oxidized glutathione.*

DEVELOPMENT OF ARTIFICIAL PHOTOSYNTHETIC SYSTEMS has been the subject of extensive research in the past two decades (*1–4*). A basic configuration of an artificial photosynthetic system is shown in Fig-

0065–2393/93/0238–0185$07.50/0

ure 1 and includes a photosensitizer (S), an electron acceptor (A), and an electron donor (D). Excitation of the photosensitizer is followed by an electron-transfer (ET) quenching process resulting in the photoproducts S^+ and A^-. Oxidation of the electron donor by the oxidized photosensitizer recycles the light-harnessing compound, and the net conversion of light energy to redox potential stored in A^- and D^+ is completed. Subsequent utilization of the intermediate redox products in chemical transformations generating fuel products or valuable chemicals is a major challenge in this area.

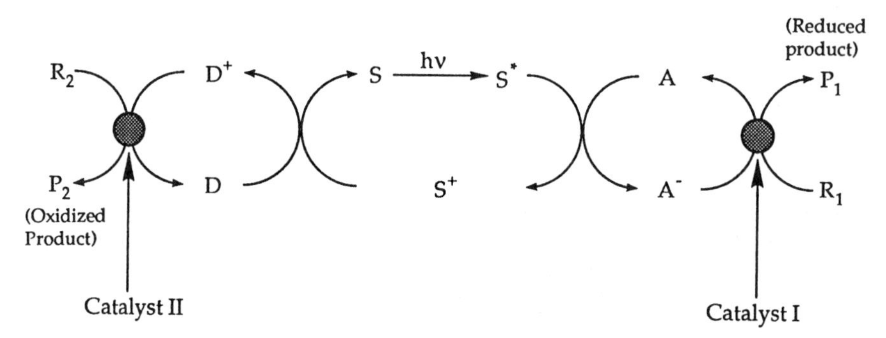

Figure 1. A general scheme of an artificial photosynthetic device. Key: S, photosensitizer; A, electron acceptor; D, electron donor; R_1 and R_2, reactants (substrates) of the reduction and oxidation processes, respectively; and P_1 and P_2, respective products of these processes.

Catalysts are essential ingredients for coupling the redox products to subsequent chemical reactions. Heterogeneous and homogeneous catalysts have been widely applied in artificial photosynthetic systems (*5–8*). Enzymes constitute nature's catalytic biomaterial and have undergone evolutionary optimization to yield very effective catalytic assemblies. Thus coupling of enzymes to artificial photosynthetic systems is an attractive route to follow (*9*).

Coupling of enzymes to artificial electron (or hole) carriers is essential for developing biocatalyzed photosynthetic systems. It is, however, very difficult to predict whether a certain electron (or hole) carrier is capable of communicating with the biocatalyst active site by accepting or donating electrons. Theory defines the energy and geometric restrictions for ET between an electron donor–acceptor pair (*10, 11*). The rate constant for ET from a donor to acceptor unit, k_{ET}, is given by the Marcus equation:

$$k_{ET} \propto e^{-\beta(d-3)} e^{(-\Delta G^\circ+\lambda)^2/4RT\lambda} \tag{1}$$

where β is a constant characterizing the donor–acceptor pair, d is the distance between the components, ΔG^o is the free energy driving force for the reaction, R is the gas constant, T is absolute temperature, and λ is the reorganization energy associated with the process. Thus for effective ET from an electron carrier to the enzyme active site (or to a hole carrier from the active site), proper distances between the two redox sites must be maintained. As a result, the charge carrier must penetrate the protein backbone in order to facilitate electron transfer. Size, electrical properties, and hydrophobicity of the charge carrier regulate its capability to penetrate the protein backbone.

In the past few years we have developed a series of artificial photosynthetic systems in which photogenerated charge relays are coupled to biocatalysts (*9*). We showed that the photogenerated 4,4′-bipyridinium radical cation, methyl viologen radical, $MV^{+\bullet}$, communicates with the enzymes lipoamide dehydrogenase, LipDH, or ferredoxin reductase, FDR, and effects the regeneration of NADH and NADPH cofactors, respectively (eqs 2 and 3).

$$2MV^{+\bullet} + H^+ + NAD^+ \xrightarrow{\text{LipDH}} 2MV^{2+} + NADH \quad (2)$$

$$2MV^{+\bullet} + H^+ + NADP^+ \xrightarrow{\text{FDR}} 2MV^{2+} + NADPH \quad (3)$$

The photogeneration of the NAD(P)H cofactors allows the utilization of numerous redox enzymes that are NAD(P)H-dependent and do not communicate with $MV^{+\bullet}$ as catalysts in the artificial systems. The photoregenerated cofactors have been coupled to various biotransformations such as reduction of ketones and keto acids, production of amino acids, and carboxylation of keto acids (*12–14*). A second method has involved the direct electron-transfer communication of photogenerated electron carriers with the biocatalyst active site (i.e., excluding the natural cofactor of the enzyme from the system). By this method reduction of CO_2 to formate proceeds in the presence of formate dehydrogenase, ForDH, and various photogenerated 4,4′-bipyridinium and 2,2′-bipyridinium radical cations, $V^{+\bullet}$ (*15*):

$$2V^{+\bullet} + H^+ + CO_2 \xrightarrow{\text{ForDH}} 2V^{2+} + HCO_2^- \quad (4)$$

In all of these systems, electron-transfer communication between the charge carrier and the biocatalyst active site proceeds by a diffusional mechanism whereby the electron carrier penetrates the protein backbone. In all of these systems, one-electron carriers, that is, bipyridinium radical cations, mediate multielectron reductive biotransformations. The conversion of the one-electron-transfer reagent to a multielectron pool occurs by discharge of electrons from the one-electron-transfer mediator to the enzyme active site. Typical pathways for electron discharge involve sequential one-electron-transfer reactions to iron–sulfur clusters acting as active sites or one-electron reduction of protein-associated cofactors, that is, flavins, followed by disproportionation of the one-electron reduced cofactor to a two-electron-relay system.

Further development in this area might include organization of biocatalytic assemblies that can directly interact with excited species. This concept is schematically outlined in Figure 2. The biocatalyst is associated with a charge-carrier component R. Within this supramolecular configuration the mutual distances of electron carrier–active site are suffi-

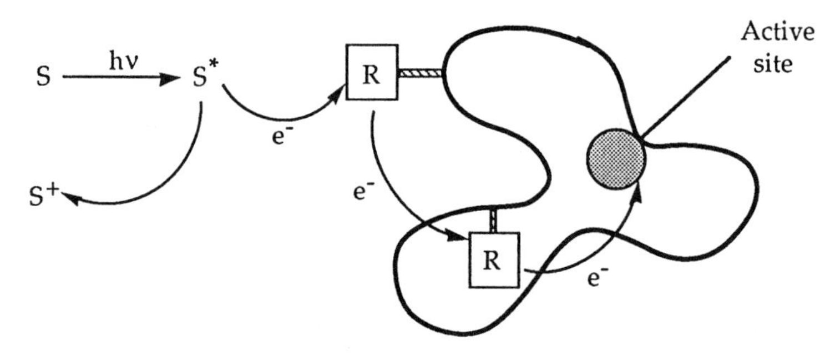

Figure 2. A scheme of an organized biocatalytic assembly capable of interacting directly with light-excited species. Key: S, photosensitizer; and R, charge carrier.

ciently short to allow effective ET rates (eq 1). As a result ET quenching of the excited species by the charge carrier followed by electron tunneling to the biocatalyst active site yield a route to communicate the shielded protein redox site with an external excited species. Such a predesigned organized biocatalytic assembly has several clear advantages over the diffusional electrical communication route:

1. Organization of the relay-active site in proper distances for effective ET abolishes the different effects controlling the permeabilities of charge carriers across proteins. Thus, a variety of charge carriers could electrically communicate with protein redox sites.

2. From a practical point of view, an artificial photosynthetic device should exhibit high turnover numbers, and its ingredients should be recycled and stored. Macromolecular electron-relay assemblies provide recyclable components.

Organization of proteins in assemblies capable of interacting electrically with their environment is a subject of growing interest in the development of biosensors (*16, 17*). Chemical modification of proteins by charge carriers (*18, 19*) or association of proteins to redox polymers (*20, 21*) leads to organized biocatalytic assemblies exhibiting electrical communication with electrode interfaces. These approaches can be further developed in relation to artificial photosynthetic devices, provided that the charge-conducting element (R) can initiate the photoinduced ET process by quenching the excited species.

In this chapter we describe a photosystem for the stepwise reduction of nitrate (NO_3^-) to ammonia (NH_3) using a multienzyme system that operates by a diffusional mechanism of the electron carrier (*22*). We also highlight organized electrically wired biocatalytic assemblies that directly interact with the excited species of the photosystem.

Photosystem for Reduction of Nitrate to Ammonia

Nitrate (as opposed to nitrogen) is the major source of the ammonia produced photosynthetically in nature. Two enzymes are involved in this reduction process: nitrate reductase catalyzes the two-electron reduction of nitrate to nitrite (eq 5), and nitrite reductase catalyzes the six-electron reduction of nitrite to ammonia (eq 6).

$$NO_3^- + 2e^- + 2H^+ \longrightarrow NO_2^- + H_2O \quad (5)$$

$$NO_2^- + 6e^- + 8H^+ \longrightarrow NH_4^+ + 2H_2O \quad (6)$$

In nature both of the enzymes are NADPH-dependent. The two enzymes can accept electrons from the synthetic electron-carrier radical cation $MV^{+\bullet}$. We therefore attempted to develop an enzyme-based artificial photosynthetic system for the reduction of nitrate to ammonia operating at ambient temperature and pressure. This goal was achieved in three complementary steps.

Step 1. The Reduction of Nitrate to Nitrite. The photosystem consists of tris(bipyridine)Ru(II), $Ru(bpy)_3^{2+}$, as the photosensitizer, $MV^{+\bullet}$ as the electron carrier, ethylenediaminetetraacetic acid (EDTA) as a sacrificial electron donor, and the enzyme nitrate reductase (EC 1.9.6.1 from *Escherichia coli* or *Aspergillus* species). Upon illumination of the system ($\lambda > 420$ nm) in the presence of the substrate nitrate, nitrite is formed, as shown in Figure 3a. The quantum yield of nitrite production is

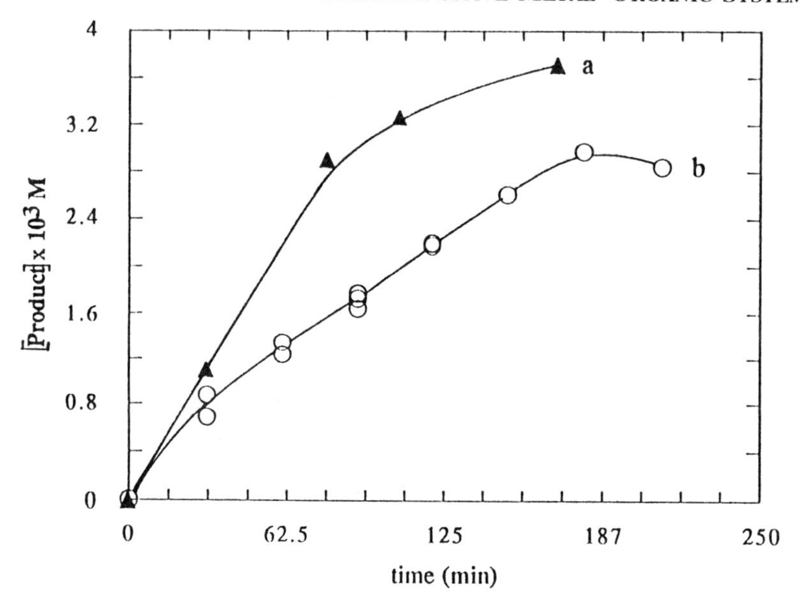

Figure 3. Rates of product formation as a function of illumination time. In all systems $[Ru(bpy)_3^{2+}] = 7.4 \times 10^{-5}$ *M,* $[Na_2EDTA] = 0.02$ *M. Curve a:* NO_2^- *formation, pH 7.0, 0.1 M Tris buffer,* 3.2×10^{-4} *M* $MV^{+\bullet}$*,* 9.9×10^{-3} *M* NO_3^-*, and 0.2 unit of nitrate reductase. Curve b:* NH_4^+ *formation, pH 8.0, 0.1 M Tris buffer,* 4.2×10^{-4} *M* $MV^{+\bullet}$*, 0.01 M* NO_2^-*, and 0.06 unit of nitrite reductase.*

$\Phi = 0.08$. The initial rate of NO_3^- reduction corresponds to 0.7 μmol/min, and reduction of nitrate proceeds until ca. 60% of the substrate is consumed.

Step 2. The Reduction of Nitrite to Ammonia. In this system, the enzyme nitrate reductase is replaced by nitrite reductase (EC 1.6.6.4 isolated from spinach leaves), and the other components of the photosystem are the same as in Step 1. Illumination of the system in the presence of the substrate nitrite leads to its reduction to ammonia, as shown in Figure 3b. The quantum yield of this process is $\Phi = 0.06$, and the rate of ammonia production levels off when ca. 28% of the nitrite has been converted to ammonia. The leveling off in ammonia production is attributed to denaturation of the biocatalyst under long-term illumination conditions.

Control experiments reveal that all the components of the two systems are required for the reduction of NO_3^- and NO_2^-, separately. Furthermore, no reduction of either substrates takes place in the dark. When the biocatalysts are excluded from either of the two photosystems,

$MV^{+\bullet}$ is accumulated. Addition of nitrate reductase to the system that includes NO_3^- and the photogenerated $MV^{+\bullet}$ or addition of nitrite reductase to the system that contains NO_2^- and the photogenerated $MV^{+\bullet}$ results in the reduction of NO_3^- and NO_2^- to nitrite and ammonia, respectively.

Step 3. The Reduction of Nitrate to Ammonia. The photosystem is composed of tris(bipyridine)Ru(II), $Ru(bpy)_3^{2+}$, as photosensitizer, EDTA as an electron donor, and $MV^{+\bullet}$ as the electron acceptor. The enzymes nitrate reductase, NitraR, and nitrite reductase, NitriR, are both included as biocatalysts for the sequential reduction of NO_3^- to NH_3. Illumination of this system results in the reduction of NO_3^- to NH_3. The progress of NH_3 formation as well as the intermediate generation of NO_2^- at time intervals of illumination are provided in Figure 4. Figure 4 shows that only after the concentration of NO_2^- reaches ca. 2.5 mM does the second reduction process become effective and lead to a higher rate of NH_3 evolution. The quantum yield for the overall reduction of nitrate to ammonia is $\Phi = 0.08$ in the integrated system, which is higher than the quantum yield of this product in the previously described

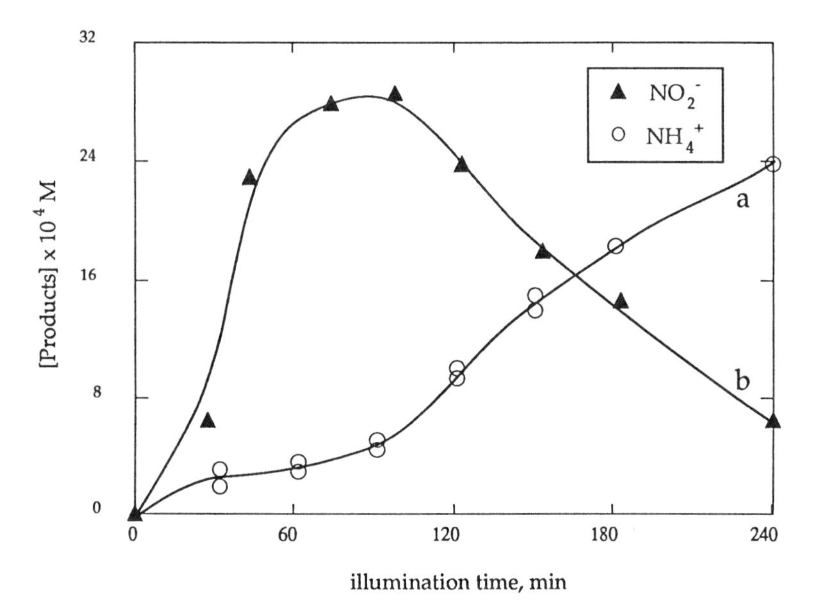

Figure 4. NO_2^- and NH_4^+ concentrations in the combined system, as a function of illumination time. Conditions: pH 8.0, 0.1 M Tris buffer, 7.4×10^{-5} M $Ru(bpy)_3^{2+}$, 0.02 M Na_2EDTA, 4.2×10^{-4} M $MV^{+\bullet}$, 0.01 M NO_3^-, 1.0 unit of nitrate reductase, and 0.35 unit of nitrite reductase.

photosystem (Φ = 0.06). This higher yield is due to the application of a higher unit content of the enzyme nitrite reductase in the integrated system (details are provided in the corresponding figure captions).

From the control experiments and the photoproducts, we suggest that the photochemically derived reduction of nitrate to ammonia proceeds through the catalytic cycles outlined in Figure 5. The artificial electron carrier $MV^{+\bullet}$ is recycled by photochemical means. It serves as an electron carrier for the two enzymes that catalyze the sequential reduction processes. The decomposition products of sacrificial electron donors, that is, EDTA, are often powerful reducing agents and might participate in electron-discharge processes to the redox enzymes. However, EDTA can be substituted in the systems by other sacrificial electron donors such as triethanolamine or 1,4-dithiothreitol. Furthermore, the rates of NO_2^- and NH_4^+ formation in the systems that apply the various electron donors follow the rates of $MV^{+\bullet}$ photogeneration in the different systems. Thus, we conclude that the sacrificial electron donors are active only in photoregeneration of $MV^{+\bullet}$, as it is the sole electron-transfer mediator for the biocatalyzed transformations.

The enzymes as well as the synthetic components comprising these systems show a considerable stability under illumination, as is evident from the turnover numbers given in Table I. Nevertheless, to further improve the stabilities of these biocatalytic assemblies and to provide a continuously operating system, organization of the components in immobilized polymer matrices is essential. This aspect will be discussed in the forthcoming sections.

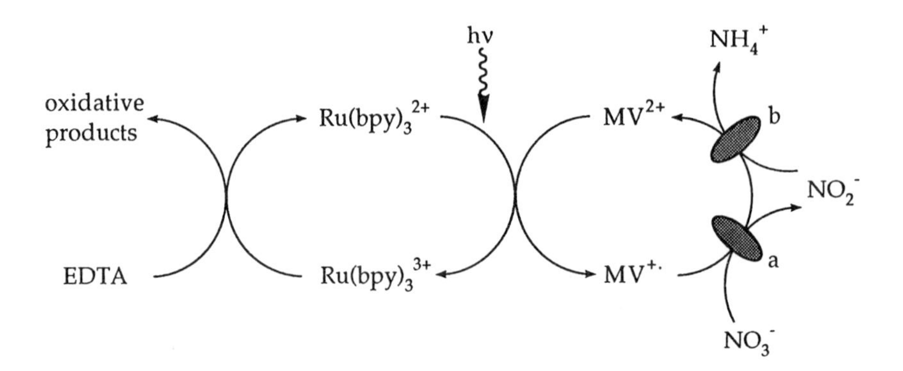

Figure 5. Biocatalyzed photosynthetic reduction of nitrate to ammonia mediated by the artificial electron carrier, $MV^{+\bullet}$, acting in a diffusive mechanism.

Table I. Turnover Numbers for Components Involved in Photosensitized Reduction of NO_3^- and NO_2^- to Ammonia

Step	$Ru(bpy)_3^{2+}$	$MV^{+\bullet}$	*Nitrate Reductase*[a]	*Nitrite Reductase*[a]
NO_3^- reduction[b]	80	18.5	6.2×10^4	
NO_2^- reduction[c]	38.5	7		2.4×10^4
NO_3^- reduction to NH_3 (combined system)[d]	32	6	9×10^4	2.1×10^4

NOTE: Turnover numbers are defined as moles of product formed divided by moles of component.

[a]Molecular weight of NitraR and NitriR is estimated as 200,000.

[b]60% conversion of NO_3^- to NO_2^-.

[c]28.5% conversion of NO_2^- to NH_4^+.

[d]23.8% conversion of NO_3^- to NH_4^+.

Photoreduction of Nitrate in a Redox Polymer–Biocatalyst Immobilized Assembly

A major problem in using enzymes in artificial photosynthetic systems is the high cost of the enzymes. It is therefore advantageous to immobilize the enzymes as well as other components of the photosystem in an organized recyclable assembly. Also, immobilization of enzymes in polymers is often accompanied by stabilization of the biocatalyst. Thus, immobilization of biocatalysts in functionalized polymers capable of interacting through electron transfer with the biocatalyst active site is an attractive route to develop further biocatalyzed artificial photosynthetic devices.

Yet, development of such systems is limited by theoretical and practical considerations. In a nonorganized assembly in which the artificial electron carrier penetrates the protein backbone by diffusion, close proximity to the active site is feasible. In contrast, in an organized rigid assembly, it is essential to consider the mutual positions of the electron carrier and biocatalyst. For effective electron-transfer rates between a polymer-bound electron carrier and the active site of an immobilized enzyme, the two components are restricted to be within a distance of ca. 20 Å (eq 1). Thus, we developed a redox polymer that contains sterically flexible redox groups that can mediate the electron transfer from a photosensitizer to the active site of the enzyme. The polymer is based on a cross-linked acrylamide backbone with pendant bipyridinium groups. The synthesis of this functionalized redox copolymer is outlined in Scheme I. Possibly, the flexible anchored bipyridinium components could lead to an adequate configurational assembly for intercomponent electron transfer within the protein backbone.

The enzyme nitrate reductase (EC 1.9.6.1 from *Aspergillus* species) is trapped inside the copolymer of acrylamide and *N*-methyl-*N'*-(3-acrylamidopropyl)-4,4'-bipyridinium (**1**). *N*-Methyl-*N'*-(3-aminopropyl)-4,4'-bipyridinium is prepared as outlined in Scheme II. Polymerization of **1** and acrylamide is initiated by adding 3-(dimethylamino)propionitrile (50% in water) and $K_2S_2O_8$ (1% in water) under anaerobic conditions. The resulting copolymer gel, exhibiting structure **2**, PV^{2+}, is ground and washed thoroughly with cold buffer solution to exclude any free bipyridinium component. The immobilized enzyme retains 55% of its native activity. The ratio of **1**:acrylamide in the copolymer is estimated as 1:35.

The photosystem is constructed by adding $Ru(bpy)_3^{2+}$ as a sensitizer and EDTA as an electron donor to an aqueous solution (pH 7.4) that includes the ground copolymer **2**, PV^{2+}, with the entrapped enzyme and NO_3^- as the substrate (*23*). Illumination (λ > 420 nm) of the immobilized system results in the reduction of NO_3^- to NO_2^-, as displayed in Figure 6 (curve a). Control experiments show that all components are essential for the photoreduction of nitrate to nitrite in this photosystem. Furthermore, during illumination the polymer becomes blue, a result revealing the generation of polymer-bound bipyridinium radical cation. The blue color persists on the polymer pieces, and no coloration of the aqueous phase is detectable. These results imply that the polymer-bound bipyridinium radical cation mediates the electron transfer to the immobilized enzyme within the copolymer assembly.

Alternatively, the reduction of NO_3^- to NO_2^- could be performed by replacing the photoactive components by a chemical reducing agent such as dithionite. Dithionite does not effect the direct reduction of nitrate. However, addition of dithionite to an aqueous solution that includes NO_3^- and the enzyme nitrate reductase immobilized in the redox copolymer results in the reduction of NO_3^- to NO_2^-.

The rate of NO_2^- formation by the chemical reductant is shown in Figure 6 (curve b) and is very similar to the rate of NO_2^- formation in the photochemical system. On the basis of the experiments described for the reduction of NO_3^- to NO_2^- in the homogeneous system using $MV^{+\bullet}$ as electron carrier, we suggest the catalytic cycle presented in Figure 7 as the mechanistic route for the reduction of nitrate in the polymer assembly. The excited photosensitizer, $^*Ru(bpy)_3^{2+}$, is quenched by the bipyridinium components anchored to the polymer backbone via an ET process. The resulting polymer-bound bipyridinium radical cation units mediate ET to the enzyme active site where the reduction of NO_3^- proceeds. Thus, the polymer-bound electron carrier units are sufficiently flexible to maintain the proper orientation and distances in respect to the active site, allowing effective ET within the rigid polymer assembly. Furthermore, as men-

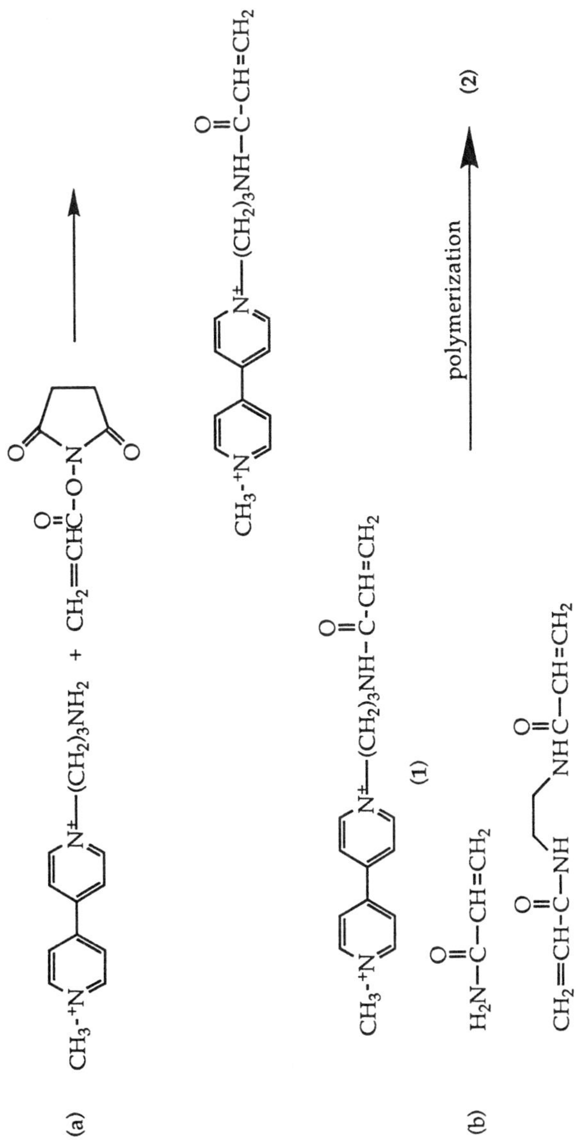

(a) $CH_3\text{-}^{+}N\langle\rangle\text{—}\langle\rangle N^{+}\text{—}(CH_2)_3NH_2 + CH_2{=}CH\overset{O}{\overset{\|}{C}}\text{-O-N}$(succinimide) ⟶ $CH_3\text{-}^{+}N\langle\rangle\text{—}\langle\rangle N^{+}\text{—}(CH_2)_3NH\text{—}\overset{O}{\overset{\|}{C}}\text{-}CH{=}CH_2$

(b) $CH_3\text{-}^{+}N\langle\rangle\text{—}\langle\rangle N^{+}\text{—}(CH_2)_3NH\text{—}\overset{O}{\overset{\|}{C}}\text{-}CH{=}CH_2$ (1)

$H_2N\text{—}\overset{O}{\overset{\|}{C}}\text{—}CH{=}CH_2$ ——polymerization——⟶ (2)

$CH_2{=}CH\text{-}\overset{O}{\overset{\|}{C}}\text{—}NH\text{—}CH_2\text{—}NH\overset{O}{\overset{\|}{C}}\text{—}CH{=}CH_2$

Scheme I. Synthesis of the functionalized redox copolymer **2**. *Reaction a: Preparation of polymerizable derivative of the bipyridinium salt (*(**1**)*). Reaction b: Copolymerization of the derivative prepared in reaction a with acrylamide and* N,N′*-methylenebis(acrylamide) as a cross-linking reagent.*

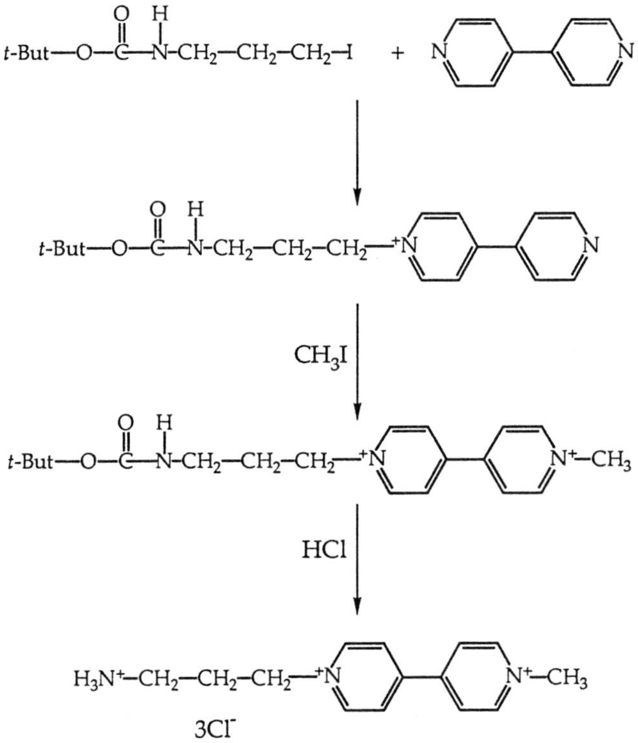

t-But—O—C(=O)—N(H)-CH_2—CH_2—CH_2—I + 4,4'-bipyridine

↓

t-But—O—C(=O)—N(H)-CH_2—CH_2—CH_2—^+N(bipyridine)N

↓ CH_3I

t-But—O—C(=O)—N(H)-CH_2—CH_2—CH_2—^+N(bipyridine)N^+—CH_3

↓ HCl

H_3N^+—CH_2—CH_2—CH_2—^+N(bipyridine)N^+—CH_3

$3Cl^-$

Scheme II. Synthetic steps in the preparation of N-*methyl*-N'-*(3-aminopropyl)-4,4'-bipyridinium. The amine end of the alkylating reagent is protected by a* tert-*butoxycarbonyl group.*

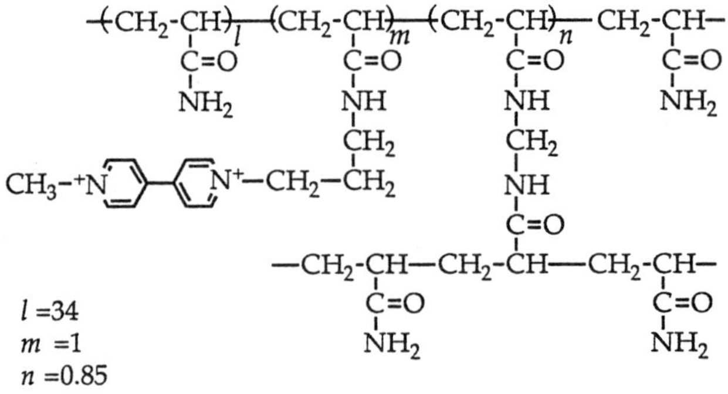

$-(CH_2\text{-}CH)_l-(CH_2\text{-}CH)_m-(CH_2\text{-}CH)_n-CH_2\text{-}CH-$

l =34
m =1
n =0.85

2

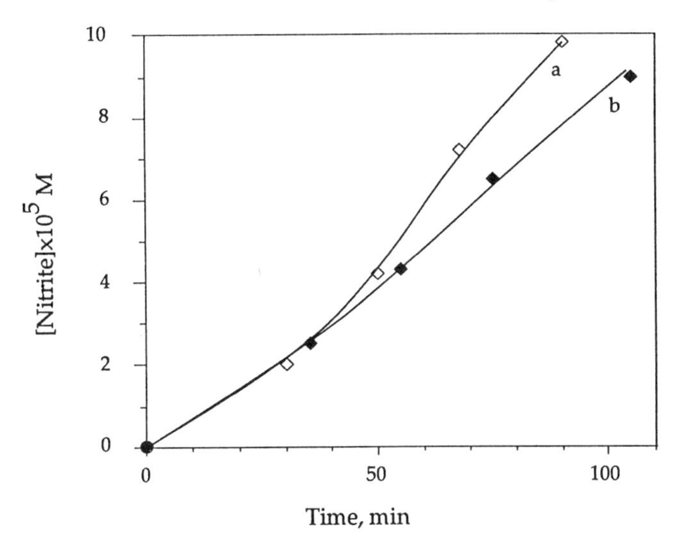

*Figure 6. Rate of NO_3^- reduction in the assembly composed of nitrate reductase (0.023 unit, based on native enzyme) immobilized in the acrylamide–**1** copolymer. In all experiments, $[NO_3^-] = 4.4 \times 10^{-3}$ M in 2.3 mL of Tris buffer, pH 7.44. Curve a: Through illumination in the presence of $Ru(bpy)_3^{2+}$ (5.5×10^{-5} M) and Na_2EDTA (4.4×10^{-3} M). Curve b: Dark reduction in the presence of 9.4×10^{-3} M sodium dithionite.*

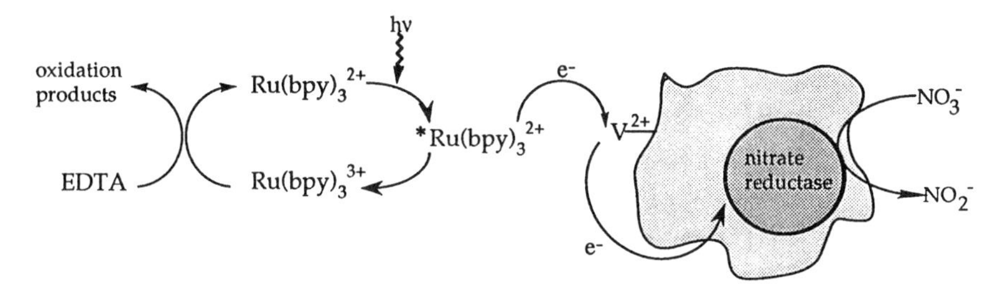

*Figure 7. Photoinduced reduction of NO_3^- to NO_2^- in an organized assembly composed of NitraR immobilized in the redox copolymer **2**. Electron transfer from the excited photosensitizer to the active site of the enzyme is mediated by the polymer-anchored bipyridinium groups.*

tioned, the rate of NO_2^- formation in the system that includes the redox polymer-immobilized enzyme is similar in the photosystem and the dark assembly including dithionite as reducing agent. These results suggest that the rate of NO_3^- reduction in the polymer assembly is limited by the ET process from the polymer to the protein and is independent in respect to the source generating the polymer-bound bipyridinium radical cations.

Insight into the rates of ET from the redox polymer to the protein could be achieved by laser flash photolysis experiments. Pulse illumination (λ = 532 nm, Nd:Yag laser) of a photosystem consisting of $Ru(bpy)_3^{2+}$ and the redox polymer (without immobilized enzyme) results in the ET products:

$$^{*}Ru(bpy)_3^{2+} + P{-}V^{2+} \longrightarrow Ru(bpy)_3^{3+} + P{-}V^{+\bullet} \quad (7)$$

These products then recombine by the thermodynamically favored back-ET process:

$$Ru(bpy)_3^{3+} + P{-}V^{+\bullet} \xrightarrow{k_b} Ru(bpy)_3^{2+} + P{-}V^{2+} \quad (8)$$

The formation of ET products and their recombination can be followed by the transient absorbance at λ = 602 nm (characteristic of polymer-bound bipyridinium radical cation) and the transient absorbance bleaching at λ = 450 nm (characteristic of $Ru(bpy)_3^{2+}$ disappearance) (Figure 8a). Excitation of the system results in the intermediate redox products that completely decay to the original reactants. From the decay transients, the back-ET reaction rate constant is calculated to be $k_b = (10 \pm 4) \times 10^6$ $M^{-1}\,s^{-1}$.

Next, a system that includes the photosensitizer, the redox polymer, and the immobilized enzyme, nitrate reductase, is analyzed. Flash illumination of this system generates again the reduced bipyridinium groups on the polymer (detected by the buildup of the absorbance at λ = 602 nm), accompanied by the decrease in the ground-state concentration of the photosensitizer (bleaching at λ = 450 nm) (Figure 8b). However, in the presence of the immobilized enzyme, the intermediate reduced polymer generated by the photoinduced ET process can decay by two possible paths. One is the back-ET reaction with the oxidized photosensitizer (eq 8). The second path involves the ET from the polymer-anchored bipyridinium radical cation to the enzyme active site:

$$V^{+\bullet}{-}P \cdots E_{ox} \xrightarrow{k_{ET}} V^{2+}{-}P \cdots E_{red} \quad (9)$$

where E_{ox} and E_{red} are the oxidized and reduced forms of the enzyme active site, respectively. This second path would result in a faster decay of the photogenerated bipyridinium radical cation as a result of the two decay processes, and only partial recovery of the absorbance at λ = 450 nm. (The recovery of the absorbance at 450 nm is due to the back reac-

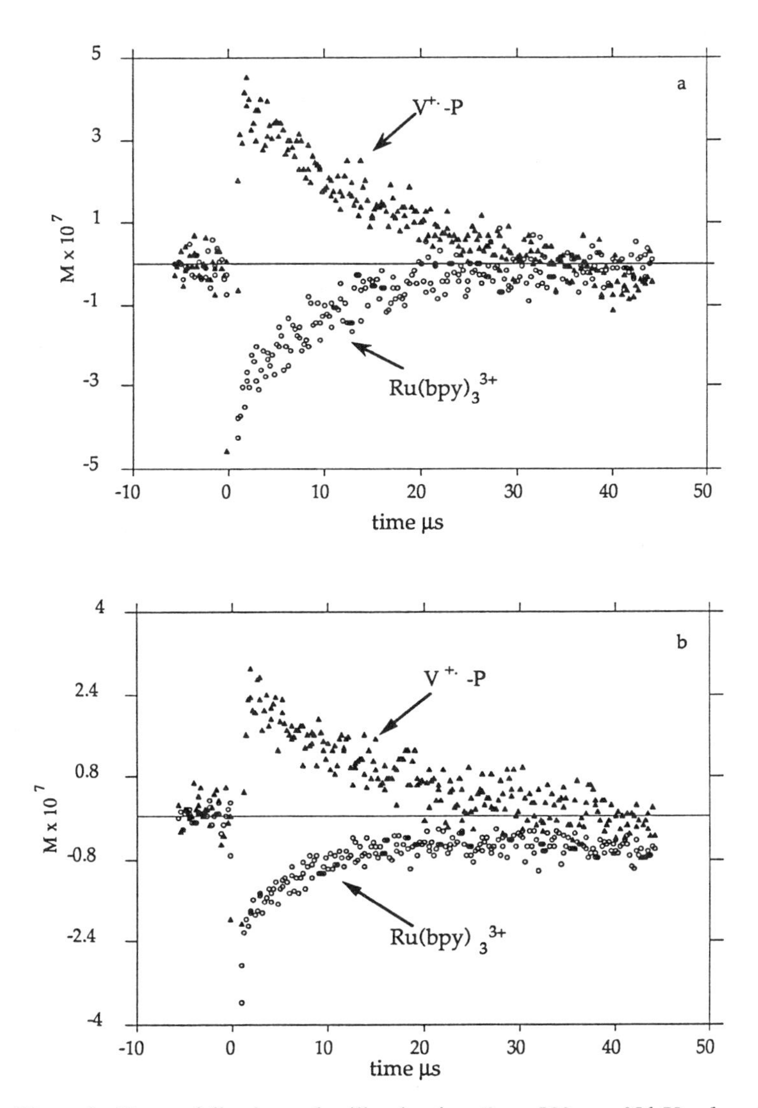

Figure 8. Events following pulse illumination (λ = 532 nm, Nd:Yag laser) of an organized photosynthetic assembly: absorbance at 602 nm, characteristic of polymer-bound bipyridinium radical cation; and absorbance bleaching at 450 nm, characteristic of $Ru(bpy)_3^{2+}$ disappearance. Part a: A system consisting of the redox copolymer and $Ru(bpy)_3^{2+}$ only. Part b: A system consisting of NitraR immobilized in the redox copolymer and $Ru(bpy)_3^{2+}$.

tion, eq 8, as in the first experiment). Figure 8b shows the decay transient of the polymer-bound bipyridinium radical cation and the transient recovery of the bleached species, $Ru(bpy)_3^{2+}$, as a result of back ET. The photogenerated polymer-bound bipyridinium radical cation decays faster than in the absence of the enzyme. On the other hand, the recovery of the bleached species, $Ru(bpy)_3^{2+}$, is not complete in the presence of the enzyme, and a steady-state concentration of $Ru(bpy)_3^{3+}$ is accumulated. This steady-state concentration of the oxidized product corresponds to the fraction of bipyridinium radical cation that reacted with the enzyme active site. (The back ET from the reduced active site to the oxidized photosensitizer is substantially slower on the experimental time scale). We thus estimate that ca. 15% of the flash-generated polymer-anchored bipyridinium radical cations transfer their electrons to the biocatalyst active site.

The decay curve of the polymer-bound bipyridinium radical cation in the presence of the enzyme can be analyzed in terms of two rate constants k_b and k_{ET} (eqs 8 and 9, respectively). Because the value of k_b is known from the system without the enzyme, and the concentration of $V^{+\bullet}$–P that recombines through back ET (eq 8) can be estimated from the transient recovery of $Ru(bpy)_3^{2+}$ in the presence of the enzyme (Figure 8b), the value of k_{ET} can be estimated. The derived value of the rate constant for ET from the polymer-anchored groups to the enzyme active site is $k_{ET} = (9 \pm 3) \times 10^5\ M^{-1}\ s^{-1}$. This value is in accordance with electron-transfer rate constants measured by Gray, Sutin and co-workers (*24*) in blue copper proteins and by Durham et al. (*25*) and McLendon et al. (*26*) in cytochromes.

Electrically Wired Enzymes by Chemical Modification of Proteins with Redox Relay Components: Direct Communication between Proteins and Excited Species

To assess the generality of using a redox polymer for electrical communication between an excited species and a redox site in proteins, other redox enzymes were incorporated in the electrically active copolymer **2**. One of these enzymes is glutathione reductase, GR. This redox enzyme is ubiquitous in nature, and its role includes the reduction of the S–S bond of oxidized glutathione, GSSG, to the reduced, GSH, using NADPH as a cofactor:

$$\text{GSSG} + \text{NADPH} + \text{H}^+ \xrightarrow{\text{GR}} 2\text{GSH} + \text{NADP}^+ \qquad (10)$$

The enzyme GR can directly communicate with the *N,N'*-dimethyl-4,4'-bipyridinium radical cation, methyl viologen, $MV^{+\bullet}$, by a diffusional mechanism, and the natural cofactor can be substituted by this artificial electron mediator. Immobilization of the enzyme GR in the redox viologen copolymer (**2**) does not lead, however, to an active biocatalytic assembly for the photoreduction of GSSG, although the immobilized enzyme maintains its native activity. Thus, electrical communication between an enzyme active site and an excited species by means of a redox copolymer matrix is not of general applicability. With nitrate reductase, effective communication is established, but the enzyme GR lacks this property. A similar difference in the ability of the two enzymes to communicate with a redox polymer is a general feature of these enzymes. When a water-soluble polymer of poly(ethyleneimine) which has been modified by attaching 4,4'-bipyridinium groups to it is used as the mediating redox polymer, the enzyme NitraR accepts electrons from it (and reduces NO_3^- to NO_2^-), whereas no communication is established with GR.

This different behavior of the two enzymes, NitraR and GR, might originate from the different positions of the active sites in the various proteins in respect to their environment, where the electron-mediating component resides. When the active site is close to the protein exterior periphery, as is the case with nitrate reductase, the resulting distances of the redox polymer and the active site allow effective ET. On the other hand, with bulky proteins that shield the active site, that is, GR, the mutual distances of the active-site ET mediator are too large and prohibit ET interactions. Thus, in order to electrically wire such bulky enzymes with their environment, the insulating shell of the protein must be transformed into a charge-conducting assembly. For this purpose, chemical modification of the protein backbone by ET mediators could generate an organized assembly where "hopping" of electrons across the electron-relay units results in the charge-conduction mechanism (Figure 9).

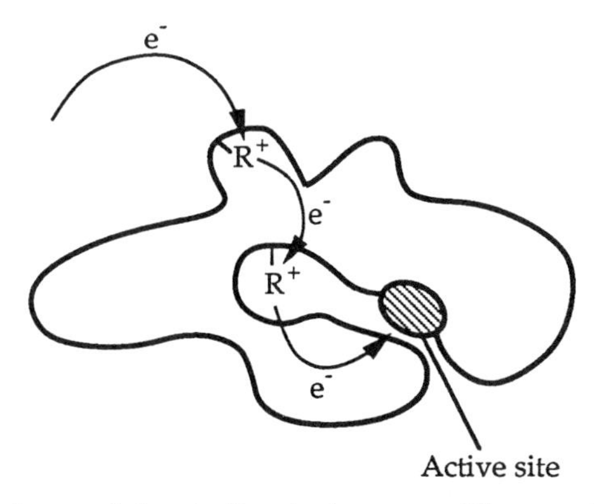

Figure 9. A scheme of electrically wired enzyme. The protein backbone is chemically modified by ET mediators, R.

We followed this approach in electrical wiring of glutathione reductase, GR (*27*). The enzyme GR (E.C 1.6.4.2 from baker's yeast, Sigmatype III) is modified by anchoring *N,N'*-bis(3-propanoic acid)-4,4'-bipyridinium, PAV (**3**), to the protein backbone. The chemical modification process is

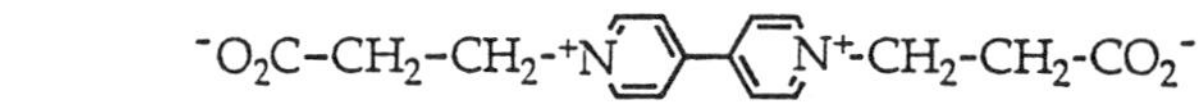

$$^{-}O_2C-CH_2-CH_2-{}^{+}N\langle\rangle-\langle\rangle N^{+}-CH_2-CH_2-CO_2^{-}$$

3

outlined in Scheme III. It involves the primary preparation of the active ester of **3** with *N*-hydroxysulfonatosuccinimide sodium salt, using 1-ethyl-3-[3-(dimethylamino)propyl]carbodiimide (EDC) as coupling agent, followed by derivatization of protein lysine residues by the active ester of PAV.

Various loading degrees of PAV on the protein are prepared by varying the concentration of PAV in the reaction medium. The modified enzyme maintains 72% of the activity of the native enzyme. Interestingly, the PAV-modified enzyme, in contrast to the native biocatalyst, does not recognize the cofactor NADPH, although it is an active protein for the reduction of GSSG, as will be discussed later. Thus, PAV-modified glutathione reductase can be considered as a novel biocatalyst for the reduction of GSSG.

The PAV-modified glutathione reductase is introduced into a photosystem that includes 6.8×10^{-5} M $Ru(bpy)_3^{2+}$ as a photosensitizer; 0.01

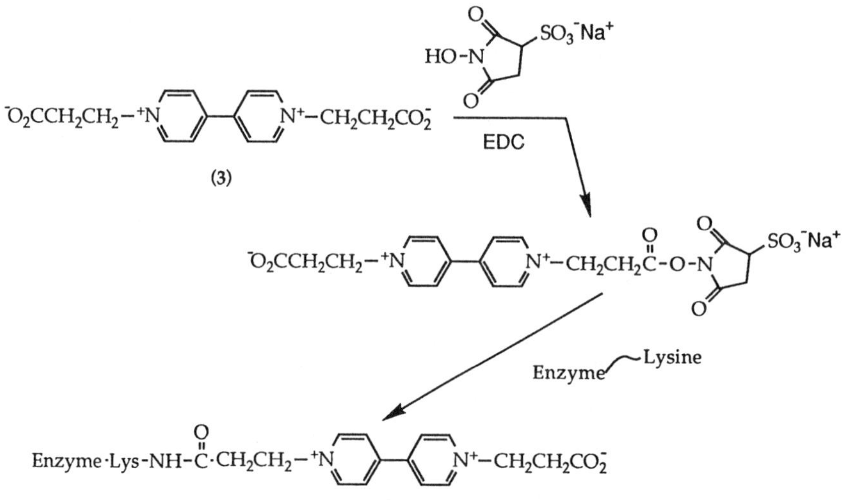

Scheme III. Chemical modification of lysine residues of an enzyme by N,N'*-bis-(3-propanoic acid)-4,4'-bipyridinium (***3***).*

M EDTA as an electron donor; and the substrate, 0.01 M GSSG. Illumination of this system, λ > 420 nm, results in the reduction of GSSG to GSH. The rate of GSSG reduction is shown in Figure 10 using biocatalysts of varying loading degrees of the PAV component. The higher the loading of the protein by PAV units, the faster is the reduction process of GSSG. Control experiments show that the biotransformation does not occur in the dark, nor does it proceed in the absence of any of the components of which the photosystem consists. The primary step in the photosensitized reduction of GSSG involves the ET-quenching of the excited photosensitizer by protein-anchored PAV units:

$$^*Ru(bpy)_3^{2+} + PAV^{+}\text{–enzyme} \xrightarrow{k_q} Ru(bpy)_3^{3+} + PVA^{\bullet}\text{–enzyme} \quad (11)$$

The quenching rate constant is determined by measuring the lifetime of the excited $Ru(bpy)_3^{2+}$ at different PAV-modified GR concentrations. The evaluated quenching rate constant corresponds to $k_q = 5.6 \times 10^9\ M^{-1}\ s^{-1}$, a value indicating that the modified enzyme effectively quenches the photosensitizer. Thus, photoreduction of GSSG in this photosystem proceeds by electrical communication of the enzyme active site and the excited species, as schematically outlined in Figure 11. The excited pho-

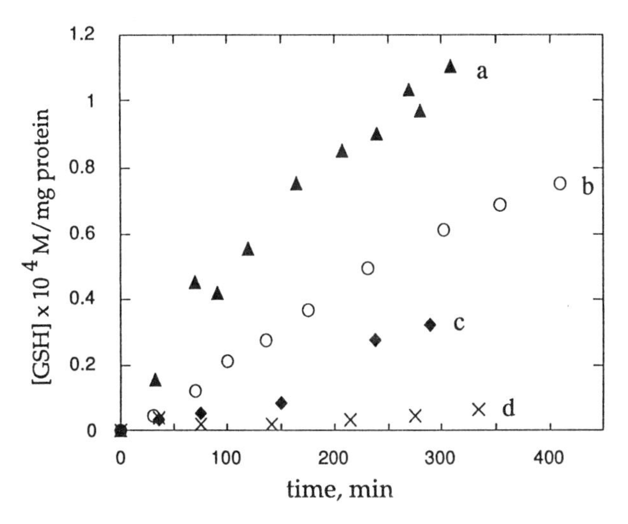

Figure 10. Rate of GSH evolution (normalized per milligram of protein) under illumination (λ > 420 nm) in a photochemical system consisting of 6.8 × 10^{-5} M $Ru(bpy)_3^{2+}$, 0.01 M EDTA, 0.01 M GSSG in 3 mL of 0.1 M phosphate buffer, pH 7.5. The loading degrees of PAV on the enzyme (mole per mole) in the various systems are (a) 3.9, (b) 1.8, (c) 1.4, and (d) 0.5.

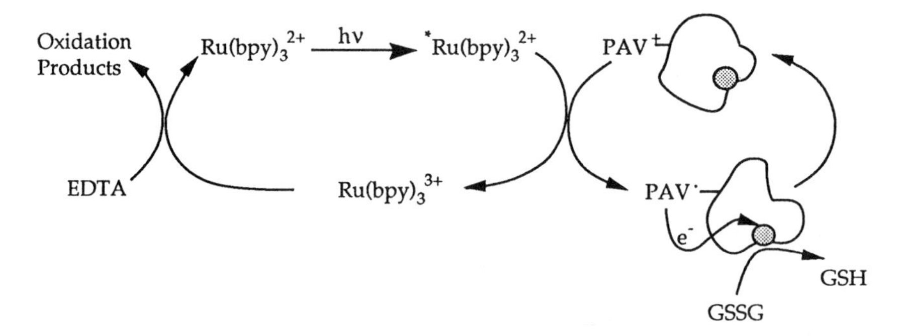

*Figure 11. Photoinduced reduction of GSSG. The biocatalyst is GR modified by the ET mediating component, PAV (**3**). The PAV groups quench the excited species and mediate ET to the active site of the enzyme.*

tosensitizer is oxidatively quenched by protein-bound PAV groups and generates reduced PAV and the oxidized photosensitizer. The oxidized photosensitizer oxidizes EDTA and recycles the light-harnessing compound. The reduced PAV groups mediate ET to the enzyme active site, where the reduction of GSSG to GSH takes place.

The loading degree of PAV on the enzyme has a marked influence on the overall reaction rate, as described earlier. The protein anchored PAV components participate in two steps of the ET process: (1) They act as ET quenchers of the excited species (eq 11), and (2) they provide the "molecular wire" and mediate ET to the active site of the enzyme. Both of these steps can, in principle, be rate-determining steps, and strongly depend on the PAV concentration. However, because the quenching rate constant (k_q) of excited $Ru(bpy)_3^{2+}$ by protein-anchored PAV is known, the fraction of $Ru(bpy)_3^{2+}$ fluorescence quenched by protein-bound PAV can be estimated at the various loading degrees.

Figure 12 correlates the fraction of excited species being quenched by proteins of different loading degrees with the overall observed reduction rate of GSSG by the respective enzymes. We realize that a linear relationship is obtained, a result implying that the rate-limiting step in the photoreduction of GSSG involves the ET quenching of the excited photosensitizer by PAV-anchored groups. Thus, the effectiveness of the electrical wiring of the biocatalyst active site is independent of the density of electrical wiring components on the protein backbone, within the range of loading degrees studied. Thus, low density of wiring units is sufficient to yield effective electrical communication with the protein active site. This conclusion was previously predicted by Heller (*18*).

The conclusion that the photoreduction of GSSG is controlled by the electron-transfer quenching process suggests that improvement of this reaction will accelerate the reduction rate. Yet, the number of bipyridi-

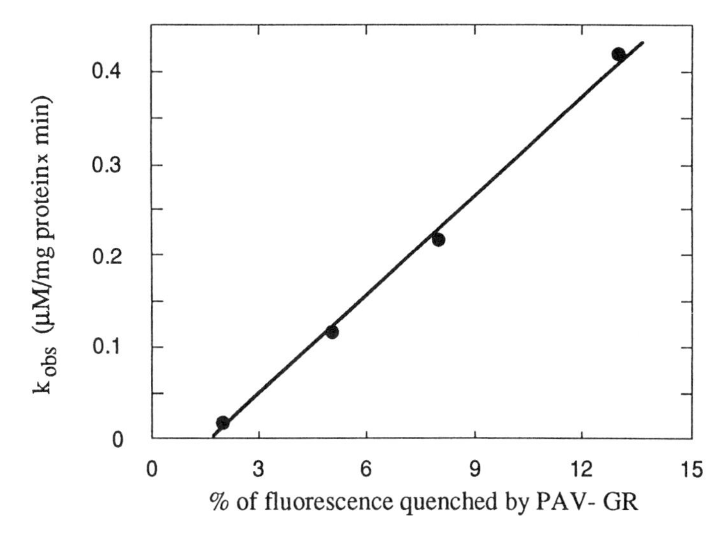

Figure 12. The observed rate of GSSG reduction versus the calculated fraction of excited species quenched by the protein-bound bipyridinium groups at the various loading degrees of PAV on the enzyme.

nium groups that can be chemically bound to the protein is limited by the number of available lysine residues. (Only lysine residues exposed to the exterior surrounding of the enzyme are relevant in this context, as the proximity of quencher to photosensitizer is essential for the ET-quenching process). Therefore, further improvement of the photosystem for reduction of GSSG by chemical manipulation of the protein is limited. Immobilization of PAV-modified GR in the redox copolymer composed of acrylamide–**1** provides, however, a path to circumvent these difficulties. The redox copolymer, **2**, includes a high density of bipyridinium groups and is anticipated to contribute to an efficient electron-transfer quenching process of the excited species, whereas the PAV-anchored groups would function only in electrical wiring of the active site to its environment.

We have immobilized (*27*) PAV-modified GR in the redox copolymer composed of acrylamide and **1**. Immobilization is performed similarly to that described for nitrate reductase. The immobilized PAV–GR biocatalyst is introduced into a photosystem that includes the photosensitizer $Ru(bpy)_3^{2+}$, the electron donor EDTA, and the substrate GSSG. Illumination of the photosystem results in the effective formation of GSH. Figure 13 compares the rate of photosensitized reduction of GSSG in the photosystem that includes PAV–GR in the redox copolymer to the most effective homogeneous photosystem that includes only the wired biocatalysts PAV–GR (Figure 10a). The photoreduction of GSSG is ca. 25-fold faster in the photosystem that includes the polymer-immobilized enzyme.

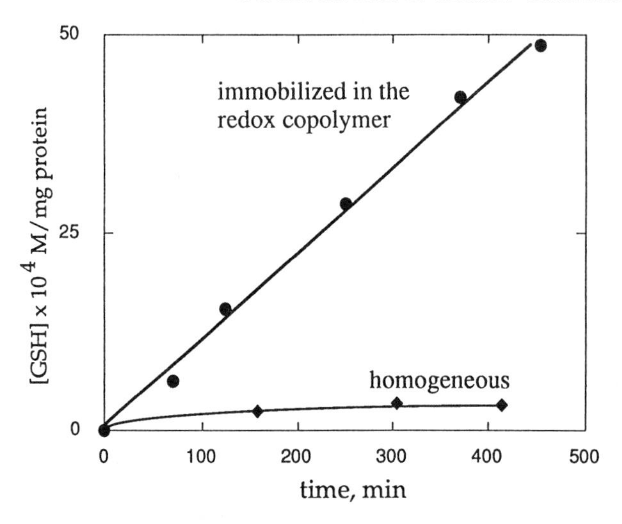

Figure 13. Rate of GSH evolution (normalized per milligram of protein) in the photosystem described in Figure 10 (a) in the system described in graph 10a; and (b) in a photosystem composed of 1.66 g of the redox polymer that contains the immobilized relay-modified glutathione reductase.

The acceleration in the reduction rate of GSSG in the presence of the redox copolymer is attributed to optimization of the reaction steps in the organized assembly (Figure 14). The redox polymer provides a highly loaded matrix of bipyridinium relay components for effective ET quenching of the excited species. The reduced polymer-bound bipyridinium components mediate ET to protein-anchored PAV units. The protein-anchored PAV reduced components act as the electrical wire to the enzyme active site, where GSSG is reduced. In addition to the acceleration of GSSG photoreduction in the system composed of PAV–GR immobilized in copolymer **2**, substantial stabilization of the biocatalyst is revealed. In the homogeneous system including PAV–GR, the evolution of GSH ceases after ca. 3 h, whereas the immobilized PAV–GR is active for prolonged illumination times and no degradation is detectable.

Conclusions and Perspectives

Proteins, that is, enzymes, can be coupled to excited species and can establish ET communication. Two different approaches to maintain electrical interactions between the enzyme active site and its environment were designed and involved immobilization in redox copolymers and chemical modification of the protein backbone by redox-relay components. Organization of the enzyme in the redox copolymer forms an organized matrix

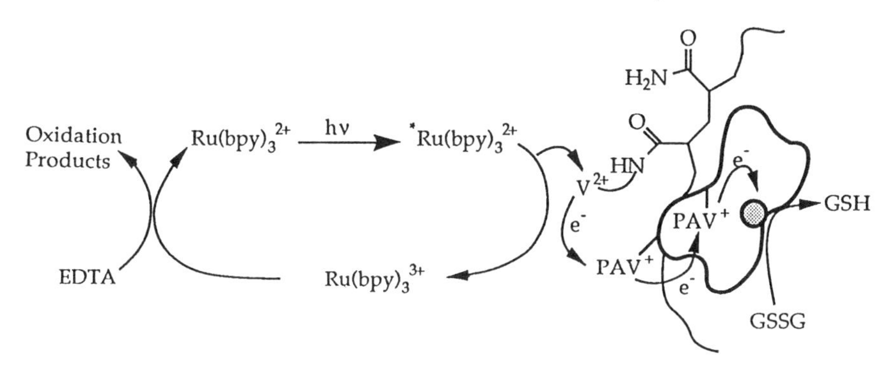

Figure 14. Photoinduced reduction of GSSG in an organized assembly composed of PAV-modified GR immobilized in the redox copolymer. The excited photosensitizer is quenched by the polymer-anchored bipyridinium groups. They mediate ET to the protein-bound PAV, and the protein-bound PAV mediates ET to the enzyme active site.

that mediates the ET process from the excited state to the biocatalyst active site. Chemical modification of the enzyme by electron-relay components generates a new biocatalyst capable of interacting with excited species. Such an electrically wired assembly allows vectorial ET from an external excited center to the biocatalyst active site across the protein backbone. Application of these approaches to other enzymes is anticipated to generate new semiartificial biocatalysts for solar-light-induced transformations. Imagination in the development of routes to electrically wired proteins can generate new catalytic biomaterials for artificial photosynthetic devices.

Yet, the subject of electrical communication between the protein active site and its environment is of broader interest. Numerous applications of electrically wired proteins in various disciplines can be envisaged. Immobilization of electrically wired enzymes onto electrodes provides the basis for an amperometric response to the biocatalytic transformation occurring at the active site. Thus, the development of biosensor devices based on these concepts seems obvious, that is, a biosensor for nitrate (a common pollutant). Furthermore, the design of electrical communication between proteins and electrodes could provide a means to convert solar light energy into electrical energy. Such a device, which is schematically outlined in Figure 15, uses the photosynthetic reaction center as the light-harnessing component and a quinone compound, Q, as the electrical wiring component for the generation of a photocurrent (or photovoltage).

Another important perspective of electrically wired enzymes involves their application as novel biocatalysts in biotechnological transformations. Numerous redox enzymes require the participation of NAD(P)H cofactors (*28, 29*). The need to regenerate and recycle these cofactors is a major

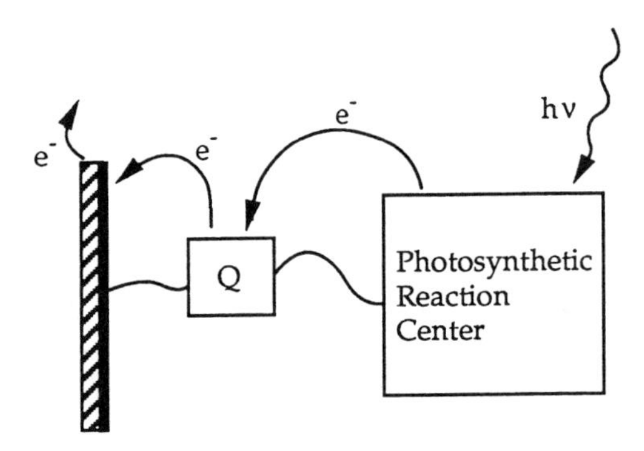

Figure 15. Proposed scheme for a photovoltaic device using the photosynthetic reaction center as the light-harnessing compound, and a quinone compound that mediates ET to the electrode.

limitation in using such enzymes for biotechnological processes. Electrically wired enzymes could provide new catalytic biomaterials that exclude cofactor requirements (*30*). Direct interaction between the reducing agent and the electrical wiring element could initiate the electron flow to the active site and effect the biotransformation. Evidently, achieving progress in various disciplines by using electrically wired protein is a challenge for future research.

Acknowledgment

Parts of this research were supported by the Fund for Basic Research administered by the Israel Academy of Sciences and Humanities and by the Ministry of Science and Technology of Israel (via a scholarship to N. Lapidot).

References

1. Bard, A. J. *Science* **1980,** *207,* 139.
2. Meyer, T. J. *Acc. Chem. Res.* **1989,** *22,* 163.
3. Willner, I.; Willner, B. In *Topics in Current Chemistry;* Mattay, J., Ed.; Springer-Verlag: Heidelberg, Germany, 1991; Vol. 159, p 153.
4. Fox, M. A. In *Topics in Current Chemistry;* Mattay, J., Ed.; Springer-Verlag: Heidelberg, Germany, 1991; Vol. 159, p 67.
5. *Energy Resources through Photochemistry and Catalysis;* Grätzel, M., Ed.; Academic Press: Orlando, FL, 1983.
6. Willner, I.; Steinberger-Willner, B. *Int. J. Hydrogen Energy* **1988,** *13,* 593.
7. Meyer, T. J. *Pure Appl. Chem.* **1986,** *58,* 1193.

8. Harriman, A. J. *Photochemistry* **1985,** *29,* 139.
9. Willner, I.; Mandler, D. *Enzyme Microb. Technol.* **1989,** *11,* 467.
10. Marcus, R. S.; Sutin, N. *Biochim. Biophys. Acta* **1985,** *811,* 265.
11. Therien, M. J.; Chang, J.; Raphael, A. L.; Bowler, B. E.; Gray, H. B. In *Structure and Bonding;* Bertrand, P., Ed.; Springer-Verlag: Heidelberg, Germany, 1991; Vol. 75, p 109.
12. Mandler, D.; Willner, I. *J. Am. Chem. Soc.* **1984,** *106,* 5352.
13. Mandler, D.; Willner, I. *J. Chem. Soc. Perkin Trans. 2* **1986,** 805.
14. Mandler, D.; Willner, I. *J. Chem. Soc. Chem. Commun.* **1986,** 851.
15. Mandler, D.; Willner, I. *J. Chem. Soc. Perkin Trans. 2* **1988,** 997.
16. Alberg, W. J.; Bartlett, P. N.; Cass, A. E. G.; Craston, D. H.; Haggett, B. G. D. *J. Chem. Soc. Faraday Trans. 1* **1986,** *82,* 1033.
17. Bartlett, P. N. In *Biosensor Technology;* Buck, R. P.; Hatfield, W. E.; Umana, M.; Bowden, E. F., Eds.; Marcel Dekker: New York, 1990; pp 95–115.
18. Heller, A. *Acc. Chem. Res.* **1990,** *23,* 128.
19. Degani, Y.; Heller, A. *J. Am. Chem. Soc.* **1988,** *110,* 2615.
20. Degani, Y.; Heller, A. *J. Am. Chem. Soc.* **1989,** *111,* 2357.
21. Gregg, B. A.; Heller, A. *Anal. Chem.* **1990,** *62,* 258.
22. Willner, I.; Lapidot, N.; Riklin, A. *J. Am. Chem. Soc.* **1989,** *111,* 1883.
23. Willner, I.; Riklin, A.; Lapidot, N. *J. Am. Chem. Soc.* **1990,** *112,* 6438.
24. Brunschwig, B. S.; DeLaive, P. J.; English, A. M.; Goldberg,M.; Gray, H. B.; Mayo, S. L.; Sutin, N. *Inorg. Chem.* **1985,** *24,* 3743.
25. Durham, B.; Pan, L. P.; Long, J. E.; Millett, F. *Biochemistry* **1989,** *28,* 8659.
26. McLendon, G. L.; Winkler, J. R.; Nocera, D. G.; Mauk, M. R.;Mauk, A. G.; Gray, H. B. *J. Am. Chem. Soc.* **1985,** *107,* 739.
27. Willner, I.; Lapidot, N. *J. Am. Chem. Soc.* **1991,** *113,* 3625.
28. Jalcovac, I. J.; Goodbrand, H. B.; Lok, K. P.; Jones, J. B. *J. Am. Chem. Soc.* **1982,** *104,* 4659.
29. Hirschbein, B. L.; Whitesides, G. M. *J. Am. Chem. Soc.* **1982,** *104,* 4458.
30. Willner, I.; Lapidot, N. *J. Chem. Soc. Chem. Commun.* **1991,** 617.

RECEIVED for review November 7, 1991. ACCEPTED revised manuscript May 28, 1992.

11

Luminescence Probes of DNA-Binding Interactions Involving Copper Complexes

David R. McMillin, Brian P. Hudson, Fang Liu, Jenny Sou, Daniel J. Berger, and Kelley A. Meadows

Department of Chemistry, Purdue University, West Lafayette, IN 47907–1393

Several chemical and spectroscopic probes of DNA and RNA structure are discussed, but the focus is on a class of complexes that allow the possibility of expanding the coordination number of the metal center while it is in the lowest energy excited state. Adduct formation with DNA can result in a dramatic increase in the luminescence intensity from such systems due to inhibition of a potent solvent-induced quenching mechanism. For example, the charge-transfer emission from $Cu(bcp)_2^+$*, where bcp denotes 2,9-dimethyl-4,7-diphenyl-1,10-phenanthroline, in aqueous methanol is dramatically enhanced when excess DNA is present in solution. Viscometric titrations suggest that partial intercalation of the complex may occur, but other types of interactions cannot be excluded. In contrast,* $Cu(dmp)_2^+$*, where dmp denotes 2,9-dimethyl-1,10-phenanthroline, is nonemissive in the presence of DNA because a solvent-accessible adduct is formed, possibly within one of the grooves of the DNA. Results of studies involving a cationic copper porphyrin are also reported. Although the free complex is nonemissive in aqueous solution, it emits in the presence of DNA as long as guanine–cytosine base pairs are present.*

SMALL MOLECULES CAN BE USED TO PROBE and modify macromolecular and microheterogeneous systems. In this regard electronic excited states of the molecular probe can be useful. Site-specific chemistry is a possibility if the probe binds tightly and preferentially to particular

0065–2393/93/0238–0211$06.50/0

types of sites and electronic excitation of the probe triggers a chemical reaction. Alternatively, study of the energetics and photophysics can provide information about the polarity and the rigidity at the binding site. Two macromolecules of interest in the present context are the polynucleotides DNA and RNA.

Overview of Polynucleotides

A polynucleotide is a linear polymer consisting of an alternating arrangement of phosphate and D-ribose (2′-deoxy-D-ribose in DNA) moieties. The information content is coded into the system via a sequence of purine and pyrimidine bases that are attached to the 1′ carbon of each sugar (Figure 1). The bases themselves are essentially planar, conjugated systems containing numerous heteroatoms that, in some cases, provide for substantial net dipole moments. The bases are capable of hydrogen bonding to each other and, as shown in Figure 1, are pairwise complementary in this regard.

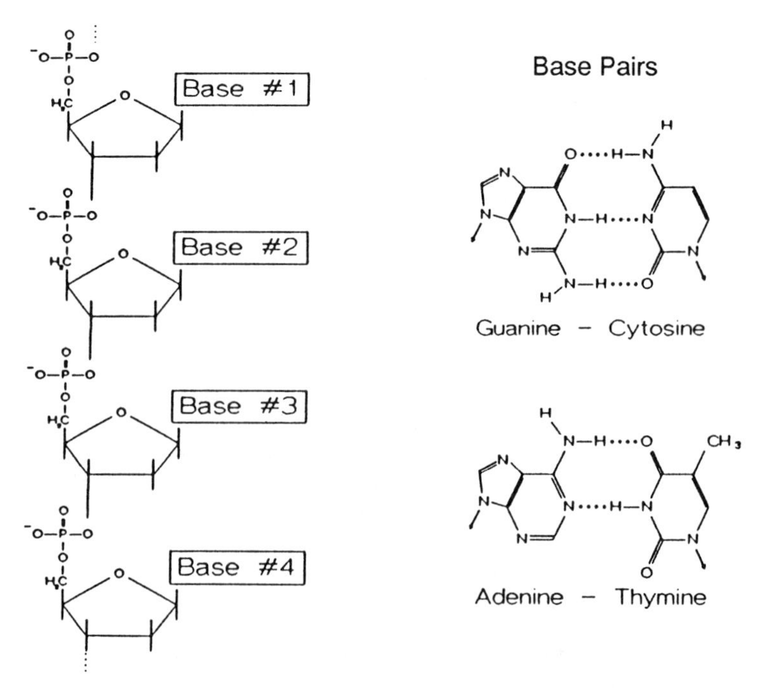

Figure 1. Structural components of DNA. Left: backbone polymer of phosphate and 2′-deoxyribose; purine and pyrimidine bases attach at C-1′. In RNA the C-2′ atom has an OH cis *to the C-3′ OH. Right: complementary base pairs of DNA and hydrogen-bonding patterns. Arrows indicate where the C-1′ carbon of the sugar attaches.*

RNA is typically a single-stranded molecule with a globular structure that is conditioned by a network of such hydrogen bonds. As illustrated schematically in Figure 2, a double-stranded motif can occur when an appropriate portion of the RNA sequence loops back on itself, and this commonly happens in naturally occurring molecules of RNA. Noncomplementary oligonucleotides such as poly(dA) tend to form helical structures with the bases stacked one upon the other. (poly(dA) denotes the polymer depicted in Figure 1 with adenine bonded to each 2′-deoxy-D-ribose unit. polyA is the corresponding polymer containing D-ribose.) The charge densities associated with individual atomic centers of the bases apparently help induce the helical twist (*1*).

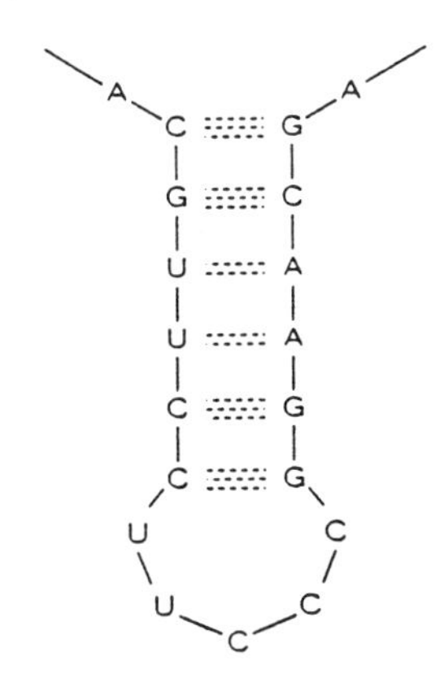

Figure 2. A hairpin loop of RNA. U denotes uracil, which is identical to thymine except there is a hydrogen in place of the methyl at C-5. The dashed lines denote hydrogen bonds.

In DNA two complementary strands are hydrogen-bonded together to form a double helix. The most prevalent form of DNA, called B-form DNA, can be described in crude terms as a twisted ladderlike structure. Complementary base pairs that are hydrogen-bonded together form the rungs, and the sides of the ladder are defined by the alternating polymer of phosphate and deoxyribose (*2, 3*). The base pairs stack one upon the other, more or less perpendicular to the long axis of the double helix, with an average vertical separation of about 3.5 Å, essentially the van der Waals thickness of an aromatic molecule.

Because the bonds to the sugars are not centrally located in the bases, two different type of grooves or indentations run along the DNA. One side of the base pairs forms the floor of what is known as the minor groove, which amounts to a narrow valley or invagination that winds around the DNA molecule. The depth of this groove is about 7.5 Å, and the width is about 5.7 Å. The floor of the major groove is defined by the opposite side of the base pairs and is more voluminous with a width of 11.7 Å and a depth of 8.5 Å. When B-form DNA is viewed from a side, the grooves alternate up and down the helix axis, as can be seen in Figure 3. Regular aspects of the structure are that the phosphate-to-phosphate distance is shorter across the minor groove than the major groove and that certain atoms are found only in one groove or the other. For example, the 3′H of the ribose lies in the major groove, while the C-1′ and C-4′ hydrogens fall in the minor groove. In B-form DNA the deoxyribose moiety exists in the C-2′-endo conformation in which the ring atoms form an envelope-shaped structure with the C-2′ carbon out of plane on the same side as C-5′ and the appended base.

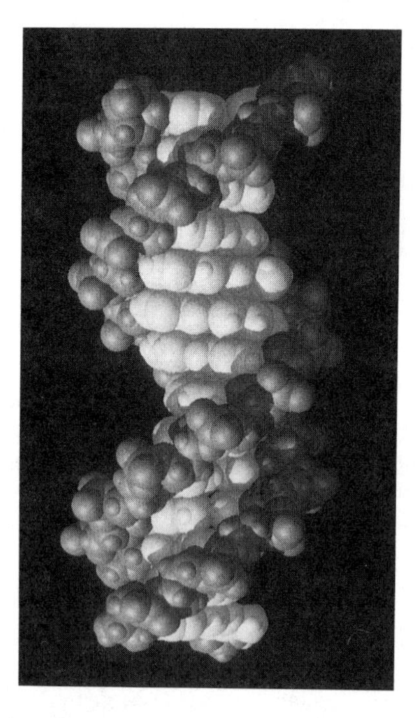

Figure 3. A segment of double-helical DNA in the ordinary B conformation.

Figure 4 is a stereoview of the local structure defined by two successive base pairs in B-form DNA. The view is down the axis of the helix, and this axis actually passes through the cytidine moiety at the point designated by a circle. The base pair depicted on top is a thymine–adenine (T•A) pair in which the filled lines designate the sigma-bonding framework. In the guanine–cytosine (G•C) base pair underneath, unfilled lines

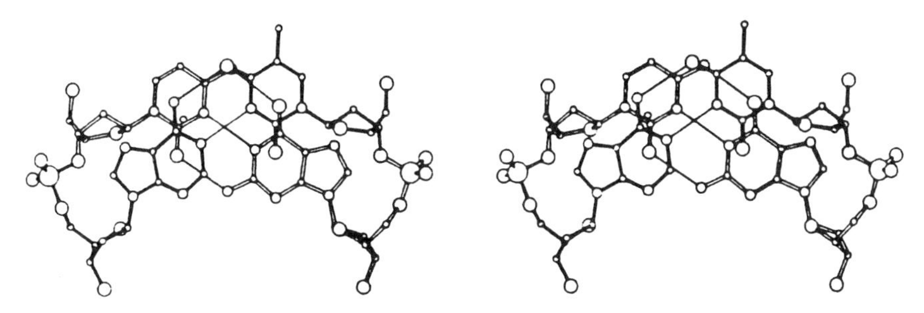

Figure 4. Stereoview of the local structure defined by two successive base pairs in B-form DNA.

represent the sigma bonds. The thin lines connecting bases represent hydrogen-bonding interactions. This drawing gives some idea of the way in which the ribose and the phosphate groups restrict access to the base pairs on the minor groove side, which falls on the bottom of the drawing. However, the width of the minor groove is actually considerably smaller than this view indicates because the shortest phosphate-to-phosphate distance between chains is not along a horizontal but along a diagonal direction, as can be appreciated from Figure 3. This happens because the next phosphate groups are displaced horizontally as well as vertically as the chains extend on account of the natural twist of the helix. (The twist angle roughly corresponds to the angle through which the hydrogen bonds have rotated in Figure 4.) By comparison with the minor groove, the major groove is a much wider channel and provides easier access to the base pairs. However, the magnitude of the coulombic potential generated by the phosphate groups is greater on average in the minor groove.

Because of a steric interaction between the C-2′ oxygen and a phosphate oxygen (*2*), the ribose group in RNA cannot adopt the same conformation. Hence in double-stranded RNA, or in a DNA–RNA hybrid, the sugar adopts the C-3′ *endo*-conformation that produces the A-form double helix. It can be viewed as a ladder that has been wound around a guide rod, rather than simply twisted (*4*). In the A-form structure the planes of the base pairs make an angle of about 70° with the helix axis, and the C-1′–C-1′ side of the base pair is quite close to the surface (*2*). In contrast, the major groove is deeply recessed and characterized by a rather narrow access. Both A-form and B-form DNA represent right-handed double helices, but there is a left-handed form as well, and it is known as Z-form DNA. It is not very common, but segments of cellular DNA could conceivably conform to this structure.

Indeed, naturally occurring DNA undoubtedly has a compacted, much less regular structure than the idealized structures just described (*5*). Bends, hairpin loops, cruciforms, and coils are structural motifs that can occur. Local structures within the DNA as well as specific sequences may be important in recognition and packaging phenomena. In principle, probes can be used to identify such features.

DNA-Binding Interactions

Binding interactions are important in DNA regulation and other processes in which recognition is important. DNA is a polyanion, so coulombic forces are clearly relevant. Indeed, they are responsible for a sheath of counterions that tends to surround the double helix. Hydrophobic cations may be drawn to the minor groove where the coulombic potential is stronger. Hydrogen-bonding interactions, for example, with atoms in the floor of the grooves, can also be important in binding, and of course stacking interactions may be important when the substrate includes a planar aromatic functionality.

The focus here will be on metal-containing binding agents. In principle, these include complex proteins with zinc fingers *(6)* or copper fists *(7)* in which multiple interactions occur, but the scope will be narrowed to small-molecule systems in which fewer specific interactions are involved. Popular systems include tethered reagents wherein the active agent is linked to a functionality specifically designed to interact with DNA. A classic system that probably should be placed in this class is the naturally occurring antibiotic bleomycin, which has a bithiazole group as an appendage that somehow interacts with DNA *(8)*. Bleomycin is a glycopeptide that can bind several metals, but the iron-containing form is probably the physiologically relevant form that cleaves DNA by a redox mechanism involving attack at a deoxyribose. Although the cobalt derivative of bleomycin is inactive in the dark, it has activity when photolyzed.

Another example of a tethered system is the totally synthetic system methidiumpropylethylenediaminetetraacetic acid (MPE)•Fe(II) (structure **1**). The methidium ring system is capable of intercalating between virtually any set of adjacent base pairs in DNA. When the methidium is bound, the iron complex on the other end of the reagent is poised next to DNA and can engage in Fenton-like chemistry that ultimately fragments a deoxyribose group *(9)*. More specific cleavage patterns can be obtained by attaching an appropriate metal complex to an oligonucleotide that is prone to bind to a specific sequence of nucleic acid *(10, 11)*.

Numerous types of untethered reagents are also known. To be useful as probes, these systems must have the inherent ability to bind to DNA as well as to effect some chemical or physical change that gives information about the binding interaction. One example is the uranyl ion UO_2^{2+}. Coulombic interactions dictate that this ion will have an affinity for the polyanion DNA, and the reactive excited state of the uranyl ion can

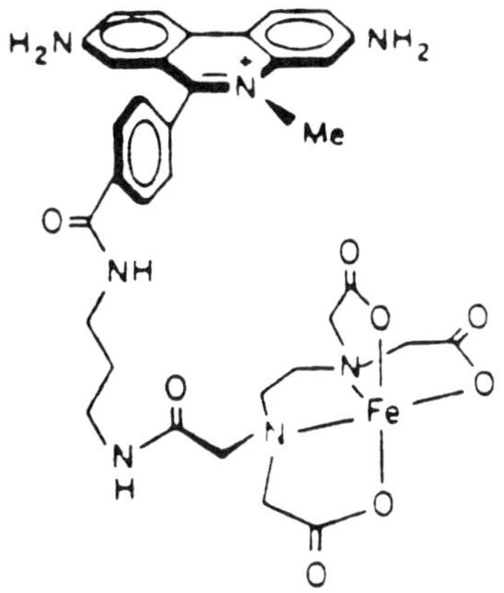

1

abstract hydrogen atoms from a variety of substrates, including alcohols. Not surprisingly, when the uranyl ion is irradiated in the presence of DNA, cleavage occurs (*12*). The mechanism probably involves hydrogen-atom abstraction from the deoxyribose, analogous in some way to the thermal chemistry seen with MPE•Fe(II).

Other interesting examples can be found in the studies of Barton and co-workers (*13–16*) who have investigated a great number of phenanthroline complexes of d^6 metal ions. The Ru(II) systems often exhibit luminescence that provides a useful handle on the DNA-binding interactions (*13*). Multiple modes of binding may exist, but partial intercalation of a phenanthroline ligand appears to play a role in the binding of many systems. A mixed-ligand system that was developed is virtually non-luminescent in aqueous media but emits in the presence of DNA (*14*).

Equally interesting are the Rh(III) analogs that also bind to DNA and are capable of photoinduced cleavage reactions (*15*). The excited states involved have not been clearly identified as yet and will no doubt be the subject of future investigations. A study (*16*) of a mixed-ligand M(III) complex in conjunction with tritium-labeled oligonucleotide suggests that the complex binds in the major groove, as the 3′H is abstracted by one of the ligands bound to the metal.

Our group's interests have centered on bis(phenanthroline)copper(I) complexes. In contrast to the d^6 systems, the d^{10} copper(I) complexes have a pseudotetrahedral coordination geometry and a +1 charge. Hence the geometric requirements and the balance that is struck between hydrophobic interactions and coulombic interactions is liable to be quite different than in the ruthenium(II) and rhodium(III) systems. The interest in the reactions between DNA and copper phenanthrolines goes back to 1979 when Sigman et al. (*17*) reported that the 1,10-phenanthroline (phen) ligand and copper(II) in combination with a thiol reagent are capable of depolymerizing poly(dA-dT)•poly(dA-dT). (The term poly(dA-dT) denotes the polymer in Figure 1 with alternating adenine and thymine bases. The analogous polymer with alternating guanine and cytosine bases is termed poly(dG-dC), and, for example, poly(dA-dT)•poly(dA-dT) denotes the hydrogen-bonded duplex.) Since then, this group and others (*18–20*) have shown that copper phenanthrolines represent a class of artificial nucleases that can be used to probe DNA and DNA complexes. Hydrogen peroxide is an important intermediate in the cleavage reaction, and a minimal reaction sequence is presented in Scheme I.

$$Cu(phen)_2^{2+} + Red \longrightarrow Cu(phen)_2^{+} + Ox \quad (1)$$

$$2Cu(phen)_2^{+} + O_2 + 2H^{+} \longrightarrow 2Cu(phen)_2^{2+} + H_2O_2 \quad (2)$$

$$Cu(phen)_2^{+} + DNA \rightleftarrows Cu(phen)_2^{+} \mid DNA \quad (3)$$

$$Cu(phen)_2^{+} \mid DNA + H_2O_2 \longrightarrow \text{oligonucleotides} + Cu(phen)_2^{2+} + OH^{-} \quad (4)$$

Scheme I.

Red and Ox denote the reduced and oxidized forms, respectively, of a sacrificial reagent such as a thiol, and $Cu(phen)_2^{+} \mid DNA$ denotes a bound form of the copper complex. The role of binding is evident from the efficiency of nucleolytic cleavage, kinetic results (*21*), as well as spectral data (*22*). On the basis of a product analysis, Sigman concluded that cleavage occurs via oxidation of the ribose moiety and that it is initiated by hydrogen-atom abstraction from the C-1′ or C-4′ carbons. Both of these hydrogen atoms are found in the minor groove, which is therefore considered to be the locus of binding (*20*).

Consistent with this reasoning, strong cutting sites in one strand correlate with strong cutting sites on the opposite strand that are offset in the 3′ direction (*18*). Sigman and co-workers (*21*) also found that the cleavage chemistry is not very sensitive to substitution at the C-5 position of the phenanthroline ligand, and they inferred that the copper complex undergoes some type of groove binding rather than intercalation. However, another group has carried out cleavage studies on oligonucleotides containing unmatched strands that result in bulges where intercalative binding is supposed to be favored (*23*). Cleavage occurs preferentially near the bulges, hence they concluded that an intercalative binding interaction may be involved. Recently, Rill and co-workers (*22*) reported viscometric data that is also consistent with intercalative binding.

No single type of experiment is likely to establish the type of binding that occurs. For example, cleavage studies can reveal information only about productive modes of binding, but more than one mode of binding may occur. Furthermore, if the reactive intermediate (a cupryl species involving the $Cu(OH)^{2+}$ moiety?) has a finite lifetime, it need not react as rapidly as it forms nor bind in the same fashion as the $Cu(phen)_2^+$ precursor. A collaborative approach involving many different types of experiments should, however, ultimately provide a sound basis for understanding these systems.

Luminescence Methods and Copper Phenanthrolines

When methyls or larger substituents are introduced into the 2- and 9-positions of the 1,10-phenanthroline ligand (structure **2**) the reduction potential of the copper complex is dramatically increased, and it is no longer an efficient catalyst for DNA cleavage (*24*). Accordingly, we have used these systems to study DNA-binding phenomena without interference from DNA-cutting reactions. The same substituents also make it possible to observe photoluminescence from the copper complexes because, in the absence of sufficiently bulky substituents, the lowest energy excited state is subject to a severe flattening distortion that effectively quenches the emission even in a low-temperature glass (*25*).

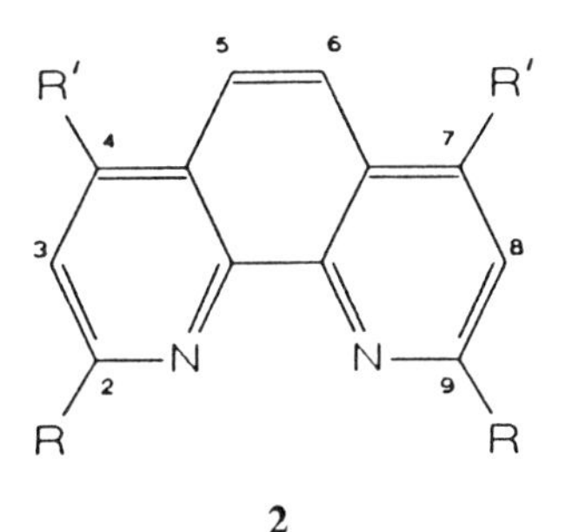

	R	R'
phen	H	H
dmp	Me	H
dip	H	Ph
bcp	Me	Ph

2

However, the excited state is subject to another potent quenching mechanism in solution whenever Lewis bases are present. Thus, the excited-state lifetime of $Cu(dmp)_2^+$, where dmp denotes 2,9-dimethyl-1,10-phenanthroline, is about 100 ns in CH_2Cl_2 at room temperature, but the lifetime is only 2 ns in CH_3CN (*26*). Donor moieties quench the excited state via nucleophilic attack at the metal center, which can formally be classified as copper(II) in the excited state. (Because there are no d–d excited states in this d^{10} system, the lowest energy excited state is a metal-to-ligand charge-transfer state.)

An adduct with an expanded coordination number has never been spectroscopically detected, but systematic exploration of steric effects (*27, 28*) and measurements of activation volumes support the proposed quenching mechanism (*29*). In addition, recent studies (*30*) involving a series of Lewis bases reveal that the activation enthalpies for quenching are uniformly negative and are consistent with reversible adduct formation. The excited-state adducts that form are rare examples of exciplexes involving coordination compounds. Exciplex formation is quite common with excited states of organic molecules, but the ligands in the primary coordination sphere of transition metal ions usually screen the d orbitals from substrates in solution. Because reasonable orbital overlap is normally necessary for effective adduct formation, the phenomenon is most likely to be observed with coordinatively unsaturated excited states such as the one that occurs in the copper systems. As will be discussed, the possibility of exciplex formation means that the complex can be used as a luminescence probe that is extremely sensitive to the local environment.

A good deal of physical data shows that $Cu(dmp)_2^+$ binds to DNA in solution. Thus, in the presence of DNA the charge-transfer absorption band of the copper complex shifts a few nanometers to the red, and there is evidence of hypochromism as well (*31, 32*). Equilibrium dialysis experiments (*33*) suggest that binding occurs, and studies involving ethidium bromide (*24*) reveal that the dmp complex binds competitively. Whatever the mode of interaction, the uptake of $Cu(dmp)_2^+$ has little influence on the rigidity or average length of double-stranded DNA in solution because the viscosity of the solution is hardly affected by the addition of the copper complex (*33*). Moreover, the emission that can be detected from $Cu(dmp)_2^+$ in aqueous media is extremely weak with or without DNA present (*32*). This weak emission suggests that the bound form of the complex is fully accessible to solvent and therefore efficiently quenched by solvent-induced exciplex formation.

The $Cu(bcp)_2^+$ system, where bcp denotes 2,9-dimethyl-4,7-diphenyl-1,10-phenanthroline, presents new problems because it is much less soluble in water, even when the chloride salt is employed. Accordingly, the spectroscopic studies were carried out in 33% methanol. As with the dmp complex, a bathochromic shift and hypochromism are evident in the charge-transfer absorption spectrum in the presence of DNA; however, two phases of binding are clearly observed (*32*). At low DNA-P:Cu ratios (where DNA-P denotes the number of phosphate, i.e., nucleotide, residues in solution and Cu denotes the amount of copper complex in solution) some type of aggregation–particulate formation occurs. However, at DNA-P:Cu ratios greater than about 10, the copper complex appears to be fully dispersed and dissolved, at least at the lower ionic strengths. The difference in character of the DNA in solution at different DNA-P:Cu ratios can be appreciated from the circular dichroic (CD) spectra

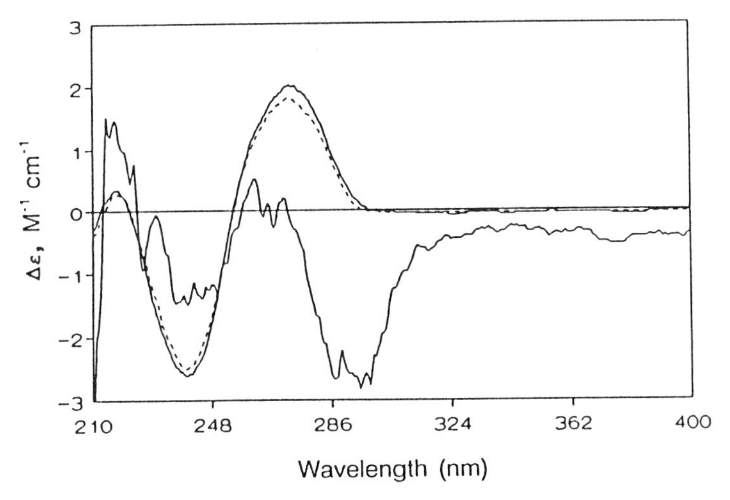

Figure 5. UV circular dichroism spectrum of salmon testis DNA with DNA-P:Cu = 0 (—, smooth trace); DNA-P:Cu = 2 (—, noisy trace); and DNA-P:Cu = 50 (- - -). The copper complex is $Cu(bcp)_2^+$, and the solvent is 1:3 by volume MeOH–pH 7.8 aqueous buffer at 20 °C. No corrections for scattering artifacts have been implemented.

presented in Figure 5. For purposes of this discussion the focus will be on the high DNA-P:Cu ratios in which the copper complex is assumed to be monomeric.

The most striking property of these solutions is that they are luminescent (*31, 32*). The charge-transfer emissions that are observed from a series of DNA solutions are depicted in Figure 6. The excitation spectra are consistent with emission from $Cu(bcp)_2^+$ molecules bound to DNA, and the emission lifetime is about 65 ns. Although the emission intensity does not vary greatly with the type of DNA used, it is more intense with poly(dA-dT)•poly(dA-dT) than with poly(dG-dC)•poly(dG-dC). On average, there is about a 10-fold enhancement of the emission intensity in the presence of DNA, and this enhancement implies that solvent-induced exciplex quenching is sharply curtailed. This curtailment suggests that the conformational flexibility of the complex is highly restricted within the DNA adduct and that the binding involves more than simple surface contact mediated by hydrophobic interactions or coulombic interactions. A more deeply embedded adduct is required. Emission polarization data are in accord with this view because they show that the rotational freedom of the complex is also highly restricted, more like that of the macromolecule in solution (*31*).

One mode of binding that would be consistent with rigid adduct formation involves partial intercalation of the complex. In the solid state, a related system, the *N*-methyl-3,5,6,8-tetramethyl-1,10-phenanthrolinium

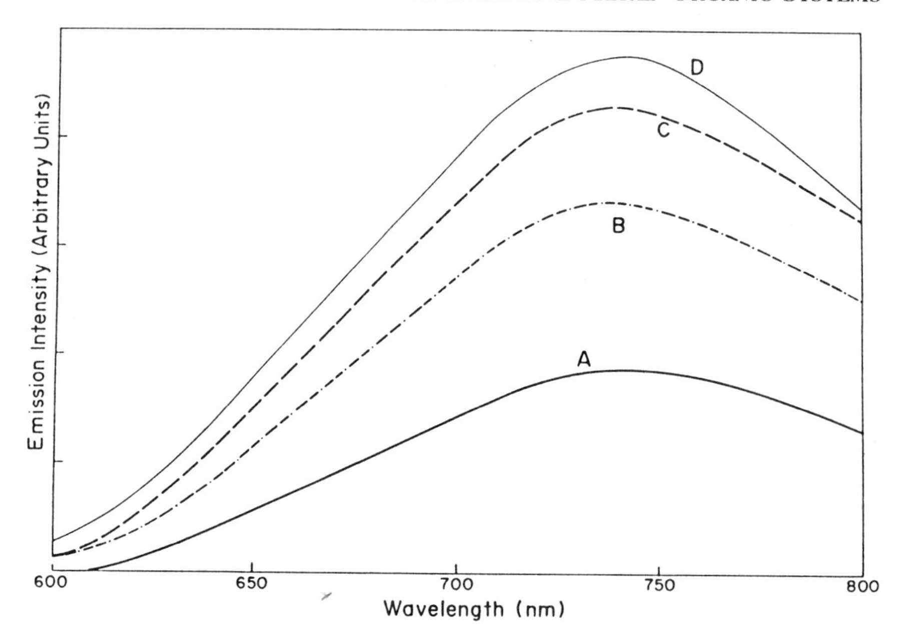

Figure 6. Corrected emission spectra from $Cu(bcp)_2^+$ in 1:3 by volume MeOH–pH 7.8 aqueous buffer at a DNA-P:Cu ratio of 20. Curves: A, poly(dG-dC)·poly(dG-dC); B, salmon testis DNA; C, poly(dA-dT)·poly(dA-dT); and D, yeast tRNA. The copper concentration was 8.7 μM, and the temperature was 20 °C. (Reproduced from reference 32. Copyright 1990 American Chemical Society.)

cation, is known to intercalate (*34*). By analogy, one might expect the copper complex to dock in the major groove with some portion of the backside of one of the bcp ligands resting between adjacent base pairs within the double helix. The extent of penetration is uncertain but will ultimately be limited by the steric influence of the bcp ligand opposite the copper. Other systems are thought to bind similarly; for example, Barton and co-workers (*35*) proposed that, in related ruthenium systems, the introduction of phenyl substituents in the 4 and 7 positions of the phenanthroline ligands promotes intercalative binding. Thus, $Ru(dip)_3^{2+}$, where dip denotes 4,7-diphenyl-1,10-phenanthroline, appears to intercalate more strongly than $Ru(phen)_3^{2+}$. Sigman (*18*) studied the DNA-cutting ability of $Cu(dip)_2^+$ and found that, in contrast to $Cu(phen)_2^+$, the cleavage pattern is much more even, that is, there appears to be less sequence specificity. This finding, too, is consistent with intercalative binding that can be a sequence-neutral process.

To explore this idea further, we made the viscometric measurements reported in Figure 7. These data show that the addition of small amounts of $Cu(bcp)_2^+$ enhances the specific viscosity of the DNA in solution.

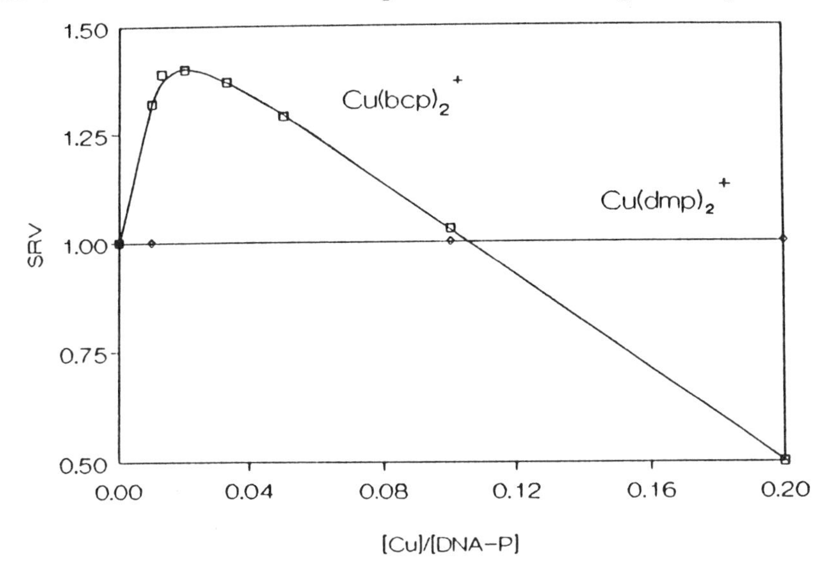

Figure 7. Specific viscosity ratios obtained with salmon testis DNA and varying amounts of $Cu(bcp)_2^+$ or $Cu(dmp)_2^+$. The ratio is the specific viscosity due to DNA with complex compared to that observed in the absence of complex. The temperature was 23 °C, and the solvent was 1:3 by volume MeOH–pH 7.8 aqueous buffer.

(The falloff that occurs at higher copper concentrations in the figure is attributed to an aggregation phenomenon; vide supra.) The viscosity increase is consistent with elongation of the helices in solution, which is a necessary consequence of intercalative binding. In contrast, $Cu(dmp)_2^+$ has virtually no effect on the specific viscosity due to the DNA in solution.

Physical studies also were carried out with RNA. The first experiments involved addition of the copper reagents to a solution containing transfer RNA (tRNA) that was isolated from yeast (*32*). As with DNA, the presence of RNA did not enhance the emission from $Cu(dmp)_2^+$, but emission enhancement did occur with $Cu(bcp)_2^+$. At first this result seems surprising because tRNA has a globular structure built out of one strand. However, similar types of binding may be possible because the structure contains stems of double-stranded structures due to the sequence looping back on itself. We also measured the effect of $Cu(bcp)_2^+$ on the specific viscosity of the synthetic, double-stranded RNA poly C•poly I, but we found virtually no enhancement. (The symbol I denotes inosine, which is guanine without the amino substituent.) Indeed, the specific viscosity was somewhat decreased in the presence of $Cu(bcp)_2^+$.

In summary, the luminescence results obtained with $Cu(dmp)_2^+$ are consistent with the binding model developed by Sigman for $Cu(phen)_2^+$. In particular, the bound form of $Cu(dmp)_2^+$ is solvent-accessible in keeping with the postulated groove binding. On the other hand, in view of the viscosity enhancements observed by Rill and co-workers (*22*), a number of questions concerning the mode of binding remain to be answered. Does the steric influence of the methyl substituents cause the dmp complex to bind differently than the phen complex? Is the viscosity increase that is associated with the phen complex due to intercalation, or does some other phenomenon account for the requisite elongation or rigidification of the double helix? For $Cu(bcp)_2^+$, the luminescence and viscometry data indicate that the complex, like $Cu(dip)_2^+$, intercalates into DNA, although charge-transfer emission can also be detected in solutions of double-stranded RNA that do not show a viscosity increase. Because of the extension of each ligand, the steric influence of the methyl substituents on the opposite ligand of $Cu(bcp)_2^+$ may be minimal; however, the steric requirements of the phenyl substituents themselves need to be considered more carefully. The problem is that the phenyl substituents are constrained to be out of the plane of the phenanthroline moiety due to steric interactions between the ortho hydrogen atoms of the phenyls and the 5- and 6-hydrogens of the phenanthroline core (*36*). Systematic studies of a copper complex of a phenanthroline ligand with a single phenyl substituent are currently underway. Because the complex of this ligand is chiral, we can also investigate whether selective binding of a particular enantiomer occurs.

Porphyrins and DNA

Porphyrins are important reagents for photodynamic therapy (*37*), in which the function apparently is sensitization of the cytotoxic agent singlet oxygen. One of the potential targets of singlet oxygen is cellular DNA; hence, there has been considerable interest in the interactions between DNA and porphyrins. The cationic metalloporphyrin TMpyP4 or 5,10,15,20-tetrakis(*N*-methylpyridinium-4-yl)porphyrin, which is shown as structure **3** with R as *N*-methylpyridinium-4-yl, and complexes thereof have been of particular interest since Fiel et al. (*38*) presented evidence that they can intercalate into DNA. The cationic nature of this porphyrin obviously favors combination with DNA, but the charge also simplifies the aqueous chemistry by minimizing self-association in solution. Several different techniques have been used to investigate this system, including circular dichroism (CD), visible spectroscopy, viscometry, and NMR spectroscopy (*39, 40*). Some of the relevant findings will be briefly summarized.

A variety of metalloporphyrins bind axial ligands; others do not, and this feature affects DNA binding. For example, the Co(III), Mn(III), and Zn(II) derivatives of TMpyP4 have axial ligands, and as a consequence

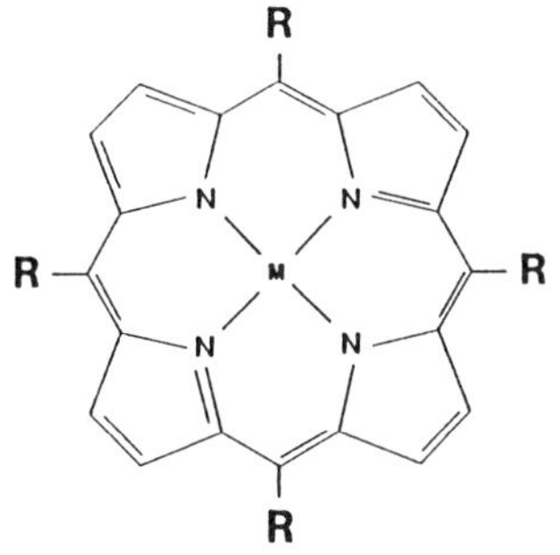

3

they are relegated to bind on the surface of DNA, possibly within one of the grooves. On the other hand, the free ligand and the Pd(II), Au(III), Ni(II), and Cu(II) derivatives typically do not have axial ligands, and they are capable of intercalating into DNA. However, intercalative binding requires guanine and cytosine bases, and there is evidence that the free ligand TMpyP4 intercalates exclusively within a 5′CpG3′ sequence [that is, where guanine follows cytosine along the chain (Figure 1)] (*41*). For an intercalator such sharply defined sequence specificity is extremely unusual.

Photophysics of Metalloporphyrins

The electronic structure and luminescence properties of metalloporphyrins, especially copper porphyrins, are of particular interest for this discussion. The electronic states associated with the ligand are fairly well understood but are complicated by the fact that the highest energy occupied orbital and the second highest energy occupied orbital have similar energies (*42*). As a consequence, the two lowest energy intraligand excited electronic configurations interact, and there are two spin-allowed π–π^* transitions with quite different intensities. The Soret band occurs at around 420 nm with an extinction coefficient of >100,000 M^{-1} cm^{-1}, while the Q band is a considerably weaker, vibronically structured transition that occurs in the neighborhood of 550 nm. In solution a free porphyrin typically exhibits π–π^* fluorescence from the lower energy singlet state; however, in most metalloporphyrins containing a transition metal, intersystem crossing to the corresponding $^3\pi$–π^* state occurs preferentially. In favorable cases this state exhibits room-temperature phosphorescence, but it deactivates by decaying into lower-energy d–d states or charge-transfer excited states when they are available.

Copper porphyrins, are a special case, for example, Cu(TPP) (TPP denotes 5,10,15,20-tetraphenylporphyrin, where R is Ph in structure **3**).

The copper center has an unpaired electron, and thus the spin-allowed transitions correspond to doublet states. These, in turn, relax to the ^2T and ^4T states that arise from the lowest energy $^3\pi$–π^* state by virtue of coupling with the unpaired electron of the copper center. In benzene, Cu(TPP) exhibits thermally equilibrated emission from the $^{2,4}T$ excited states with a lifetime of 29 ns (*43*). Kim et al. (*44*) reported that the emission is quenched when strong donors such as pyridine are incorporated into the solution by virtue of formation of a five-coordinate form of the complex. On the basis of iterative extended Hückel calculations, they concluded that the emission from the ^2T and ^4T excited states was quenched by a lower energy charge-transfer excited state of the five-coordinate adduct. The possibility of expanding the coordination number suggests that exciplex interactions could be important in the excited-state relaxation. Accordingly, we have begun to investigate the luminescence properties of Cu(TMpyP4) in solution and in the presence of DNA. Preliminary results are described in the following section.

Luminescence Studies of Cu(TMpyP4) in the Presence of DNA

Although Cu(TMpyP4) itself exhibits no detectable emission in aqueous solution, a broad, unstructured emission can be observed in the presence of DNA (Figure 8). In a corrected emission spectrum the maximum occurs at about 800 nm, and the emission can be assigned to the ^2T and ^4T excited states by analogy with the Cu(TPP) system. As indicated in Figure 8, the emission depends on the DNA-P:Cu ratio, but the focus will be on solutions with a high DNA-P:Cu ratio in which the emission has reached a maximum value. Because the emission intensity increases with the guanine and cytosine content, we assign the emission to the intercalated form of the complex. Other spectral data, including electronic absorption and CD results as well as viscometry studies, are in full accord with this interpretation.

A model that successfully accounts for the luminescence results can be presented with the help of Figure 9. Two different forms of Cu(TMpyP4) are represented across the diagram. On the right the porphyrin complex is depicted as a four-coordinate complex, the most stable form of the complex in aqueous solution. Opposite this is the higher energy five-coordinate form with a water bound at one of the axial positions. The diagrams also present the energies of the ^2T and ^4T excited states, which are expected to be about the same regardless of the coordination number. The more interesting state is the d–d excited state associated with the $d_{z^2} \rightarrow d_{x^2-y^2}$ transition. As is well known from studies of the

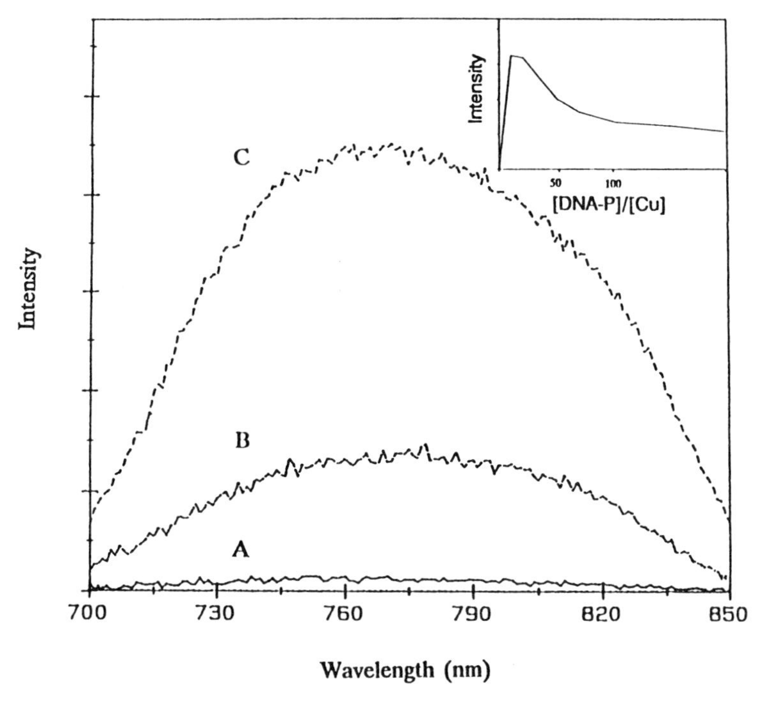

Figure 8. Uncorrected emission spectra of Cu(TMpyP4) at DNA-P:Cu ratios of 50:1. Curves: A, poly(dA-dT)·poly(dA-dT); B, salmon testis DNA; and C, poly(dG-dC)·poly(dG-dC). The conditions are μ = 0.2 M NaCl, pH 7.8, and T = *25 °C. Inset: emission intensity versus the DNA-P:Cu ratio for salmon testis DNA.*

pentaammine effect (*45*), this state is expected to drop to lower energy in the five-coordinate form, and we propose that it falls below the $^{2,4}T$ manifold. This condition would result in the quenching of the emission because the d–d excited states of copper(II) systems are extremely short-lived. In essence the idea is that the emission from free Cu(TMpyP4) is quenched in aqueous solution because of solvent-induced exciplex formation. In this case the exciplex is short-lived because it readily deactivates via a d–d state. On the other hand, emission from the $^{2,4}T$ states can be observed when the complex is intercalated into DNA because the axial positions of the complex are protected from solvent attack.

Summary and Conclusions

DNA is a complex macromolecule capable of forming many different types of local structures. Transcription and regulation of DNA molecules

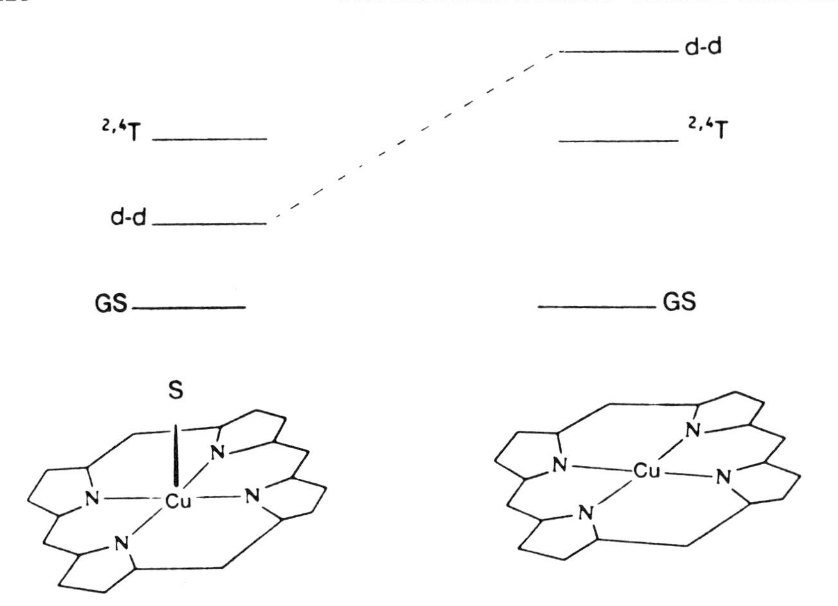

Figure 9. State diagram for four-coordinate versus five-coordinate Cu(TMpyP4). S denotes a solvent molecule.

depend on recognition phenomena that may involve multiple types of binding interactions. Coulombic forces, hydrogen-bonding interactions, and hydrophobic and steric effects are all factors that come into play. There are so many possibilities that understanding the binding of even simple ligands in solution requires information from many different experiments. We have illustrated a luminescence method that can be applied when binding to DNA interferes with a potent solvent-induced quenching process involving exciplex formation. This type of complex formation with nucleophilic solvents requires effective orbital overlap, and this overlap is possible in general when the excited state is coordinatively unsaturated.

The copper phenanthrolines, $Cu(dmp)_2^+$ and $Cu(bcp)_2^+$, have been investigated as analogs of the artificial nucleases, $Cu(phen)_2^+$ and $Cu(dip)_2^+$, respectively. Although $Cu(dmp)_2^+$ binds to DNA, no significant induction of luminescence occurs, and we conclude that some type of solvent-accessible adduct is formed that may involve groove binding. On the other hand, when $Cu(bcp)_2^+$ is combined with an excess of DNA or RNA, a significant charge-transfer luminescence signal can be observed. The results are consistent with intercalative binding, but other types of deeply buried binding may also explain the results. All that is required is that the binding be secure enough to preclude the structural reorganization that is necessary for nucleophilic attack at the metal center.

The results with $Cu(bcp)_2^+$ reveal how profound an influence the introduction of hydrophobic substituents can have on the DNA-binding interaction. The methyl substituents attached at the 2- and 9-positions should be inconsequential in this regard, but the possibility of a steric influence on the DNA-binding interaction cannot be totally dismissed. This caveat aside, our results strongly suggest that intercalative binding does not occur with $Cu(dmp)_2^+$. By inference it is unlikely to be significant for $Cu(phen)_2^+$ either. As the nuclease chemistry of new derivatives of $Cu(phen)_2^+$ unfolds and as new types of systems are probed by $Cu(phen)_2^+$ and the like, luminescence studies of appropriate analogs can be an important aid in the effort to understand the complexities of the interactions, especially the fundamental binding interactions.

Finally, we presented new results pertaining to a copper porphyrin, Cu(TMpyP4). In aqueous solution the intraligand phosphorescence of the complex is completely quenched because attack by solvent at an axial position induces deactivation via a low-lying d–d state. However, the emission can be observed from certain types of DNA adducts. Because the emission intensity correlates with the GC content of the DNA, the emissive form is assigned to be the intercalated complex, which is incapable of accepting axial ligands.

Acknowledgments

This work was supported by the National Science Foundation through Grant No. CH–9024275. J. Sou and B. Hudson were the recipients of undergraduate summer fellowships from Eli Lilly and the Department of Chemistry of Purdue University, respectively.

References

1. Sarai, A.; Mazur, J.; Nussinov, K.; Jernigan, R. L. *Biochemistry* **1988,** *27,* 8498–8502.
2. Dickerson, R. E. *Sci. Amer.* **1983,** *249,* 94–111.
3. Kennard, O.; Hunter, W. N. *Quart Rev. Biophys.* **1989,** *22,* 327–379.
4. Jain, S.; Zon, G.; Sundaralingam, M. *Biochemistry* **1991,** *30,* 3567–76.
5. Livolant, F. *J. Mol. Biol.* **1991,** *218,* 165–181.
6. Berg, J. M. *J. Biol. Chem.* **1990,** *265,* 6513–6516.
7. Nakagawa, K. H.; Inouye, C.; Hedman, B.; Karin, M.; Tullius, T. D.; Hodgson, K. O. *J. Am. Chem. Soc.* **1991,** *113,* 3621–3623.
8. Stube, J.; Kozarich, J. W. *Chem. Rev.* **1987,** *87,* 1107–1136.
9. Dervan, P. B. *Science (Washington, D.C.)* **1986,** *232,* 464–471.
10. Strobel, S. A.; Dervan, P. B. *Nature* **1991,** *350,* 172–174.

11. Francois, J. C.; Saison-Behmoaras, T.; Barbier, C.; Chassignol, M.; Thuong, N. T.; Hélène, C. *Proc. Natl. Acad. Sci. U.S.A.* **1989,** *86,* 9702–9706.
12. Nielsen, P. E.; Mlegaard, N. E.; Jeppesen, C. *Nucleic Acids Res.* **1990,** *18,* 3847–3851.
13. Basile, L. A.; Barton, J. K. *Met. Ions. Biol. Syst.* **1989,** *25,* 31–103.
14. Friedman, A. E.; Chambron, J. C.; Sauvage, J. P.; Turro, N. J.; Barton, J. K. *J. Am. Chem. Soc.* **1990,** *112,* 4960–4962.
15. Pyle, A. M.; Morii, T.; Barton, J. K. *J. Am. Chem. Soc.* **1990,** *112,* 9432–9434.
16. Long, E. C.; Absalon, M. J.; Stubbe, J.; Barton, J. K. *J. Inorg. Biochem.* **1991,** *43,* 436.
17. Sigman, D. S.; Graham, D. R.; D'Aurora, V.; Stern, A. M. *J. Biol. Chem.* **1979,** *254,* 12269–12272.
18. Sigman, D. S. *Acc. Chem. Res.* **1986,** *19,* 180–186.
19. Sigman, D. S.; Chen, C.-H. B. *Annu. Rev. Biochem.* **1990,** *59,* 207–236.
20. Sigman, D. S.; Chen, C.-H. B. In *Metal–DNA Chemistry;* Tullius T. D., Ed.; ACS Symposium Series 402; American Chemical Society: Washington, DC, 1989; pp 24–47.
21. Thederahn, T. B.; Kuwabara, N. D.; Larsen, T. A.; Sigman, D. S. *J. Am. Chem. Soc.* **1989,** *111,* 4941–4946.
22. Veal, J. M.; Rill, R. L. *Biochemistry,* **1991,** *30,* 1132–1140.
23. Williams, L. D.; Thivierge, J.; Goldberg, I. H. *Nucleic Acids Res.* **1988,** *16,* 11607–11615.
24. Tamilarasan, R.; McMillin, D. R.; Liu, F. In *Metal–DNA Chemistry;* Tullius T. D., Ed.; ACS Symposium Series 402; American Chemical Society: Washington, DC, 1989; pp 48–58.
25. Everly, R. M.; Ziessel, R.; Suffert, J.; McMillin, D. R. *Inorg. Chem.* **1991,** *31,* 559–561.
26. Palmer, C. E. A.; McMillin, D. R.; Kirmaier, C.; Holten, D. *Inorg. Chem.* **1987,** *26,* 3167–3170.
27. Dietrich-Buchecker, C. O.; Marnot, P. A.; Sauvage, J. P.; Kirchhoff, J. R.; McMillin, D. R. *J. Chem. Soc., Chem. Commun.* **1983,** *24,* 513–515.
28. McMillin, D. R.; Kirchhoff, J. R.; Goodwin, K. V. *Coord. Chem. Rev.* **1985,** *64,* 83–92.
29. Crane. D. R.; DiBenedetto, J.; Palmer, C. E. A.; McMillin, D. R.; Ford, P. C. *Inorg. Chem.* **1988,** *27,* 3698–3700.
30. Stacy, E. M.; McMillin, D. R. *Inorg. Chem.* **1990,** *29,* 393–396.
31. Tamilarasan, R.; Ropartz, S.; McMillin, D. R. *Inorg. Chem.* **1988,** *27,* 4082–4084.
32. Tamilarasan, R.; McMillin, D. R. *Inorg. Chem.* **1990,** *29,* 2798–2802.
33. Graham, D. R.; Sigman, D. S. *Inorg. Chem.* **1984,** *23,* 4188–4191.
34. Jain, S. C.; Bhandary, K. K.; Sobell, H. M. *J. Mol. Biol.* **1979,** *135,* 813–840.
35. Goldstein, B. M.; Barton, J. K.; Berman, H. M. *Inorg. Chem.* **1986,** *25,* 842–847.
36. Klemens, F. K.; Fanwick, R. E.; Bibler, J. K.; McMillin, D. R. *Inorg. Chem.* **1989,** *28,* 3076–3079.
37. Diamond, I.; Granelli, S. G.; McDonagh, A. F. *Biochem. Med.* **1977,** *17,* 121–127.
38. Fiel, R. J.; Datta-Gupta, N.; Mark, E. H.; Howard, J. C. *Cancer Res.* **1981,** *41,* 3543–3545.

39. Pasternack, R. K.; Gibbs, E. J. In *Metal–DNA Chemistry;* Tullius, T. D., Ed.; ACS Symposium Series 402; American Chemical Society: Washington, DC, 1989; pp 59–73.
40. Marzilli, L. G. *New J. Chem.* **1990,** *14,* 409–420.
41. Marzilli, L. G.; Banville, D. L.; Zon, G.; Wilson, W. D. *J. Am. Chem. Soc.* **1986,** *108,* 4188–4192.
42. Gouterman, M. In *The Porphyrins;* Dolphin, D., Ed.; Academic: Orlando, FL, 1978; Vol. III, Part A, pp 1–165.
43. Asano, M.; Kaizu, Y.; Kobayashi, H. *J. Chem. Phys.* **1988,** *89,* 6567–6576.
44. Kim, D.; Holten, D.; Gouterman, M. *J. Am. Chem. Soc.* **1984,** *106,* 2793–2798.
45. Hathaway, B. J.; Tomlinson, A. A. G. *Coord. Chem. Rev.* **1970,** *5,* 1–43.

RECEIVED for review November 7, 1991. ACCEPTED revised manuscript May 4, 1992.

12

Molecular Models for Semiconductor Particles

Luminescence Studies of Several Inorganic Anionic Clusters

Thomas Türk[1,2], Arnd Vogler[1], and Marye Anne Fox[2]*

[1]Department of Chemistry, Institut für Anorganische Chemie, Universität Regensburg D-8400 Regensburg, West Germany

[2]Department of Chemistry, University of Texas, Austin, TX 78712

In the transition region from a bulk semiconductor to a molecular cluster, the optical and photocatalytic properties of a semiconductor drastically change as the size of the crystallite is decreased. Several inorganic cluster molecules with a well-defined particle size and structure, (that is, $Zn_4(SPh)_{10}^{2-}$, $Cd_4(SPh)_{10}^{2-}$, $Cd_{10}S_4(SPh)_{16}^{4-}$, and $Zn_4O(OAc)_6$) were synthesized and compared to the analogous mononuclear complexes $Zn(SPh)_4^{2-}$ and $Cd(SPh)_4^{2-}$. These clusters showed structured absorption spectra and a red-shift of their absorption edges with increasing crystallite size. The observed extinction coefficients suggest that the absorption bands can be best ascribed to ligand-to-metal charge-transfer (LMCT) transitions, which can be thought of as molecular cluster analogs of valence-to-conduction band transitions in bulk ZnS and CdS. Luminescence lifetimes and Stokes shifts provide further information about the nature of the optical transitions.

THE UTILITY OF SEMICONDUCTOR SUSPENSIONS as photocatalysts for the oxidative degradation of a wide range of organic compounds (*1*, *2*)

*Corresponding author

0065-2393/93/0238-0233$06.00/0

derives from the collective versus molecular properties of the semiconductor cluster. Thus, band-gap excitation of the semiconductor cluster promotes an electron from the valence band to the conduction band and thereby generates an electron–hole pair that can be trapped by interfacial electron transfer. This species can then react with adsorbed species or exchange with lattice oxides to form other oxy radicals. Adsorbed oxygen acts as an effective electron trap, thus forming surface-bound superoxide. Many organic molecules or water can function as effective single-electron donors to trap a photogenerated hole. Subsequent reactions occurring between these redox-activated species or between these radical ions and other traps present in solution initiate oxygenation and oxidative cleavage, which ultimately lead to degradation and, in some cases, to complete mineralization.

The thermodynamics of these conversions is governed by the energetic positions of the valence and conduction bands. In both metal oxide and metal chalcogenide bulk semiconductor particles, the band positions are dependent on the medium; both bands shift 59 mV per pH unit in aqueous solutions. In nonaqueous suspension, for example in acetonitrile, the band-gap positions can be specified on a standard electrochemical scale. For titanium dioxide, for example, the valence band edge lies at about +2.4 V versus the saturated calomel electrode (SCE), and that for the conduction band lies at about –0.8 V. The potential for the conduction band is almost isoenergetic with the reduction potential of oxygen and thus permits facile electron trapping, but the valence band edge is highly oxidizing and thus permits single-electron oxidation of virtually any organic compound that bears either a lone pair or any conjugation.

The band gap of a given semiconductor is also dependent on particle size. The electronic properties of a given semiconductor cluster depend on the periodic arrangement of many atoms or molecules in a crystal lattice, and hence the shrinking diameter of a given semiconductor crystallite causes a gradual shift from a cluster exhibiting bulk semiconductor properties to one exhibiting well-spaced, discrete orbital levels. This phenomenon, referred to in the literature (*3–8*) as size quantization, is accompanied by pronounced effects on both the optical characteristics and photocatalytic efficiency of the particles. As the cluster size becomes smaller and smaller, the band gap widens, with a dramatic blue shift from the absorption onset observed for the bulk particle.

A number of techniques have been employed to generate size-quantized clusters in stable environments. The in situ preparation of small clusters has been accomplished by the following techniques:

- ion exchange into spatially defined cavities such as polymers (*9–12*), vesicles (*13, 14*), and zeolites (*15–18*)

- surface modification by chemisorption or physisorption of capping reagents (*19–25*)
- biosynthesis (*26*)
- layer formation within Langmuir–Blodgett films (*27*)
- separation of size-disperse mixtures by size exclusion chromatography (*28–31*)

The in situ preparation methods are typically induced by ion exchange of the semiconductor cation into a defined cavity with cation-exchange ability (*32, 33*). Surface techniques can then be used to characterize both the phase of the crystallite (*34*) and the photocatalytic activity of the resulting included material (*32*).

By employing an ion-dilution technique, ultrasmall particles are formed by a method analogous to formation of an inverted micelle within a microemulsion (*12*). In this approach, the cation-to-ionomer cluster ratio is controlled by diluting the exchange solution with an inert ion, for example, Ca^{2+}. The absorption onset for the resulting particles can be tuned over a range of more than 3 eV. Analogous spectral shifts are observed when layers of size-quantized particles, with dimensions smaller than 50 Å, are prepared by exposure of Langmuir–Blodgett films of cadmium arachidate to H_2S to yield semiconductor clusters of cadmium sulfide particles held within a lattice of layered arachidate anions (*27*). As in the clusters generated by ion-dilution of perfluorinated ionomer membranes (Nafion) (*12*), the onset of absorption of this layer is significantly blue-shifted from that of the bulk semiconductor.

Despite the utility of these techniques in forming small particle sizes, all nonetheless give a relatively broad particle size distribution, which complicates the quantitative correlation of the physical properties of the observed semiconductor cluster with particle size. This problem can be overcome if the synthesis of a monodisperse cluster incorporating the atomic subunits of semiconductor particles is undertaken. These synthetic clusters provide models for conventional semiconductors but possess a well-defined size and shape (*35–39*). We describe in this chapter our characterization of the optical and electrochemical properties of three cadmium benzenethiolate clusters (Türk, T.; Resch, U.; Fox, M. A.; Vogler, A., unpublished results) and two zinc benzenethiolate clusters (Türk, T.; Resch, U.; Fox, M. A.; Vogler, A., unpublished results) studied as a function of cluster size. We are particularly interested in characterizing the gradual transition from molecular to semiconductor properties, as had been attempted in our previous effort to characterize the photocatalytic activity and spectroscopy of heteropolyoxyanions in comparison with metal oxide clusters and powders (*40*). Thus, we undertook a search for parallel photocatalytic activity and optical similarities between these clusters and bulk semiconductor particles.

The available crystal structures (*37*) of these metal chalcogenide clusters imply that they are reasonable molecular models for CdS and ZnS in that the coordination environments of both the metal and the sulfur atoms, as well the bond distances and angles, are similar to those observed in the bulk semiconductor. We seek in this study to determine whether the organic capped clusters may bear similar analogy to quantized inorganic clusters. This chapter compares the spectroscopic properties of more detailed studies of each of these families, which are as yet unpublished.

Spectroscopy

The absorption spectra of $Cd(SPh)_4{}^{2-}$, $Cd_4(SPh)_{10}{}^{2-}$, $Cd_{10}S_4(SPh)_{16}{}^{4-}$, $Zn(SPh)_4{}^{2-}$, and $Zn_4(SPh)_{10}{}^{2-}$ show characteristic absorption bands in the ultraviolet region (Türk, T.; Resch, U.; Fox, M. A.; Vogler, A., unpublished results). In the cadmium complex containing only one Cd atom, only a single symmetrical absorption band at 282 nm can be observed, whereas two overlapping bands are present for both the tetranuclear and decanuclear complexes, respectively, at 249 and 275 nm and at about 250 and 280 nm. This additional band presumably derives from the existence of two types of thiolate ligands at bridging and terminal positions. The absorption of benzenethiolate itself (λ = 303 nm, ϵ (molar absorptivity) = 13,600 M^{-1} cm^{-1}) appears at a position well-resolved from these bands. As has been shown for many oxyanions and carbanions, coordination at the negatively charged site with a metal or alkyl group typically induces a pronounced blue shift on the observable absorption band (*41*). The red shift observed for the decanuclear cluster similarly parallels that expected as a ligand-to-metal charge-transfer (LMCT), whereby metal association in the higher molecular weight complex causes a lowering of the metal-centered antibonding orbital. The assignment of this transition as a LMCT transition is also parallel to that expected for the semiconductor CdS, in which the valence band is largely composed of filled sulfide 3p valence orbitals, whereas the conduction band is principally composed of empty Cd 5s orbitals (*39, 42*).

Similar considerations also apply to the zinc complexes. For example, for $Zn(SPh)_4{}^{2-}$, the absorption band at 273 nm is presumably an intraligand transition, because simple monometallic zinc complexes (bound to halide or hydroxide ligands) have been so characterized (*39, 43*), presumably because the zinc 4s orbital lies at too high an energy to be an accessible acceptor orbital for a LMCT transition. Although the energy of this level will come down somewhat in $Zn_4(SPh)_{10}{}^{2-}$, the transition observed is probably a composite of intraligand and LMCT transitions, as has been suggested for a tetranuclear zinc oxocluster $Zn_4O(OAc)_6$ (*39*).

Emission Spectra and Singlet Lifetimes

The multinuclear cadmium complexes exhibit luminescence spectra that are independent of excitation wavelength. As with the absorption spectra, the emission of the decanuclear complex (λ_{max} = 545 nm) is broad and red-shifted from that of the tetranuclear complex (λ_{max} = 500 nm). The substantial Stokes shifts from the absorption maxima just discussed indicate that appreciable geometric distortion occurred upon photoexcitation. Large Stokes shifts have also been previously reported for other tetranuclear clusters of d^{10} metals (*39, 44–47*). Solid samples of the Cd_4 and Cd_{10} powders similarly display intense, broad emissions, with lifetimes of approximately 1 ms at 77 K being obtained by transient diffuse reflectance spectroscopy. Unlike the emission observed in solution, these emissions are not ordered with respect to cluster size as would be predicted for a metal-to-ligand charge-transfer (MLCT) transition and are instead assigned as spectroscopically forbidden intraligand transitions. The much shorter lifetimes associated with the red shift of the emission band and with increasing cluster size for this complex in solution at room temperature allow differentiation of this emission (as an allowed MLCT band) from that observed in the solid state at low temperature.

Similarly, the mononuclear zinc complex gives no detectable emission, but the tetranuclear zinc complex shows weak luminescence at 360 nm. The short lifetime of this emission (~35 ps) permits its assignment as an allowed MLCT transition. The excitation spectrum for this emission does not parallel its absorption spectrum, a result further supporting the previous assignment of the direct absorption as deriving at least in part from intraligand transitions. As with the cadmium clusters, a large Stokes shift (almost 10,000 cm^{-1}) argues for appreciable geometrical excited-state distortion that, as was mentioned with respect to the cadmium clusters, is consistent with a LMCT transition deriving from population of metal antibonding orbitals upon photoexcitation.

Cluster Electrochemistry

Both the oxidation and reduction peak potentials for the cadmium clusters are sensitive to cluster size. The oxidation peak potential shifts from +0.88 V vs. SCE in acetonitrile for the mononuclear cluster to +0.77 V for the tetranuclear cluster to +0.68 V in the decanuclear cluster. Similarly, a reduction wave is absent from the mononuclear complex (lying more negative than –2.8 V vs. the Ag–AgCl electrode), whereas the tetranuclear complex shows a well-defined wave at 2.47 V and the decanuclear complex at –2.02 V. $Zn(SPh)_4^{2-}$ displays a broad oxidation wave at

+0.05 V vs. Ag–AgCl, whereas $Zn_4(SPh)_{10}^{2-}$ exhibits an oxidation peak potential at +0.79 V. Although both oxidative potentials are completely irreversible, their positions are seemingly strongly affected by cluster size in parallel to the effects observed in the cadmium clusters. Neither zinc cluster showed discernible reduction waves within the solvent window (at potentials less negative than –2.8 V vs. Ag–AgCl).

Formation of Charge-Transfer Complexes

A 1:1 mixture of the multinuclear cadmium complexes with methyl viologen in acetonitrile results in the appearance of a new band at about 470 nm for the Cd_4 complex and at 440 nm for the Cd_{10} complex. This blue shift seems to be related to the increasing anionic charge density in the resulting charge-transfer complex, as has been observed for bands formed between methyl viologen and electron-rich naphthalene derivatives (*48*). With the monomolecular cadmium cluster, no evidence for a stable ground-state charge-transfer complex was available from absorption spectroscopy, although the intense blue color of the reduced methyl viologen monocation radical was obvious upon mixing. Presumably the enhanced driving force for thermal electron transfer accounts for this observation.

A broad charge-transfer absorption band is also formed upon mixing in a 1:1 molar ratio $Zn_4(SPh)_{10}^{2-}$ and methyl viologen. The resulting band, centered at 445 nm, could be bleached by flash laser excitation in which the reduced methyl viologen radical, absorbing at 395 nm and 605 nm (*49*), was obvious.

Photosensitivity of the Clusters

UV illumination of any of the cadmium clusters causes disappearance of the UV absorption bands as a new broad emission band at 355 and a shoulder at 440 nm appear. Neither the rate nor the spectral shape of this new emission band is sensitive to oxygen. The rate at which the bleaching occurred was, however, dependent on cluster size, and the degradation of the mononuclear cluster occurs about 5 times as fast as the tetranuclear cluster, which decomposed about twice as fast as the Cd_{10} cluster. Mass spectroscopic analysis shows a product distribution consistent with the formation of thianthrene, benzothiophene, and benzenethiol. Completely parallel reactivity is observed with the zinc clusters, with a strong bleaching of the absorption band occurring upon UV radiation, while an intensely luminescent band appears with a maximum at 355 and shoulder at 440 nm. $Cd(SPh)_4^{2-}$ is roughly 10 times more photosensitive than the tetranuclear cluster. As with the zinc clusters, the rate of the photobleaching is independent of the presence of oxygen, and the same distribution of photoproducts, as indicated by gas chromatography–mass spectrometry, was obtained upon photolysis of these clusters as with band-gap irradiation of a CdS suspension in the presence of an acetonitrile solution of benzenethiol.

Conclusions

Increasing cluster size in the Zn_n and Cd_n anionic clusters bearing benzenethiolate ligands causes shifts in their absorption spectra that reflect increasing LMCT character. Substantial Stokes shifts observed in the multinuclear clusters are consistent with appreciable excited-state geometrical distortion. Electrochemical oxidation and reduction peak potentials are also size-dependent, the oxidations and reductions becoming easier with increasing cluster size. Charge-transfer complexation of the multinuclear complexes with methyl viologen indicates enhanced redox activity. Photosensitivity of the complexes is also consistent with enhanced LMCT with a product mixture resulting from subsequent reactions of a surface-bound benzenethiyl radical. The optical, electrochemical, and photochemical properties of these clusters thus represent reasonable transition models for the development of colligative properties in moving from very small molecules to the bulk semiconductor.

Acknowledgements

We are grateful to the German Federal Government (Department of Technology) for support of the student fellowship that permitted T. Türk's work in Austin and to the U.S. Department of Energy, Office of Basic Energy Sciences, for financial support in Texas. We thank F. Sabin for assistance in the synthesis of the clusters and Jon Merkert for assistance in the electrochemical measurements. The reported lifetimes were measured at the Center for Fast Kinetics Research, a facility jointly supported by the National Institutes of Health and the University of Texas. We are grateful to D. J. Kiserow and S. M. Hubig for assistance in the time-resolved measurements.

References

1. Fox, M. A. *Top. Curr. Chem.* **1991,** *159,* 68.
2. Fox, M. A. In *Photocatalysis;* Serpone, N.; Pelizzetti, E., Eds.; Academic Press: Orlando, FL, 1989; Chapter 13.
3. Henglein, A. *Top. Curr. Chem.* **1988,** *143,* 113.
4. Henglein, A. *Chem. Rev.* **1989,** *89,* 1861.
5. Brus, L. E. *J. Phys. Chem.* **1986,** *90,* 2555.
6. Steigerwald, M. L.; Brus, L. E. *Accts. Chem. Res.* **1990,** *23,* 183.
7. Wang, Y.; Herron, N. *J. Phys. Chem.* **1991,** *95,* 525.
8. Stucky, G. D.; MacDougall, J. E. *Science* **1990,** *47,* 669.
9. Krishnan, M.; White, J. R.; Fox, M. A.; Bard, A. J. *J. Amer. Chem. Soc.* **1983,** *105,* 7002.
10. Kuczynski, J. P.; Miloslavjevic, B. H.; Thomas, J. K. *J. Phys. Chem.* **1984,** *88,* 980.
11. Wang, Y.; Suna, A.; Mahler, W.; Kasowski, R. *J. Chem. Phys.* **1987,** *87,* 7315.
12. Smotkin, E. S.; Brown, R. M., Jr.; Rabenberg, L. K.; Salomon, K.; Bard, A. J.; Campion, A.; Fox, M. A.; Mallouk, T. E.; Webber, S. E.; White, J. M. *J. Phys. Chem.* **1990,** *94,* 7543.
13. Watzke, H. J.; Fendler, J. H. *J. Phys. Chem.* **1987,** *91,* 854.
14. Chang, A. C.; Pfeiffer, W. F.; Buillaume, B.; Baral, S.; Fendler, J. H. *J. Phys. Chem.* **1990,** *94,* 4284.
15. Wang, Y.; Herron, N. J. *J. Phys. Chem.* **1988,** *92,* 4988.
16. Herron, N.; Wang, Y.; Eddy, M. M.; Stucky, G. D.; Cox, D. E.; Moller, K.; Bein, T. *J. Am. Chem. Soc.* **1989,** *111,* 530.
17. Persaud, L.; Bard, A. J.; Campion, A.; Fox, M. A.; Mallouk, T. E.; Webber, S. E.; White, J. M. *Inorg. Chem.* **1987,** *26,* 3825.
18. Pettit, T. L.; Fox, M. A. *J. Org. Chem.* **1985,** *50,* 5013.
19. Dannhauser, T.; O'Neal, M.; Johansson, K.; Whitten, D.; McLendon, G. *J. Phys. Chem.* **1986,** *90,* 6074.
20. Spandel, L.; Haase, M.; Weller, H.; Henglein, A. *J. Am. Chem. Soc.* **1987,** *109,* 5649.
21. Wang, Y.; Suna, A.; McHugh, J.; Hilinski, E. F.; Lucas, P. A.; Johnson, R. D. *J. Chem. Phys.* **1990,** *92,* 6927.
22. Steigerwald, M. L.; Alivasatos, A. P.; Gibson, J. M.; Harris, T. D.; Kortan, R.; Muller, A. J.; Thayer, A. M.; Duncan, T. M.; Douglass, D. C.; Brus, L. E. *J. Am. Chem. Soc.* **1988,** *110,* 3046.
23. Nosaka, Y.; Yamaguchi, K.; Miyama, H. *Chem. Lett.* **1988,** 605.
24. Kortan, A. R.; Hull, R.; Opila, R. L.; Bawendi, M. G.; Carroll, P. J.; Brus, L. E. *J. Am. Chem. Soc.* **1990,** *112,* 1327.
25. Herron, N.; Wang, Y.; Eckert, H. *J. Am. Chem. Soc.* **1990,** *112,* 1322.
26. Dameron, C. T.; Reese, R. N.; Mehra, R. K.; Kortan, A. R.; Carroll, P. J.; Steigerwald, M. L.; Brus, L. E.; Winge, D. R. *Nature* **1989,** *338,* 596.
27. Smotkin, E. S.; Lee, C. L.; Bard, A. J.; Campion, A.; Fox, M. A.; Mallouk, T.; Webber, S. E.; White, J. M. *Chem. Phys. Lett.* **1988,** *152,* 265.
28. Fischer, Ch. H.; Lilie, J.; Weller, H.; Katsikas, L.; Henglein, A. *Ber. Bunsenges Phys. Chem.* **1989,** *93,* 61.
29. Fischer, Ch. H.; Weller, H.; Katsikas, L.; Henglein, A. *Langmuir* **1989,** *5,* 429.

30. Eychmüller, A.; Katsikas, L.; Weller, H. *Langmuir* **1990,** *6,* 1605.
31. Fischer, Ch. H.; Weller, H.; Foijtik, A.; Lume-Pereira, C.; Jonata, E.; Henglein, A. *Ber. Bunsenges. Phys. Chem.* **1986,** *90,* 46.
32. Kakuta, N.; White, J. M.; Campion, A.; Bard, A. J.; Fox, M. A.; Webber, S. E. *J. Phys. Chem.* **1985,** *89,* 48.
33. Mau, A. W. H.; Huang, C. B.; Kakuta, N.; Bard, A. J.; Campion, A.; Fox, M. A.; White, J. M.; Webber, S. E. *J. Am. Chem. Soc.* **1984,** *106,* 6537.
34. Kakuta, N.; White, J. M.; Campion, A.; Bard, A. J.; Fox, M. A.; Webber, S. E. *J. Phys. Chem.* **1985,** *89,* 48.
35. Lacelle, S.; Stevens, W. C.; Kurtz, D. M.; Richardson, J. W.; Jacobsen, R. A. *Inorg. Chem.* **1984,** *23,* 930.
36. Dean, P. A. W.; Vittal, J. J. *Inorg. Chem.* **1986,** *25,* 514.
37. Dance, I. G.; Choy, A.; Scudder, M. L. *J. Am. Chem. Soc.* **1984,** *106,* 6285.
38. Brennan, J. G.; Siegrist, T.; Stuczynski, S. M.; Steigerwald, M. L. *J. Am. Chem. Soc.* **1990,** *112,* 9233.
39. Kunkely, H.; Vogler, A. *J. Chem. Soc., Chem. Commun.* **1990,** 1204.
40. Cardona, R.; Gaillard, E.; Fox, M. A. *J. Am. Chem. Soc.* **1987,** *107,* 6347.
41. Fox, M. A. *Chem. Rev.* **1979,** *79,* 253.
42. Bahnemann, D. W.; Kormann, C.; Hoffmann, M. R. *J. Phys. Chem.* **1987,** *91,* 3789.
43. Bird, B. D.; Day, P. *J. Chem. Soc., Chem. Commun.* **1967,** 741.
44. Vogler, A.; Kunkely, H. *J. Am. Chem. Soc.* **1986,** *108,* 7211.
45. Vogler, A.; Kunkely, H. *Chem. Phys. Lett.* **1989,** *158,* 74.
46. Vogler, A.; Kunkely, H. *Chem. Phys. Lett.* **1988,** *150,* 135.
47. Kunkely, H.; Vogler, A. *Chem. Phys. Lett.* **1989,** *164,* 621.
48. Hubig, S. M. *J. Lumin.* **1991,** *47,* 137.
49. Hubig, S. M.; Dionne, B. C.; Rodgers, M. A. J. *J. Phys. Chem.* **1986,** *90,* 5873.

RECEIVED for review November 7, 1991. ACCEPTED revised manuscript May 11, 1992.

13

Polyoxometalates in Catalytic Photochemical Hydrocarbon Functionalization and Photomicrolithography

Excited-State Lifetimes and Subsequent Thermal Processes Involving $W_{10}O_{32}^{4-}$

Craig L. Hill[1], Mariusz Kozik[2,3], Jay Winkler[2,4], Yuqi Hou[1], and Christina M. Prosser-McCartha[1]

[1]Department of Chemistry, Emory University, Atlanta, GA 30322

[2]Department of Chemistry, Brookhaven National Laboratory, Upton, NY 11973

The energetic and mechanistic features of catalytic photochemical oxidation of organic substrates by polyoxometalates and then the applications of this chemistry to catalytic alkane functionalization, microlithography, and catalytic dehalogenation are succinctly reviewed. Three sets of experiments that affect these areas are presented, and the future development of catalytic photoredox processes effected by polyoxometalates is discussed. The excited state of decatungstate, $W_{10}O_{32}^{4-}$ *(**1**), and the conventional ground-state radical,* tert-*BuO·, have similar reactivities and lead to similar products upon reaction with various organic substrates. The simultaneous photooxidation of cyclooctane and tetrahydrofuran leads to some cross-coupling product and a complex organic product distribution that is consistent to a large degree*

[3]Current address: Department of Chemistry, Canisius College, Buffalo, NY 14208

[4]Current address: Beckman Institute, California Institute of Technology, Pasadena, CA 91125

0065–2393/93/0238–0243$06.00/0

with intermediate freely diffusing radicals. Laser flash photolysis measurements (355-nm frequency tripled Nd:YAG output) of the heretofore unreported emission of **1** *(λ_{max} = 615 nm) establish that the lifetime of the excited state* (**1***) *is 21 ± 3 ps at 25 °C in 9:1 acetonitrile–water solution.*

EARLY TRANSITION METAL–OXYGEN ANION CLUSTERS (*1*), which we will refer to as polyoxometalates, form a large class of inorganic compounds with great molecular diversity and significant potential applications in a range of areas. The sizes, shapes, charges, ground- and excited-state redox potentials, solubilities in polar and nonpolar media, and other properties of polyoxometalates can be systematically varied to a considerable extent. In this chapter we review a range of past and ongoing research involving the interactions of light and polyoxometalate derivatives, including systems for the catalytic functionalization of saturated hydrocarbons, 260-nm photomicrolithography, and catalytic dehalogenation of environmentally undesirable chloro- and bromocarbons. Furthermore, we present new data that address key issues regarding the photophysical and photochemical properties of polyoxometalate-based photooxidation processes and that affect all of these areas.

The isopolyoxometalate complex decatungstate, $W_{10}O_{32}^{4-}$ (**1**), whose structure in polyhedral and bond notation is given in Figure 1, is the principal focus of this work. The first two studies presented here further define the nature of the step involving attack of excited-state polyoxometalate, **1***, on various substrates and subsequent steps involving organic radicals, and the final studies, time-resolved laser flash photolysis measurements of **1**, report the first emission from **1** and delimit the lifetime of the luminescent excited state.

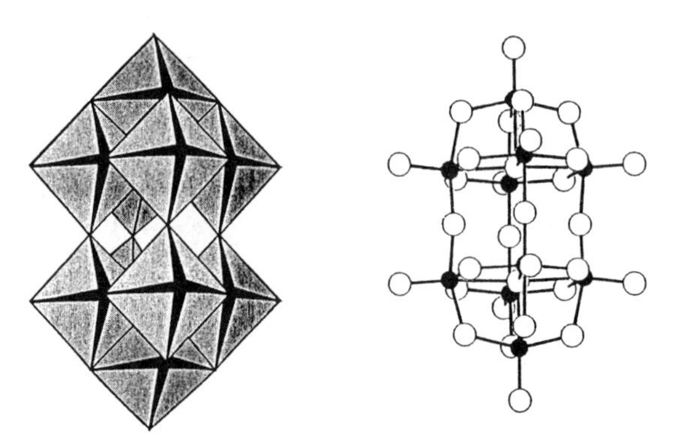

Figure 1. Structure of decatungstate, $W_{10}O_{32}^{4-}$, **1**, *in polyhedral (left) and bond (right) notation.*

Background

Catalytic Photochemical Oxidation of Organic Materials by Polyoxometalates: General Features. Many polyoxometalates composed primarily of V^{V}, Mo^{VI}, W^{VI}, or combinations of these ions that contain principally type I MO_6 octahedra, photooxidize a wide range of organic substrates, $OrgH_2$, with blue or UV light (eqs 1 and 2) (*2–19*).

$$P_{ox} + h\nu \longrightarrow P_{ox}^{*} \tag{1}$$

$$P_{ox}^{*} + OrgH_2 \longrightarrow Org + 2H^{+} + P_{red} \tag{2}$$

where P_{ox} is the oxidized form of the polyoxometalate (all transition metal centers are usually in the d^0 electronic configuration); P_{red} is the reduced form of the polyoxometalate; the * indicates the excited state; and $OrgH_2$ is organic substrate such as alcohols, acids, ethers, amides, and hydrocarbons.

Type I octahedra are those whose central d^0 transition metal ions have only one terminal oxo group. The lowest unoccupied molecular orbitals (LUMOs) of type I MO_6 units are less M–O antibonding than the LUMOs of MO_6 units with two or more terminal oxo groups. One consequence of this property is that whereas thermal or photochemical reduction of polyoxometalates with the type I MO_6 units tends to proceed readily and reversibly, reduction of polyoxometalates with the type II MO_6 units tends to proceed at higher potentials and irreversibly either chemically or electrochemically (*1*). Reoxidation of the reduced polyoxometalates, P_{red}, either by H_2 evolution or O_2 reduction, leads to the net processes catalytic in polyoxometalate in equations 3 and 4, respectively.

$$OrgH_2 + h\nu \xrightarrow{P_{ox}\ \text{(catalyst)}} Org + H_2 \tag{3}$$

$$OrgH_2 + \frac{1}{2}O_2 + h\nu \xrightarrow{P_{ox}\ \text{(catalyst)}} Org + H_2O \tag{4}$$

Equation 3 is photosynthetic, converting substantial light into chemical energy, particularly when the organic substrates, $OrgH_2$, are hydrocarbons. Equation 4 provides an entry into catalyzed O_2-based organic oxidations that do not proceed by the usual autoxidation mechanisms dominated by radical chains. Substrate oxidation and polyoxometalate reduction (eq 2) can be separated from reduced polyoxometalate reoxidation by O_2; therefore, the substrate oxidation process in equation 4 does not involve attack by alkoxy and other oxy radicals but rather by the excited-state polyoxometalate, and O_2 serves only to reoxidize the reduced polyoxometalate.

The absorption and photoredox action spectra of polyoxometalates (eqs 1 and 2) can be moved toward the visible regime or the terrestrial solar spectrum both by substitution of d^0 ions that absorb more in the visible region (e.g., V^V substituted for W^{VI}) or by medium effects (*7, 13*). Electron donor–acceptor complexes between polyoxometalates and electron-rich organic substrates have been characterized in both the solid state and solution and can give rise to bathochromic or red shifts of several hundred nanometers in the absorption and photoredox action spectra (*13*).

Catalytic Modification of Hydrocarbons. Before 1985 the only organic substrates photooxidized by polyoxometalates were alcohols and classes of compounds with fairly low potentials, but we believed that the excited states of some polyoxometalates should have adequate energies to overcome anticipated substantial kinetic barriers (overpotentials) associated with oxidizing more difficult and interesting types of materials, including alkanes and hydrocarbon polymers. The excited-state potential for the common Keggin polyoxotungstate, $\alpha\text{-}PW_{12}O_{40}^{3-}$, should be $\sim$+3.0 V versus the normal hydrogen electrode (NHE) (the measured ground-state potentials of $\sim$+0.1 V in acetonitrile + light energy at 420 nm, the approximate location of the 0,0 transition, of $\sim$+2.95 V). After our initial report (*14*) of alkane photooxidation by $\alpha\text{-}PW_{12}O_{40}^{3-}$ under anaerobic conditions and a subsequent detailed study of this system (*15*), Yamase and Usami (*16*) in Japan reported oxidation of alkenes under similar conditions. Although substantial chemistry is seen under aerobic conditions, generally these conditions are avoided for initial studies because conventional autoxidation chemistry (radical chain oxidation by O_2) initiated by radical intermediates generated by the polyoxometalate (eq 2) can obscure the inherent chemistry between the organic substrates and the excited polyoxometalates themselves.

Although the photooxidation of readily oxidized or electron-rich classes of organic materials by polyoxometalates probably proceeds by initial electron transfer, considerable evidence from the first studies indicated that the alkanes and alkenes are photooxidized by initial atom transfer (H abstraction). It appeared reasonable that organic radicals generated from hydrocarbon substrates in equation 2 might be susceptible to high-efficiency trapping, oxidation, or even reduction, a condition rendering the overall hydrocarbon functionalization chemistry considerably broader in scope and potentially more useful. This condition proved to be the case. Systematic subsequent studies led to the development of polyoxometalate-based systems that couple the initial photochemical generation of radicals with rapid subsequent thermal processes. The more oxidizing polyoxotungstate systems such as the Keggin heteropolytungstates, $X^{n+}W_{12}O_{40}{}^{(8-n)-}$ (where X^{n+} is one of several main-group or transition metal ions situated in the central T_d cavity of the molecule), effect alkane functionalization via radical generation and subsequent radical oxidation. Products derived from carbonium ions are observed: principally the most substituted or thermodynamic olefins, and in acetonitrile solvent, *N*-alkylacetamides, for example, equations 5 and 6 (*14, 17*).

$$\text{2-methylpentane} \xrightarrow[h\nu,\ CH_3CN,\ Ar]{\alpha\text{-}H_3PW_{12}O_{40}\cdot nH_2O} \text{2-methyl-2-pentene} + H_2 \quad (5)$$

~100% selectivity

$$(CH_3)_3C\text{–}H \xrightarrow[h\nu,\ CH_3CN,\ Ar;\ \text{reaction aided by Pt(0)}]{\alpha\text{-}H_3PW_{12}O_{40}\cdot nH_2O} (CH_3)_3C\text{–}NHCOCH_3 + H_2 \quad (6)$$

~95% selectivity

Isobutane gives an exceptionally high ratio of *N*-alkylacetamide to alkene product as only tertiary and primary C–H centers are present (only a terminal alkene, one with a CH_2 group, can be formed). With more reducing polyoxometalates such as **1**, radical oxidation is not seen. In these cases, radical cage escape and subsequent radical–radical reactions are observed. These give rise preferentially to the least substituted or nonthermodynamic olefins, which represent the kinetic products of radical–radical disproportionation (eq 7) (*17*).

$$\text{2-methylpentane} \xrightarrow[h\nu,\ CH_3CN,\ Ar]{(n\text{-}Bu_4N)_4W_{10}O_{32}} \text{2-methyl-1-pentene} + H_2 \quad (7)$$

~95% selectivity

Attack on alkanes in halogenation and most conventional radical chain oxidation systems do not give rise to these interesting products as other radical-trapping processes interfere.

Alkane photooxidation by **1** under conditions where substantial reduced polyoxometalate, P_{red}, accumulates results in reduction of alkyl radical intermediates to carbanions (*18*). The carbanion intermediates are then rapidly trapped by electrophiles such as D^+ sources (e.g., D_2O) or nitriles to generate deuterated alkane or alkyl methyl ketones, respectively.

The ketones arise from hydrolysis of imine intermediates *(18)*. The mechanisms for these processes are summarized in Figure 2. The rate law, isotope effect, product distribution, and spectroscopic studies are presented in references 17–19.

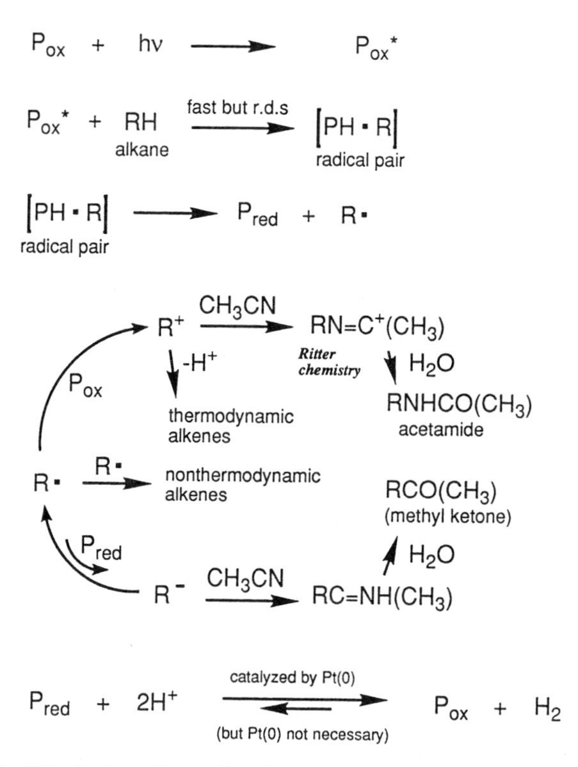

Figure 2. Principal pathways involved in the photochemical functionalization of alkanes by polyoxometalates based on rate law, isotope effect, inorganic and organic product distribution, and spectroscopic studies.

Subsequent detailed studies *(19)* have shown that alteration of the ground- and excited-state redox potentials of **1** by protonation completely changes the chemistry from that characterized by radical–radical reactions and radical reduction to that characterized by radical oxidation and carbonium ions. Both the kinetics and the chemistry of the protonated form of **1** are completely separable from that of the unprotonated form. Equation 8 gives a fairly complete steady-state rate law for photoreduction of **1** (P) (production of P_{red}); the first term involves the protonated form of **1** (solely applicable in the presence of >2.5 equiv of acid), and the second term involves the unprotonated form of **1** (solely applicable in the absence of acid). At intermediate values of pH both forms and both chemistries are applicable *(19)*.

$$\frac{+\mathrm{d}[\mathrm{P_{red}}]}{\mathrm{d}t} = \left(\frac{\mathrm{H}_k\mathrm{P}^{(4-k)-}}{[\mathrm{P}^{4-}]_0}\right)\left(\frac{\Phi_k I_{\mathrm{aT}} k_k[\mathrm{RH}]}{k_{qk}[\mathrm{RH}] + k_{dk}}\right) + \left(\frac{[\mathrm{P}^{4-}]_0 - \mathrm{H}_k\mathrm{P}^{(4-k)-}}{[\mathrm{P}^{4-}]_0}\right)\left(\frac{\Phi_0 I_{\mathrm{aT}} k_0[\mathrm{RH}]}{k_{q0}[\mathrm{RH}] + k_{d0}}\right) \quad (8)$$

where P is the decatungstate moiety; superscripts designate charge; $+\mathrm{d}[\mathrm{P_{red}}]/\mathrm{d}t$ is the rate of decatungstate reduction, including thermal reduction by alkyl radical intermediates; $[\mathrm{P}^{4-}]_0$ is the initial concentration of $W_{10}O_{32}^{4-}$; I_{aT} is the the amount of light absorbed by all the decatungstate species; Φ_k is the quantum yield for production of the photochemically active excited state; k_k is the rate constant for reductive quenching of excited state; k_{dk} is the rate constant for nonradiative decay of excited state; and k_{qk} is the total alkane-quenching rate constant. Here $k = 0$ for the nonprotonated species, and $k = 1$ or 2 for the reactive protonated species. Reference 19 gives a full discussion of the terms, derivation, and applicability of equation 8.

Microlithographic Applications. Photochemical processes, like electrochemical processes, are far less common than thermal processes in industrial chemical synthesis, but photochemical processes are common in high technology. One area in which photochemistry currently has a sizable commercial role is in the area of microlithography, the technology used to fabricate high-resolution circuit patterns on ultrahigh-purity semiconductor substrates such as Si (*20–22*). A general scheme for the lithographic process is shown in Figure 3. After selective irradiation, the irradiated areas are worked up by baking, etching, and various development procedures to yield the circuit pattern. A negative tone image is shown as the ultimate product of this multifaceted development process in Figure 3.

Three goals currently being sought with respect to microlithographic technology are (1) photoresist systems that will be photosensitive in the deep UV region (250–260-nm light), (2) chemical amplification systems for maximizing the differential solubility or reactivity between the irradiated and nonirradiated areas (photoresist systems that generate H^+ are the most common category here; such systems are under intense development) (*21*), and (3) pattern transfer or etching via gases or plasmas rather than liquids (*20, 22*).

Various polyoxometalate derivatives are of interest in microlithography as they potentially fulfill all three desired goals. First, the maximum in the absorption and photoredox action spectra for many polyoxotungstates is ~260 nm. Second, equation 2 can be used to photochemically generate acid, and considerable control over this process can be achieved. Third, solubilization of resist material achieved by photooxidation and/or acid generation followed by baking could be followed by removal of tungsten as volatile WF_6 after treatment with HF gas. In general polyoxometalate-based photoresist materials should offer reasonable sensitivity, contrast, and resolution, but it is far too early to have ranges in these parameters or much knowledge about how readily they can be rationally and systematically altered. Work in this area is just beginning (*23, 24*).

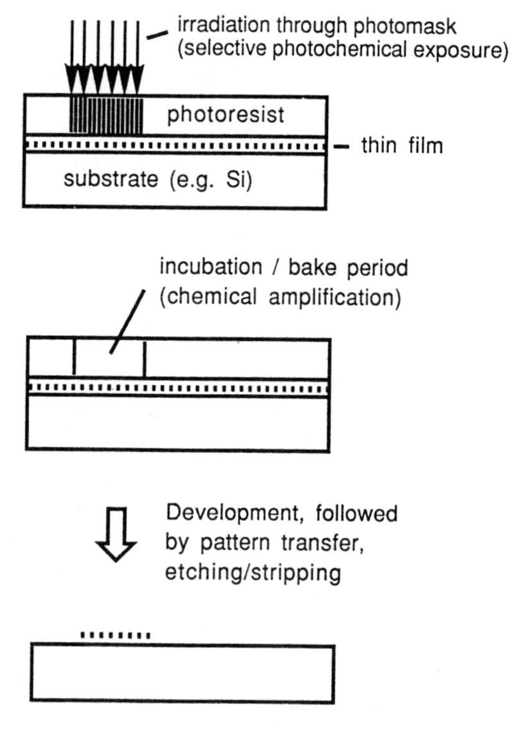

Figure 3. Scheme for the general steps in photochemical microlithography.

Catalytic Photochemical Dehalogenation of Halocarbons. The use of halocarbons as solvents is becoming increasingly less viable in industrial processes as concern about the toxicity and carcinogenicity of this class of compounds grows (*25–28*). Some radicals effect C–X (X is bromine or chlorine) bond cleavage (*29*), for example,

$$(CH_3)_2\dot{C}OH + RX \longrightarrow (CH_3)_2CO + R\cdot + X^- + H^+ \qquad (9)$$

We reasoned that combining the catalytic and sustained generation of radicals from less toxic substrates such as 2-propanol using polyoxometalate photocatalysis (eq 2 with the appropriate polyoxometalates including **1**) with equation 9 would facilitate the catalytic dehalogenation of halocarbons. This proved to be the case (*27*). One example, the photodechlorination of CCl_4 by 2-propanol, is given in equation 10.

$$3CCl_4 + 3(CH_3)_2CHOH + W_{10}O_{32}{}^{4-} + h\nu \longrightarrow CHCl_3 + C_2Cl_6 + 3H^+ + 3Cl^- + 3(CH_3)_2CO + H_2W_{10}O_{32}{}^{4-} \quad (10)$$

Furthermore, superoxide, a species known to dehalogenate halocarbons (*30*) generated in the reoxidation of reduced polyoxometalates with O_2, also participates in catalytic homogeneous dehalogenation by polyoxometalates. A third mode of dehalogenation involves direct reaction of some reduced polyoxometalates with halocarbons themselves (*27, 31*).

Experimental Section

Materials and Methods. The different salts of **1**, $(n\text{-}Bu_4N)_4W_{10}O_{32}$ (*32*) and $Na_4W_{10}O_{32}\cdot nH_2O$ (*17*), were prepared by literature methods. The acetonitrile solvent was reagent grade or Burdick and Jackson glass-distilled grade. The tetrahydrofuran (THF) was Burdick and Jackson glass-distilled grade and was run through a column of activity-grade-I neutral alumina just prior to use. The organic substrates were reagent grade from Wiley, Fluka, Aldrich, and Pfaltz and Bauer and had purities >99.9% by gas chromatography (GC) except for the di-*tert*-butyl peroxide (DTBPO), which had a purity of 98%. The gas chromatographic (GC), GC–mass spectrometric (GC–MS) and spectroscopic (NMR, IR, and UV–visible) data were collected as previously described (*15, 17*).

Kinetic Studies (Table I). All the relative rates for both the reactions of the polyoxometalate excited state (**1***) and the authentic localized ground-state radical, *tert*-BuO•, were determined under competitive kinetic conditions because the reproducibility was the highest under these conditions. The reactions were monitored by the chromophore of reduced form of **1** (λ_{max} = 630 nm, ϵ (molar absorptivity) = 12,500 cm^{-1} M^{-1} in CH_3CN) and the initial rate method (substrate conversions were <2% in all cases), and optically dense conditions were used throughout. A 1000-W Xe arc lamp coupled with a 290-nm interference filter and 1-cm path length quartz cuvette were used.

In the reactions with **1***, the concentrations of the substrates and polyoxometalate were 1 M and 2.0 mM, respectively, unless noted otherwise. Error due to reoxidation of reduced polyoxometalate (either by H_2 evolution or O_2 reduction) under the conditions of measurement was assessed and found to be negligible.

Table I. Relative Rates of Oxidation of Five Organic Substrates by the *tert*-Butoxy Radical and the Excited State of $(n\text{-}Bu_4N)_4W_{10}O_{32}$ (1*)

Reacting Species	cyclohexane (S)	cyclohexene	cyclohexanol (OH, S)	cyclohexanecarboxylic acid (COOH, S)	cyclooctane (S)
1*	1.0	3.0	12	2.1	5.8
tert-BuO·	1.0	6.8	2.8	1.3	3.5

NOTES: For the reactions of **1***, stirred acetonitrile solutions of the substrates (1.0 M in each) and **1** (2.0 mM) under Ar at ~25 °C were irradiated by using a 1000-W Xe arc lamp with a 290-nm interference filter. For reactions of *tert*-BuO·, stirred acetonitrile solutions of the substrates (0.1 M in each) and the *tert*-butoxy radical precursor, di-*tert*-butyl peroxide, under Ar at ~25 °C were irradiated with the same lamp but with a 280-nm cutoff filter for 1.0 h.

An "S" inside the structures of the products designates saturated ring systems; for both sets of reactions, the rate of oxidation of the cyclohexane is taken as 1.0. The absolute value for the rate (not rate constant) of reaction of **1*** with cyclohexane is 2.0×10^{-7} M s^{-1}; the absolute value for the rate of reaction of *tert*-BuO· with cyclohexane was not determined. Competitive kinetics and the initial rate method were used in all rate determinations. The products were identified and quantified in all reactions by GC and GC–MS.

In the reactions with *tert*-BuO·, cyclohexane (0.1 M), the second substrate (0.1 M), and the radical precursor (0.5 M di-*tert*-butyl peroxide), were irradiated in acetonitrile solution under argon for 1 h with a 1000-W Xe lamp but now using a 280-nm cutoff filter. This filter was chosen to provide an appropriate level of irradiance for this rate study. Although parity of substrate conversion relative to the reactions with **1*** was desired, this proved to be difficult to achieve experimentally with *tert*-BuO·; final conversions in these reactions ranged from 10 to 60%. Rates were determined in the latter reactions by quantification of the organic products with time by gas chromatography (GC): 5% phenylmethylsilicon capillary column, nitrogen carrier gas, temperature programming, and flame ionization detection (FID). Inasmuch as *tert*-BuO· absorbs weakly at 320 nm, control reactions were conducted by using both light of λ >280 and >400 nm. The same relative rates were produced in both cases, consistent with a minimal contribution from reactions of excited-state *tert*-BuO·. All reactions were repeated four times, and the reproducibility was ±15%.

Product Distribution Studies in Table II. To be able to quantify cross-coupling products, these reactions were run at varying but generally high concentrations of both cyclooctane and tetrahydrofuran substrates. The relative quantities of both these substrates and acetonitrile solvent are given in Table II (column 2). In a typical reaction, the concentration of **1** was 0.0026 M, the total solution volume was 10 mL, and 3 mg of powdered Pt(0) was suspended in the stirred solution during irradiation (550-W medium-pressure mercury lamp with

Table II. Simultaneous Catalytic Anaerobic Photochemical Oxidation of Cyclooctane and Tetrahydrofuran by $(n\text{-}Bu_4N)_4W_{10}O_{32}$–Pt(0)

Substrates and Solvent[b]	*H_2 Product*[c] *(moles)*	*Turnovers*[d]	*Amount of Product (moles per mole of $W_{10}O_{32}^{4-}$)*[a]									*Other Organic Products*
			[dihydrofuran structure]	[alcohol structure, OH]	[cyclooctene structure]	[lactone structure]	[THF dimer structure]	*Bu_3N*	[acetylcyclooctane structure]	[THF–cyclooctane structure]	[bicyclooctyl structure]	
1 (7:67:26)	7.0×10^{-4}	29.3	6.5	0.4	0.2	6.2[e]	6.9 (15)[f]	1.0[g]	0.1	1.3 (3)[h]	1.3 (8)[i]	5.4 (16)
2[j] (7:67:26)	1.2×10^{-3}	34.6	10.0	0.3	0.2	5.8[e]	6.2 (19)[f]	1.0[g]	0.1	1.2 (3)[h]	1.3 (7)[i]	8.5 (25)
3 (7:7:86)	1.6×10^{-3}	16.7	5.4	0.1	1.1	0.9[e]	0.1 (2)[f]	1.4[g]	1.6	1.7 (7)[h]	1.2 (6)[i]	3.2 (23)
4[k] (7:7:86)	3.2×10^{-7}	0	0	0	0	0	0	0	0	0	0	0
5[l] (7:7:86)	6.6×10^{-6}	0	0	0	0	0	0	0	0	0	0	0

NOTE: Reaction conditions were as follows: 10 mL of a 0.0026 M solution of $W_{10}O_{32}^{4-}$ catalyst in substrates and solvent and 3 mg of suspended Pt(0) powder under Ar at ~57 °C was irradiated with a 550-W Hg lamp in a Pyrex Schlenk flask for 24 h with stirring. Products were identified by GC and GC–MS.

[a] The number of discrete isomers or products combined to produce the yields reported are given in parentheses after the yield.

[b] The numbers are the volume percent ratios of cyclooctane, THF substrates, and acetonitrile (AN) solvent.

[c] The amount of hydrogen produced after 24 h of irradiation.

[d] Turnovers equals the total amount of organic products divided by the amount of catalyst, both in moles.

[e] The yield includes the amount of 2-acetyltetrahydrofuran product because the two products coinject on the GC.

[f] The number of products includes all THF coupling products.

[g] Bu_3N is derived from the tetrabutylammonium cation of the catalyst.

[h] The number of products includes all THF–cyclooctane cross-coupling products.

[i] The number of products includes all cyclooctane coupling products.

[j] The same reaction conditions as reaction 1 except that hydrogen was removed from the reaction after 8 and 16 h of irradiation.

[k] No $W_{10}O_{32}^{4-}$ catalyst or Pt(0) was present.

[l] Pt(0) was present but not $W_{10}O_{32}^{4-}$ catalyst.

Pyrex 280-nm cutoff) under an argon atmosphere at ~57 °C. The products were identified and quantified by GC as described and by GC–MS.

Picosecond Laser Flash Photolysis Studies. The laser flash photolysis apparatus was described previously (*33*). The picosecond time-resolved emission spectra were recorded following excitation with a 30-ps pulse from the third harmonic (355 nm) of a flash-lamp-pumped, actively–passively mode-locked Nd:YAG laser. The repetition rate of the laser system was 1 Hz. Emitted light was dispersed by a spectrograph and directed to the entrance slit of a streak camera. The response of the streak camera was corrected for nonuniform response of the camera and for dark current. The data in Figures 4 and 5 were obtained by using a 2 mM solution of the sodium salt of **1** in 9:1 acetonitrile–water (v/v) after averaging of 1000 laser shots. Similar results were obtained by using the lithium salt of **1** in 9:1 acetonitrile–water and the sodium salt of **1** in 100% dry acetonitrile. The signal-to-noise ratio on the sodium salt data was not as good as on the lithium salt data, however, as a consequence of solubility limitations of the complex in this medium.

Results and Discussion

Comparative Kinetics and Product Distribution Studies Involving the Excited State of 1 and the *tert*-Butoxy Radical, *tert*-BuO•. Although considerable evidence, as already discussed, is consistent with attack on alkane by the excited state of polyoxotungstates (eq 2, where $OrgH_2$ is an alkane and P_{ox} is a polyoxotungstate) being H-atom abstraction, we seek to compare reactivities of such excited states (e.g., **1***) with those of conventional ground-state radicals. Table I gives the relative reactivities of the excited state of **1** (or **1***) and the localized ground-state *tert*-butoxy radical, *tert*-BuO•, with representative organic substrates. Reaction conditions are given in the table footnote and in the Experimental Section. The substrates are cyclohexane, cyclohexene (the corresponding cycloalkene with allylically activated C–H bonds), cyclohexanol (the corresponding alcohol), cyclohexanecarboxylic acid, and cyclooctane.

In Table I the relative rates for the reactions of both **1*** and *tert*-BuO• are reported relative to the parent hydrocarbon. The rates for the different radical-like species are not normalized with respect to each other. The rates for all the reactions of **1*** are reported at low conversions (<2%) to minimize the obscuring effects of subsequent reactions. Although the relative rates are not identical for the two types of reactive species, the trends in reactivities are similar, a result further indicating H-atom abstraction character in the reactions of **1*** with these substrates.

The initial products of all reactions were also determined. Reaction of **1*** with cyclohexane, cyclohexene, cyclohexanol, cyclohexanecarboxylic

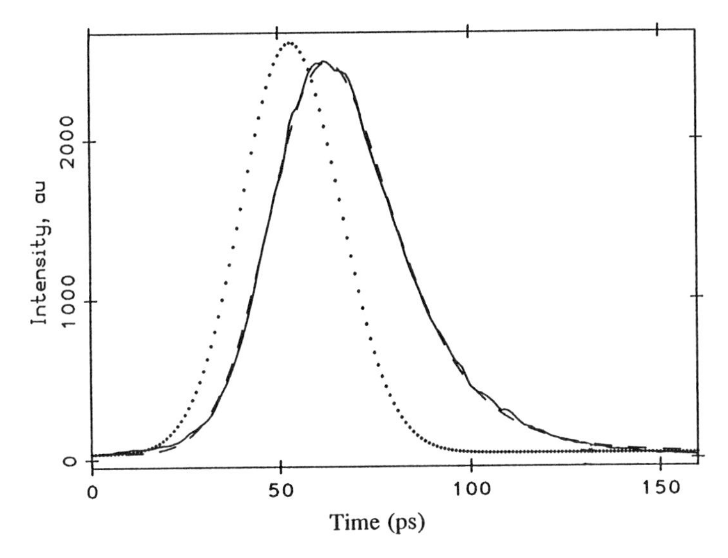

Figure 4. Plots as a function of time in picoseconds: emission intensity (combined intensity from 600 to 640 nm) (—); single exponential fit to the data from deconvolution (- - -) (the lifetime inferred from the data is 21 ± 3 ps); and instrument response function (35 ps) determined using Raman scattering from water at 405 nm (• • •).

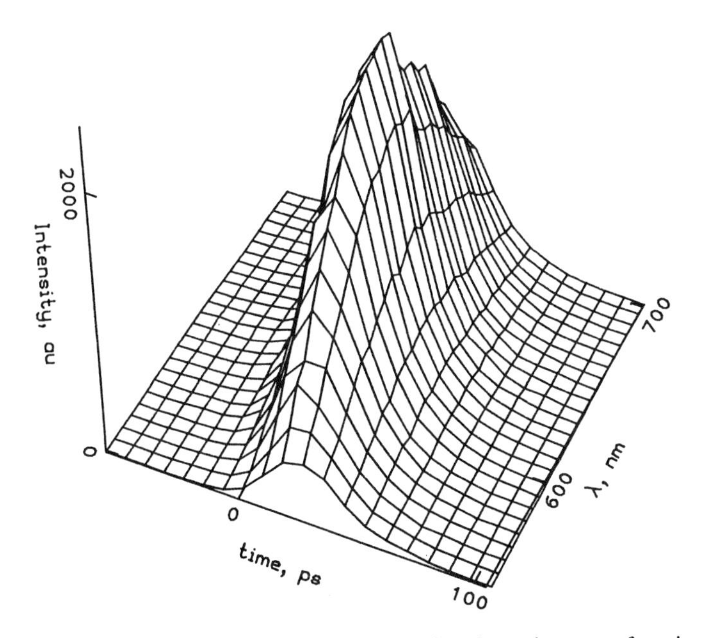

Figure 5. Three-dimensional plot of emission intensity as a function of time and wavelength.

acid, and cyclooctane gave cyclohexene, 3,3′-dicyclohexenyl, cyclohexanone, cyclohexane, and cyclooctene, respectively, all in ~90% selectivity. All these products are also consistent with initial radical H abstraction.

The ability of **1** to catalyze the photochemical cross coupling of alkanes with other types of organic reactants was also examined in this investigation. The impetus for examining such reactions is twofold: they would provide further mechanistic information, and they might be applicable to cross-linking and solubilization reactions in irradiated photoresist materials. The products generated by the simultaneous photooxidation of two representative substrates, cyclooctane and the ether, tetrahydrofuran (THF), by **1**, are given in Table II. Reaction conditions are given in the table footnote and in the Experimental Section. The reactions were carried out under pseudo-first-order conditions (substrates present in great excess), and thus low conversions (~1% for THF and ~2% for cyclooctane) to minimize the effect of secondary reactions on the product distributions. A catalytic amount of the hydrogen-evolution catalyst, Pt(0) powder, was added to facilitate reoxidation of reduced **1** and thus increase turnovers. The presence of Pt(0) minimally perturbed the initial kinetic product distributions, its main effect being to increase the amount of the cyclooctyl methyl ketone relative to the other products. Dark control reactions with and without **1** and Pt(0) were conducted; none gave any organic oxidation products detectable by gas chromatography.

Several points follow from the product distribution data in Table II. First, several dehydrogenation and coupling products are formed, and their relative quantities in the different reactions are consistent with their formation via dominant if not exclusive disproportionation and coupling of intermediate 2-tetrahydrofuranyl and cyclooctyl radicals. Second, cross-coupling products (isomers of 2-tetrahydrofuranylcyclooctane) are formed and in roughly statistical amounts, judged from the other products derived from the 2-tetrahydrofuranyl and cyclooctyl radicals. Third, the mechanism of generation by butyrolactone under these anaerobic conditions is not clear at the present time, but oxidation of 2-tetrahydrofuranyl radical to the corresponding resonance-stabilized carbonium ion and subsequent capture by the few equivalents of water present is most likely. Both tributylamine and 1-butanol are derived from the tetra-*n*-alkylammonium counterions of **1**. Fourth, the simultaneous presence of H_2 and Pt(0) has minimal effect on the product distributions (cf., reactions 1 and 2 in Table II). This minimal effect is surprising, as these reducing conditions were shown to be kinetically competent in other polyoxometalate systems examined earlier to alter the quantity of the largely carbanion-derived alkyl methyl ketone products (*15, 34*).

All the rate and product studies presented here are consistent with the thermal reactions of intermediate caged or freely diffusing radicals rather than the photochemical generation of these species as the primary factors behind the tremendous diversity of documented polyoxometalate-catalyzed hydrocarbon functionalization processes.

Lifetime of and Emission from 1*. Time-resolved measurements of **1** in the picosecond time domain were instigated for several reasons. Only recently have time-resolved laser flash photolysis studies of polyoxometalate systems of any kind been reported and none on the systems with the potent photooxidation or extensive thermal redox capabilities such as **1** (*13, 35, 36*). Furthermore, recent work (*13*) has made it clear that the time domains for the photoredox events in polyoxotungstates are very fast indeed, with substantial activity in the subnanosecond regime. Second, the conventional wisdom has been that redox-active polyoxotungstates are nonluminescent, and statements to this effect have repeatedly been made by other groups (*7*). We were not comfortable with such proclamations because weak orange emission from **1** had been noted by our group. The purity of the complexes and the presence of impurities in the solvent can have a significant effect on emission quantum yields of polyoxometalate excited states, although no substantive investigation addressing this subject has yet been published. Third, time-resolved laser flash photolysis data on redox-active polyoxometalates could, in principle, answer a number of questions regarding the photophysics and photochemistry of these compounds.

Irradiation of **1** (λ_{max} = 322 nm) with a frequency-tripled Nd:YAG laser system (355-nm excitation) results in a pale orange emission with a large Stokes shift (λ_{max} = 615 nm). Similar results are observed with the sodium or lithium salts of **1** in 9:1 acetonitrile–water (v/v) or in dry acetonitrile. Figure 4 plots the emission intensity for the sodium salt in dry acetonitrile as a function of time (as channel number; 1 ps per channel) along with a single exponential fit to the data and the instrument response function. A lifetime for the emission of 21 ± 3 ps is extracted from the data. Figure 5 is a three-dimensional plot of emission from the sodium salt of **1** as a function of time and wavelength.

Acknowledgements

Research performed at Emory University was supported by National Science Foundation (Grant No. CHE–9022317) and the Army Research Office (Grant No. DAAL03–91–G–0021), and research performed at Brookhaven National Laboratory was carried out under Contract No. DE–AC02–76CH00016 with the U.S. Department of Energy and supported by its Division of Chemical Sciences, Office of Basic Energy Sciences.

References

1. Pope, M. T. *Heteropoly and Isopoly Oxometalates;* Springer-Verlag: Berlin, Germany, 1983.
2. Akid, R.; Darwent, J. R. *J. Chem. Soc., Dalton Trans.* **1985**, 395.
3. Attanasio, D.; Suber, L. *Inorg. Chem.* **1989,** *28,* 3781.
4. Hill, C. L.; Bouchard, D. A. *J. Am. Chem. Soc.* **1985,** *107,* 5148.
5. Chambers, R. C.; Hill, C. L. *Inorg. Chem.* **1990,** *112,* 8427.
6. Nomiya, K.; Miyazaki, T.; Maeda, K.; Miwa, M. *Inorg. Chim. Acta* **1987,** *127,* 65.
7. Papaconstantinou, E. *Chem Soc. Rev.* **1989,** *18,* 1.
8. Savinov, E. N.; Saidkhanov, S. S.; Parmon, V. N.; Zamaraev, K. I. *Doklady, Phys. Chem. USSR* **1983,** *272,* 741.
9. Shul'pin, G. B.; Kats, M. M. *Zhurn. Obshch. Khim.* **1989,** *59,* 2738.
10. Ward, M. D.; Brazdil, J. F.; Mehandu, S. P.; Anderson, A. B. *J. Phys. Chem.* **1987,** *91,* 6515.
11. Yamase, T.; Watanabe, R. *J. Chem. Soc., Dalton Trans.* **1986,** 1669.
12. Yamase, T.; Suga, M. *J. Chem. Soc., Dalton Trans.* **1989,** 661.
13. Hill, C. L.; Bouchard, D. A.; Kadkhodayan, M.; Williamson, M. M.; Schmidt, J. A.; Hilinski, E. F. *J. Am. Chem. Soc.* **1988,** *110,* 5471.
14. Renneke, R. F.; Hill, C. L. *J. Am. Chem. Soc.* **1986,** *108,* 3528.
15. Renneke, R. F.; Hill, C. L. *J. Am. Chem. Soc.* **1988,** *110,* 5461.
16. Yamase, T.; Usami, T. *J. Chem. Soc., Dalton Trans.* **1988,** 183.
17. Renneke, R. F.; Pasquali, M.; Hill, C. L. *J. Am. Chem. Soc.* **1990,** *112,* 6585.
18. Prosser-McCartha, C. M.; Hill, C. L. *J. Am. Chem. Soc.* **1990,** *112,* 3671.
19. Renneke, R. F.; Kadkhodayan, M.; Pasquali, M.; Hill, C. L. *J. Am. Chem. Soc.* **1991,** *113,* 8357.
20. Reichmanis, E.; Thompson, L. F. *Chem. Rev.* **1989,** *89,* 1271.
21. Reichmanis, E.; Houlihan, F. M.; Nalamasu, O.; Neenan, T. X. *Chem. Mater.* **1991,** *3,* 394.
22. MacDonald, S. A.; Schlosser, H.; Ito, H.; Clecak, N. J.; Willson, C. G. *Chem. Mater.* **1991,** *3,* 435.
23. Yoshimura, T.; Ishikawa, A.; Okamoto, H.; Miyazaki, H.; Sawada, A.; Tanimoto, T.; Okazaki, S. *Microelectron. Eng.* **1991,** *13,* 97.
24. Carls, J. C.; Argitis, P.; Heller, A. *J. Electrochem. Soc.* **1992,** *139,* 786.
25. Ollis, D. F.; Pelizzetti, E.; Serpone, N. In *Photocatalysis: Fundamentals and Applications;* Serpone, N.; Pelizzetti, E., Eds.; Wiley: New York, 1989.
26. Pelizzetti, E.; Carlin, V.; Minero, C.; Grätzel, M. *New J. Chem.* **1991,** *15,* 351.
27. Sattari, D.; Hill, C. L. *J. Chem. Soc. Chem. Commun.* **1990,** 634.
    Sattari, D.; Hill, C. L. *J. Am. Chem. Soc.* **1993,** *115,* in press.
28. Maldotti, A.; Bartocci, C.; Amadelli, R.; Carassiti, V. *J. Chem. Soc., Dalton Trans.* **1989,** 1197.
29. Brault, D.; Bizet, C. Morliere, P.; Rougee, M.; Land, E. J.; Santus, R.; Swallow, A. J. *J. Am. Chem. Soc.* **1980,** *102,* 1015.
30. Sugimoto, H.; Matsumoto, S.; Sawyer, D. T. *J. Am. Chem. Soc.* **1987,** *109,* 8081.
31. Eberson, L.; Ekström, M. *Acta Chem. Scand.* **1988,** *B42,* 113.

32. Chemseddine, A.; Sanchez, C.; Livage, J.; Launay, J. P.; Fournier, M. *Inorg. Chem.* **1984,** *23,* 2609.
33. Winkler, J. R.; Netzel, T. L.; Creutz, C.; Sutin, N. *J. Am. Chem. Soc.* **1987,** *109,* 2381.
34. Renneke, R. F. Ph.D. Thesis, Emory University, Atlanta, GA, 1989.
35. Kraut, B.; Ferraudi, G. *Inorg. Chem.* **1989,** *28,* 2692.
36. Kraut, B.; Ferraudi, G. *Inorg. Chem.* **1990,** *29,* 4834.

RECEIVED for review November 7, 1991. ACCEPTED revised manuscript May 18, 1992.

14

Photoredox Chemistry of Metal Complexes in Microheterogeneous Media

Robin Cowdery-Corvan, Susan P. Spooner, George L. McLendon, and David G. Whitten*

Department of Chemistry, University of Rochester, Rochester, NY 14627–0216

This chapter reports on the photoreactivity of metal-based reactants constrained in microheterogeneous media. First, the discussion focuses on the photoredox reactions of colloidal CdS with organic electron donors such as amino alcohols. By using the CdS particles as photoexcited acceptors, the characteristic photooxidative cleavage of amino alcohols was observed, but with overall characteristics that are medium-dependent, for particles stabilized within Aerosol-OT (dioctyl sodium sulfosuccinate) reversed micelles, carboxymethylamylose polymer in water, and acetonitrile solutions. Second, the reactivity of some amphiphilic cobalt(III)–carboxylate complexes is discussed. These complexes, which contain a trans-*stilbene chromophore and can be incorporated as structural components in a variety of media including Langmuir–Blodgett assemblies, micelles, and reverse micelles, undergo intramolecular electron transfer followed by complex decomposition in overall reactions that are strongly medium-dependent.*

THE REACTIVITY OF MOLECULES constrained or incorporated into interfaces is a topic of many current investigations. Of particular interest is photoinduced charge transfer across interfaces and the factors that con-

*Corresponding author

0065–2393/93/0238–0261$06.00/0

trol, limit, or facilitate such processes. Interfacial electron-transfer processes might involve two species on opposite sides of an interface, reactions between an interfacial component molecule and a molecule in the contacting medium, or intramolecular charge transfer within a molecule that is either a component or embedded into an interface in which there is a steep gradient of charge.

Another theme of interest is often referred to as supramolecular chemistry or the chemical properties and reactivities of species that by some assembling process exhibit behavior beyond the "conventional" molecular scale. Specifically, in this case questions are posed as to how "molecular" reactivity is modified at an interface. Among the possibilities are (1) selective aggregation, enhanced or limited by the organizing interface or medium topology; (2) selective exclusion or concentration of reagents, particularly by charged interfaces; and (3) modification of unimolecular reactivity either by rapid expulsion of geminate fragments or by enhanced "cage" lifetimes.

In this chapter we will focus on two studies in our laboratory in which these two aspects of interfacial reactivity are combined, that is, a photoinduced charge-transfer process involving a specific role of the interface and the participation in this process of a species that may be regarded as a "supermolecule". In most of the examples, the medium is a microheterogeneous assembly formed by the self-organization of surfactant or amphiphile units either in solution or at the air–water interface.

Confined Semiconductor Microcrystallites as Photooxidants for Fragmentable Amines

Many studies (*1–15*) have shown that growth of semiconductor particles such as cadmium sulfide can be limited when carried out in the presence of potential hosts that have limited domains in which the microcrystallite particles can grow. These hosts include zeolites, reversed micelles, various soluble polymers, and a number of other microheterogeneous media. The characteristics of stabilized particles are very dependent on the specific host medium employed, but some studies (*1, 16, 17*) demonstrated that the atomic spacing in these particles for CdS is identical to that in bulk crystalline CdS. In most cases, however, these particles absorb at higher than band-gap energies and emit at relatively long wavelengths. Many of the chemical and photophysical properties of these particles have been and are being explored from a number of perspectives, and one of the most interesting is an evaluation or comparison of their photochemical and photophysical behavior relative to that of more conventional molecular excited states. Previous studies (*18–25*) demonstrated that excited states of the microcrystallites can be quenched by both ground-state donors and acceptors in what appeared to be fairly conventional excited-state single-electron-transfer (SET) quenching processes, although the rates and efficiencies of these processes appear to be quite system-dependent.

In this chapter we discuss particles stabilized by three relatively different host media: Aerosol-OT (AOT, dioctyl sodium sulfosuccinate) reversed micelles with heptane as the continuous phase (Figure 1), amylose or carboxymethylamylose (CMA) polymers solubilized in water, and acetonitrile solutions. These three media have in common the property of solubilizing CdS to form reasonably stable subband-gap size particles that absorb light at relatively short wavelengths and emit in the visible region (*7, 19, 21, 26*).

The Aerosol-OT–heptane–water reversed-micelle medium has proven to be an extremely versatile host for preparation of a variety of microcrystallite particles of CdS and other semiconductors (*2, 7–10, 27–30*). In this study, we have generally kept the Aerosol-OT concentration at 0.5 M

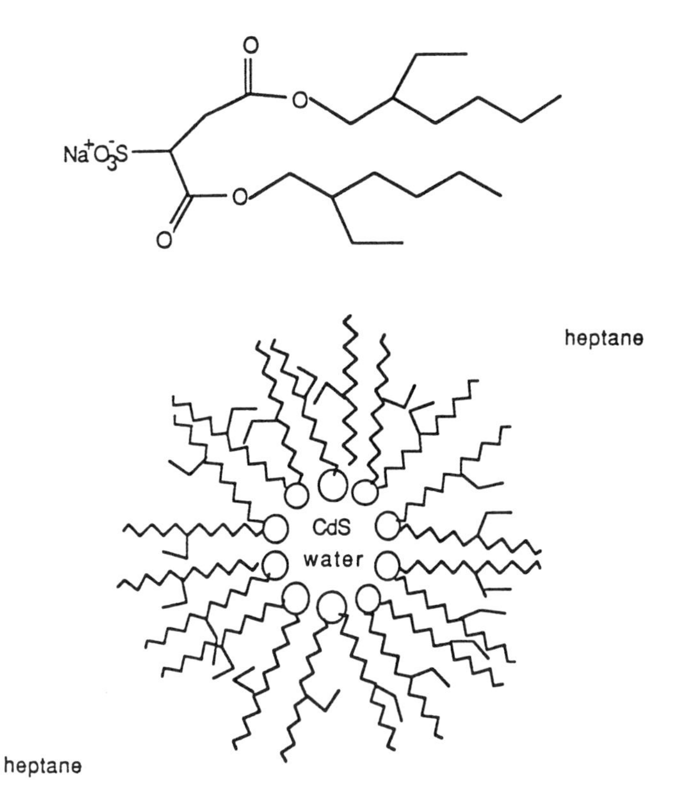

Figure 1. Schematic representations of AOT (top) and an AOT-reversed micelle (bottom).

and varied the amount of water present; the reversed-micelle solution is thus characterized by the mole ratio of water to surfactant (ω). For the reversed micelle itself, as the value of ω increases in the range 1–30, the size of the reversed-micelle "water pool" increases, and the nature of the water contained within the reversed micelle changes substantially (*31, 32*). For the very small reversed micelles (ω = 1–10) virtually all of the water is "interfacial" and strongly associated with the charged head groups of the Aerosol-OT amphiphile. As ω increases, the water pool is composed of both interfacial and "bulk" water.

We found that stable CdS particles could be obtained in the range ω = 1–30, but with ω >32, bulk CdS forms upon standing for more than a few hours, and stable particles cannot be formed in solutions with ω <1. Figure 2 shows absorption spectra and emission spectra of CdS particles stabilized in AOT–water–heptane reversed micelles for ω = 1, 5, and 10. A characteristic shift of both absorption and emission spectra to longer wavelengths is observed with increasing ω. Particle size correlates with absorption wavelength (*1, 17, 33–36*), and hence the particle diameters can be estimated from the absorption spectrum.

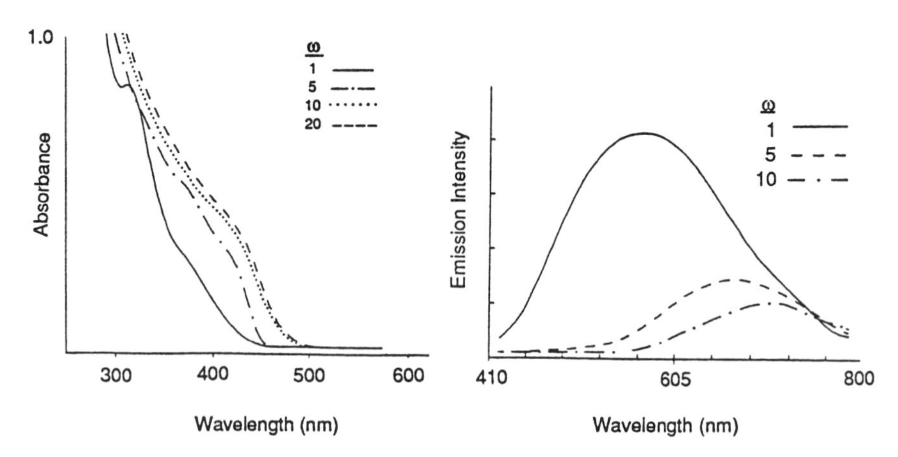

Figure 2. Absorption and luminescence of CdS in AOT–water–heptane reversed micelles.

Table I gives a comparison of the water pool size measured for "empty reversed micelles" (*31, 32*) with the particle diameters estimated from the absorption onset (*37, 38*). These data indicate that the particle size increases more slowly than the water pool diameter. The emission, which has been attributed to electron–hole recombination (*3, 19, 26, 35, 36, 39, 40*), is red-shifted from the absorption and exhibits the same general trends as absorption with red shifts with increasing ω. The relatively large Stokes shift is a nearly constant 1.15 eV in the range ω = 1–10.

Table I. Correlation of Water Pool Size with Particle Size

ω = [water]:[AOT]	*Water Pool Diameter (Å)*	*Particle Diameter (Å)*
1	10	14
5	36	23
10	48	27
20	76	30

Emission from the CdS particles is typically multiexponential; we have been able to fit the decay reasonably well to a triple exponential in which one component is in the subnanosecond range, the second is in the range 1–10 ns, and the third, 30–300 ns. In most cases the longest component is the major species in terms of numbers of quanta. The Aerosol-OT reversed-micelle stabilized CdS particles are fairly stable over a period of weeks, but on prolonged storage in the dark, the particles undergo changes that are followed by a bathochromic shift in the absorption spectrum; these changes are attributed to an "Ostwald ripening" or particle agglomeration (*1, 40, 41*). Upon irradiation of AOT-stabilized CdS particles, a shift of the absorption spectrum to the blue is observed and is attributed to "photoanodic dissolution" (*42*).

Amylose and carboxymethylamylose (CMA) are polymeric materials that can be used in CdS particle stabilization. Amylose is a starch consisting of a long chain of glucose residues; CMA is amylose that has been carboxylated at some of the hydroxyl positions. Both amylose and CMA in dimethyl sulfoxide (DMSO)–water 50:50 and in water, respectively, form helical structures into which linear rodlike amphiphiles can be incorporated (*43–48*). The initial impetus for using CMA was to explore the possibility of forming linear strands of semiconductors with possible two-dimensional confinement. Although CMA in water was found to support the formation of stable microcrystallite particles of CdS, transmission electron micrographs (TEMs) of the particles revealed that they were spherical and not cylindrical in nature. In fact the TEMs suggest that the polymer is simply coated to the particle surface in a fashion similar to that reported for hexametaphosphate and other polymeric stabilizers (*1, 35*). The indicated particle diameters are 35–50 Å for 4×10^{-4} M CdS in a 1.0% CMA–water sample: similar particles are obtained in 50:50 DMSO–water using 1.0% amylose as the stabilizing polymer. These particle diameters measured by transmission electron microscopy are very similar to those measured on the basis of absorption spectra (40–45 Å), as shown in Figure 3.

In the CMA in water, the particle size of CdS, as indicated by absorption species, can be controlled effectively by changing the ratio of the

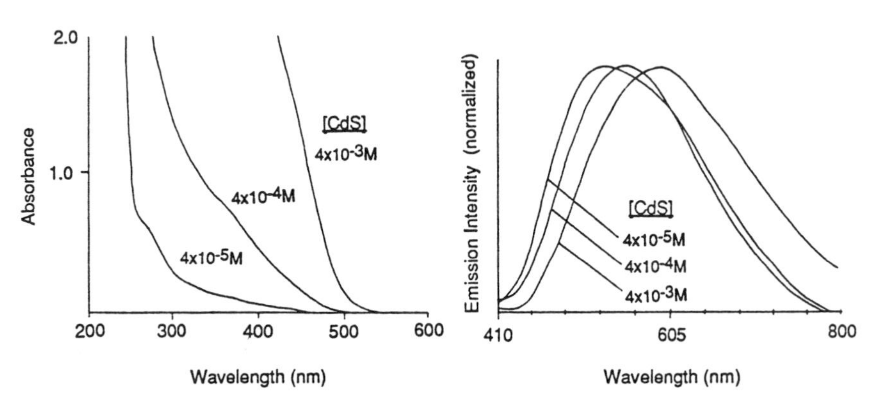

Figure 3. Absorption and luminescence of CdS in 1:1 DMSO:water CMA polymer medium.

molecular semiconductor to polymer concentrations, lower ratios yielding smaller particles. Polymer concentrations higher than 5.0% are not soluble in water or DMSO–water mixtures, and therefore small particles (10–50 Å) are generated at very low molecular semiconductor concentrations (i.e., $\leq 4 \times 10^{-4}$ M).

As with particles stabilized by other means, CMA-stabilized CdS exhibits visible luminescence that in turn exhibits multiexponential decay. In comparison with the Aerosol-OT reversed-micelle stabilized particles, the Stokes shifts decrease as the particles become larger. The luminescence lifetimes are similar to those observed for CdS particles entrapped in reversed micelles. CMA-stabilized CdS stored in the dark showed minimal changes in absorption spectra over periods of up to 5 days, but changes under irradiation similar to those observed with the AOT-stabilized CdS were observed upon radiation. As with the AOT-stabilized particles, removal of the solvent by lyophilization followed by redispersion of the dried particles led to identical absorption and emission spectra, a result indicating that the particles remain dispersed during this process.

CdS colloids can also be generated in acetonitrile or 2-propanol by using the method of "arrested precipitation". Stable solutions could be obtained at CdS concentrations of 4×10^{-4} M, which were argon-degassed at room temperature. Particle sizes in acetonitrile increase upon the addition of water (Figure 4) (*5, 19*). Once again the emission observed in acetonitrile or 2-propanol solutions shows a substantial Stokes shift and exhibits multiexponential decay. The long component lifetime in 2-propanol is somewhat shorter (<100 ns) than in the other two media.

In some ways, then, the microcrystallite particles of CdS stabilized by the three media, as described, all possess low-lying luminescent excited states with energies and photophysical properties comparable to those of molecular excited states of commonly used "photosensitizers" in conven-

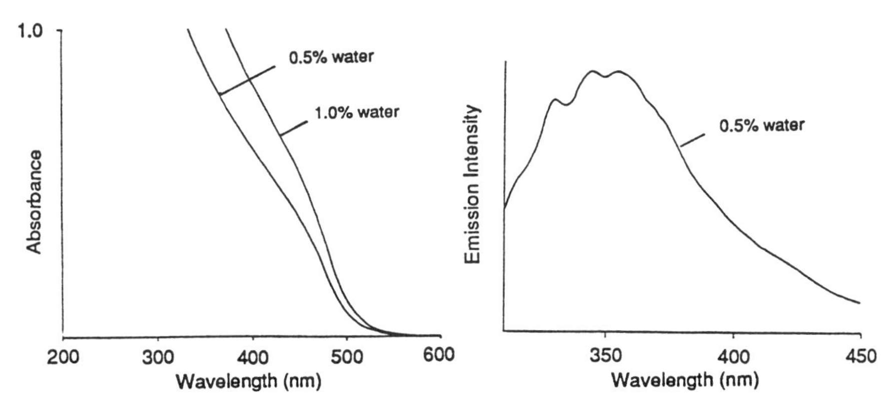

Figure 4. Absorption and luminescence of CdS in acetonitrile–water.

tional photoreactions such as triplet sensitization or single-electron-transfer processes. The fact that these particle excited states can be quenched by both potential single-electron acceptors such as quinones or viologens, as well as by potential electron donors such as amines or thiols, suggests that these "supermolecules" may behave very similarly to molecular excited states in their photophysical and photochemical processes (*21, 35, 39, 49, 50, 52*). To this end, we investigated the photoreactivity of the microcrystallite particles in biomolecular photochemical reactions involving fragmentable electron donors.

The fragmentable electron donors we studied with the microcrystallites behaving as excited electron acceptors are the amino alcohols shown as **1** and **2**.

C_6H_5–CH(OH)–CH(NC_4H_8O)–C_6H_5 **1**

C_6H_5–CH(OH)–CH_2–NC_4H_8 **2**

These were previously shown (*52–55*) to undergo relatively clean excited electron-acceptor-mediated fragmentation upon irradiation of acceptors such as thioindigo, 9,10-dicyanoanthracene (DCA), and 2,6,9,10-tetracyanoanthracene (TCA). These studies (*52–55*) showed that irradiation of these acceptors in the presence of **1** or **2** leads to efficient quenching of the acceptor excited singlet state and to clean fragmentation of the amine as shown in equation 1 for **1**.

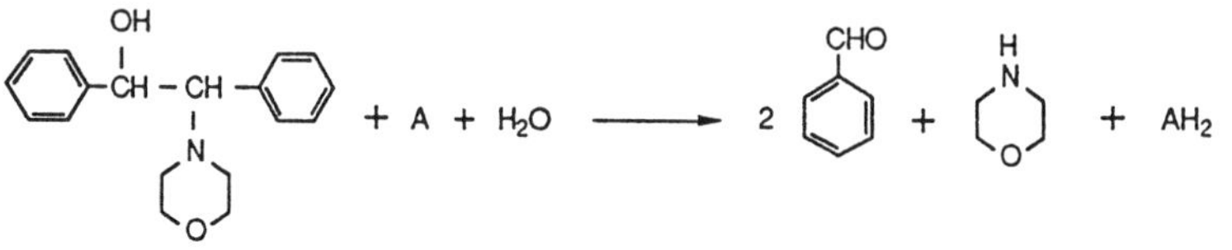

$$+ \ A + H_2O \longrightarrow 2\ (\text{CHO}) + (\text{morpholine}) + AH_2 \qquad (1)$$

The mechanism proposed for this reaction in solvents of low to moderate polarity under conditions in which free ion formation is relatively unimportant is given in Scheme I. Under these conditions, although the reaction occurs in very high chemical yield, the observed quantum efficiencies are relatively low because of the unfavorable competition between back electron transfer, k_{bet}, and fragmentation, k_{frag}. Values of k_{frag} (eq 2) depend on solvent, the basicity of the acceptor radical anion A^{-}, and the particular amine.

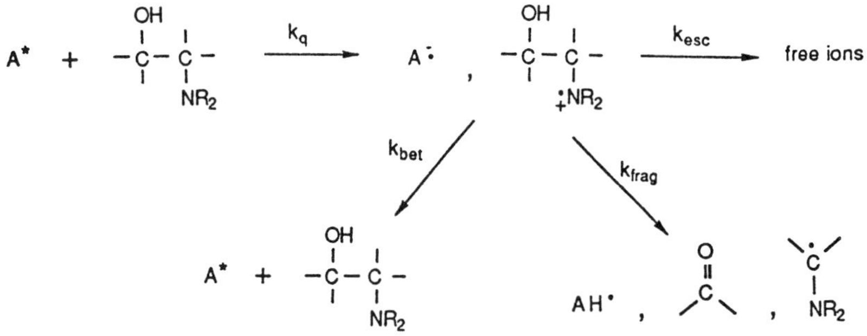

Scheme I.

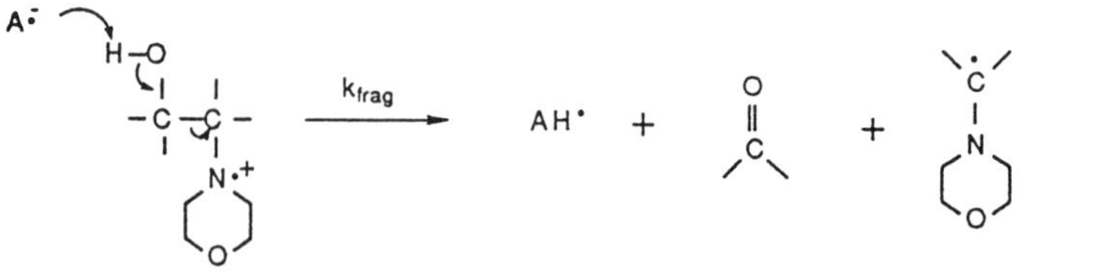

(2)

For acceptors whose radical anions are relatively basic such as thioindigo, moderate quantum efficiencies (quantum yield Φ = 0.01–0.03) are observed, especially in relatively nonpolar solvents in which the ion radical pair should be a closely coupled unsolvated contact radical ion pair or exciplex.

As the polarity increases, the quantum yield decreases, a property indicating that solvent-separated radical ion pairs are much less reactive; much lower quantum efficiencies are observed when acceptors whose conjugate anion radicals are not very basic are employed. The rate constants for acceptor anion-radical-assisted fragmentation are in the range of 10^6

s^{-1}. Back-reaction rate constants for relatively exothermic process are estimated to be $\geq 10^{11}$ s^{-1}. Under conditions in which free ions are generated either by cosensitization or by direct generation via nonphotochemical means, fragmentation is observed. It is at present unclear exactly what the rate constant for unassisted fragmentation should be (eq 3); however, probably the unassisted process has a rate constant lower than 10^6 s^{-1}.

$$\text{S–O(H):} \;\; \text{H–O–}\underset{|}{\overset{|}{\text{C}}}\text{–}\underset{\overset{+}{:}NR_2}{\overset{|}{\text{C}}}\text{–} \xrightarrow{k_{frag}} \text{S–}\overset{+}{\text{O}}\text{H}_2 + \text{>C=O} + \text{>}\dot{\text{C}}\text{–NR}_2 \quad (3)$$

Irradiation of microparticulate CdS, prepared using all three methods of stabilization as described, in the presence of amines **1** or **2** results both in quenching of the CdS luminescence and in fragmentation of the amine, with no change in CdS absorption spectra upon amine addition, a result suggesting no change in particle size. However, immediate interpretation of photophysical and photochemical behavior observed on selective excitation of CdS in the presence of the amines is not so simple because in several cases (Figure 5) irradiation of CdS at low concentrations of amine leads to "antiquenching", as previously reported (*56, 57*). This antiquenching is attributed to initial binding of the amine to the microparticle surface, which results in removal of "trap sites" that could otherwise lead to nonradiative decay in competition with luminescence. The removal of trap sites actually leads to an enhancement and spectral blue shift in the overall luminescence and has been observed for several different combinations of amine and type of particle preparation (*56*).

In most cases, further addition of amine ultimately results in quenching of the particle luminescence, as detailed in Figure 5. The quenching does not follow simple Stern–Volmer relationships and indicates that the particles have varied accessibilities to the quencher. In any case, as listed in Table II, the irradiation of the CdS particles in the presence of both amines leads to moderately efficient fragmentation but with quantum yields that differ substantially for the differently stabilized particles. The overall quantum yields measured reflect in another complicated way the local concentration of amine in the presence of the microcrystallite particle, the balance between quenching and antiquenching activity of the amine, and the photoreactivity of the "quenched particle-amine" photoproduct produced in the quenching act itself.

As data in Table III indicate, the actual photofragmentation efficiencies, α, seem to vary in a relatively straightforward way. (The photofragmentation efficiency, α, is defined as the quantum yield of amine fragmentation divided by the quantum yield of particle emission quenching, where

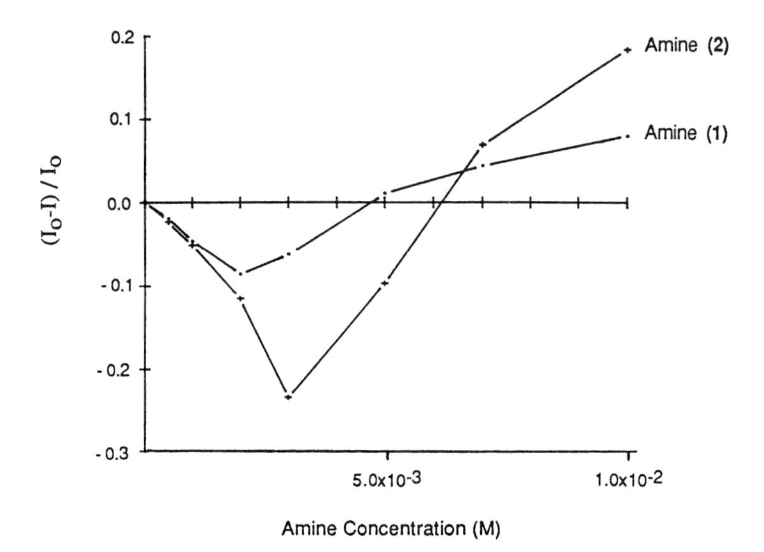

Figure 5. Amine quenching of CdS emission in AOT (ω = 3) reversed micelles. The quantum yield of particle emission quenching $\Phi_{em} = (I_0 - I_q)/I_0$; I_0 is the CdS emission intensity in the absence of quencher, and I_q is the CdS emission intensity in the presence of quencher.

Table II. Amine Photofragmentation Quantum Yields (Φ)

	CH_3CN		AOT: ω = 10		CMA,
$[Amine]_{initial}$ (M)	**1**	**2**	**1**	**2**	**2**
5.0×10^{-3}	0.27	0.25	0.06	0.09	0.16
7.5×10^{-3}	0.32	0.30	0.09	0.12	0.46
1.0×10^{-2}	0.27	0.37	0.17	0.13	0.91

Table III. Amine Photofragmentation Efficiencies (α)

	CH_3CN		AOT: ω = 10		CMA,
$[Amine]_{initial}$ (M)	**1**	**2**	**1**	**2**	**2**
5.0×10^{-3}	≥1.0	1.0	0.14	0.11	0.26
7.5×10^{-3}	≥1.0	1.0	0.15	0.14	0.59
1.0×10^{-2}	≥1.0	1.0	0.33	0.15	1.0

the quantum yield of particle emission quenching $\Phi_{em} = (I_0 - I_q)/I_0$; I_0 is the CdS emission intensity in the absence of quencher, and I_q is the CdS emission intensity in the presence of quencher.) For the acetonitrile-stabilized particles, it appears that for both amines **1** and **2** the α is effectively unity. This result is perhaps not too surprising, but nonetheless noteworthy, because it contrasts greatly with the much lower α observed with "molecular" electron acceptors. The most likely reasons for this striking difference are either that k_{bet} is relatively small as a result of electron dispersal over the microparticle or that cage escape k_{esc} is relatively large in the rather polar and fluid acetonitrile medium or both.

Because of the relatively low steady-state light intensity and consequent intermediate low concentrations, once free ion radicals are formed, their unimolecular reactions (eq 3) should dominate any bimolecular return electron transfer. Somewhat lower α values are observed with the CMA-stabilized particles, although again the quantum efficiencies here are much higher than those observed with molecular electron acceptors. The somewhat reduced quantum efficiencies may be consistent with a less polar or more hydrophobic surface region in the polymer-stabilized CdS that inhibits somewhat free ion formation compared to acetonitrile. Alternatively the increase of reaction quantum yield as amine concentration is increased for CMA-stabilized CdS may also indicate that reaction of a second molecule of amine serving as a base can enhance fragmentation near the particle surface in competition with otherwise limiting back electron transfer, k_{bet}.

For the reversed-micelle-solubilized CdS, the α value is somewhat lower and also shows a small increase with increase in amine concentration, even after the quenching is factored out. It is tempting to conclude that the lower α value observed for the two amines in the reversed micelle is simply due to the fact that even though the medium is a highly polar reversed-micelle water pool, the product "ion radicals" are generated in a highly restricted environment that inhibits a true cage escape to free ions and enhances back electron transfer relative to the high escape yields suggested for acetonitrile.

In summary, the studies with the different stabilized CdS and fragmentable amines show that CdS particles do behave as molecular excited states as far as electron-transfer reactions are concerned and that overall higher quantum efficiencies in mediating electron-transfer processes occur in all the cases studied, compared to corresponding reactions in which "molecular" excited singlet states having comparable lifetimes are quenched. These studies suggest that fruitful future investigations might involve a series of electron-transfer reactions involving either donors or acceptors whose corresponding ion radicals react with an independent "clock" such that a more precise estimate of the important photophysical parameters k_{bet} and k_{esc} can be better assessed.

Photoredox Reactions of Amphophilic Stilbenecarboxylate–Cobalt(III) Complexes in Microheterogeneous and Homogeneous Media

An important class of photosensitive metal–organic systems involves metal complexes in which photoinduced electron transfer induces a change in the metal oxidation level and consequently a dramatic alteration in the ability of the metal to bind ligands. A rather interesting example of this alteration in binding involves *trans*-stilbenecarboxylatopentaamineco-balt(III) studied several years ago by Vogler and co-workers (*58, 59*). Irradiation of this complex (**3**) results in redox decomposition as shown in equation 4.

$COO\text{-}Co(NH_3)_5^{2+}$ $\xrightarrow[313\ nm]{h\nu}$

$COO\cdot + Co^{2+} + 5\,NH_3$

CHO

(4)

The initial photochemical step is believed to be an electron transfer from stilbene ligand to cobalt that labilizes the cobalt center and leads to release of ammonia and in a subsequent, much more complicated reaction, to further oxidation of the *trans*-stilbenecarboxylate radical to benzal-dehyde (*59*). The subsequent steps are fairly complicated and show an increase of the quantum yield with increase in the concentration of **3** due to the metal-complex-mediated oxidation of the intermediate radical. Because of our interest in the photophysics and photochemistry of stilbene and related polyenes in microheterogeneous media, we prepared and studied the surfactant *trans*-stilbenecarboxylate **4** and its cobalt complex **5**. Both **4** and **5** are relatively good amphiphiles that can be incorporated into a number of different microheterogeneous environments, including Langmuir–Blodgett assemblies, micelles, and reversed micelles.

$COO\text{-}Co(NH_3)_5^{2+}$

3

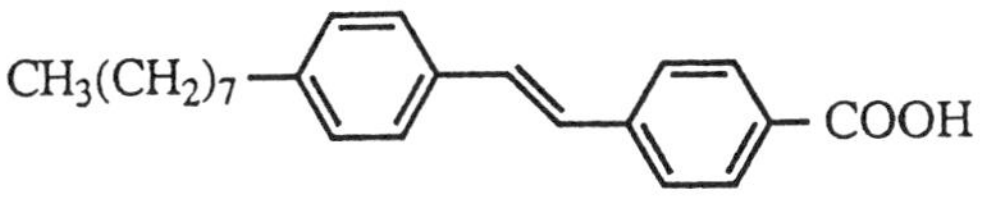

$CH_3(CH_2)_7$–C_6H_4–CH=CH–C_6H_4–COOH

4

$CH_3(CH_2)_7$–C_6H_4–CH=CH–C_6H_4–$COO\text{-}Co(NH_3)_5^{2+}$

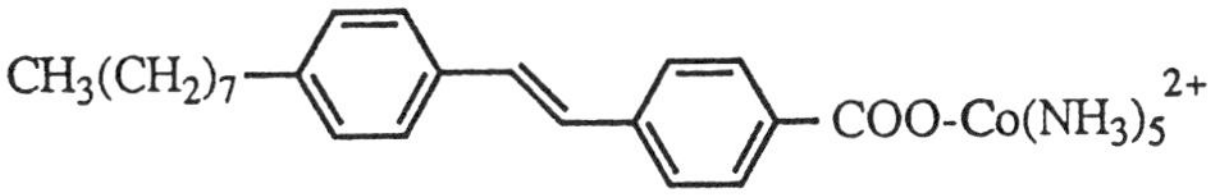

5

In our studies of **4** and **5** as both photophysical and photochemical probes, we uncovered a number of interesting aspects of their excited-state behavior. First, in both homogeneous solution and microheterogeneous media, the parent amphiphile has a typical *trans*-stilbene absorption and a very strong fluorescence. The fluorescence is particularly strong in highly restricted environments such as Langmuir–Blodgett multilayers in which the chromophore is in a restricted semi-rigid environment that inhibits excited-state decay via the isomerization path.

In all media in which it has been studied thus far, cobalt complex **5** shows an almost identical absorption spectrum to that of **4** but virtually no fluorescence (Figure 6). Incorporation of **5** into fluid microheterogeneous media such as acetonitrile or acetonitrile–water solutions or into aqueous micelles (Triton X-100 (octoxynol) micelles or cetyltrimethylammonium chloride (CTAC) micelles) results in photoreactivity apparently similar to that previously observed by Vogler, albeit with results that are strongly medium-sensitive (*59*). In each case, irradiation results in the formation in octylbenzaldehyde apparently due to ligand-to-metal electron transfer induced by excitation of the stilbene chromophore.

The efficiencies observed in acetonitrile or 9:1 acetonitrile–water are much lower than those reported by Vogler, and so a direct comparison of the reactivity has not yet been possible (*59*). However, reactivity in the two apparently homogeneous solutions is much lower than that observed when **5** is incorporated into either neutral (Triton X-100) or cationic (CTAC) micelles. Reactivity is 5 times greater in the Triton X-100 micelles than in acetonitrile and more than 50 times greater than in the cationic (CTAC) micelles. In contrast, when **5** is incorporated into transferred Langmuir–Blodgett multilayers supported on glass or quartz, irradiation results in no fluorescence and no photochemical reaction whatsoever. This finding suggests then that the photoreactivity of **5** is best accounted for by the diagram presented in Scheme II.

The absorption spectrum is dominated by the intraligand transition and suggests that the first excited state populated is the stilbene localized excited singlet that is strongly fluorescent in compound **4**. In cobalt complex **5**, this excited state decays rapidly to a ligand-to-metal charge-transfer

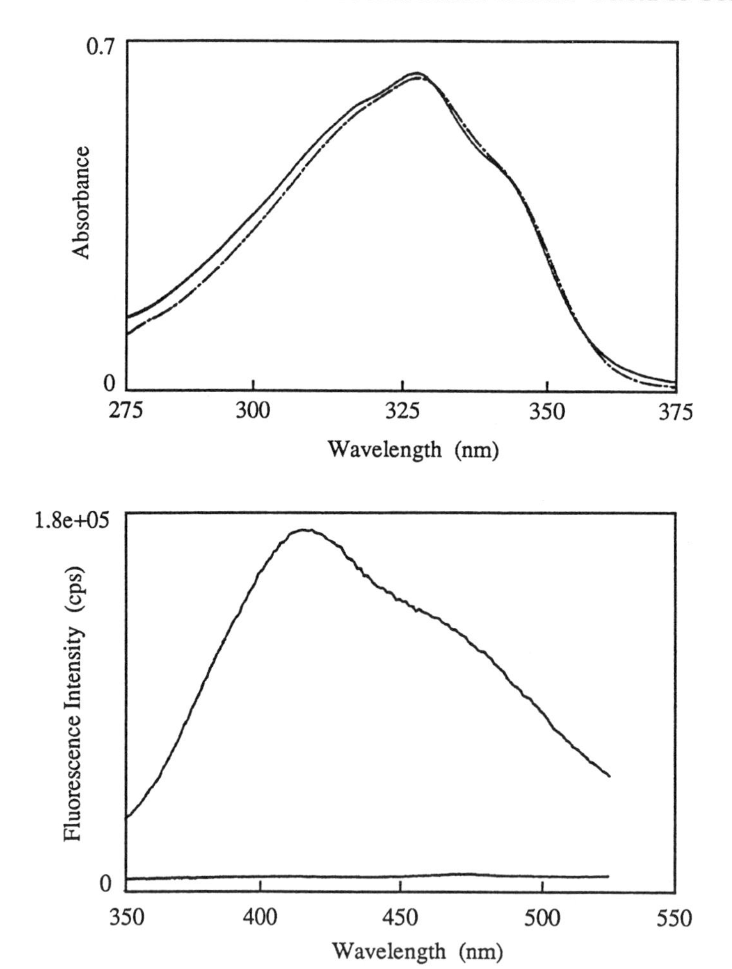

Figure 6. Top: Absorption spectra of **4** *(—) and* **5** *(- -) in acetonitrile solution. Bottom: Fluorescence intensity of* **4** *(—) and* **5** *(- -) in multilayer assemblies.*

(LMCT) state that can subsequently decompose to give a neutral fragment containing the stilbenecarboxylate and the cationic cobalt species. In Langmuir–Blodgett assemblies, most likely relatively little cleavage of the carboxylate to cobalt bond occurs, and perhaps only decay via the LMCT state occurs in dry assemblies. Nonetheless, this decay is sufficient to completely deactivate the complex nonradiatively.

In homogeneous solution and in aqueous micelles, the formation of octylbenzaldehyde suggests that the decay of the LMCT state to the radical pair occurs and that subsequent reactivity occurs with varying efficiency. The enhanced reactivity in Triton X-100 micelles could be ascribed to a "hydrophilic–hydrophobic sorting" of the products formed by decay of the LMCT state. This mechanism is certainly supported by the greatly enhanced reactivity observed in the cationic CTAC micelles in which the expulsion of the cationic cobalt fragment should be assisted near the cationic micellar surface.

The cobalt complex **5** has turned out to be an extremely useful probe quencher in Langmuir–Blodgett assemblies. We investigated a number of Langmuir–Blodgett assemblies containing the surfactant stilbenes such as **6–9**, which show typical stilbene absorption spectra slightly blue-shifted from those of **4** and **5**. In one study we examined the fluorescence-quenching ability of **5** compared to simple cobalt complex with stilbene **6** in assemblies, as shown in Figure 7. Not surprisingly, the fluorescence of **6** is unquenchable by simple cobalt complexes such as **10** in which a hydrophobic spacer of 18 carbons separates the chromophore from the cobalt center. Replacement of **10** by **5** as shown in the figure results in fairly effective quenching of the fluorescence from a bilayer of **6**. We interpret this result as quenching of the majority of the fluorescence from the proximal monolayer and effectively no quenching from the distal monolayer of the stilbene bilayer because virtually no energy transfer from stilbene to stilbene occurs as reported elsewhere (*60*).

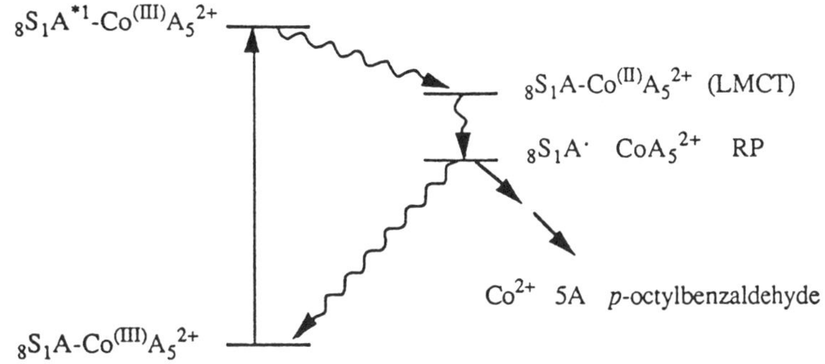

$_8S_1A^{*1}\text{-}Co^{(III)}A_5^{2+}$

$_8S_1A\text{-}Co^{(II)}A_5^{2+}$ (LMCT)

$_8S_1A^{\cdot}$ CoA_5^{2+} RP

Co^{2+} 5A *p*-octylbenzaldehyde

$_8S_1A\text{-}Co^{(III)}A_5^{2+}$

Scheme II.

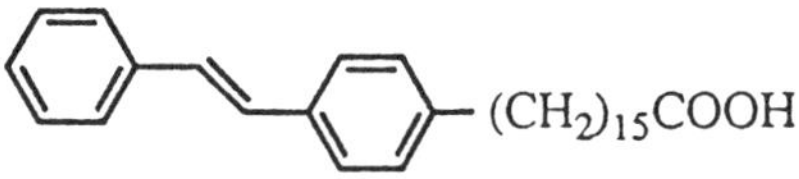

$(CH_2)_{15}COOH$

6

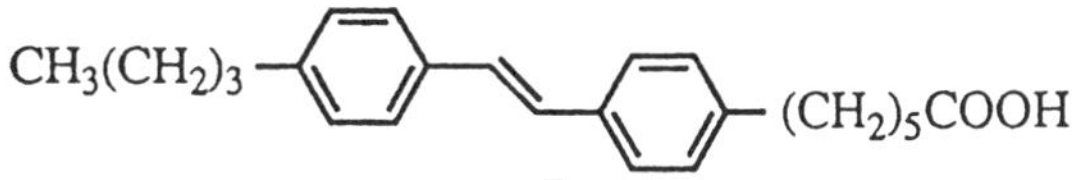

$CH_3(CH_2)_3$ $(CH_2)_5COOH$

7

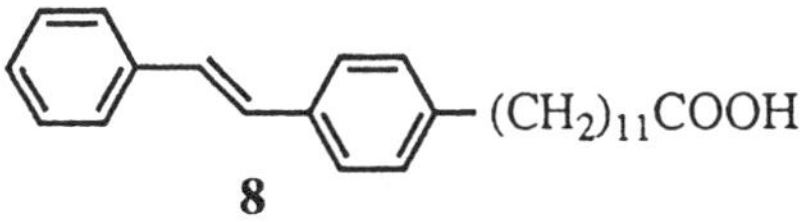

$(CH_2)_{11}COOH$

8

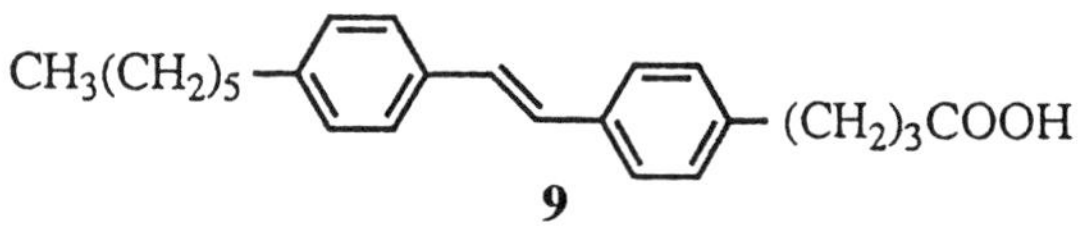

$CH_3(CH_2)_5$–C₆H₄–CH=CH–C₆H₄–$(CH_2)_3COOH$

9

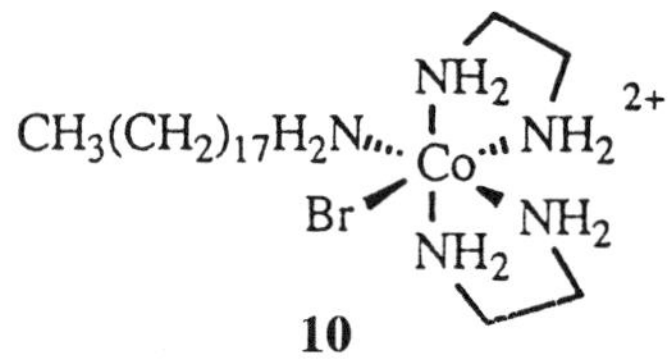

$CH_3(CH_2)_{17}H_2N$–Co(NH_2)$_4$Br $^{2+}$

10

Key

spacer

6

10

5

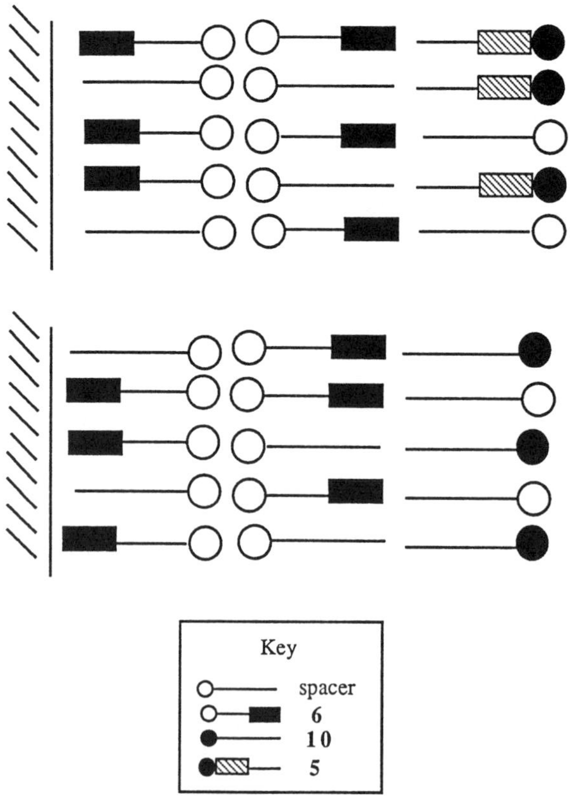

Figure 7. Architecture of Langmuir–Blodgett assemblies containing cobalt complexes **5** *and* **10** *as potential quenchers for a bilayer of stilbene* **6***. Compared to a slide containing two layers of* **6***, the addition of a layer of* **5** *(top) results in a 45% reduction in the fluorescence intensity, but the addition of a layer of* **10** *(bottom) results in little if any reduction in fluorescence intensity.*

Even more interesting results occur in mixtures of stilbenes **7**–**9** in the presence and absence of the stilbenecarboxylate and stilbenecarboxylate–cobalt complexes **4** and **5**. As detailed in Figure 8, multilayers of the stilbene mixture give evidence for efficient energy migration, and not surprisingly the cobalt complex **5** is able to quench efficiently fluorescence from up to a five-layer assembly of the stilbene mixture. However, incorporation of **4** into the stilbene mixture results in a change of the fluorescence from that of the stilbene mixture aggregate to predominantly that of **4**, a result indicating that the lower energy excited singlet of **4** behaves effectively as an energy localizer. This behavior in fact inhibits migration of the excitation through the multilayer and inhibits somewhat the quenching of **6** in the multilayer, as indicated in the figure. We interpret our results to indicate that stilbene **6** behaves as both an energy- and electron-transfer quencher in these assemblies and thus can serve as an

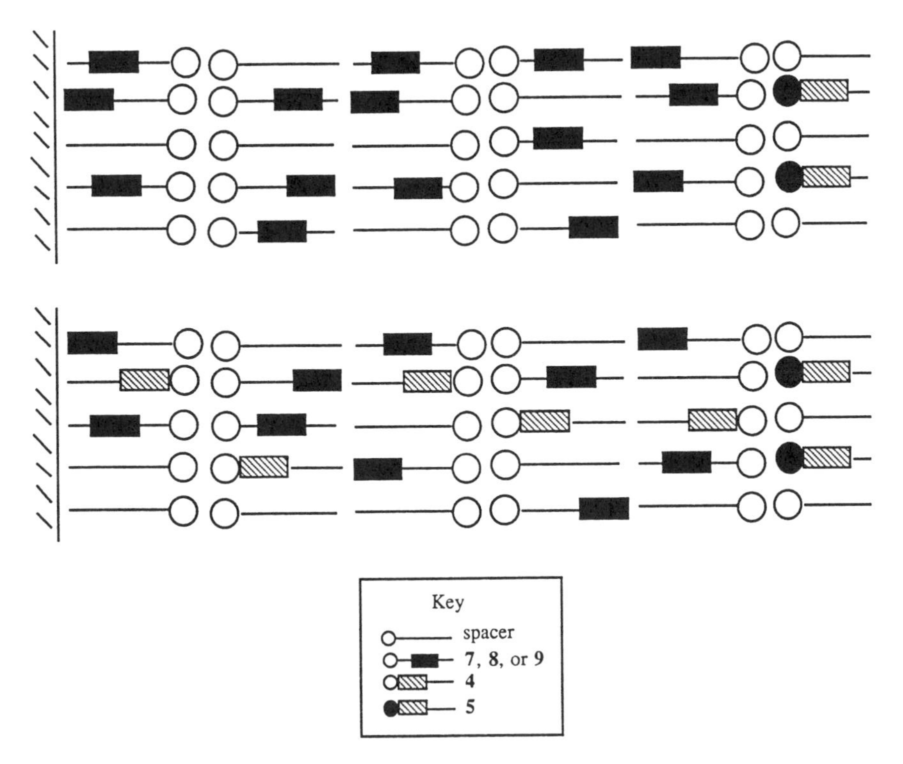

Figure 8. Molecular arrangement in multilayer assemblies containing **7**, **8**, *and* **9** *without surfactant* **4** *(top) and with* **4** *(bottom). Top: The addition of a layer of* **5** *as a quencher results in a reduction in the fluorescence intensity of 41% in a slide containing five layers of the* **7**–**8**–**9** *mixture, indicating substantial energy delocalization. Bottom: In contrast, only a 30% reduction in fluorescent intensity is observed when a layer of* **5** *is added.*

important component in the construction of multilayer assemblies that can promote electron flow over relatively long (>150 Å) distances. We are currently extending these investigations.

Acknowledgment

We are grateful to the National Science Foundation for support of this research through Grant No. CHE–86I6361.

References

1. Brus, L. E.; Rossetti, R.; Ellsion, J. L.; Gibson, J. M. *J. Chem. Phys.* **1984,** *80,* 4464.
2. Brus, L. E.; Steigerwald, M. L.; Alivisatos, A. P.; Harris, A. L.; Levinos, N. J. *J. Chem. Phys.* **1988,** *89,* 4001.
3. Henglein, A.; Fojtik, A.; Weller, H. *Chem. Phys. Lett.* **1985,** *120,* 552.
4. Gratzel, M.; Moser, J. *J. Am. Chem. Soc.* **1984,** *106,* 6557.
5. Kamat, P. V.; Dimitrijevic, N. M.; Fessenden, R. W. *J. Phys. Chem.* **1988,** *95,* 2324.
6. Fox, M. A.; Nosaka, Y. *Langmuir,* **1987,** *3,* 1147.
7. Fendler, J. H. *Chem. Rev.* **1987,** *87,* 877.
8. Thomas, J. K.; Lianos, P. *Chem. Phys. Lett.* **1986,** *125,* 299.
9. Thomas, J. K.; Lianos, P. *J. Colloid. Interface Sci.* **1987,** *117,* 505.
10. Backer, C. A.; Cowdery-Corvan, J. R.; Spooner, S. P.; Armitage, B.; McLendon, G. L.; Whitten, D. G. *J. Surf. Sci. Tech.* **1990,** *6,* 59.
11. Wang, Y.; Suna, A.; McHugh, J.; Hilinski, E. F.; Lucas, P.; Johnson, R. D. *J. Chem. Phys.* **1990,** *92,* 6927.
12. Bard, A. J., Becker, W. G. *J. Phys. Chem.* **1983,** *87,* 4888.
13. Bard, A. J.; Fox, M. A.; Krishnan, M.; White, J. R. *J. Am. Chem. Soc.* **1983,** *105,* 7002.
14. Bard, A. J.; Enea, O. *J. Phys. Chem.* **1986,** *90,* 301.
15. Anpo, M.; Aikawa, N.; Kubokawa, Y.; Che, M.; Luois, C.; Giamello, E. *J. Phys. Chem.* **1985,** *89,* 5017.
16. Rossetti, R.; Nakahara, S.; Brus, L. E. *J. Chem. Phys.* **1983,** *79,* 1086.
17. Brus, L. E. *J. Phys. Chem.* **1986,** *90,* 2555.
18. Brus, L. E.; Possetti, R.; Beck, S. M. *J. Am. Chem. Soc.* **1984,** *106,* 980.
19. Gratzel, M.; Ramsden, J. J.; Webber, S. E. *J. Phys. Chem.* **1985,** *89,* 2740.
20. Thomas, J. K. *J. Colloid. Interface Sci.* **1984,** *100,* 116.
21. Thomas, J. K. *J. Phys. Chem.* **1987,** *91,* 267.
22. Rossetti, R.; Brus, L. E. *J. Phys. Chem.* **1986,** *90,* 558.
23. Moser, J.; Gratzel, M. *J. Am. Chem. Soc.* **1983,** *105,* 6547.
24. Duonghong, D.; Ramsden, J.; Gratzel, M. *J. Am. Chem. Soc.* **1982,** *104,* 2977.
25. Nosaka, Y.; Fox, M. A. *Langmuir* **1987,** *3,* 1147.
26. Brus, L. L.; Chestnoy, N.; Harris, T. D.; Hull, R. *J. Phys. Chem.* **1986,** *90,* 3393.

27. Fendler, J. H.; Chang, A.-C.; Pfeirner, W. F.; Guillaume, B.; Baral, S. *J. Phys. Chem.* **1990,** *94,* 4284.
28. Fendler, J. H.; Watzke, H. J. *J. Phys. Chem.* **1987,** *91,* 854.
29. Fendler, J. H.; Meyer, M.; Wallberg, C.; Kurihara, K. *J. Chem. Soc. Chem. Commun.* **1984,** 90.
30. Thomas, J. K.; Kuczynski, J. *J. Phys. Chem.* **1983,** *87,* 5498.
31. Thomas, J. K. *The Chemistry of Excitation at Interfaces;* ACS Monograph 181; American Chemical Society: Washington, DC, 1984.
32. Eiche, H.-F.; Zulauf, M. *J. Phys. Chem.* **1979,** *83(4),* 480.
33. Brus, L. E. *J. Chem. Phys.* **1984,** *80,* 4403.
34. Fojtik, A.; Weller, H.; Koch, U.; Henglein, A. *Ber Bunsenges. Phys. Chem.* **1984,** *88,* 969.
35. Ramsden, J. J.; Gratzel, M. *J. Chem. Soc., Faraday Trans. I* **1984,** *80,* 919.
36. Nozik, A. J.; Williams, F.; Nenadovic, M. T.; Rajh, T.; Micic, O. I. *J. Phys. Chem.* **1985,** *89,* 397.
37. Weller, H.; Schmidt, H. M.; Koch, U.; Fojtik, A.; Baral, S.; Kunad, W.; Weiss, K.; Dieman, E.; Henglein, A. *Chem. Phys. Lett.* **1986,** *124,* 557.
38. Wang, Y.; Heron, N. *Phys. Rev. B.* **1990,** *42(11),* 7253.
39. Rossetti, R.; Brus, L. *J. Phys. Chem.* **1982,** *56,* 4470.
40. Fojtik, A.; Weller, H.; Koch, U.; Henglein, A. *Ber Bunsenges. Phys. Chem.* **1984,** *88,* 969.
41. Ostwald, W. *Die Welt, der Vemachlassigten Dimensionen, Aufl.;* Th. Steinkopff: Dresden, Germany, 1920.
42. Henglein, A.; Fojtik, A.; Weller, H. *Ber. Bunsenges. Phys. Chem.* **1987,** *91,* 441.
43. Rundle, R. E.; French, D. *J. Am. Chem. Soc.* **1943,** *65,* 558.
44. Rao, V. S. R.; Foster, J. F. *Biopolymers* **1965,** *3,* 185.
45. Brant, D. A.; Dubin, P. L. *Macromolecules* **1975,** *5,* 831.
46. Pfannemuller, B.; Mayerhofer, H.; Schulz, R. C. *Biopolymers* **1971,** *10,* 243.
47. French, A. D.; Zobel, H. F. *Biopolymers* **1967,** *5,* 457.
48. Hui, Y.; Gai, Y. *Macromol. Chem.* **1988,** *189,* 1287.
49. Henglein, A. *J. Phys. Chem.* **1982,** *56,* 2291.
50. Gopidas, K. R.; Kamat, V. *Langmuir* **1989,** *5,* 22.
51. Meyer, G. J.; Lisensky, G. C.; Ellis, A. B. *J. Am. Chem. Soc.* **1988,** *110,* 4914.
52. Ci, X.; Lee, L.; Whitten, D. G. *J. Am. Chem. Soc.* **1987,** *109,* 2536.
53. Ci, X.; Whitten, D. G. *J. Am. Chem. Soc.* **1987,** *109,* 7215.
54. Ci, X.; Whitten, D. G. *J. Phys. Chem.* **1991,** *95,* 1988.
55. Ci, X.; Kellett, M. A.; Whitten, D. G. *J. Am. Chem. Soc.* **1991,** *113,* 91.
56. Dannhauser, T.; O'Neil, M.; Johansson, K.; Whitten, D. G.; McLendon, G. L. *J. Phys. Chem.* **1986,** *90,* 6074.
57. Lisensky, G. C.; Penn, R. L.; Murphy, C. J.; Ellis, A. J. *Science* **1990,** *288,* 840.
58. Adamson, A. W.; Vogler, A.; Lantzke, I. *J. Phys. Chem.* **1969,** *73,* 4183.
59. Vogler, A., Kern, A. *Z. Naturforsch.* **1979,** *388,* 271.
60. Mooney, W. F.; Whitten, D. G. *J. Am. Chem. Soc.* **1986,** *108,* 5712.

RECEIVED for review November 7, 1991. ACCEPTED revised manuscript May 5, 1992.

15

Heterogeneous Photocatalyzed Oxidation of Phenol, Cresols, and Fluorophenols in TiO_2 Aqueous Suspensions

Nick Serpone[1], Rita Terzian[1], Claudio Minero[2], and Ezio Pelizzetti[2]

[1]Laboratory of Pure and Applied Studies in Catalysis, Environment, and Materials, Department of Chemistry and Biochemistry, Concordia University, Montreal, Canada H3G 1M8

[2]Dipartimento di Chimica Analitica, Università di Torino, 10125 Torino, Italy

The photocatalyzed total oxidation of phenol (PhOH), cresols (2-, 3-, and 4-MePhOH), and fluorinated phenols (2-, 3-, and 4-FPhOH and 2,4- and 3,4-F_2PhOH) to CO_2 takes place in irradiated, air-equilibrated aqueous TiO_2 (anatase) dispersions. Major hydroxylated intermediate substrates, produced along the temporal course of the oxidation process, were identified and confirmed in most cases by high-performance liquid chromatographic (HPLC) methods. The photooxidation of phenol, a model and the parent compound of the cresols and fluorophenols, was re-examined and is herein treated in some detail, as it provides useful apparent kinetic data uninfluenced by additional ring substituents. The oxidation was carried out at pH 3 (H_2SO_4), in distilled water (pH ~6.5), and at pH 7 (phosphate buffer). The evolution of CO_2 proceeds via hydroquinone and catechol intermediates; catechol was not detected at pH 7, and both intermediates were undetected in 0.01 M NaOH solutions. Apparent kinetics of disappearance of the initial aromatic substrate, formation and subsequent degradation of the intermediate products, as well as formation of CO_2 were obtained (computer-fitting) by using a simple phenomenological model that implicates consecutive and parallel reactions. The analytical form of the rate of product evolution is reminiscent of a Langmuirian type rate expression that cannot delineate between the various operational mechanisms.

0065–2393/93/0238–0281$09.50/0

TODAY'S HIGHLY INDUSTRIALIZED ENVIRONMENT is charged with a multitude of potentially toxic agents. Contamination of water resources by a variety of organic substances is an increasingly common problem. The nine general classes of environmental contaminants include phenol and the cresols (*1*). Common uses of these phenolic substrates range from fumigants and insecticides to wood preservatives and disinfectants (*2*). Although the mono- and difluorinated phenols have yet to appear in wastewaters and their impact on the environment has yet to be assessed, their presence in ecosystems is not distant because of the biological and nonbiological degradation of pharmaceuticals, foodstuff preservatives, and high-quality polymers (*3*).

In this chapter, we examine the temporal course of the oxidation of cresols and the mono- and difluorinated phenols, and re-examine the photocatalyzed oxidation of phenol in air-equilibrated aqueous TiO_2 suspensions under AM1-simulated sunlight irradiation (except for the cresols). (AM1 stands for air-to-mass ratio of 1.) The emphasis is on process kinetics and plausible mechanistic routes that lead to the end product CO_2. This follows our earlier report (*4*) on the photodegradation of phenol in a thin-film TiO_2 reactor operating in a continuous recirculation mode.

Why Photooxidations?

The behavior and fate of aromatic and aliphatic hydrocarbons in the environment are of primordial importance in processes to detoxify wastewater and groundwater. The photochemical transformation of these environmental substances, whether or not mediated by suitable catalysts, is an attractive alternative nonbiological method to natural biodegradational phenomena (*5*). Inconveniences and detrimental factors in direct solar photolysis are the lack of sunlight absorption by these substrates, attenuation of the sunlight, and the relatively shallow penetration depth of sunlight in natural aquatic bodies (several contaminants are adsorbed on sediments in lakes and rivers; e.g., polychlorinated biphenyls). Thus, a catalyst that can absorb solar photons and mediate the oxidation (mineralization) of the substrates, or at least their conversion to environmentally less harmful substances, is desirable.

Detoxification processes designed to remove toxic wastewater constituents must consider the use of other raw materials, newer processes, and newer technologies to minimize the amounts of pollutants in the waste effluents. At present, the various means of treating industrial wastewaters, which contain a mixture of pollutants, are a combination of ultrafiltration, extraction, phase transfer (air stripping, evaporation, and adsorption), incineration, and oxidation by ozone or hydrogen peroxide (*6*). In the latter oxidation methods, only partial contaminant destruction occurs. New methods that are being developed involve photochemical destruction of the contaminant(s) (*7*). The simultaneous coupling of light and an oxidant (O_3, H_2O_2, and O_2), often referred to as advanced oxidation processes (AOPs), has demonstrated (*8*) the complete transformation of organic carbon to CO_2, and hence, the central advantage of photooxidative approaches to water treatment.

In earlier studies (*9*), we examined a variety of materials (metal oxides and others) under UV–visible irradiation to act as photocatalysts. Anatase titania, TiO_2, has so far met the criteria imposed on these materials (light absorption, photochemical and chemical stability, cost, and availability). In each instance, total oxidation was demonstrated by following the temporal evolution of the products CO_2 (and/or HX for halogenated organic compounds) along with the concomitant disappearance of the original substrate. Systematic kinetic studies on TiO_2-mediated photooxidations on some organic substrates have been undertaken (*4, 10–14*).

Significance of Heterogeneous Photocatalysis in Photooxidations

Catalysis, by definition, implicates a catalytic entity that participates and accelerates the chemical transformation of a substrate, itself remaining unaltered at the end of each catalytic cycle. Demonstration of a major excess of reactant conversion, defined by monolayer equivalents of the heterogeneous catalyst, without any measurable solid dissolution or other change has been suggested (*15*) as proof of a truly catalytic phenomenon. An early review (*15*) that used this criterion noted a number of catalytic examples in the literature.

A suitable definition of homogeneous and heterogeneous photocatalysis has been a matter of debate for some time. An excellent account is given by Kisch (*16*). For a truly ideal heterogeneous photocatalyst used in a chemical transformation of reactants to products (*17*),

1. The catalytic entity (here TiO_2) remains unaltered at process completion.
2. The reaction is exoergic and is accelerated only by the entity.
3. Electrons and holes, generated on the photocatalyst, are needed by the process.
4. As will be evident in this chapter, products are formed with high specificity and are different from those obtained in the homogeneous phase.

The distinction between photosensitization and photocatalysis is noted in Figure 1 (*17, 18*). Photosensitization implicates a photosynthetic process

PHOTOSENSITIZATION

$$Ox_1 + Red_2 \rightleftarrows Red_1 + Ox_2$$

$$\Delta G^0 > 0$$

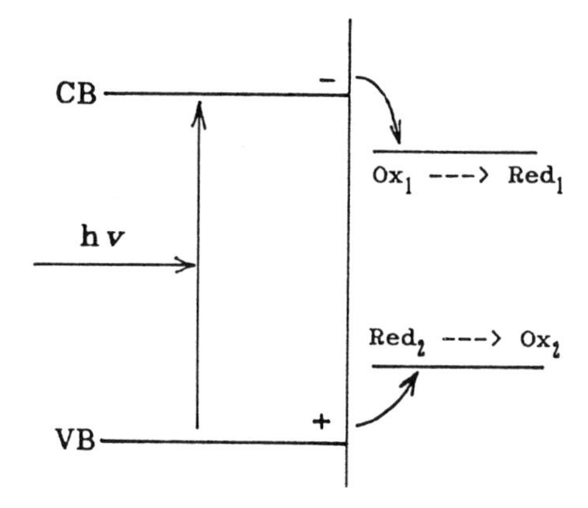

Semiconductor/Electrolyte

PHOTOCATALYSIS

$$Ox_1 + Red_2 \rightleftarrows Red_1 + Ox_2$$

$$\Delta G^0 < 0$$

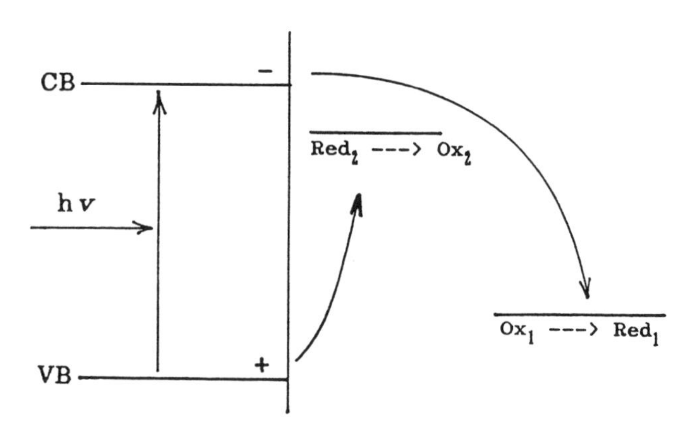

Semiconductor/Electrolyte

Figure 1. Distinction between photosensitization and photocatalysis using semiconductor particulates. Abbreviations: CB, conduction band; VB valence band; Red, electron donor; and Ox, electron acceptor.

(conformational free energy $\Delta G^o > 0$) whereby light induces formation of two products in which photon energy is stored. In photocatalysis, no energy is stored; there is merely an acceleration of a slow event by a photon-assisted process ($\Delta G^o < 0$).

Photocatalysis with irradiated semiconductor dispersions leads to highly effective, spatially controlled oxidations and reductions of organic and inorganic substrates, respectively. TiO_2 particulates have proven excellent catalysts in the photooxidation of a variety of aromatic substrates such as phenol (*4, 19–21*), chlorinated phenols (*4, 9a, 10, 22, 23*), dioxins and polychlorinated biphenyls (PCBs) (*24, 25*), DDT (*26*), surfactants (*11, 27–33*), saturated aliphatic hydrocarbons (*34*), and *s*-triazine (*35*); they have been used extensively in our laboratories and elsewhere (*36–40*). The success of TiO_2-assisted photooxidative degradations of organic compounds is based on several factors:

1. The process occurs under ambient conditions.
2. The formation of photocyclized intermediate products, unlike direct photolysis techniques, is avoided.
3. Oxidation of the substrates to CO_2 is complete.
4. The photocatalyst is inexpensive and has a high turnover.
5. TiO_2 can be supported on suitable reactor substrates (*4, 41, 42*).
6. The process offers great potential as an industrial technology to detoxify wastewaters (*8*).

Nature of the Reactive Species

The present evidence supports the notion that the OH radical (•OH) is the major oxidizing species in the photooxidation of most of the organic compounds examined (*38, 43–45*). Such evidence stems from the observation of highly hydroxylated intermediate products (*4*) that bear close resemblance to the products obtained by oxidation with Fenton's reagent (*46*). In the oxidation of phenol by photoactivated TiO_2 in aqueous media, Okamoto and co-workers (*19*) identified hydroquinone (HQ) and catechol (CC) as the major components for 16% conversion, together with smaller quantities of pyrogallol (PG), 1,2,4-trihydroxybenzene (HHQ), and hydroxybenzoquinone (HBQ).

Similar hydroxylated species (3-fluorocatechol, fluorohydroquinone, 4-fluorocatechol, and 1,2,4-trihydroxybenzene) were identified in the photooxidation of 3-fluorophenol (*14*). Added support for •OH as the primary oxidant in aqueous media originates with a recent kinetic deuterium isotope experiment (*47*) that showed that the rate-limiting step in the photooxidation of 2-propanol on TiO_2 is formation of active oxygen species through a reaction involving the solvent water. Electron paramagnetic resonance (EPR) studies also identified •OH as the sole radical by spin-trap methods in aqueous solutions (*48, 49*), and in gas–solid systems (*50*) of irradiated TiO_2, under ambient conditions.

In principle, two oxidizing species are possible when a semiconductor particle is illuminated with photon energies greater or equal to its band-gap energy, E_{bg}. For a metal chalcogenide semiconductor, the species are the photogenerated valence band holes, $h_{VB}{}^{+}$, and the •OH (discussed later). Recent studies (*51–53*) addressed the issue of direct hole versus OH radical oxidations. A thorough discussion of the views presented here is treated elsewhere (*54*). Briefly, no distinction exists between a photogenerated trapped hole and an •OH formed on the particle surface (*52, 53*).

Several routes lead to formation of •OH on an illuminated semiconductor particle surface (*44, 45, 55–58*). The primary photochemical act, subsequent to near-UV light absorption by TiO_2 particles (wavelengths <380 nm), is the generation of electron–hole pairs whose separation into conduction band electrons ($e_{CB}{}^{-}$) and valence band holes ($h_{VB}{}^{+}$) is facilitated by the electric-field gradient in the space-charge region (eq 1):

$$TiO_2 + h\nu\,(\geq E_{bg}) \xrightarrow{\Phi I_a{}^n} e^- + h^+ \qquad (1)$$

where I_a refers to the rate of absorption of light at low photon flux (n = 1) or at higher photon flux (n = 0.5). Subsequently, the charge carriers can recombine:

$$e^- + h^+ \xrightarrow{k_{rec}} TiO_2 + \text{heat (and/or } h\nu) \qquad (2)$$

where k_{rec} is the rate constant for electron–hole recombination. Competing with reaction 2 is charge-carrier migration (*17, 59*) to the surface (Figure 2); there the charge carriers are ultimately trapped by intrinsic subsurface energy traps ($Ti^{IV}–O^{2-}–Ti^{IV}$) for the hole and by surface traps ($–Ti^{IV}–$) for the electrons (eqs 3 and 4) (*60*),

$$(Ti^{IV}-O^{2-}-Ti^{IV})_{subsurface} + h_{VB}{}^{+} \longrightarrow (Ti^{IV}-O^{-}\bullet-Ti^{IV})_{subsurface} \qquad (3)$$

$$(-Ti^{IV}-)_{surface} + e_{CB}{}^{-} \longrightarrow (-Ti^{III}-)_{surface} \qquad (4)$$

and/or by extrinsic surface traps via interfacial electron transfer with adsorbed electron donors $(Red_2)_{ads}$ and acceptors $(Ox_1)_{ads}$, respectively (eqs 5 and 6).

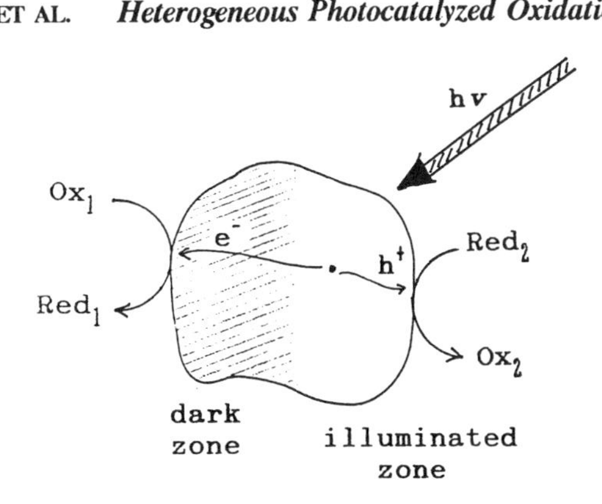

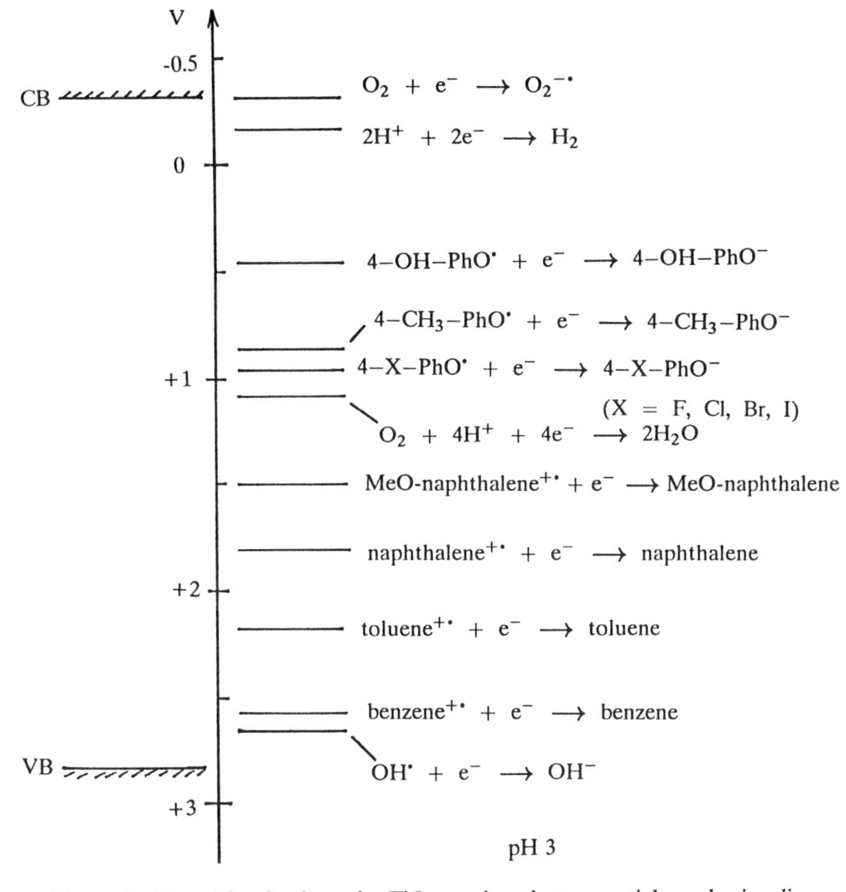

Figure 2. Top: Simple view of a TiO_2 semiconductor particle under irradiation and the ensuing redox reactions. Bottom: Redox potential scale showing the potentials (vs. NHE) of the valence and conduction bands of anatase TiO_2 and the redox potentials of some redox couples in homogeneous phase.

$$h_{VB}^{+} + (Red_2)_{ads} \longrightarrow (Ox_2)_{ads} \quad (5)$$

$$e_{CB}^{-} + (Ox_1)_{ads} \longrightarrow (Red_1)_{ads} \quad (6)$$

The event embodied in equation 2 necessitates that Red_2 and Ox_1 be adsorbed prior to light excitation of the TiO_2 photocatalyst. Adsorbed redox-active solvents can also act as electron donors and acceptors. For a hydrated and hydroxylated TiO_2 anatase surface, hole trapping by interfacial electron transfer yields surface-bound •OH (*48–50, 59–65*). It is a major route of formation of •OH (eqs 7a and 7b).

$$(Ti^{IV}-O^{2-}-Ti^{IV})-OH^{-} + h_{VB}^{+} \underset{k_{-7}}{\overset{k_7}{\rightleftarrows}} (Ti^{IV}-O^{2-}-Ti^{IV})-{\cdot}OH \quad (7a)$$

$$(Ti^{IV}-O^{2-}-Ti^{IV})-OH_2 + h_{VB}^{+} \underset{k_{-7}}{\overset{k_7}{\rightleftarrows}} (Ti^{IV}-O^{2-}-Ti^{IV})-{\cdot}OH + H^{+} \quad (7b)$$

where k_7 refers to the rate constant for the oxidation of surface OH^- and/or H_2O, and k_{-7} denotes the corresponding rate constant for the reverse reaction. In principle, OH radicals may also be formed from H_2O_2 via the superoxide radical anion O_2 (eqs 8 and 9) (*19b*), and by photolysis of H_2O_2 (eq 10).

$$O_2(ads) + e^{-} \longrightarrow O_2^{-}{\cdot}(ads) \quad (8)$$

$$H_2O_2 + O_2^{-}{\cdot} \longrightarrow {\cdot}OH + OH^{-} + O_2 \quad (9)$$

$$H_2O_2 + h\nu \longrightarrow 2{\cdot}OH \quad (10)$$

However, when irradiation is done with AM1-simulated sunlight (λ >340 nm), reactions 9 and 10 are of little consequence because the quantity of H_2O_2 formed is small (*66*).

Some years ago, we reported (*22a*) that both O_2 and H_2O are essential species in the photomineralization of 4-chlorophenol in the presence of irradiated TiO_2; no photodegradation occurs in the absence of either O_2 or H_2O, or both (*22*).

Even trapped electrons and trapped holes can rapidly recombine on the particle surface. To obviate recombination, electrons are scavenged by preadsorbed (and photoadsorbed) molecular oxygen to give O_2(ads) (eq 8) which can be further reduced to the peroxide dianion, O_2^{2-}(ads) (eq 11).

$$O_2^{-}{\cdot}(ads) + e^{-} \longrightarrow (O_2^{2-})_{surface} \quad (11)$$

Alternatively, surface peroxo species can be formed (*67–70*) either by hydroxyl radical (hole) pairing (eq 12), or by sequential two-hole capture by the same OH group (eq 7 followed by eq 14), or by dismutation of O_2 (eq 15).

$$2(\cdot OH)_{surface} \longrightarrow (H_2O_2)_{surface} \quad (12)$$

$$(H_2O_2)_{surface} \rightleftarrows (O_2^{2-})_{surface} + 2H^+ \quad (13)$$

$$(Ti^{IV}-O^{2-}-Ti^{IV})-\cdot OH + h_{VB}^+ \xrightarrow{-H^+} (O)_{surface} \longrightarrow \frac{1}{2}O_2(ads) \quad (14)$$

$$2O_2^{-}\cdot(ads) \longrightarrow O_2(ads) + O_2^{2-}(ads) \quad (15)$$

In acidic media (pH 3), O_2 exists as the hydroperoxide radical, $HO_2\cdot$ (pK_a 4.88, ref. 71). Other plausible reactions that might occur on the TiO_2 particle surface and that are solvent-related (i.e., water-related) are summarized in equations 16–19.

$$2HO_2\cdot \longrightarrow H_2O_2 + O_2 \quad (16)$$

$$H_2O_2 + e^- \longrightarrow \cdot OH + OH^- \quad (17)$$

$$H_2O_2 + h^+ \longrightarrow O_2 + 2H^+ \quad (18)$$

$$O_2^{-}\cdot + HO_2\cdot \longrightarrow O_2 + HO_2^- \xrightarrow{H^+} H_2O_2 \quad (19)$$

Many of these reactions have not been assessed under the photocatalytic conditions used in these photooxidations. They are inferred from radiation chemical studies in homogeneous phase.

The energies of the valence and conduction bands (pH 3) on a redox scale (vs. the normal hydrogen electrode or NHE) for TiO_2 anatase, as well as the homogeneous phase potentials of relevant redox couples (*45, 72, 73*) are illustrated in Figure 2. A photogenerated valence band hole, redox potential E_{redox} ~+2.9 V, can in principle oxidize OH^- to $\cdot OH$, and can oxidize various organic substrates to give the corresponding radical cations and phenoxyl species. No inference is made here regarding the magnitude of the redox potentials of the couples in heterogeneous media. No doubt, physisorption or chemisorption of species to a heterogeneous surface will alter their redox potentials. For instance, the redox potential of the chemisorbed $\cdot OH/OH^-$ couple on TiO_2 recently (*54*) was estimated as ~+1.5 V; the equivalent redox potential for this $\cdot OH/OH^-$ couple in homogeneous aqueous media is ~+2.62 V (pH 3). Similar data for other species adsorbed on a heterogeneous surface are desirable.

Experimental Procedures

Phenol (Carlo Erba, >95% pure), and high-performance liquid chromatographic (HPLC) grade methanol and *o*-phosphoric acid were used as received. Titanium dioxide was Degussa P 25 (Brunauer–Emmett–Teller (BET) surface area, 55 m^2/g; particle size, 20–30 nm); it was used without prior treatment. Water was doubly distilled throughout.

Irradiation of air-equilibrated aqueous solutions of phenolic samples (20 mL, 20 ppm (mg/L), 212 μM) was carried out with AM1-simulated sunlight (CO.FO.MEGRA, Milan, Italy; 1500-W Xe lamp equipped with a 340-nm cutoff filter). The catalyst TiO_2 loading was 2 g/L. The photodegradation process was examined at three pH values: pH 3 (adjusted with H_2SO_4), natural pH of 6.5 (unbuffered twice-distilled water), and pH 7 (0.01 M phosphate buffer).

Analyses of reactants and intermediate products were carried out by HPLC methods on 3-mL aliquots collected at selected intervals from the irradiated TiO_2 solutions. The aliquots were centrifuged and filtered over Millipore GSTF filters (pore size 0.22 μm) prior to analysis on a Perkin-Elmer Series 2 liquid chromatograph equipped with a UV detector; the detection wavelength was 280 nm, and the stationary phase was a Bondapak C18 reverse-phase column. The mobile phase consisted of a 40.0:59.9:0.1 mixture of methanol–water–*o*-phosphoric acid. The CO_2 temporal evolution was monitored by gas chromatographic (GC) techniques on a HP-5890 gas chromatograph, and confirmed by the quantity of $BaCO_3$ produced from bubbling the evolved gas into a $Ba(OH)_2$ solution.

Details of the experimental procedures used in the photooxidation of cresols and fluorinated phenols were presented elsewhere (*13, 14*); they are similar to those given here with minor variations (e.g., for the cresols, the illumination was carried out using a 900-W Hg–Xe lamp at λ >320 nm).

Results and Discussion

Photooxidation of Phenol. The photooxidation of phenol in irradiated TiO_2 suspensions was examined earlier (*4, 20, 21*) under a variety of conditions. In addition to hydroquinone and catechol, 1,2,4-benzenetriol, pyrogallol, and 2-hydroxy-1,4-benzoquinone were identified, albeit in smaller quantities (*19*). Elsewhere, only catechol and benzoquinone were detected (*20a*). Herein, we report the identification of hydroquinone and catechol. These differences arise from variations in experimental conditions, the type, source, and any prior treatment of the TiO_2 used. Okamoto and co-workers (*19*) used anatase pretreated by reduction in a stream of hydrogen at 520 °C for 6 h (surface area, 38 m^2/g). Augugliaro et al. (*20a*) employed TiO_2 (BDH Inc., anatase; surface area, 10 m^2/g). In our work, we use untreated TiO_2 (Degussa P-25, largely anatase; surface area, 55 m^2/g). The photocatalytic activity of TiO_2 and other semiconductor photocatalysts depends on the pretreatment to which they are subjected.

Rutile TiO_2 appears inactive in the degradation of phenol, and different batches of anatase show different activities (*20*). The nature of the intermediates formed must depend on the nature of the catalyst and on the experimental conditions. The role of surface properties of TiO_2 particles, although an unknown parameter, is significant (*54*). The TiO_2 P 25 (~80% anatase, ~20% rutile) consists of 99.5% TiO_2, <0.3% Al_2O_3, <0.3% HCl, <0.2% SiO_2, and <0.01% Fe_2O_3. These impurities, segregated on the particle surface, could strongly influence the kinetics and the thermodynamics of interfacial electron transfer between the photocatalyst and the various substrates.

Oxidation of phenol by the UV–H_2O_2 advanced oxidation process (AOP) yields dihydroxy and trihydroxy intermediates and ultimately CO_2 (*74*). By contrast, oxidation by UV–O_3 yields nine intermediate products, the principal ones being formic acid, glyoxal, glyoxylic acid, and oxalic acid (*75*).

Irradiation of an air-equilibrated aqueous phenolic–TiO_2 suspension with AM1-simulated sunlight leads to the disappearance of phenol. The temporal course of the process was examined at pH 3 (adjusted with H_2SO_4; Figure 3), in distilled water (pH ~6.5), and at pH 7 (phosphate buffer). At [phenol] = 212 μM, only hydroquinone (HQ; in distilled water and at pH 3 and 7) and catechol (CC; pH 3) were detected. Trace

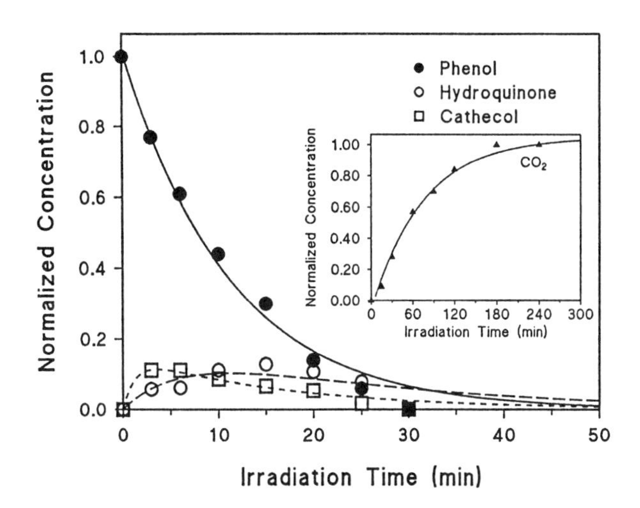

Figure 3. Photodegradation of an air-equilibrated solution of phenol (212 μM) in the presence of irradiated TiO_2 (2 g/L) at pH 3 (H_2SO_4). The curves were obtained from the computer fit of the experimental data. Inset: evolution of CO_2.

quantities of CC (1–2 μM; i.e., 0.1–0.2 ppm, the sensitivity limits of the analytical method employed) were found in distilled water. By contrast, neither HQ nor CC was detected in 0.01 M NaOH (pH ~12) solutions. In all cases, disappearance of phenol follows reasonably good first-order kinetics for about two or three half-lives; formation of carbon dioxide is stoichiometric (eq 20).

$$C_6H_5OH + 7O_2 \xrightarrow[TiO_2]{h\nu} 6CO_2 + 3H_2O \quad (20)$$

Phenol Mineralization: pH 3 (H_2SO_4). Oxidation of phenol by irradiation in the presence of TiO_2 (52% after 10 min; apparent rate of decomposition k_{dec} = 0.091 + 0.006 min^{-1}; half-life t = 7.6 min) yields HQ and CC with comparable chemical yields (11% and 9%, respectively), but at different times in the process (~3 min for CC and ~15 min for HQ). The apparent rate of formation of HQ is k_f = 0.030 min^{-1}, and its rate of mineralization is 0.21 min^{-1} (Figure 3); curves represent first-order computer fits of the data. The corresponding apparent kinetics for CC are $2k_f$ = 0.060 min^{-1} and k_{dec} = 0.76 min^{-1}. The total degradation of phenol and its intermediates HQ and CC occurred at about 30 min into the process. The evolution of CO_2 is also shown in Figure 3 (inset); it is nearly quantitative after ~3 h of irradiation. The apparent kinetic parameters for both HQ and CC yield a reasonably good fit for up to 30 min; the rate of generation of CO_2 is k_{CO_2} = 0.013 min^{-1}. The error in the curve fitting and the reproducibility of the experimental data are about ±10% (or better).

Phenol Mineralization: Distilled Water (pH ~6.5). After 10 min of irradiation, 74% of phenol degraded (155 μM; k_{dec} = 0.16 ± 0.02 min^{-1}; t = 4.4 min), and 13% of HQ formed (27 μM; k_f = 0.13 min^{-1}). By 20 min, there was no trace of phenol remaining in solution, and more than 95% of HQ had decomposed (k_{dec} = 0.18 + 0.03 min^{-1}; t = 3.9 min). This quantity accounted for only 66% of the CO_2 produced (~832 μM). The quantity of CC detected was within the experimental "noise". After 30 min of irradiation, 10% (or ~20 μM) of the initial phenol concentration was converted to intermediate products and ~90% to CO_2. The expected stoichiometric quantity (reaction 20; 1260 μM) of CO_2 evolved was reached after about 60 min, by which time all the intermediate species had degraded. Curve-fitting the quantity of CO_2 evolved gave an apparent rate constant k_{CO_2} = 0.050 ± 0.012 min^{-1}.

Phenol Mineralization: pH 7 (Buffer). After 10 min of irradiation of the phenol–TiO_2 suspension, 151 μM (72%) of phenol degraded (k_{dec} = 0.13 ± 0.01 min^{-1}; t = 5.3 min), 20% of which (~42 μM) was converted to HQ; the apparent rate of formation of HQ was k_f = 0.13 min^{-1}, and its apparent rate of decomposition was k_{dec} = 0.16 min^{-1} (t = 4.4 min). No other aromatic intermediate was detected. We presume that the remaining 52% of phenol was converted to aliphatic carboxylates (discussed later) and/or to carbon dioxide.

Photooxidation of Cresols. Liquid chromatograms showed only two clearly detectable intermediates formed for each of the three cresols examined. Hydroxylation of *p*-cresol may yield two intermediates: 4-methylresorcinol and 4-methylcatechol. By contrast, both *o*- and *m*-cresol can in principle yield four isomeric species. Common to both the *p*- and *m*-cresol is 4-methylcatechol (4-MCC), and the common intermediate in *o*- and *m*-cresol was identified as methylhydroquinone (MHQ) (Figure 4). Another intermediate seen in the reaction of *p*-cresol was described as 4-methylresorcinol (*13*); understandably, this species did not form in the photooxidation of *o*-cresol because hydroxylation at position meta to the OH substituent is unlikely. Formation of 4-methylcatechol in the oxidation of the cresols was confirmed by diffuse reflectance spectra.

None of the cresols examined degrade in the dark in the presence of TiO_2. Direct irradiation of aqueous solutions of the cresols with UV–visible light, but in the absence of TiO_2, leads to very small decreases in the concentration of the cresols: 3–5% after ~6 h of irradiation (*see* Figures 4A–4C). For 20 mg/L of the cresols, the rate constants are 6.3 × 10^{-5} min^{-1} (*o*-cresol), 18 × 10^{-5} min^{-1} (*m*-cresol), and 9.9 × 10^{-5} min^{-1} (*p*-cresol), about 1–2 orders of magnitude smaller than the catalyzed processes (0.4–1.3 × 10^{-2} min^{-1}).

The photocatalyzed oxidation of the three cresols in the presence of TiO_2 is also illustrated in Figures 4A–4C. Both *o*- and *p*-cresol (20 mg/L) degrade via first-order kinetics; *m*-cresol (also 20 mg/L) decomposes by zero-order kinetics under otherwise identical experimental conditions of light source and initial pH. The formation and subsequent degradation of the two intermediates are also shown. In each case, total disappearance of the original cresol and decomposition of the intermediate species occurs in less than 4 h of irradiation. Product analysis showed that the photocatalyzed oxidation followed the expected stoichiometry (eq 21):

$$CH_3C_6H_4OH + \frac{17}{2}O_2 \xrightarrow[TiO_2]{h\nu} 7CO_2 + 4H_2O \qquad (21)$$

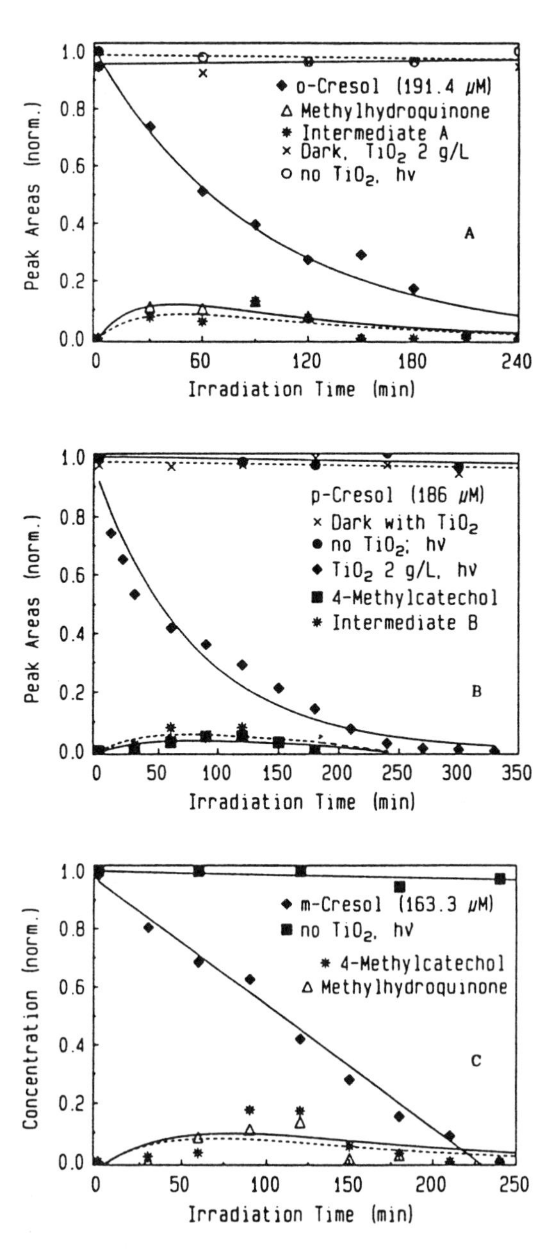

Figure 4. Plots of normalized peak areas (or concentration) as a function of irradiation time showing the degradation of the three cresols (ca. 20 mg/L) and formation and decomposition of two intermediates in the photomineralization process with irradiated TiO_2 present (2 g/L) at pH 3. The behavior of the cresols under dark conditions but in the presence of the TiO_2, and under direct photolysis (no TiO_2 present) are also indicated. The curves are computer fits to equations noted in the text. (Reproduced with permission from reference 13. Copyright 1991.)

Approximately 32.5 μmol of CO_2 was expected for an initial cresol concentration of 20 mg/L. Four hours of irradiation, when all the cresols and the aromatic intermediate species had decomposed, produced only ~65–70% of CO_2; 80–85% of CO_2 evolved after about 7 h. Other, slow-degrading intermediates (aliphatic) formed. Starvation of the suspension of needed oxygen was not precluded (*22*).

Consequently, the effect of the oxygen concentration on the kinetics of decomposition was carried out on *m*-cresol under conditions in which the suspension was saturated with O_2. The decomposition of this cresol in air-equilibrated and in oxygen-saturated suspensions is compared in Figure 5A. Apparent zero-order kinetics are also evident. The corresponding

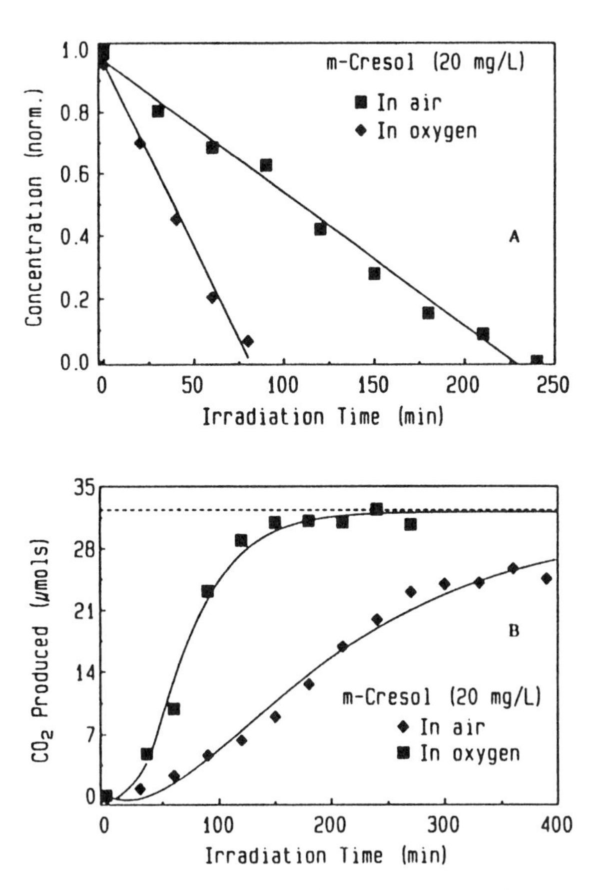

Figure 5. Top: Plots showing the photodegradation of 20 mg/L of m-*cresol in the presence of 2 g/L TiO_2 in air-equilibrated and in oxygen-saturated suspensions; initial pH 3. Bottom: Plots showing the corresponding temporal evolution of CO_2 from the photomineralization of 20 mg/L of* m-*cresol in air-equilibrated and in oxygen-saturated suspensions of TiO_2. (Reproduced with permission from reference 13. Copyright 1991.)*

parameters are, air versus O_2, respectively: k_{dec}, 4.0×10^{-3} and 12×10^{-3} min^{-1}; initial rates, 0.74 and 2.2 μM/min; and t(app), 174 and 59 min. In the presence of excess oxygen, decomposition of m-cresol occurs in less than 1.5 h. Comparison of the evolution of carbon dioxide from air-equilibrated and oxygen-saturated suspensions is made in Figure 5B. For the oxygen-saturated suspensions, about 60% of CO_2 evolved after the cresol had decomposed. Near quantitative formation of CO_2 is seen in ~2.5 h; k_{CO_2} ~9.5×10^{-3} min^{-1} (air) and 32×10^{-3} min^{-1} (O_2). The slower evolution of products cannot be due solely to the lack of oxygen, but also to the intermediacy of relatively stable, slower degrading intermediates.

Photooxidation of Fluorophenols. *Monofluorophenols.* Illumination of air-equilibrated aqueous suspensions containing 50 mg/L of TiO_2 and 2.0×10^{-4} M of monofluorinated phenols with AM1-simulated sunlight leads to rapid disappearance of the phenols in short time (~15–40 min; Figures 6 and 7). Both 2-fluorophenol (2-FPh) and 4-fluorophenol (4-FPh) decompose via good (apparent) first-order kinetics that are nearly identical within experimental error: 0.104 ± 0.005 and 0.106 ± 0.004 min^{-1}, respectively (t = 6.7 min). Oxidation of 3-fluorophenol (3-FPh) follows reasonably good first-order kinetics to yield an apparent $k_{dec} = 0.074 \pm 0.007$ min^{-1} (t = 9.4 min). No photooxidation occurs in the absence of TiO_2; in the dark, but in the presence of TiO_2, no changes were noted in the concentrations of the three phenols examined.

The decomposition of 2-fluorophenol; the formation of F^-; and the formation of the ultimate product CO_2, which confirms the complete oxidation of the initial organic substrate, are depicted in Figure 7. The dotted line represents the expected stoichiometric quantity of carbon dioxide that in this case is achieved after ~4 h of irradiation. Oxidation of 2-FPh occurs with nearly concomitant formation of fluoride, whether the quantity of TiO_2 was 50 or 100 mg/L: $k_{dec} = 0.10 \pm 0.01$ and 0.17 ± 0.01 min^{-1}, and the apparent rate of formation of F^-, k_F, $= 0.081 \pm 0.002$ and 0.11 ± 0.02 min^{-1}, respectively. The corresponding rate constant for the formation of CO_2 is 0.016 ± 0.002 min^{-1} (t = 42 min). Similar patterns were observed for the other two monofluorinated phenols. Formation of fluoride reaches a maximum at ~70–80%, which corresponds to the fractional quantity of fluoride in solution. The remaining 20–30% is chemisorbed on the TiO_2 particle surface.

The formation and subsequent decomposition of the intermediate products, fluorohydroquinone (FHQ), 3-fluorocatechol (FCC), and catechol (CC) obtained from the decomposition of 2-fluorophenol, along with the formation of fluoride, are shown in Figure 6. After 25 min into the process, 2-FPh has degraded while both FCC and CC have formed and

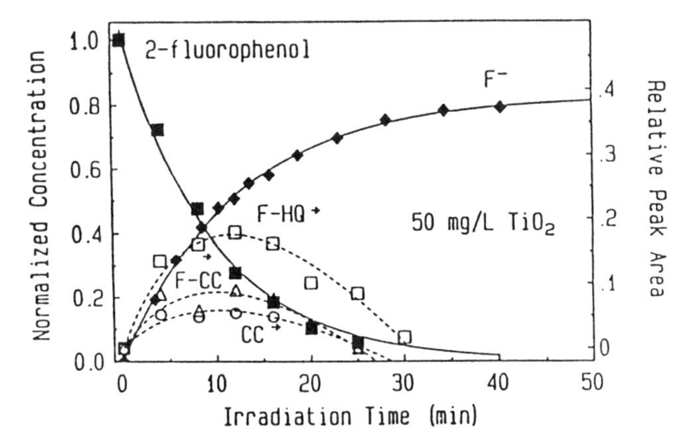

Figure 6. Plots showing the photodecomposition of 2-fluorophenol and the nearly concomitant formation of fluoride as a function of irradiation time at constant concentration of TiO_2 at 50 mg/L. The relative peak areas for three of the identified intermediates (fluorohydroquinone, fluorocatechol, and catechol) as a function of irradiation time are also indicated. (Reproduced from reference 14. Copyright 1991 American Chemical Society.)

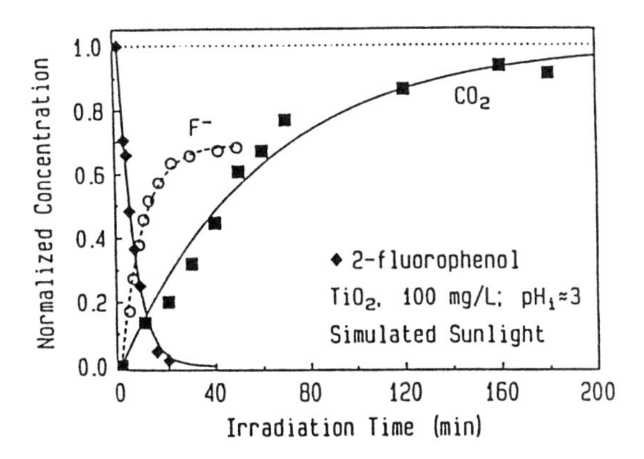

Figure 7. First-order plots showing the decomposition of 2-fluorophenol, the formation of the F^- anion (or HF) and formation of carbon dioxide (the product of total photomineralization) as a function of irradiation time. (Reproduced from reference 14. Copyright 1991 American Chemical Society.)

decomposed; FHQ also reaches a maximum after ~10 min and is totally decomposed after 30 min of irradiation. By this time, approximately 20% of the phenol has been converted to carbon dioxide. The other 80% consists of aliphatic intermediates that ultimately also degrade, but more slowly, to yield the final oxidation products. Formation of CO_2 is faster for 3-FPh and 4-FPh (compare t = 42 vs. 25 min).

The apparent rate of decomposition of 4-fluorophenol in a 100-mg/L TiO_2 aqueous suspension is 0.16 ± 0.01 min^{-1}. Two of the intermediates are hydroquinone (HQ) and 4-fluorocatechol. Curve-fitting the data for HQ gives k_{dec} = 0.16 min^{-1}, identical with the value of 0.16 ± 0.01 min^{-1} obtained for the degradation of a pure sample of HQ under otherwise identical conditions.

Figure 8 illustrates the effect of varying the concentration of the TiO_2 on the observed rate constant for the disappearance of 4-fluorophenol. The curve-fitted $k_{obs} = A[TiO_2]/(B + [TiO_2])$, with $A = k_{obs}(lim) = 0.49 \pm 0.05$ min^{-1} and $B = 1/K = 0.43 \pm 0.17$ g/L (K denotes the photoadsorption coefficient and is $\sim 3 \times 10^8$ M^{-1} in terms of concentration of TiO_2 particles).

Difluorophenols. The decomposition of a difluorophenol (2,4-DFPh) in suspensions containing 100 mg/L of TiO_2 is shown in Figure 9; the process follows good first-order kinetics (0.24 ± 0.01 min^{-1}) for several half-lives. It is complete by 12 min. Under identical conditions of concentration of catalyst, the decomposition of 2,4-DFPh and 3,4-DFPh are nearly identical.

The formation of fluoride and CO_2 in the photocatalyzed oxidation of 2,4-DFPh is also presented in Figure 9. Formation of F^- follows the degradation of the parent substrate 2,4-DFPh: $k_{dec} = 0.24 \pm 0.01$ min^{-1} and $k_F = 0.14 \pm 0.02$ min^{-1}. Similar to the observations noted earlier, $[F^-]$ in solution reaches a maximum value of ~75% of the stoichiometric value. Formation of CO_2, which confirms the total destruction of 2,4-DFPh, is nearly complete after 90 min of irradiation ($k_{CO_2} = 0.036 \pm 0.002$ min^{-1}; t = 19 min).

Identification of Intermediates. Liquid chromatographic techniques identified the intermediate products from the photodegradation of 2-FPh: fluorohydroquinone (FHQ), 3-fluorocatechol (3-FCC), and catechol. The 4-fluoro analog gives 4-fluorocatechol (4-FCC) and hydroquinone, and 3-fluorophenol yields fluorohydroquinone, 3-fluorocatechol, and 4-fluorocatechol. For the difluorophenols, only one aromatic intermediate was identified: fluorohydroquinone.

Apparent Kinetics. As will be evident later, the presence of TiO_2 particulates presents a formidable challenge in describing the kinetics in heterogeneous photocatalysis. The difficulty originates with the heterogeneous surface, which interacts with solvent and substrates to various degrees of complexities. Some of these interactions have yet to be

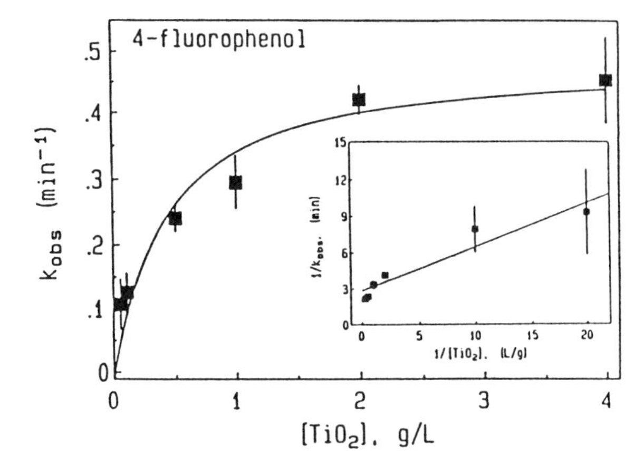

Figure 8. Dependence of k_{obs} *on the concentration of the photocatalyst* TiO_2 *for constant concentration (200 μM) of 4-fluorophenol. The inset shows the linear transform of the adsorptionlike isotherm. (Reproduced from reference 14. Copyright 1991 American Chemical Society.)*

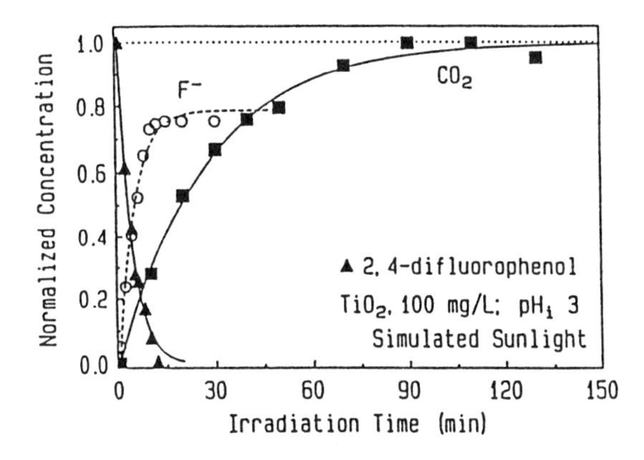

Figure 9. Plots showing the photodecomposition of 2,4-difluorophenol, the concomitant formation of fluoride, and formation of the final product CO_2 *in the photomineralization of the fluorophenol. The concentration of the photocatalyst was 100 mg/L. (Reproduced from reference 14. Copyright 1991 American Chemical Society.)*

assessed in photooxidations. Complicating the picture is the effect photons have on such surface properties as, for example, adsorption–desorption equilibria and the nature of the catalytic sites, among others. Process kinetics in heterogeneous photocatalysis must be considered, for the moment, as apparent kinetics until such time as more details are understood from the semiconductor–solution interfacial dynamics.

The photocatalytic process described in the following section for the oxidation of phenol, cresols and the fluorinated phenols, on the basis of the constraints just given, has been modeled in its most simplistic form by the phenomenological consecutive and parallel reactions in Scheme I:

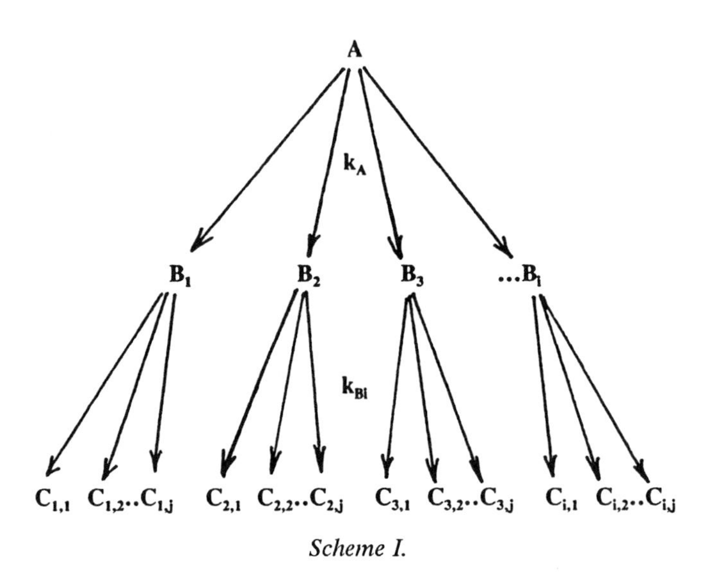

Scheme I.

where A is the initial substrate, B_i are intermediate products, and $C_{i,j}$ are subsequent intermediate species towards complete oxidation; k_A represents $\sum_i k_{Ai}$ for $i = 1, 2, \ldots n$ and n is the number of possible sites available on the aromatic ring for attack by the primary oxidizing species, the •OH radical (discussed later); and k_B denotes the macroscopic rate constant for consecutive and parallel reactions of the disappearance of the various stable B_i intermediate products. When the initial organic substrate is a haloaromatic, we use k_X to represent the observed rate constant of formation of the halide, irrespective of its form (X^- or HX) and its source. The implicit assumption has also been made (*11*) that k_{Ai} are nearly identical, irrespective of the position of attack of the ring by the •OH. The disappearance of the initial substrate, A, together with the formation and disappearance of B can be described by (*76*):

$$[A_{(t)}] = [A]_0 \exp(-k_A t) \tag{22a}$$

$$[B_{i(t)}] = \frac{k_A [A]_0}{k_B - k_A} \left[\exp(-k_A t) - \exp(-k_B t) \right] \tag{22b}$$

where t is time. The formation of X^- and CO_2 follows in nearly all cases simple exponential growth kinetics: [product] = constant $[1 - \exp(-k_P t)]$.

Although the phenols examined here are only slightly adsorbed to the particle surface in the dark, the extent of photoadsorption is an inaccessible parameter depending on adsorption–desorption rates and on whatever other process(es) that take place subsequent to adsorption. In describing a rate expression for the photooxidation process, the implicit assumption is often made that there is a constant fraction, however small, of the organic substrate on the catalyst's oxidative active surface sites, $Ti^{IV}|\cdot OH$ (*77*). Following ·OH attack on the aromatic ring of the substrate S_{ads}, one or more intermediates (I_{ads} and/or I_{sol}) that form subsequently or nearly simultaneously undergo dehalogenation and fragmentation to aliphatic species, which ultimately also degrade to produce stoichiometric quantities of CO_2 (eq 23).

$$\overbrace{(\text{site}) + S \underset{k_d}{\overset{k_a}{\rightleftarrows}} S_{ads}}^{A} \xrightarrow{k_S} \overbrace{[I_{ads} + I_{sol}]}^{B} \xrightarrow{k_I} \overbrace{\cdots \longrightarrow\longrightarrow\longrightarrow CO_2}^{C} \qquad (23)$$

$$S_{ads} \searrow \{X^-\} \qquad [I_{ads} + I_{sol}] \searrow \{X^-\}$$

where k_S denotes Σ(rate constants) in the formation of various intermediate species and k_I represents a sum of rate constants in the fragmentation of these intermediates.

Employing the methods used earlier (*12, 13*), the rate of formation of product will be given by (*78*):

$$\text{rate} = \frac{k_S K_S N_S[S]}{1 + \alpha K_S[S]} \qquad (24)$$

where $K_S = k_a/(k_d + k_S)$ is the photoadsorption coefficient for the various substrates, N_S is the number of oxidative active sites, and $\alpha = (k_S + k_I)/k_I$. Expression 24 is reminiscent of a Langmuirian rate equation (*79*). Further modification of equation 24 is needed to take account of the formation of the reactive ·OH species described by reactions 7a and 7b.

The quantum yield of ·OH formation is $\Phi_{OH} = \Phi I_a{}^n k_7 \tau$ (at low light fluxes $n = 1$), where τ is the lifetime of the photogenerated h^+ [$\tau = 1/(k_{rec} + k_7)$]. If $k_{rec} > k_7$ (*80*), then $\Phi_{OH} \sim \Phi I_a \beta A_p k_7/k_{rec}$, where βA_p is the fraction of the irradiated surface and A_p is the particle surface area. As well, the rate of formation of products will depend on the lifetime of the bound ·OH species, τ_{OH}. This dependence leads to:

$$\text{rate} = \frac{\left(\dfrac{\Phi I_a{}^n \beta A_p k_7 \tau_{OH}}{k_{rec}}\right) k_S K_S N_S[S]}{1 + \alpha K_S[S] + K_w[H_2O] + K_I[I] + K_{ions}[ions]} \times \frac{K_{O_2}[O_2]}{1 + K_{O_2}[O_2]} \tag{25}$$

Here, k_{rec} is the rate constant for electron–hole recombination events (radiative and nonradiative). The additional terms in the denominator indicate the influence of the solvent water, the intermediates, and the various anions and cations present in the system; through adsorption, any one or all of these can act as inhibitors by blocking some of the active surface sites on the photocatalyst (*81*). The adsorption isotherm expression for oxygen is included in equation 25 to consider the effect of molecular oxygen on the rate of product formation (*13*).

Expression 25 has the same analytical form as the equations reported by Okamoto and co-workers (*19*) and more recently by Turchi and Ollis (*77*). Turchi and Ollis noted that, with minor variations, the expression for the rate of photooxidation of organic substrates on irradiated TiO_2 presents the same saturation-type kinetic behavior as portrayed by the Langmuir–Hinshelwood rate law (*79*), and this behavior occurs whether or not (1) reaction occurs while both reacting species are adsorbed, (2) a "free" radical species reacts with the adsorbed substrate, (3) a surface-bound radical reacts with a free substrate in solution, and (4) the reaction takes place between the free reactants in solution. Assigning an operational mechanism for reactions taking place in heterogeneous media to a Langmuir-type process, to an Eley–Rideal pathway, or to an equivalent type process is not possible on the basis of observed kinetics alone (*82, 83*). For example, an equation analogous to the Langmuirian expression is obtained in homogeneous media for the reduction of *cis*-$Ru(NH_3)_4Cl_2{}^+$ by Cr^{2+}(aq) (*84*). Clearly, although the analytical expression obtained for the rate of photooxidation may be analogous to the Langmuir–Hinshelwood relationship, nothing can be concluded about the operational mechanism in a heterogeneous photocatalysis experiment. Other experiments must be undertaken to unravel the complex events that take place. The events that occur on an illuminated semiconductor alone present a formidable challenge.

Photooxidation Mechanisms. In aqueous acidic media (pH 3) O_2 protonates to form the hydroperoxide radical $HO_2\cdot$, which is subsequently converted to H_2O_2 (*38, 43, 44b, 55–57*). Absorption of UV light by hydrogen peroxide or its reaction with e^- can break down H_2O_2 to produce •OH (*19, 45, 55, 56*). However, the major source of •OH is reactions 7a–7b or their equivalent (*43, 58*). Direct oxidation of the organic substrates by the valence band holes of the photocatalyst is probably minor (*77*). Our studies are silent on the possible direct involvement of O_2 anion radicals (or HO_2•) in the photodegradation via either oxidation or reduction of the fluorinated phenols and organic compounds in general. However, O_2 can oxidize aromatic compounds; for example, methylhydroquinone is oxidized by the superoxide radical anion with $k \sim 1.7 \times 10^7$ M^{-1} s^{-1} (*85*).

Photocatalyzed oxidations of organic substrates are typically bimolecular. Classically, heterogeneously photocatalyzed oxidations have been viewed as occurring via two principal routes: the Eley–Rideal (ER) and the Langmuir–Hinshelwood (LH) pathways. In the ER pathway, the reaction is viewed as occurring between a surface-adsorbed oxidant and the organic substrate in solution, but in the LH mechanism, reaction involves the two reactants adsorbed on the particle surface on which they can either compete for the same available sites or they are adsorbed on two different types of available sites. Appropriate rate expressions that deal with these two possibilities were presented elsewhere (*4*). The LH route predominates in several photooxidative conversions (*82*), but examples in which both the ER and LH pathways contribute to the overall reactions have been noted (*83*) (however, *see* the following paragraphs).

Two sites can be identified on the surface of the TiO_2 photocatalyst particles: lattice Ti^{III} sites and surface Ti^{IV}–OH^- sites. Organic compounds adsorb on surface hydroxyls (*86*), but molecular O_2 (and other electron acceptors) adsorbs on the Ti^{III} sites (*87*), which it oxidizes to form the superoxide radical anion, O_2 (*60*).

Although •OH radicals are formed at the Ti^{IV}–OH^- surface sites, they may, in principle, diffuse away from the particle surface and react with a solution substrate (*77*). Under the conditions used for the photooxidation of the phenols, the reaction–diffusion modulus for •OH (*77*) suggests that the diffusion length for •OH is about 200 Å (*88, 89*). A recent spin-trap–EPR study at ambient temperature noted that OH radicals (and no other) are present on the particle surface, and that reaction between •OH species and an organic substrate occurs at the surface (*90*). A thorough recent discussion on these issues is presented elsewhere (*55*). The photooxidation of phenols (and other organic compounds) by irradiated TiO_2 occurs between two adsorbed substrates (•OH species and phenols) (*52, 53*). We consider this point next to describe the kinetics in greater detail.

The intermediates that have been identified by HPLC methods, together with the present status of knowledge of events on semiconductor particle surfaces, at least for TiO_2, permit the inference of a mechanistic pathway. Taking phenol as a model substrate in the present context, a tentative view of the more significant steps is illustrated in Scheme II. This scheme presents our views on the most plausible events in a typical photocatalyzed oxidation process.

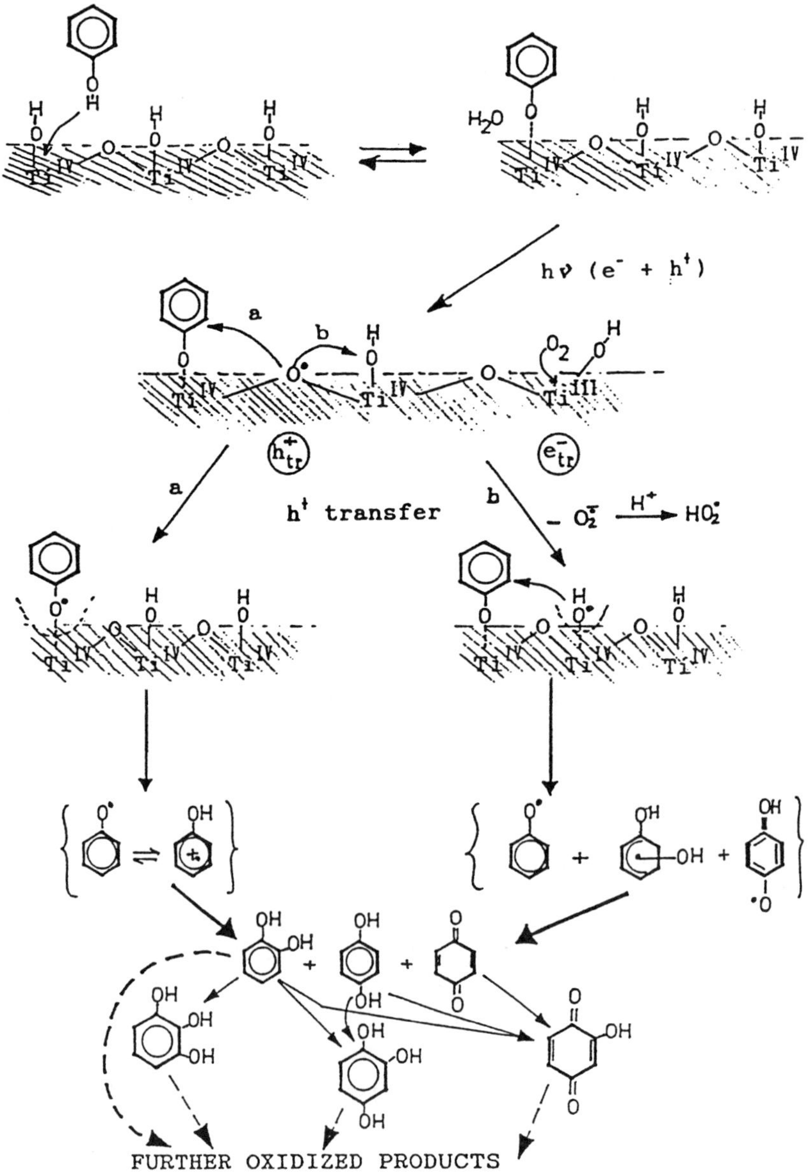

hν (e⁻ + h⁺)

h⁺ transfer

FURTHER OXIDIZED PRODUCTS

Scheme II.

Phenol is dissociatively chemisorbed on a surface Ti^{IV}, displacing an OH^- surface group. Following irradiation, the photogenerated e^- and h^+ are subsequently trapped: the electron as a surface Ti^{III} (e_{tr}^-) and the hole as a subsurface lattice Ti^{IV}–O–Ti^{IV} entity (h_{tr}^+), and this in competition with e^-–h^+ recombination. Adsorbed molecular O_2 scavenges the surface-trapped electron to give O_2, which yields the $HO_2\cdot$ radical in acid media. The h_{tr}^+ species may now oxidize directly either the chemisorbed phenoxide to give the phenoxyl radical (step a of Scheme II) (*91*) or the surface Ti–OH^- groups to give a chemisorbed •OH (step b), which in turn may produce the phenoxyl, the dihydroxycyclohexadienyl, and the semiquinone radicals, all of which have been observed for pentachloro- and pentabromophenols (*92*). The available data suggest that this primary oxidation event occurs at the particle surface; unfortunately nothing is known about whether the subsequent steps are surface related or take place in solution. This notwithstanding, Scheme II also notes that hydration of the phenoxyl radical to the same intermediate products [catechol, hydroquinone, benzoquinone, and others (*19*)] as those from •OH radical oxidation of the Ti^{IV}–O^-–Ph moiety, establishes that product analysis is silent as to the course of events in photooxidation.

Continued oxidation of these intermediate products ultimately yields carbon dioxide. Of interest, OH radical (produced by irradiation of NO_3^- ion in aqueous media) oxidation of phenol in homogeneous phase yields the intermediate products shown in equation 26, plus four others containing NO_2 and NO group substituents (*93*). For the cresols, no doubt the details will differ but the overall path will be similar.

OH (phenol) + $NO_3^{-\bullet}$ $\longrightarrow$

OH, OH + OH, OH + OH, OH + O, O + OH, OH, OH + O, O, OH (26)

Scheme III summarizes in more detail the various presumed steps in the formation of the intermediate products in the photooxidation of phenol by •OH (*19, 94*), irrespective of whether it occurs on the surface or in solution. At some point along the photooxidation path, the phenyl ring cleaves to form various aliphatic aldehydes and carboxylate species. By analogy with results obtained from the ozone oxidation of 4-chlorophenol (*94*) in neutral aqueous media, the nature of the species that may also form in the present context are noted in Scheme IV.

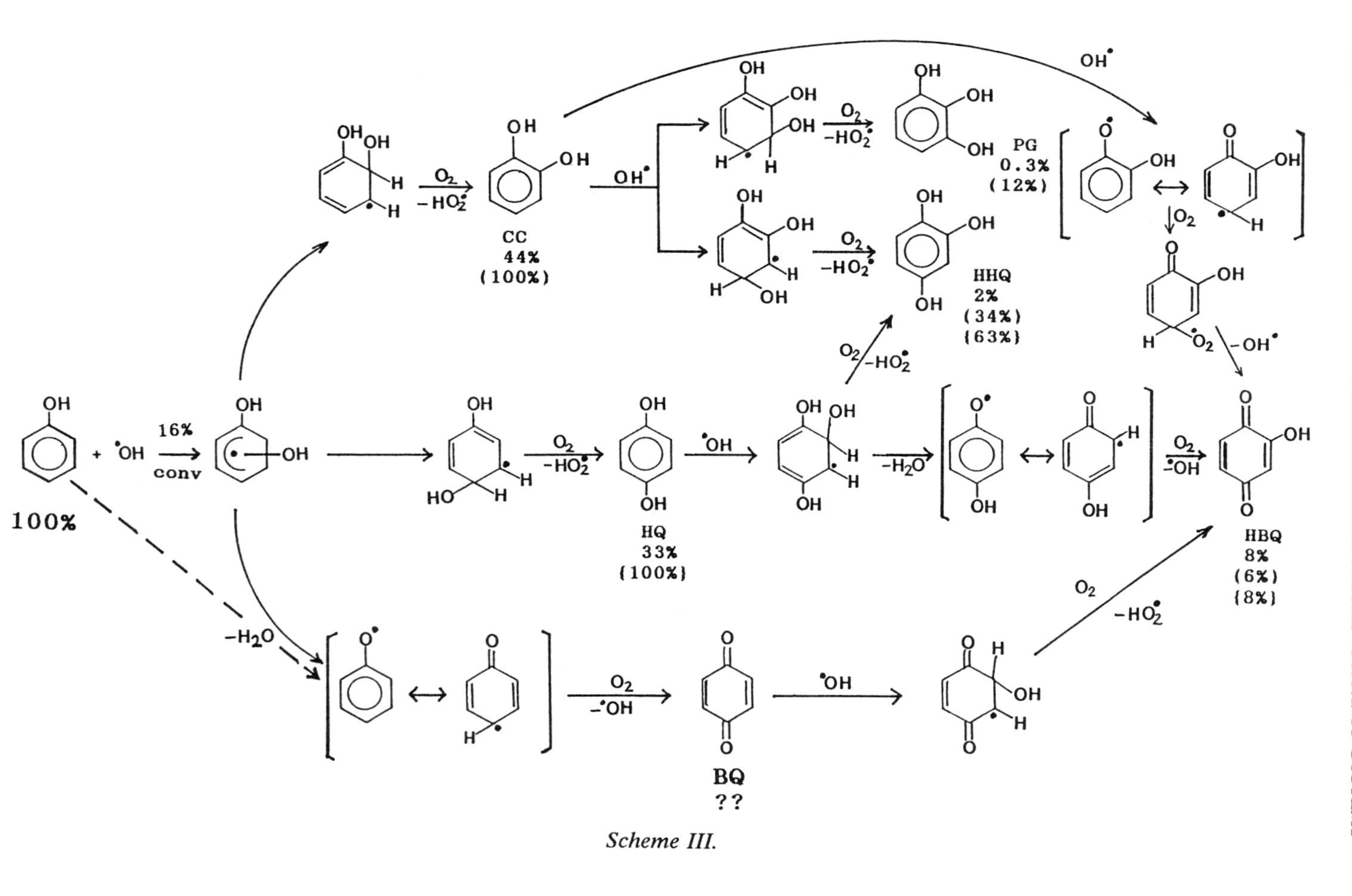

Scheme III.

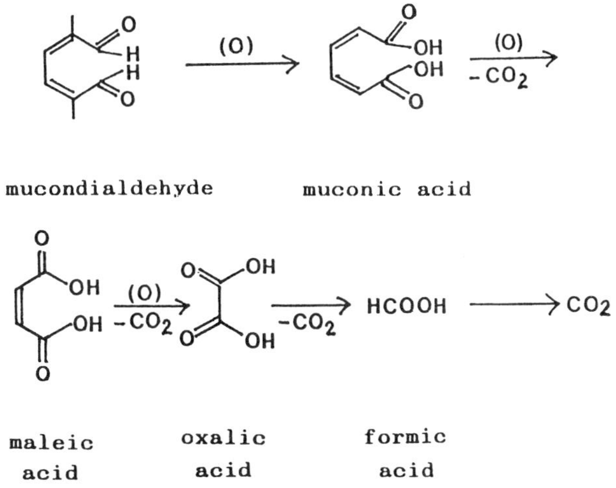

mucondialdehyde muconic acid

maleic acid oxalic acid formic acid

HCOOH → CO_2

Scheme IV.

The greater number of intermediate aromatic products identified for the fluorinated phenols also permits a description of the plausible pathways for their ultimate photooxidation to carbon dioxide and fluoride (*14*). Schemes V and VI summarize the events by analogy with our recent pulse radiolytic study on pentahalophenols (*92*).

The first step (a) in Scheme V yields a fluorodihydroxycyclohexadienyl radical (*92, 95*) that, through loss of H_2O (step 1), forms a phenoxyl radical intermediate; this intermediate can interact with the $HO_2\cdot$ radical (*see* Scheme II) to give back the original substrate **A**. Further attack by $\cdot OH$ on the phenoxyl radical species produces **G** and/or **F**.

Loss of fluoride occurs in the formation of **G** to give benzoquinone for the monofluorinated phenols (observed for 4-fluorophenol). The expected products from intermediate **F** are fluorocatechol and fluorohydroquinone, as seen for the monofluorophenols but not for the difluorinated analogs. Loss of fluoride by the cyclohexadienyl radical (step 4) yields a fluorohydroxyphenoxyl radical that interacts further with $\cdot OH$ to give **E** (step 5) and **D** (step 6), while reaction with $HO_2\cdot$ gives **C** (step 7). Species **E** is not a possible intermediate for the monofluorophenols, but difluorophenols would produce hydroxybenzoquinone (not observed). For the monofluoro analogs, **D** gives trihydroxybenzene, observed for both 2-FPh and 4-FPh. No fluorinated trihydroxybenzenes were identified (*14*); these species must be thermally unstable, as are the trihydroxybenzenes, which rapidly oxidize and fragment to aliphatic species. When **C** forms, the expected intermediate products are hydroquinone (4-FPh) and catechol (2-FPh). For the difluorinated phenols, both fluorohydroquinone and fluorocatechol are possible and have been identified (*14*).

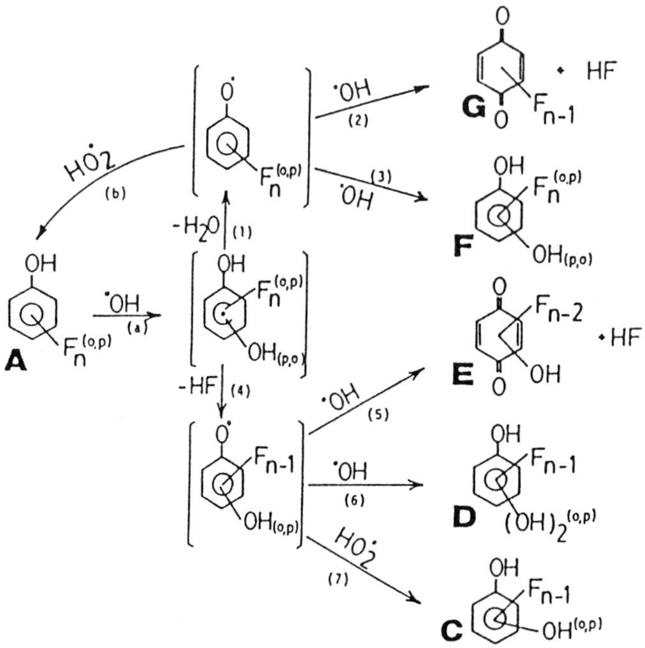

Scheme V.

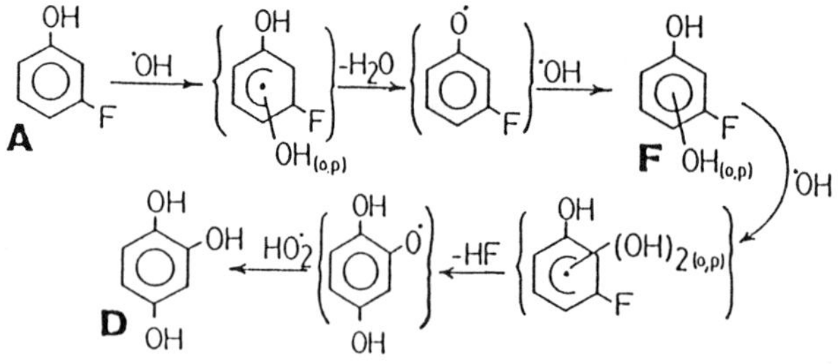

Scheme VI.

The pathway for 3-fluorophenol is described separately in Scheme VI, in which intermediates **F** and **D** are relevant. Reaction of this phenol with •OH gives the cyclohexadienyl radical that, through loss of water, gives the fluorophenoxyl species. Further •OH attack yields **F** (3-FCC, 4-FCC, and FHQ) and subsequently another cyclohexadienyl radical; loss of fluoride and reaction with HO_2• ultimately yields the trihydroxybenzene species **D**.

Conclusions

The complete photocatalyzed oxidation of phenol, cresols, and fluorinated phenols in anatase TiO_2 dispersions irradiated with AM1-simulated sunlight occurs with relatively good photochemical efficiencies, despite the low quantity of sunlight radiation absorbed ($\leq 3\%$) by the photocatalyst. Degradation of the initial phenol, at concentrations of ~20 mg/L normally found in polluted waters, is relatively rapid (20–30 min for phenol, ~4–5 h for cresols, and ~20–30 min for the fluorophenols) depending on conditions, but stoichiometric evolution of carbon dioxide, which manifests total oxidation (mineralization), requires longer times (1–3 h for phenol, ~10 h for cresols, and ~1.5–4 h for fluorophenols).

In spite of past and some recent reports that show interest only in the disappearance of the initial substrate in proposing heterogeneous photocatalytic processes (treated here) to detoxify wastewaters, it is the demonstrated formation of the ultimate oxidation product CO_2 that is relevant. Practical applications of heterogeneous photocatalysis must also consider the potential of forming stable intermediate products that may be as or more toxic than the parent substrate. Any economic analysis must be based on CO_2 evolution rates (thus the interest in the phenomenological kinetic aspects of the present work) or to the formation of some stable, nontoxic product, for example, cyanuric acid in the photooxidation of atrazine (*35*). Heterogeneous photocatalysis is rapidly maturing toward commercial application to detoxify wastewaters containing phenolic substances.

Acknowledgments

Our work in Montreal is supported by the Natural Sciences and Engineering Research Council of Canada (to N. Serpone), and in Torino it is generously supported by the Consiglio Nazionale delle Ricerche, the Ministero della Publica Istruzione, the Regione Piemonte, ENI Ricerche, and the European Economic Community (to E. Pelizzetti). We are particularly grateful to the North Atlantic Treaty Organization for an exchange grant (NATO Grant No. CRG 890746) between our two laboratories and those of Pierre Pichat (Lyon, France) and Marye Anne Fox (Austin, Texas).

N. Serpone acknowledges the kind hospitality of Pierre Pichat during a 1990–1991 sabbatical at the Laboratoire de Photocatalyse, Catalyse et Environnement, Ecole Centrale de Lyon, 69130 Ecully, France.

References

1. Callahan, M. A.; Slimak, M.; Gbel, N.; May, I.; Fowler, C.; Freed, R.; Jennings, P.; Dupree, R.; Whitemore, F.; Maestri, B.; Holt, B.; Gould, C. *Water Related Environmental Fate of 129 Priority Pollutants;* United States Environmental Protection Agency: Washington, DC, 1979; Report No. EPA–44014–79–029 a and b.
2. (a) *Dangerous Prop. Ind. Mater. Rep.* **1985,** *5,* 30, and references therein. (b) *Dangerous Prop. Ind. Mater. Rep.* **1986,** *6,* 41, and references therein.
3. Seidel, H. *Chem. Ind.* **1988,** *11,* 62.
4. Al-Ekabi, H.; Serpone, N. *J. Phys. Chem.* **1988,** *92,* 5726.
5. Leifer, A. *The Kinetics of Environmental Aquatic Photochemistry: Theory and Practice;* American Chemical Society: Washington, DC, 1988; Chapters 3 and 4.
6. Martinetz, D. In *Chemical Waste: Handling and Treatment;* Bromley, J.; Fraquhar, J. T.; Gidley, P. T.; James, S.; Martinetz, D.; Robin, A.; Schomaker, N. B.; Stephens, R. T.; Walters, D. B., Eds.; Springer-Verlag: Berlin, Germany, 1986; p 113, and references therein.
7. Ollis, D. F.; Pelizzetti, E.; Serpone, N. In *Photocatalysis: Fundamentals and Applications;* Serpone N.; Pelizzetti, E., Eds.; Wiley-Interscience: New York, 1989; pp 603–637.
8. Ollis, D. F.; Pelizzetti, E.; Serpone, N. *Environ. Sci. Technol.* **1991,** *25,* 1523.
9. (a) Barbeni, M.; Pramauro, E.; Pelizzetti, E.; Borgarello, E.; Serpone, N. *Chemosphere,* **1985,** *14,* 195. (b) Borello, R.; Minero, C.; Pramauro, E.; Pelizzetti, E.; Serpone, N.; Hidaka, H. *Environ. Toxicol. Chem.* **1989,** *8,* 997. (c) Pelizzetti, E.; Minero, C.; Borgarello, E.; Tinucci, L.; Serpone, N., unpublished.
10. Al-Ekabi, H.; Serpone, N.; Pelizzetti, E.; Minero, C.; Fox, M. A.; Draper, R. B. *Langmuir,* **1989,** *5,* 250.
11. Pelizzetti, E.; Minero, C.; Maurino, V.; Sclafani, A.; Hidaka, H.; Serpone, N. *Environ. Sci. Technol.* **1989,** *23,* 1385.
12. Terzian, R.; Serpone, N.; Minero, C.; Pelizzetti, E.; Hidaka, H. *J. Photochem. Photobiol. A Chem.* **1990,** *55,* 243.
13. Terzian, R.; Serpone, N.; Minero, C.; Pelizzetti, E. *J. Catal.* **1991,** *128,* 352.
14. Minero, C.; Aliberti, C.; Pelizzetti, E.; Terzian, R.; Serpone, N. *Langmuir,* **1991,** *7,* 3081.
15. Childs, L. P.; Ollis, D. F. *J. Catal.* **1980,** *66,* 393.
16. Kisch, H. In *Photocatalysis: Fundamentals and Applications;* Serpone N.; Pelizzetti, E., Eds.; Wiley-Interscience: New York, 1989; Chapter 1.
17. Bahnemann, D. W. In *Photochemical Conversion and Storage of Solar Energy;* Pelizzetti, E.; Schiavello, M., Eds.; Kluwer Academic: Dordrecht, Netherlands, 1991; pp 251–276.
18. Fox, M. A. In *Photocatalysis: Fundamentals and Applications;* Serpone, N.; Pelizzetti, E., Eds.; Wiley-Interscience: New York, 1989; Chapter 13.
19. (a) Okamoto, K.; Yamamoto, Y.; Tanaka, H.; Tanaka, M.; Itaya, A. *Bull. Chem. Soc. Jpn.* **1985,** *58,* 2015. (b) Okamoto, K.; Yamamoto, Y.; Tanaka, H.; Itaya, A. *Bull. Chem. Soc. Jpn.* **1985,** *58,* 2023.

20. (a) Augugliaro, V.; Palmisano, L.; Sclafani, A.; Minero, C.; Pelizzetti, E. *Toxicol. Environ. Chem.* **1988,** *16,* 89. (b) Sclafani, A.; Palmisano, L.; Schiavello, M. *J. Phys. Chem.* **1990,** *94,* 829. (c) Augugliaro, V.; Davi, E.; Palmisano, L.; Schiavello, M.; Sclafani, A. *Appl. Catal.* **1990,** *65,* 101. (d) Sclafani, A.; Palmisano, L.; Davi, E. *New J. Chem.* **1990,** *14,* 265. (e) Sclafani, A.; Palmisano, L.; Davi, E. *J. Photochem. Photobiol. A Chem.* **1991,** *56,* 113.
21. Tseng, J.; Huang, C. P. *Emerging Technologies in Hazardous Waste Management;* Tedder, D. W.; Pohland, F. G., Eds.; ACS Symposium Series 422; American Chemical Society: Washington, DC, 1990; pp 12–39.
22. (a) Barbeni, M.; Pramauro, E.; Pelizzetti, E.; Borgarello, E.; Gratzel, M.; Serpone, N. *Nouv. J. Chim.* **1984,** *8,* 547. (b) Shimamura, Y.; Misawa, H.; Oguchi, T.; Kammo, T.; Sakuragi, H.; Tokumaru, K. *Chem. Lett.* **1983,** 1691.
23. Barbeni, M.; Morello, M.; Pramauro, E.; Pelizzetti, E.; Vincenti, M.; Borgarello, E.; Serpone, N. *Chemosphere,* **1987,** *16,* 1165.
24. Barbeni, M.; Pramauro, E.; Pelizzetti, E.; Borgarello, E.; Serpone, N.; Jamieson, M. A. *Chemosphere,* **1986,** *15,* 1913.
25. Pelizzetti, E.; Borgarello, M.; Minero, C.; Pramauro, E.; Borgarello, E.; Serpone, N. *Chemosphere,* **1988,** *17,* 499.
26. Borello, R.; Minero, C.; Pramauro, E.; Pelizzetti, E.; Serpone, N.; Hidaka, H. *Environ. Technol. Chem.* **1989,** *8,* 997.
27. Hidaka, H.; Kubota, H.; Gratzel, M.; Serpone, N.; Pelizzetti, E. *Nouv. J. Chim.* **1985,** *9,* 67.
28. Hidaka, H.; Kubota, H.; Gratzel, M.; Pelizzetti, E.; Serpone, N. *J. Photochem.* **1986,** *35,* 219.
29. Hidaka, H.; Fujita, Y.; Ihara, K.; Yamada, S.; Suzuki, K.; Serpone, N.; Pelizzetti, E. *J. Jpn. Oil Chem. Soc.* **1987,** *36,* 837.
30. Hidaka, H.; Ihara, K.; Fujita, Y.; Yamada, S.; Pelizzetti, E.; Serpone, N. *J. Photochem. Photobiol. A Chem.* **1988,** *42,* 375.
31. Hidaka, H.; Yamada, S.; Suenaga, S.; Kubota, H.; Serpone, N.; Pelizzetti, E.; Gratzel, M. *J. Photochem. Photobiol. A Chem.* **1989,** *47,* 103.
32. Hidaka, H.; Yamada, S.; Suenaga, S.; Kubota, H.; Serpone, N.; Pelizzetti, E. *J. Mol. Catal.* **1990,** *59,* 270.
33. Hidaka, H.; Zhao, J.; Suenaga, S.; Serpone, N.; Pelizzetti, E. *J. Jpn. Oil Chem. Soc.* **1990,** *39,* 963.
34. Pelizzetti, E.; Minero, C.; Maurino, V.; Hidaka, H.; Serpone, N.; Terzian, R. *Ann. Chim.* **1990,** *80,* 81.
35. Pelizzetti, E.; Maurino, V.; Minero, C.; Carlin, V.; Pramauro, E.; Zerbinati, O.; Tosato, M. L. *Environ. Sci. Technol.* **1990,** *24,* 1559.
36. Pelizzetti, E.; Pramauro, E.; Minero, C.; Serpone, N.; Borgarello, E. In *Photocatalysis and Environment;* Schiavello, M., Ed.; Kluwer Academic: Dordrecht, Netherlands, 1988; p 469, and references therein.
37. Pelizzetti, E.; Minero, C.; Maurino, V. *Adv. Colloid Interfac. Sci.* **1990,** *32,* 271.
38. Matthews, R. W. In *Photochemical Conversion and Storage of Solar Energy;* Pelizzetti, E.; Schiavello, M., Eds.; Kluwer Academic: Dordrecht, Netherlands, 1991; pp 427–439, and references therein.

39. Serpone, N.; Lawless, D.; Terzian, R.; Minero, C.; Pelizzetti, E. In *Photochemical Conversion and Storage of Solar Energy;* Pelizzetti, E.; Schiavello, M., Eds.; Kluwer Academic: Dordrecht, Netherlands, 1991; pp 451–475, and references therein.
40. Ollis, D. F. In *Photochemical Conversion and Storage of Solar Energy;* Pelizzetti, E.; Schiavello, M., Eds.; Kluwer Academic: Dordrecht, Netherlands, 1991; pp 593–622, and reference therein.
41. Serpone, N.; Borgarello, E.; Harris, R.; Cahill, P.; Borgarello, M.; Pelizzetti, E. *Sol. Energy Mater.* **1986,** *14,* 121.
42. (a) Matthews, R. W. *J. Catal.* **1986,** *97,* 565. (b) Matthews, R. W. *Water Res.* **1986,** *20,* 569.
43. Matthews, R. W. *J. Chem. Soc. Faraday Trans. 1* **1984,** *80,* 457.
44. (a) Izumi, I.; Dunn, W. W.; Willbourn, K. O.; Fan, F. F.; Bard, A. J. *J. Phys. Chem.* **1980,** *84,* 3207. (b) Izumi, I.; Fan, F. F.; Bard, A. J. *J. Phys. Chem.* **1981,** *85,* 218.
45. Fujihira, M.; Satoh, Y.; Osa, T. *Bull. Chem. Soc. Jpn.* **1982,** *55,* 666.
46. Barbeni, M.; Minero, C.; Pelizzetti, E.; Borgarello, E.; Serpone, N. *Chemosphere,* **1987,** *16,* 225.
47. Cunningham, J.; Srijaranai, S. *J. Photochem. Photobiol. A Chem.* **1988,** *43,* 329.
48. Ceresa, E. M.; Burlamacchi, L.; Visca, M. *J. Mater. Sci.* **1983,** *18,* 289.
49. Jaeger C. D.; Bard, A. J. *J. Phys. Chem.* **1979,** *83,* 3146.
50. Anpo, M.; Shima, T.; Kubokawa, Y. *Chem. Lett.* **1985,** 1799.
51. Draper, R. B.; Fox, M. A. *J. Phys. Chem.* **1990,** *94,* 4628.
52. Lawless, D.; Serpone, N.; Meisel, D. In *Proceedings of the Symposium on Semiconductor Photoelectrochemistry;* Koval, C., Ed.; The Electrochemical Society: Pennington, NJ, 1991.
53. Lawless, D.; Serpone, N.; Meisel, D. *J. Phys. Chem.* **1991,** *95,* 5166.
54. Serpone, N.; Lawless, D.; Terzian, R.; Meisel, D. In *Electrochemistry in Colloids and Dispersions;* McKay, R.; Texter, J., Eds.; VCH: New York, 1992, pp 399–416.
55. Cundall, R. B.; Rudham, R.; Salim, M. S. *J. Chem. Soc. Faraday Trans. 1* **1976,** *72,* 1642.
56. Harvey, P. R.; Rudham, R.; Ward, S. *J. Chem. Soc. Faraday Trans. 1* **1983,** *79,* 1381.
57. Herrmann J.-M.; Pichat, P. *J. Chem. Soc. Faraday Trans. 1* **1980,** *76,* 1138.
58. Kormann, C.; Bahnemann, D. W.; Hoffmann, M. R. *Environ. Sci. Technol.* **1988,** *22,* 798.
59. Bahnemann, D. W.; Henglein, A.; Spanhel, L. *Faraday Discuss. Chem. Soc.* **1984,** *78,* 151.
60. Howe, R. F.; Gratzel, M. *J. Phys. Chem.* **1987,** *91,* 3906.
61. Avudaithai, M.; Kutty, T. R. N. *Mat. Res. Bull.* **1988,** *23,* 1675.
62. Völz, H. G.; Kaempf, G.; Fitzky, H. G.; Klaeren, A. *Photodegradation and Photostabilization of Coatings;* Pappas, S. P.; Winslow, F. H., Eds.; ACS Symposium Series 151; American Chemical Society: Washington, DC, 1981; pp 163–182.
63. Gratzel, M.; Howe, R. F. *J. Phys. Chem.* **1990,** *94,* 2566.

64. Maldotti, A.; Amadelli, R.; Bartocci, C.; Carassiti, V. *J. Photochem. Photobiol. A Chem.* **1990,** *53,* 263.
65. Howe, R. F. *Adv. Coll. Interfac. Sci.* **1982,** *18,* 1.
66. Serpone, N.; Borgarello, E.; Barbeni, M.; Pelizzetti, E.; Pichat, P.; Herrmann, J.-M.; Fox, M. A. *J. Photochem.* **1987,** *36,* 373.
67. Markham, M. C.; Laidler, J. K. *J. Phys. Chem.* **1953,** *57,* 363.
68. (a) Oosawa, Y. O.; Gratzel, M. *J. Chem. Soc. Faraday Trans. 1* **1988,** *84,* 197. (b) Thampi, K. R.; Rao, M. S.; Schwarz, W.; Gratzel, M.; Kiwi, J. *J. Chem. Soc. Faraday Trans. 1* **1988,** *84,* 1703. (c) Gu, B.; Kiwi, J.; Gratzel, M. *Nouv. J. Chim.* **1985,** *9,* 539.
69. Muraki, H.; Saji, T.; Fujihira, M.; Aoyagui, S. *J. Electroanal. Chem.* **1984,** *169,* 319.
70. Gratzel, C. K.; Jirousek, M.; Gratzel, M. *J. Mol. Catal.* **1990,** *60,* 375.
71. Sawyer, D. T.; Gibian, M. *J. Tetrahedron,* **1979,** *35,* 1471.
72. Fox, M. A.; Chen, C.-C.; Younathan, J. N. N. *J. Org. Chem.* **1984,** *49,* 1969.
73. Fujihira, M.; Satoh, Y.; Osa, T. *J. Electroanal. Chem.* **1981,** *126,* 277.
74. Ho, P. C. In *Management Hazard: Toxic Wastes Process;* Ind. [Int. Congr.]; Kolaczkowski, S. T.; Crittenden, B. D., Eds.; Elsevier: London, 1987; pp 563–573; *Chem. Abstr.* **1987,** *109,* 11146x.
75. Takahashi, N.; *Kogai* **1988,** *23,* 341; *Chem. Abstr.* **1988,** *111,* 139828e.
76. Laidler, K. J. *Chemical Kinetics,* 3rd ed.; Harper and Row: New York, 1987; pp 278–281.
77. Turchi C.; Ollis, D. F. *J. Catal.* **1990,** *122,* 178.
78. Reference 76, pp 400–406.
79. Reference 76, pp 243–251.
80. Rothenberger, G.; Moser, J.; Gratzel, M.; Serpone, N.; Sharma, D. K. *J. Am. Chem. Soc.* **1985,** *107,* 8054.
81. Abdullah, M.; Low, G. K.-C.; Matthews, R. W. *J. Phys. Chem.* **1990,** *94,* 6820.
82. Pichat, P.; Herrmann, J.-M. In *Photocatalysis: Fundamentals and Applications;* Serpone, N.; Pelizzetti, E., Eds.; Wiley-Interscience: New York, 1989; Chapter 8, pp 217–250.
83. Al-Ekabi, H.; Serpone, N. In *Photocatalysis: Fundamentals and Applications;* Serpone, N.; Pelizzetti, E., Eds.; Wiley-Interscience: New York, 1989; Chapter 14, pp 457–488.
84. Movius, W. G.; Linck, R. G. *J. Am. Chem. Soc.* **1970,** *92,* 2677.
85. Rao, P. S.; Hayon, E. *J. Phys. Chem.* **1975,** *79,* 397.
86. Nagao, M.; Suda, Y. *Langmuir,* **1987,** *3,* 786.
87. (a) Munuera, G.; Rives-Arnau, V.; Sancedo, A. *J. Chem. Soc. Faraday Trans. 1* **1979,** *75,* 736. (b) Harbour, J. R.; Hair, M. L. *J. Phys. Chem.* **1979,** *83,* 652. (c) Howe, R. F.; Gratzel, M. *J. Phys. Chem.* **1985,** *89,* 4495. (d) Munuera, G.; Gonzalez-Elipe, A. R.; Rives-Arnau, V.; Navio, A.; Malet, P.; Soria, J.; Conesa, J. C.; Sanz, J. In *Adsorption and Catalysis on Oxide Surfaces;* Che, M.; Bond, G. C., Eds.; Elsevier: Amsterdam, Netherlands, 1985.
88. The reaction–diffusion modulus for OH radicals (*77*) is given by $\Phi' = k_{OH}[S]/(D/L^2)$ where Φ' is 1 for equal reaction and diffusion rates; k_{OH} is 1.4 $\times$ 10^{10} M^{-1} s^{-1} (*84*); [S] is 2.10 $\times$ 10^{-4} M; D is the liquid diffusion coefficient, approximately 10^{-5} cm^2/s; and L is the diffusion length, which is calculated to be 184 Å.

89. (a) Adams, G. E.; Michael, B. D. *Trans. Faraday Soc.* **1967,** *63,* 1171. (b) Land, E. J.; Ebert, M. *Trans. Faraday Soc.* **1967,** *63,* 1181.
90. (a) Bolton, J. R. *Proceedings of the 13th DOE Solar Photochemistry Research Conference;* U.S. Department of Energy, Argonne National Laboratory: Argonne, IL, 1989; Vol. 13. (b) Bolton, J. R. *Symposium on Advanced Oxidation Processes for Treatment of Contaminated Water and Air;* Wastewater Technology Center: Burlington, Ontario, Canada, 1990; *Chem. Abstr.* **1990,** *116,* 90854.
91. Draper, R. B.; Fox, M. A. *Langmuir,* **1990,** *6,* 1396.
92. Terzian, R.; Serpone, N.; Draper, R. B.; Fox, M. A.; Pelizzetti, E. *Langmuir,* **1991,** *7,* 3081.
93. Lemaire, J.; Boule, P.; Sehili, T.; Richard, C. In *Photochemical Conversion and Storage of Solar Energy;* Pelizzetti, E.; Schiavello, M., Eds.; Kluwer Academic: Dordrecht, Netherlands, 1991; pp 477–495.
94. (a) Getoff, N.; Solar, S. *Radiat. Phys. Chem.* **1986,** *28,* 443. (b) Getoff, N.; Solar, S. *Radiat. Phys. Chem.* **1988,** *31,* 121.
95. Draper, R. B.; Fox, M. A.; Pelizzetti, E.; Serpone, N. *J. Phys. Chem.* **1989,** *93,* 1938.

RECEIVED for review November 7, 1991. ACCEPTED revised manuscript April 27, 1992.

16

Homogeneous Metal-Catalyzed Photochemistry in Organic Synthesis

Robert G. Salomon, Subrata Ghosh, and Swadesh Raychaudhuri

Department of Chemistry, Case Western Reserve University, Cleveland, OH 44106

The scope of synthetically useful transformations of organic substrates that may be achieved by homogeneous metal-catalyzed photochemical reactions is examined. Recent examples of alkane C–H bond activation, oxidation of alcohols, oxidative cleavage of C–C bonds, carbon–heteroatom bond activation, and additions to C=C bonds are presented with an emphasis on the role of the metal and, where applicable, with comparisons to the metal-free photochemistry of the same substrates. The compatibility of Cu(I)-catalyzed intramolecular $2\pi + 2\pi$ photocycloadditions with various functional groups and the utility of those groups for facilitating useful transformations of the photoproducts is thoroughly surveyed. Besides applications to the total synthesis of natural products, effective new Cu(I)-catalyzed photochemical methods for construction of various carbon ring systems are reported.

Metal-Catalyzed Photochemical Reaction Types

All metal-catalyzed photochemical reactions may be divided into two categories: photogenerated catalysis or catalyzed photolysis. Photogenerated catalysis involves generation of a thermally active catalyst in a photochemical reaction (Figure 1). Such reactions may exhibit an induction period during which the catalyst is being generated. They also may continue after termination of the irradiation (i.e., postphotochemical reaction) and, therefore, may exhibit quantum yields $\Phi > 1$ mol/einstein. Any of these characteristics is diagnostic for photogenerated catalysis, but their

0065–2393/93/0238–0315$06.00/0

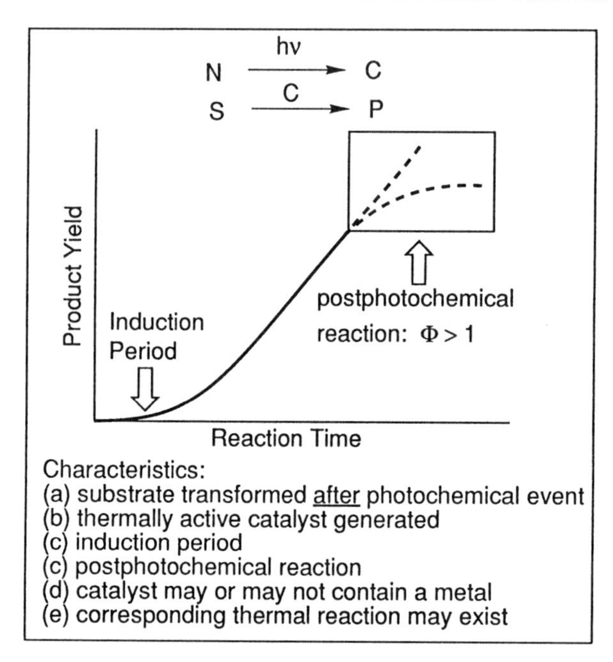

Figure 1. Photogenerated catalysis. Symbols: N, nominal catalyst; C, thermally active catalyst; S, substrate; and P, product.

absence does not exclude such a mechanism. The same catalyst can often be produced nonphotochemically; hence a corresponding thermal reaction may exist. The synthetically important advantages of photogeneration might include the convenience of preparing or storing the catalyst precursor or the thermally mild conditions under which highly unsaturated metal catalysts can be produced.

In contrast, catalyzed photolysis involves transformation of the organic substrate during the photochemical event (Figure 2). Therefore, neither an induction period nor a postphotochemical reaction is possible and $\Phi \leq 1$. Perhaps most importantly, catalyzed photochemical reactions are unique. No corresponding thermal reaction exists. This fact and the relevance of ground-state metal–substrate interactions for predicting the reaction course are synthetically important features of catalyzed photolyses.

Irradiation of a metal catalyst–substrate complex leading directly to product is only one possible mechanism for a catalyzed photolysis. Other possibilities include generation of an intermediate excited-state complex that affords a photoproduct or an excited state of the substrate in a subsequent step (Scheme 1). Photoexcitation of the substrate might also occur by a classical photosensitization mechanism involving the metal catalyst as photosensitizer.

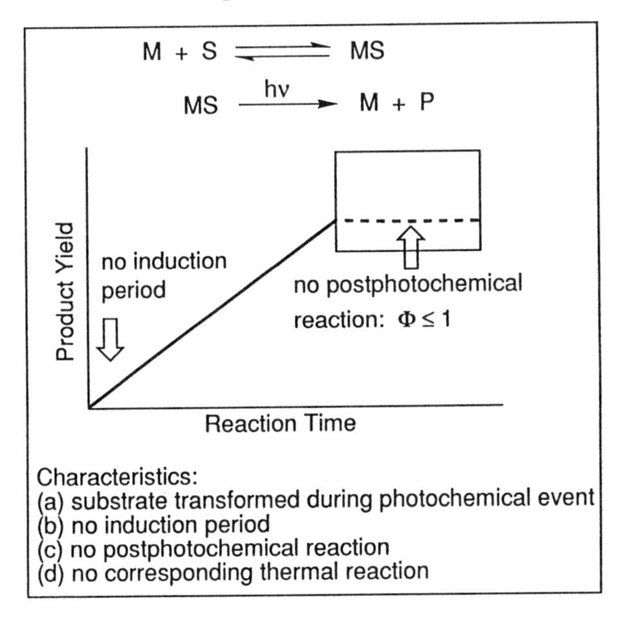

Figure 2. Catalyzed photolysis. Symbols: M, metal catalyst; S, substrate; and P, product.

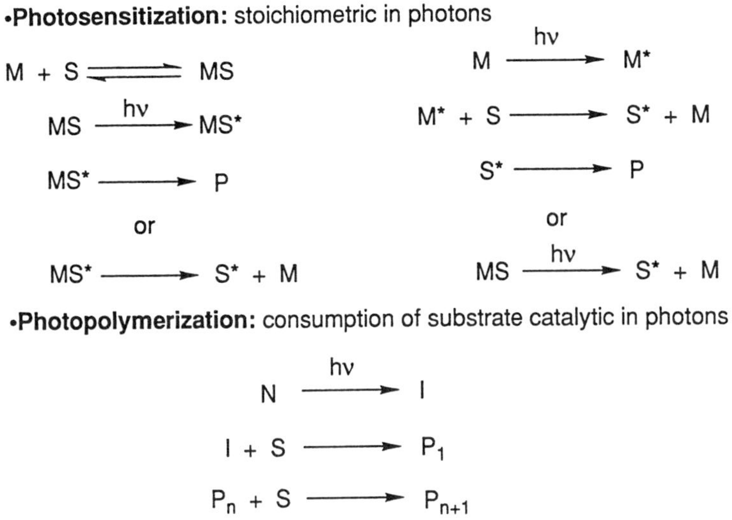

•**Photosensitization:** stoichiometric in photons

$$M + S \rightleftharpoons MS \qquad\qquad M \xrightarrow{h\nu} M^*$$

$$MS \xrightarrow{h\nu} MS^* \qquad\qquad M^* + S \longrightarrow S^* + M$$

$$MS^* \longrightarrow P \qquad\qquad S^* \longrightarrow P$$

or   or

$$MS^* \longrightarrow S^* + M \qquad\qquad MS \xrightarrow{h\nu} S^* + M$$

•**Photopolymerization:** consumption of substrate catalytic in photons

$$N \xrightarrow{h\nu} I$$

$$I + S \longrightarrow P_1$$

$$P_n + S \longrightarrow P_{n+1}$$

Scheme 1. Photosensitization and photopolymerization. Symbols: M, metal catalyst; S, substrate; MS, excited state of MS complex; S*, excited state of substrate; M*, excited state of metal catalyst; P, product; N, nominal catalyst; I, polymerization initiator; and P_n, polymer from n molecules of monomeric substrate.*

One type of catalyzed photolysis, initiation of a photopolymerization, results in a consumption of substrate that is catalytic in photons (Scheme 1). Nevertheless, the quantum yield of product, that is, polymer molecules, $\Phi \leq 1$. Here a metal could be involved as a catalyst in the photogeneration of an initiator I from its precursor N or the precursor N itself might be a metal complex.

Recent Examples of Homogeneous Metal-Catalyzed Photochemical Reactions

In a review published in 1983 (*1*), Salomon delineated the scope of synthetically useful transformations of organic substrates that may be achieved by homogeneous metal-catalyzed photochemical reactions. To update that review for this chapter, we chose examples from the recent work of several groups not represented at the symposium upon which this book is based.

Rh-Catalyzed Terminal Carbonylation of Alkanes. Highly selective insertion of a carbonyl group into a terminal C–H bond of *n*-alkanes **1** (Scheme 2) occurs and affords aldehydes **2** upon irradiation with CO in the presence of $(Me_3P)_2RhCl(CO)$ (*2*). The corresponding *n*-alkenes **3** and acetaldehyde are also produced in a secondary, non-metal-catalyzed, photofragmentation of **2**. The insertion presumably involves photogeneration of a thermally active catalyst, **5**, from the nominal catalyst, **4**, by decarbonylation. Steric factors may account for the regioselectivity for insertion into a terminal C–H bond to produce a Rh(III)-alkyl, **6**. Coordination of a carbonyl ligand and 1,2-alkyl shift in the resulting Rh(III) carbonyl **7** produces Rh(III) acyl **8** from which **9** is generated by reductive elimination. Ligand exchange then releases aldehyde **2** and regenerates the nominal catalyst. Because the nominal catalyst is regenerated in the catalytic cycle, the quantum yield for such a photogenerated catalytic reaction will be $\Phi \leq 1$.

Rh-Catalyzed Oxidation–Decarbonylation of Primary Alcohols. A recent example of homogeneous metal-catalyzed oxidation of an alcohol is especially interesting because it is accompanied by decarbonylation of the resulting aldehyde. Thus, irradiation of ethanol in the presence of $(iso\text{-}Pr_3P)_2RhH(CO)$ affords methane, carbon monoxide, and hydrogen (*3*). This transformation presumably involves photogeneration of a thermally active catalyst, **13**, by decarbonylation of the nominal catalyst, **12** (Scheme 3). If this coordinatively unsaturated Rh(I) complex

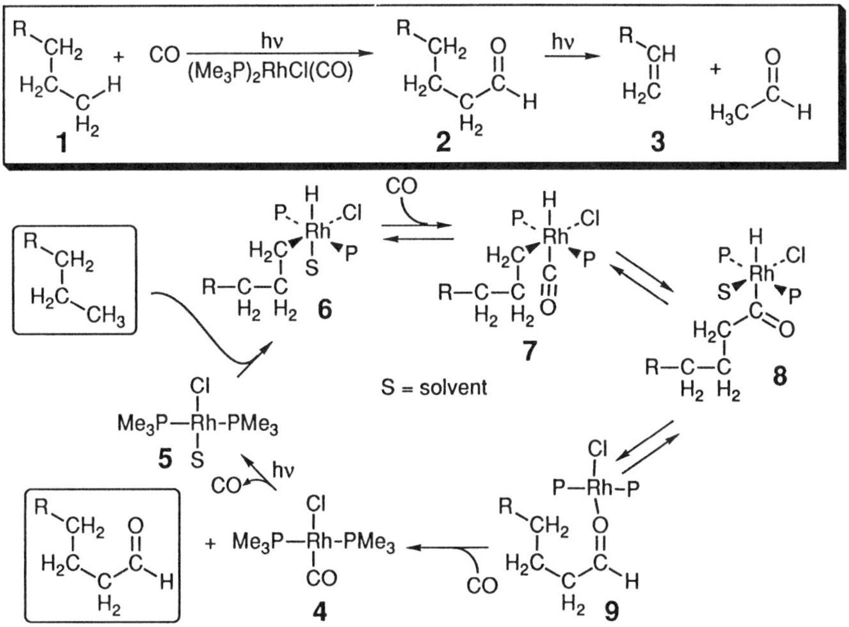

Scheme 2. *Rh-catalyzed terminal carbonylation of alkanes.*

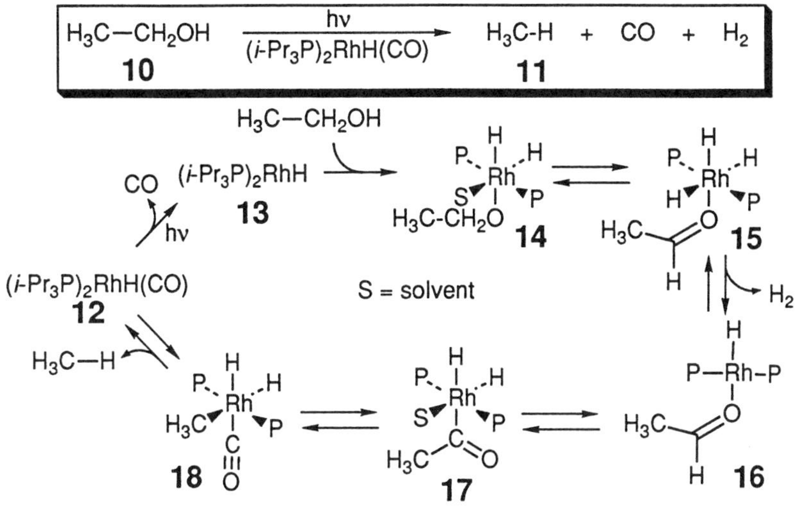

Scheme 3. *Rh-catalyzed oxidation–decarbonylation of primary alcohols.*

oxidatively adds ethanol to give Rh(III) complex **14**, then β-hydride elimination can produce **15**. If reductive elimination of hydrogen produces **16**, then decarbonylation could generate methane through intermediates **17** and **18** in steps corresponding to the reverse of the terminal carbonylation just described (*see* Scheme 2). The harsh conditions required to effect thermal decarbonylation of metal carbonyls such as **12** limit the synthetic utility of the corresponding thermally induced decarbonylations. Furthermore, the thermal reactions often require stoichiometric amounts of expensive organometallic reagents. If the catalytic cycle outlined in Scheme 3 is effective for primary alcohols in general, it would be a uniquely valuable reaction for organic synthesis.

Metal-Catalyzed Photolytic Oxidative Cleavage of Vicinal Diols. The contrasting reactions that occur upon irradiation of glucose (**19**, Scheme 4) in the presence of $FeCl_3$ (*4*) or $TiCl_4$ (*5*) are interesting. Regioselective cleavage of the 1,2-C–C bond occurs upon irradiation in the presence of $FeCl_3$. Acetylation of the photoproduct affords **20** (Scheme 4). The catalytic cycle presumably involves photoinduced ligand-to-metal electron transfer in a Fe(III) complex, **21**, that produces **22** and Fe(II). Regeneration of Fe(III) by nonphotochemical reaction with oxygen completes the catalytic cycle. The proclivity of Fe(III) toward coordination of the hemiacetal hydroxyl at position 1 in **19** provides regioselectivity. In remarkable contrast, regioselective cleavage of the 5,6-C–C bond occurs upon irradiation in the presence of $TiCl_4$. Acetylation of the photoproduct affords **24** (Scheme 5). The catalytic cycle presumably involves photoinduced ligand-to-metal electron transfer in a Ti(IV) complex, **25**, that produces **26** and Ti(III). Regeneration of Ti(IV) by nonphotochemical reaction with oxygen completes the catalytic cycle. The proclivity of Ti(IV) toward coordination of the least sterically congested terminal vicinal diol at positions 5 and 6 in the furanose form of **19** provides regioselectivity.

Ag-Promoted Addition of Phenacyl Chloride to Alkenes. The Ag-promoted photolytic addition of phenacyl chloride (**27**) to alkenes (*6*) is exemplified by the reaction with 2-methyl-2-butene (**28**) that produces tetralone **29** (Scheme 6). The addition is presumably initiated by Ag-promoted photoactivation of the C–Cl bond in **27**. Markovnikov addition of the resulting carbocation to the C=C bond in **28** would afford **31** from which **29** would arise by electrophilic aromatic substitution. In the absence of Ag(I), an entirely different photoreaction occurs between **27** and **28**. A Paterno–Büchi $2\pi + 4\pi$ cycloaddition generates the oxetane **32**.

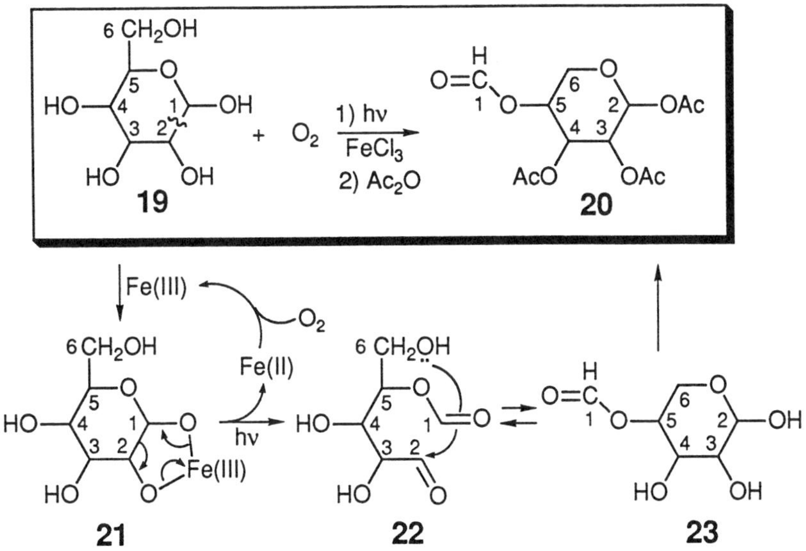

Scheme 4. Fe-catalyzed oxidative cleavage of vicinal diols.

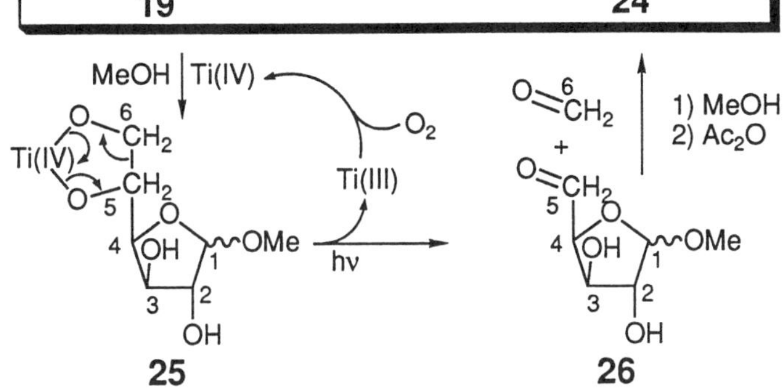

Scheme 5. Ti-catalyzed oxidative cleavage of vicinal diols.

Cu-Promoted Photoactivation of C–O Bonds. Polymerization of tetrahydrofuran (**33**, Scheme 7) occurs upon irradiation in the presence of copper(I) trifluoromethanesulfonate (CuOTf). Presumably, catalyzed photolytic activation of a C–O bond initiates this photopolymerization (*7*). Thus, ligand-to-metal electron transfer in an excited complex, **35a**, generates a cation radical, **35b** (Scheme 7). The consequently enhanced nucleofugacity of oxygen in **35b** favors alkylation of a second molecule of **33**, leading to an oxonium intermediate **35c**. Further trans-alkylations of additional molecules of **33** delivers polytetrahydrofuran (**34**).

A related mechanism could account for Cu(I)-catalyzed photodimerization of allyl alcohol (*8*). Thus, photoinduced ligand-to-metal electron transfer in an excited complex derived from **37** generates a cation radical **38** (Scheme 8). The enhanced nucleofugacity of the OH group in **38** favors alkylation of a second molecule of allyl alcohol (**36**), leading to an oxonium intermediate **39**. Deprotonation then delivers diallyl ether (**41**).

Synthetic Applications of Cu(I)-Catalyzed $2\pi + 2\pi$ Photocycloadditions

Diallyl ether (**41**) undergoes further reaction upon irradiation in the presence of CuOTf. An intramolecular photocycloaddition generates 3-oxabicyclo[3.2.0]heptane (**43**). Numerous observations support the key intermediacy of chelated Cu(I) complex **42**. The corresponding *n*-butyl-substituted diallyl ether (**44**) is transformed into *exo*-2-butyl-3-oxabicyclo-[3.2.0]heptane (**45**) upon irradiation in the presence of CuOTf (*9*) (Scheme 9). The stereoselectivity of this reaction results from a preference for the equatorially substituted complex **44** over the more sterically congested complex **46** that is required to generate the *endo*-butylated stereoisomer **47** of **45**.

A strong preference for generating the bicyclo[3.2.0] ring system is evident in the Cu(I)-catalyzed photobicyclizations of various allyl dienyl ethers. Thus, generation of **49** rather than **51** (*9*) is understandable in terms of a preference for the bidentate complex **48** over the isomeric complex **50** (Scheme 10). Similarly the exclusive formation of **53** rather than **55** (*10*) results from a preference for the bidentate complex **52** over the isomeric complex **54** (Scheme 11).

The photobicyclization of diallyl ethers can be used to produce a variety of multicyclic carbon networks (*9*) such as **57** or **60** from **56** or **59**, respectively (Scheme 12). The ether functionality in these photoproducts facilitates synthetically useful transformations such as the regioselective oxidations of the ethers **57** and **60** into butyrolactones **58** or **61**, respectively (*9*).

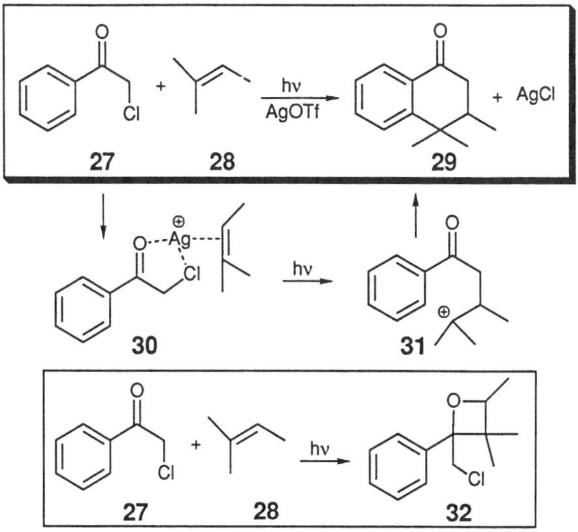

Scheme 6. Ag-promoted addition of phenacyl chloride to alkenes.

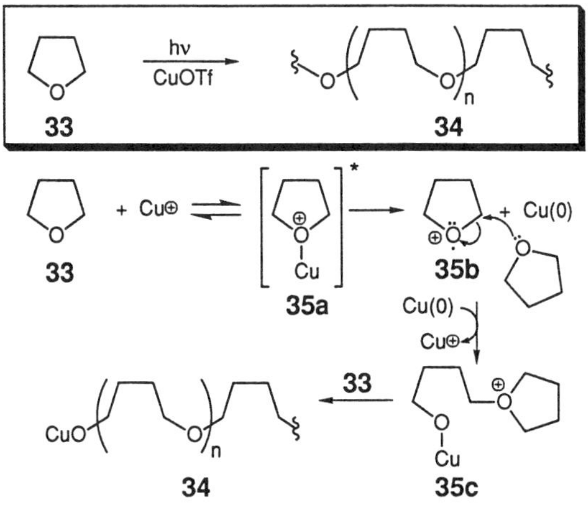

Scheme 7. Cu(I)-catalyzed photopolymerization of THF.

A factor that limits the functionality that can be present in substrates for Cu(I)-catalyzed photobicyclizations is the ultraviolet absorption spectrum of the functional group. For example, irradiation of di-2-propenamides (diallyl amides) **62a** or **62b** fails to generate any bicyclization products **63a** or **63b**, but the corresponding carbamate **62c** gives **63c** in good yield (Scheme 13) (*11*). The Cu(I) complex **62c** has an ultraviolet λ_{max} = 233.4 nm, and the dipropenylcarbamate ligand is nearly tran-

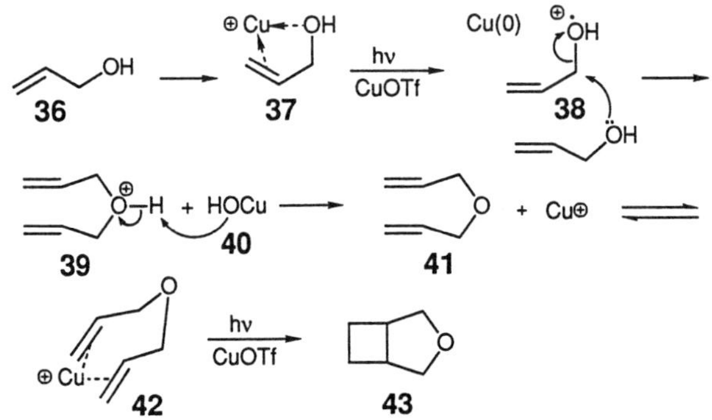

Scheme 8. Cu(I)-catalyzed photodimerization of allyl alcohol.

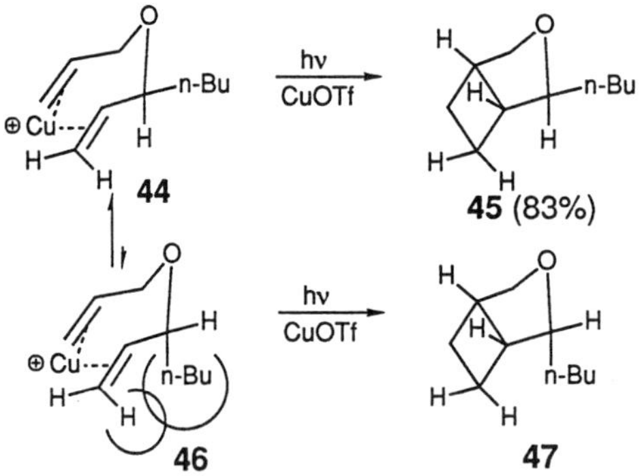

Scheme 9. Cu(I)-catalyzed photobicyclization of diallyl ethers.

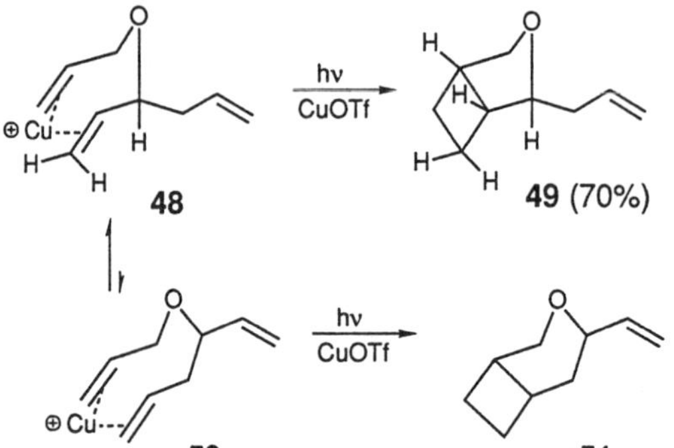

Scheme 10. A preference for the bicyclo[3.2.0] ring system.

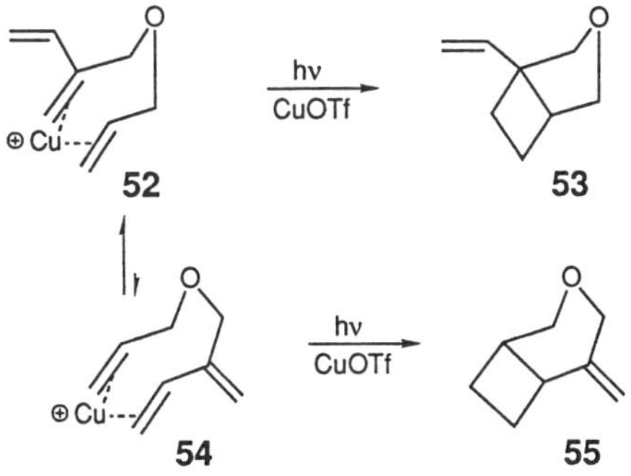

Scheme 11. Another example of a preference for the bicyclo[3.2.0] ring system.

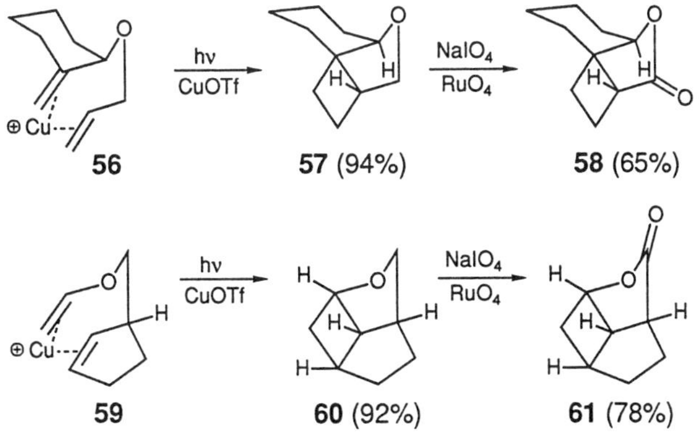

Scheme 12. Synthesis of multicyclic carbon networks by photobicyclization.

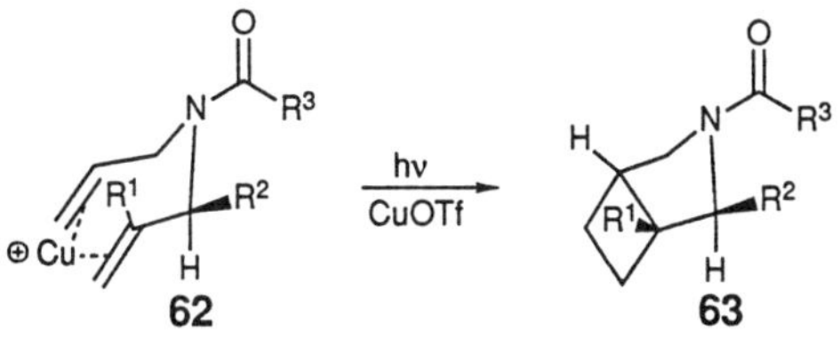

Amide	R^1	R^2	R^3	Reaction Time (h)	Yield (%)	ε^b
a	H	H	Me	24	0^a	192
b	H	H	H	24	0^a	231
c	H	H	OEt	24	74	15
d	H	Me	OEt	48	60	- -
e	Me	H	OEt	22	76	- -

(a) Nearly quantitative recovery of starting N,N-diallylamide.
(b) Molar extinction at 233.4 nm.

Scheme 13. Cu(I)-catalyzed photobicyclization of N,N-*diallylcarbamates.*

sparent at that wavelength. In contrast, the di-2-propenamide (diallyl amide) ligands in **62a** and **62b** absorb strongly at that wavelength and thereby interfere with the requisite photoexcitation of the corresponding Cu(I) complexes.

The Cu(I)-catalyzed photobicyclization of myrcene (**64**) again demonstrates the preference for formation of the bicyclo[3.2.0] ring system in **66** rather than the [3.1.1] or [2.1.1] ring systems of **67** or **68** (Scheme 14) (*10*). Interestingly, triplet sensitized irradiation of **64** produces the bicyclo[2.1.1] product **68** exclusively, and direct irradiation delivers both **67** and **68**. Only the Cu(I)-catalyzed photoreaction produces **66**, presumably by photoexcitation of an intermediate bidentate complex **65**.

Cu(I)-catalyzed photobicyclization of 3-hydroxy-1,6-heptadiene favors generation of *endo*-2-hydroxy product **73** over the *exo* epimer **71** (Scheme 15) (*12*). This stereoselectivity undoubtedly reflects a preference for the tridentate chelation in complex **72** over a bidentate complex, **70**. Where steric factors do not favor tridentate chelation, as in **76**, preferential generation of an *exo*-2-hydroxy photobicyclization product **75** is favored over the *endo* epimer **77** (Scheme 16). For 4-hydroxy-1,6-heptadiene, a tridentate chelate, **80**, and a bidentate complex, **81**, are similarly favorable, resulting in similar yields of *endo*- and *exo*-3-hydroxybicyclo[3.2.0]heptanes **79** and **82** (Scheme 17) (*8*).

Cu(I)-catalyzed photobicyclization is clearly useful for construction of multicyclic carbon networks that incorporate the bicyclo[3.2.0] ring system such as the panasinsene sesquiterpenes **87** and **88**. These were assembled efficiently by photobicyclization of the cyclohexane **83** (Scheme 18) (*13*). Oxidation of the photoproduct **84** delivered tricyclic ketone **86**. The synthetic utility of the photobicyclization route is underscored by the complete failure of attempts to generate **86** by an alternative non-metal-catalyzed photochemical route, photocycloaddition of enone **85** with isobutylene.

Bicyclo[3.2.0]heptanes containing structure **92** were proposed for dimethyl ether derivatives of the sesquiterpene phenol robustadials on the basis of NMR and mass spectral studies (*14*). We accomplished an efficient construction of this ring system by Cu(I)-catalyzed photobicyclization of diene **89** (Scheme 19) (*15*). After generation of the pyran ring in **91**, formylation delivered three of the four possible diastereomers of **92**. None of these corresponded to the natural products, so our synthesis disproved the proposed structure. Our subsequent synthetic studies demonstrated that the natural products incorporate the bicyclo[3.1.1]-heptylpinane ring system **93** rather than a bicyclo[3.2.0]heptyl ring system (*16*).

The compatibility of Cu(I)-catalyzed photobicyclization with various functional groups raises the possibility of exploiting those groups to facilitate conversions of the bicyclo[3.2.0]heptane moiety of photoproducts into

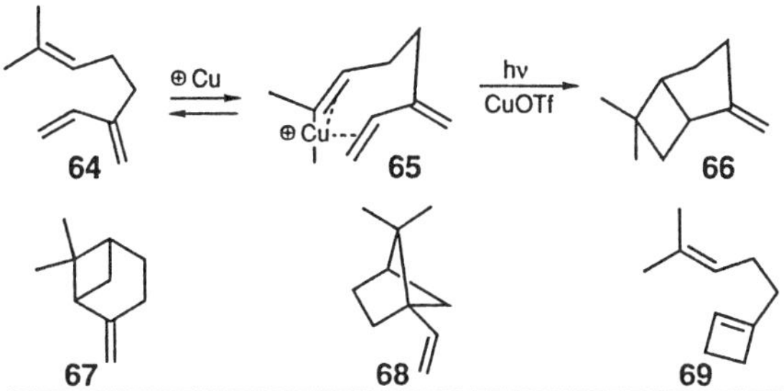

reaction type	66	67	68	69
	------ products (%) ------			
triplet sensitized[a]			75	
direct irradiated[b]		20	12	52
CuOTf catalyzed[c]	20	0	0	35

(a) β -acetonaphthone sensitizer, pyrex immersion well.
(b) Vycor filter.
(c) Quartz immersion well, 0.011 M CuOTf.

*Scheme 14. Cu(I)-catalyzed photobicyclization of myrcene (**64**).*

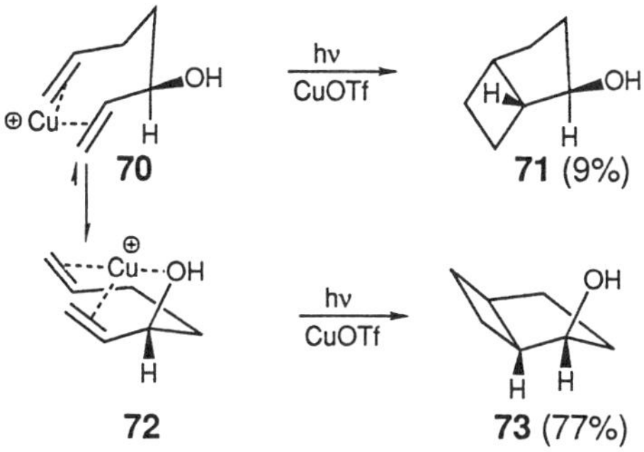

Scheme 15. Cu(I)-catalyzed photobicyclization of 3-hydroxy-1,6-heptadienes.

other ring systems. For example, Cu(I)-catalyzed photobicyclization can be exploited for stereocontrolled construction of cyclobutanes. A synthesis of the terpene (±)-grandisol (**98**) was accomplished starting with the high-yielding conversion of diene **94** into bicyclo[3.2.0]heptane **95** (Scheme 20) (*17*). Cleavage of the cyclopentane ring was accomplished by fragmentation of oxime **96**. Reduction of the resulting acid **97** delivered **98**.

With the expectation that a strategically placed trimethylsilyl substituent could facilitate cleavage of the furan ring in 3-oxabicyclo[3.2.0]-heptanes, we attempted to generate **101** by Cu(I)-catalyzed photobicyclization the silyl diallyl ether **99** (Scheme 21) (Salomon, R. G.; Raychaudhuri, S. R., unpublished data). Unexpectedly, irradiation of **99** in the presence

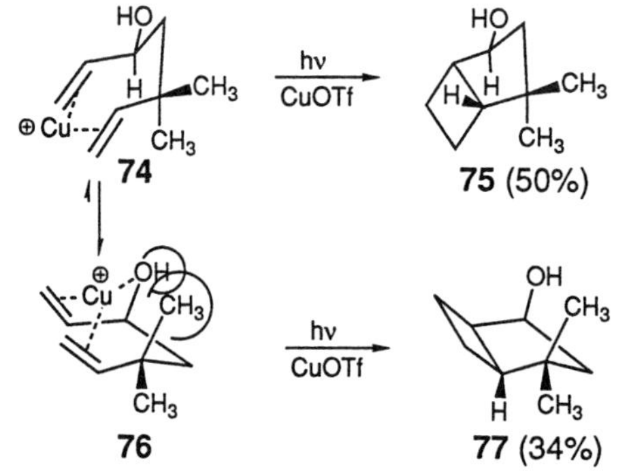

Scheme 16. The exception that proves the rule: preferential generation of exo-*2-hydroxyl photobicyclization product.*

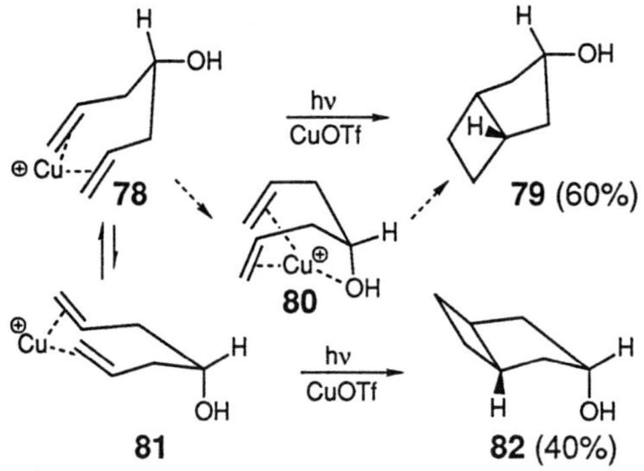

Scheme 17. Cu(I)-catalyzed photobicyclization of 4-hydroxy-1,6-heptadienes.

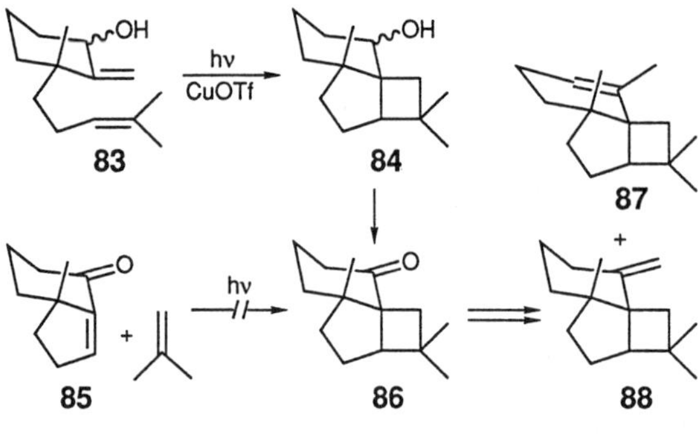

Scheme 18. Total synthesis of panasinsene sesquiterpenes.

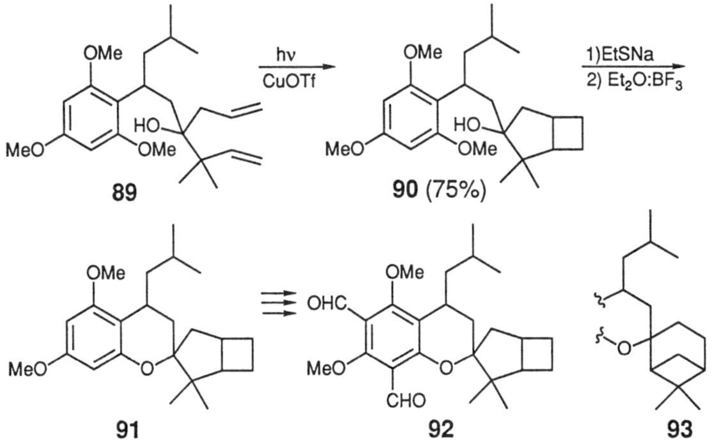

Scheme 19. Total synthesis of putative (±)-robustadial sesquiterpene phenols.

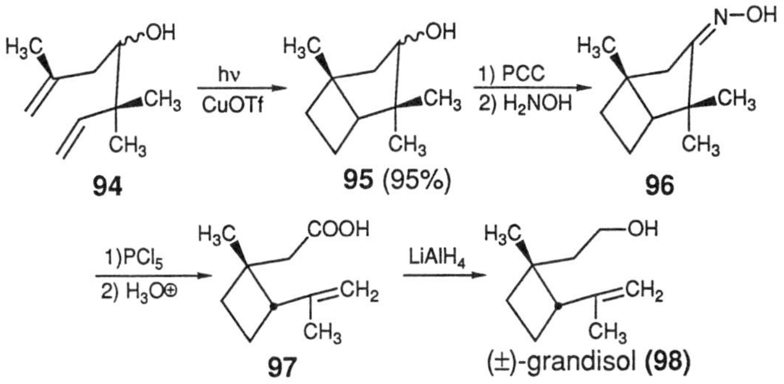

Scheme 20. Total synthesis of (±)-grandisol.

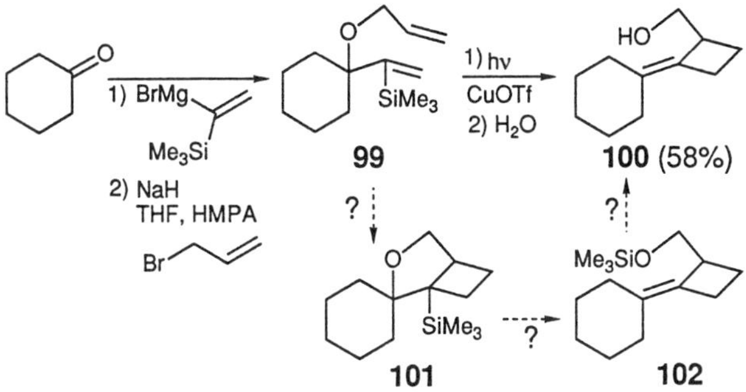

Scheme 21. Synthesis of 2-alkylidene-1-hydroxymethylcyclobutanes.

of CuOTf delivered 2-cyclohexylidene-1-hydroxymethylcyclobutane (**100**) directly. Apparently the β-silyl ether **101** formed initially undergoes β-elimination to produce **102** under the photolysis reaction conditions. Hydrolysis of **102** during workup delivers **100**.

Photochemical energy stored as ring strain in the cyclobutyl portion of the photocycloaddition products provides the driving force for synthetically valuable transformations of the bicyclo[3.2.0] ring system. Thus, $2\sigma + 2\sigma$ cycloeliminations can be exploited to generate 2-cyclopenten-1-ones from bicyclo[3.2.0]heptyl ketones that are readily available from acyclic precursors by Cu(I)-catalyzed photobicyclization (*12*). For example, Cu(I)-catalyzed photobicyclization of the 3-hydroxy-1,6-heptadiene **103** delivers bicyclo[3.2.0]heptanol **104** in excellent yield (Scheme 22). Pyrolysis of the derived ketone **105** provides cyclopentenone **106** in good overall yield.

The relief of ring strain can also foster carbon skeletal rearrangements. For example, 7-hydroxynorbornanes are available by acid-catalyzed rearrangement of 2-hydroxybicyclo[3.2.0]heptane photoproducts (*18*). This protocol provides an efficient route to *exo*-1,2-polymethylene-7-hydroxynorbornanes **113** from allyl cycloalkanones **107** (Scheme 23). Photobicyclization of an epimeric mixture of intermediate dienes **108** delivers hydroxybicyclo[3.2.0]heptane epimers **109**. Owing to a rapid equilibrium between the epimeric carbocations **111** and alkene **110**, a single epimeric cation **112** is generated upon treatment of **109** with acid, and this treatment affords epimerically pure **113** stereoselectively.

A different carbon skeletal rearrangement is accomplished by treatment of 1-ethoxy-3-oxabicyclo[3.2.0]heptanes with acid. For example, Cu(I)-catalyzed photobicyclization of 2-ethoxy diallyl ether, **114**, affords **115**. Acid treatment induces rearrangement with expansion of the cyclobutane ring, delivering cyclopentanone **117** in excellent yield (Scheme 24) (Ghosh, S.; Salomon, R. G., unpublished observations). The same strategy can be exploited for the construction of bridged multicyclic cyclopentanones. For example, 2-methoxy diallyl ether **119** affords **120** that rearranges with expansion of the cyclobutane ring delivering the bridged multicyclic cyclopentanone **122** upon treatment with acid (Scheme 25) (Ghosh, S.; Salomon, R. G., unpublished observations).

Conclusions

That Cu(I)-catalyzed photobicyclizations should be valuable for organic synthesis was suggested by the facts that these reactions (1) are often clean and high yielding, (2) generate two C–C bonds, (3) generate energy-rich cyclobutane derivatives, and (4) are compatible with several functional groups. This review shows how the interplay of these factors provides a

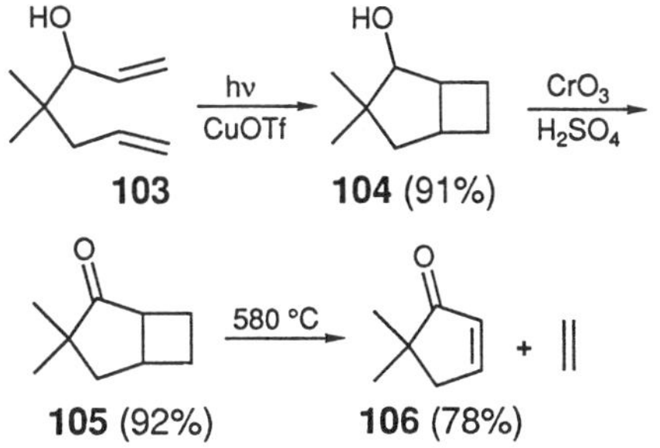
HO
hν
CuOTf
HO
CrO_3
H_2SO_4
103
104 (91%)
O
580 °C
O
+
105 (92%)
106 (78%)

Scheme 22. Synthesis of 2-cyclopenten-1-ones.

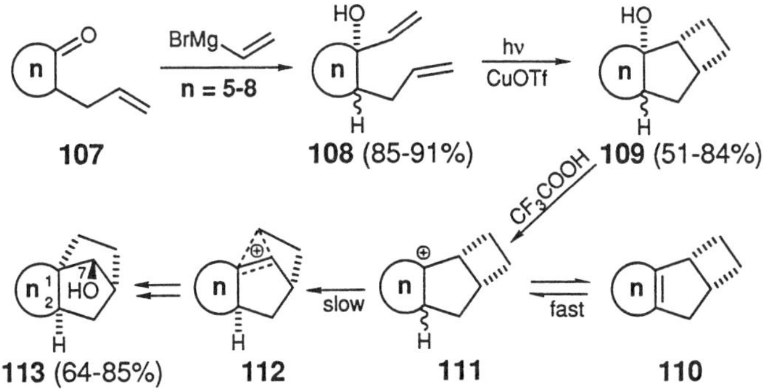
O
n
BrMg
n = 5-8
HO
n
H
hν
CuOTf
HO
n
H
107
108 (85-91%)
CF_3COOH
109 (51-84%)
n
1
2
7
HO
H
n
⊕
H
slow
n
⊕
H
fast
n
113 (64-85%)
112
111
110

Scheme 23. Synthesis of exo-*1,2-polymethylene-7-hydroxynorbornanes.*

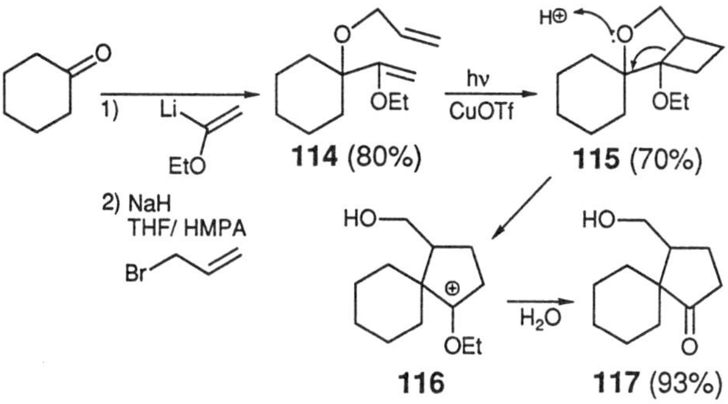
O
1) Li
EtO
2) NaH
THF/ HMPA
Br
O
OEt
114 (80%)
hν
CuOTf
H⊕
:O
OEt
115 (70%)
HO
⊕
OEt
116
H_2O
HO
O
117 (93%)

Scheme 24. Synthesis of 3-hydroxymethylcyclopentanones.

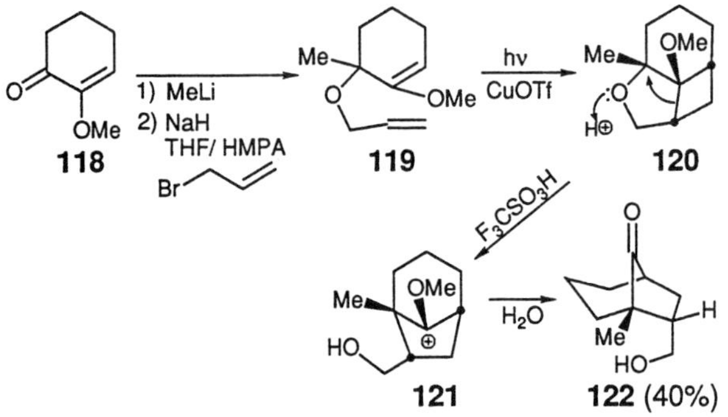

Scheme 25. Synthesis of bridged multicyclic cyclopentanones.

wide variety of synthetic applications. The pace of discovery of new metal-catalyzed photochemical reactions of organic molecules is strong. For these new reactions to be widely used in organic synthesis, it is imperative to determine their compatibility with the various functional groups present in synthetic targets or required to facilitate transformations of synthetic intermediates.

Acknowledgments

Financial support for our early studies on homogeneous metal-catalyzed photochemistry came from the Petroleum Research Fund. More recent support was provided by the National Science Foundation. Our collaborators in the studies reported herein include Kamlakar Avasthi, Daniel Coughlin, Subrata Ghosh, Kasturi Lal, Thomas Mirante, Swadesh Raychaudhuri, Wiley Youngs, Michael Zagorski, and Anthony Zarate.

References

1. Salomon, R. G. *Tetrahedron* **1983,** *39,* 485.
2. Sakakura, T.; Hayashi, T.; Tanaka, M. *Chem. Lett.* **1987,** 859–862.
3. Delgado-Lieta, E.; Luke, M. A.; Jones, R. F.; Cole-Hamilton, D. J. *Polyhedron* **1982,** *1,* 839–840.
4. Ichikawa, S.; Tomita, I.; Hosaka, A.; Sato, T. *Bull. Chem. Soc. Jpn.* **1988,** *61,* 513–520.
5. Sato, T.; Takahashi, K.; Ichikawa, S. *Chem. Lett.* **1983,** 1589.
6. Sato, T.; Tamura, K. *Tetrahedron Lett.* **1984,** *25,* 1821–1824.
7. Woodhouse, M. E.; Lewis, F. D.; Marks, T. J. *J. Am. Chem. Soc.* **1978,** *100,* 996–998.

8. Evers, J. Th. M.; Mackor, A. *Tetrahedron Lett.* **1978,** 821–824.
9. Ghosh, S.; Raychaudhuri, S. R.; Salomon, R. G. *J. Am. Chem. Soc.* **1987,** *109,* 83–90.
10. Avasthi, K.; Raychaudhuri, S. R.; Salomon, R. G. *J. Org. Chem.* **1984,** *49,* 4322–4324.
11. Salomon, R. G.; Ghosh, S.; Raychaudhuri, S. R.; Miranti, T. S. *Tetrahedron Lett.* **1984,** *25,* 3167–3170.
12. Salomon, R. G.; Coughlin, D. J.; Ghosh, S.; Zagorski, M. G. *J. Am. Chem. Soc.* **1982,** *104,* 998.
13. McMurry, J. E.; Choy, W. *Tetrahedron Lett.* **1980,** *21,* 2477–2480.
14. Xu, R.; Snyder, J. K.; Nakanishi, K. *J. Am. Chem. Soc.* **1984,** *106,* 734.
15. Lal, K.; Zarate, E. A.; Youngs, W. J.; Salomon, R. G. *J. Org. Chem.* **1988,** *53,* 3673–3680.
16. Salomon, R. G.; Mazza, S. M.; Lal, K. *J. Org. Chem.* **1989,** *54,* 1562–1570.
17. Rosini, G.; Geier, M.; Marotta, E.; Petrini, M.; Balini, R. *Tetrahedron* **1986,** *41,* 6027–6032.
18. Avasthi, K; Salomon, R. G. *J. Org. Chem.* **1986,** *51,* 2556–2562.

RECEIVED for review November 7, 1991. ACCEPTED revised manuscript June 8, 1992.

17

Photooxidation of Metal Carbynes

Lisa McElwee-White, Kevin B. Kingsbury, and John D. Carter

Department of Chemistry, Stanford University, Stanford, CA 94305

Radical cations formed upon photooxidation of the metal carbyne complexes (η^5-C_5H_5)$L_1L_2M{\equiv}CR$ (M is Mo or W; L_1 and L_2 are CO or $P(OMe)_3$; and R is alkyl or aryl) were found to be extremely reactive. Depending on the reaction conditions, either the radical cations exhibited the ligand-exchange and atom-abstraction processes that are typical of metal radicals, or the carbyne ligands underwent a series of highly unusual rearrangements to yield organic products. When organic products were generated, they were formed with good stereochemical and regiochemical control.

REACTIVE ORGANIC FRAGMENTS CAN BE STABILIZED with transition metals so that they can be used for synthesis under ordinary reaction conditions. One example is the extensive utilization of metal carbenes in organic synthesis (*1*). It is thus somewhat surprising that there are so few reported cases of generation of organic products from the closely related metal carbynes. Although the chemistry of metal carbynes has been explored (*2*), most of the reported reactions involve conversion into other carbyne complexes by exchange of ancillary ligands; interconversion of carbyne, carbene, and vinylidene complexes; or [2 + 2] cycloaddition with alkenes and alkynes followed by metathesis or polymerization. With the exception of polymers and metathesis products, organic species derived from the carbyne ligand are almost never generated.

One method for inducing new modes of reactivity in metal complexes is one-electron (1-e^-) oxidation. The differences in reaction rates between 18-e^- complexes and their 17-e^- counterparts can be quite dramatic (*3–5*). Reaction manifolds that are not available to an 18-e^- complex can thus become accessible to its corresponding 17-e^- radical cation. This chapter

0065–2393/93/0238–0335$06.00/0

provides some background on one-electron oxidation of organometallic species and then discusses our application of this strategy to the activation of metal carbyne complexes by photooxidation. Radical cations produced via photochemical electron transfer from the carbynes exhibit unprecedented reaction pathways, including rearrangement and loss of the carbyne ligand to yield organic products.

Photooxidation of Organometallic Complexes

To place the photooxidation of metal carbynes in perspective, a few examples of photochemical electron transfer from other organometallic species will be discussed. Photoinduced electron transfer involving organometallic complexes is of considerable interest, as evidenced by a 1988 review (*6*). When the excited-state complex serves as a reductant, the most common electron acceptors are the halogenated solvents CCl_4, $CHCl_3$, and CH_2Cl_2. Not only are the halocarbons readily reduced by excited-state organometallic compounds, but the electron-transfer step is irreversible because the reduced solvent undergoes a rapid fragmentation (*7*). Irreversible electron transfer is advantageous for obtaining chemical reactions by the oxidized species because back transfer, if possible in the system, can be the most rapid process available to the newly generated ion pair.

In most cases, the excited states that give rise to electron transfer are of the metal-to-ligand charge transfer (MLCT) or charge-transfer-to-solvent (CTTS) variety. As an example, electron transfer from the MLCT states of the hexakisarylisocyano complexes, $M(CNPh)_6$ (M is Mo or W), to $CHCl_3$ resulted in formation of the seven-coordinate products $[M(CNPh)_6Cl]^+[Cl]^-$ by a net 2-e^- oxidation process (*8*). Transfer via CTTS states is somewhat more common and is exemplified by the photooxidation of $(\eta^5\text{-}C_5H_5)_2Fe$ in CCl_4 (*9*). Fragmentation of the reduced solvent in this system has been demonstrated by addition of nitroxide spin traps to the reaction mixtures, whereupon adducts of the resulting radical are observed along with the organometallic product $(\eta^5\text{-}C_5H_5)_2FeCl$.

The chemical reactions induced by photooxidation of metal complexes are generally dominated by the facile ligand-exchange and atom-abstraction processes that are characteristic of metal radicals (*3–5*). Examples of the increased substitutional lability of 17-e^- cationic complexes relative to their neutral 18-e^- counterparts are well documented for electrochemical and chemical oxidation (*3–5, 10*). This effect is also seen in cases of photoinduced electron transfer such as the irradiation of $Ni(phen)(S_2C_2Ph_2)$ in $CHCl_3$ (*11*). The resulting cationic species then undergoes rapid ligand exchange to give the symmetric complexes $Ni(S_2C_2Ph_2)_2$ and $[Ni(phen)_2]^{2+}$.

Halogen-atom abstraction is also a common result of irradiation of metal complexes in CCl_4 or $CHCl_3$. In the photooxidation of $M(CNPh)_6$ (M is Mo or W) just discussed, the $M(CNPh)_6$ radical cation is postulated to abstract a Cl atom within the caged ion pair to yield the final product, $[M(CNPh)_6Cl]^+Cl^-$ (*8*). Similar atom-abstraction reactions are also observed for the photochemically generated radicals $\cdot Mn(CO)_5$ and $(\eta^5\text{-}C_5H_5)Fe(CO)_2\cdot$ (*12, 13*).

Another consequence of photochemical electron transfer is the initiation of chain processes. The reduction of halogenated solvents produces radical species (*7*), so reactions that occur via radical chain propagation can result. An example is the oxidative addition of CH_2Cl_2 to Pt[2-(2-thienyl)pyridine]$_2$ following CTTS excitation (*14*). The $\cdot CH_2Cl$ radicals from solvent fragmentation are carriers in a chain process for oxidative addition with an average chain length of about 40.

Chains that propagate by redox reactions can also be initiated by photochemical electron transfer. One such case is the photosubstitution of PPh_3 for CH_3CN in $[(CH_3CN)Re(CO)_3phen]^+$ (*15*). Electron transfer in this system is from PPh_3 to the complex, resulting in a 19-e^- species that undergoes rapid ligand exchange to form $[(PPh_3)Re(CO)_3phen]^0$. This zero-valent intermediate then reduces the starting material to propagate the chain, as evidenced by quantum yields of approximately 20.

Reactions within the Ligands of Metal Radicals

For the systems described, photoinitiated electron transfer triggers the characteristic chemical reactions of metal radicals. The ligand-exchange, halogen-abstraction, and chain processes that result are interpretable in terms of this model. The examples discussed share a common feature: The chemistry has occurred at the metal atom.

In order to convert photooxidized metal carbyne complexes into organic products derived from the carbyne ligand, reactions of the cation radical must take place within that ligand. Chemistry at an organic ligand occurs in metal radicals, although it is much less common than reactions at the metal atom. Most cases of ligand-centered radical reactivity occur in 19-e^- species or in cationic 17-e^- complexes (*16*). For example, Arbuzov reaction of $P(OMe)_3$ occurs during the reaction of the 19-e^- species $(\eta^5\text{-}C_5H_5)(\eta^6\text{-}C_6H_6)Fe$ with $P(OMe)_3$ (*17*). As a 17-e^- example, the intermediate alkynyl complex $(\eta^5\text{-}C_5H_5)(dppe)Fe\text{–}C{\equiv}CMe^{\cdot +}$ formed upon oxidation and deprotonation of $(\eta^5\text{-}C_5H_5)(dppe)Fe{=}C{=}CHMe^+$ undergoes carbon–carbon coupling at C_β of the alkynyl ligand to yield a vinylidene dimer (*18*). (dppe is 1,2-bis(diphenylphosphino)ethane.) A related process involves coupling at C_β to form $[(\eta^5\text{-}C_5H_5)[P(OMe)_3]_2\text{-}Mo{\equiv}CCH\text{-}t\text{-}Bu]_2$ following 1-e^- oxidation of the anionic vinylidene $[(\eta^5\text{-}C_5H_5)[P(OMe)_3]_2Mo{=}C{=}CH\text{-}t\text{-}Bu]^-$ by ferrocinium ion (*19*).

Conjugated π ligands such as the cyclopentadienyl ring are particularly prone to radical processes. If the photochemical electron transfer from ferrocene to CCl_4 is carried out in the presence of ethanol, ring-substituted products are obtained (*20*). These are believed to arise from addition of the $\cdot CCl_3$ radical to the ring of the odd electron species $(\eta^5\text{-}C_5H_5)_2Fe^+$ followed by solvolysis of the $-CCl_3$ group. In a related process, homolytic loss of the benzyl radical from $(\eta^5\text{-}C_5H_5)Fe(CO)_2(\eta^1\text{-}CH_2C_6H_5)$ to give the 17-e^- species $(\eta^5\text{-}C_5H_5)Fe(CO)_2$ is followed by coordination of a third CO and recombination of the benzyl radical with the $(\eta^5\text{-}C_5H_5)$ ring to give an 18-e^- complex (*21*). Ring–ring radical coupling was also observed in the dimerization of several odd-electron species, including $(\eta^5\text{-}C_5H_5)Rh(CO)(PPh_3)^+$ (*22*), $(\eta^5\text{-}C_6H_7)Fe(CO)_3$ (*23*), and $(\eta^5\text{-}C_5H_5)Co(\eta^5\text{-}C_7H_9)$ (*16*).

The case of $(\eta^5\text{-}C_5H_5)Rh(CO)(PPh_3)^+$ is particularly interesting. The PPh_3-substituted complex undergoes the ligand-centered ring-coupling process (*22*), but its analogs $(\eta^5\text{-}C_5H_5)Rh(CO)L^+$, where L is PMe_3 or $P(OPh)_3$, dimerize by formation of a metal–metal bond (*24*). This system is a rare example of closely related metal radical species exhibiting different modes of reactivity, one at the metal and the other at a ligand. A case in which a single species shows reactivity at both sites is $(CO)_5W\text{–}C{\equiv}N\text{–}CR_2\cdot$ (R is mesityl) (*25*). In the presence of Br_2, it yields $Br(CO)_4WCNCR_2$ in competition with dimerization through the ligands to give $(CO)_5W\text{–}C{\equiv}N\text{–}CR_2CR_2\text{–}N{\equiv}C\text{–}W(CO)_5$.

Chemical Oxidation of Metal Carbynes

Support for the idea that photooxidation of metal carbynes could produce organic products derived from the carbyne ligand is found in early reports by Fischer (*26*) on the treatment of the chromium carbyne complex $Br(CO)_4Cr{\equiv}CPh$ (I) with chemical oxidants. As shown in Scheme I, a variety of organic products could be obtained, depending on the oxidizing agent. Cl_2 and Br_2 led to formation of the benzotrihalides, and oxidation in nucleophilic solvents such as water and methanol gave products in which the solvent had been incorporated. Coupling of the phenylcarbyne ligands to produce $PhC{\equiv}CPh$ was also observed when Ce(IV) served as oxidant.

Different pathways were reported by Green (*27*) for the reactions of $(\eta^5\text{-}C_5H_5)[P(OMe)_3]_2Mo{\equiv}CCH_2\text{-}t\text{-}Bu$ (**II**, Scheme II) with potential 1-e^- oxidants. Treatment of **II** with $[4\text{-}FC_6H_4N_2][BF_4]$ afforded the vinylidene complex, $(\eta^5\text{-}C_5H_5)(N_2C_6H_4\text{-}4\text{-}F)[P(OMe)_3]Mo{=}C{=}CH\text{-}t\text{-}Bu$, and reaction with CF_3I yielded both the related vinylidene complex $(\eta^5\text{-}C_5H_5)(I)[P(OMe)_3]_2Mo{=}C{=}CH\text{-}t\text{-}Bu$ (III), and the unusual chelate carbene, $(\eta^5\text{-}C_5H_5)(I)[P(OMe_3)]Mo[{=}C(CH_2\text{-}t\text{-}Bu)P(O)(OMe)_2]$ (IV) (*see*

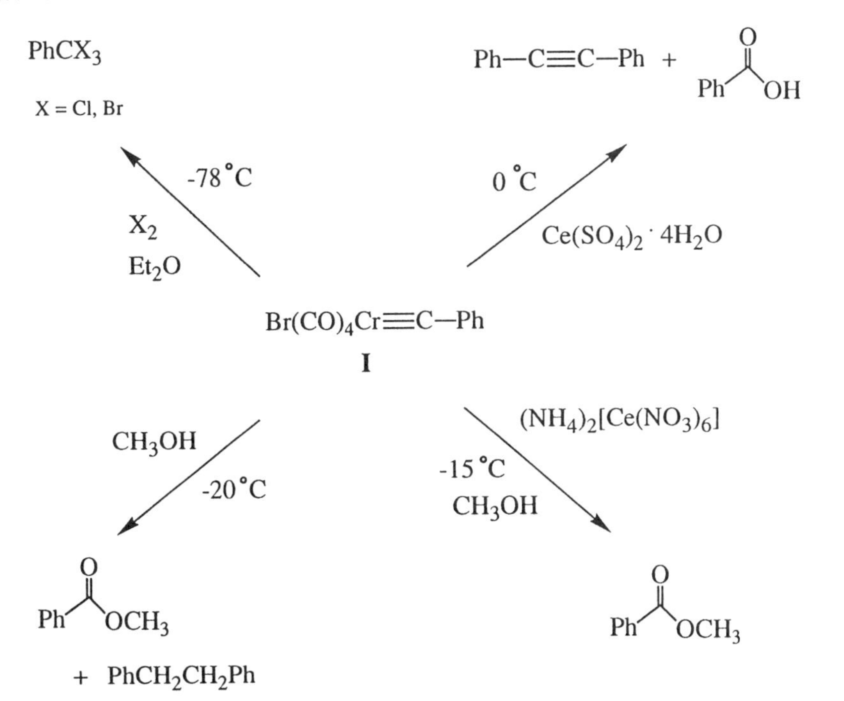

Scheme I. Treatment of chromium carbyne complex **I** *with chemical oxidants. (Reproduced from reference 26.)*

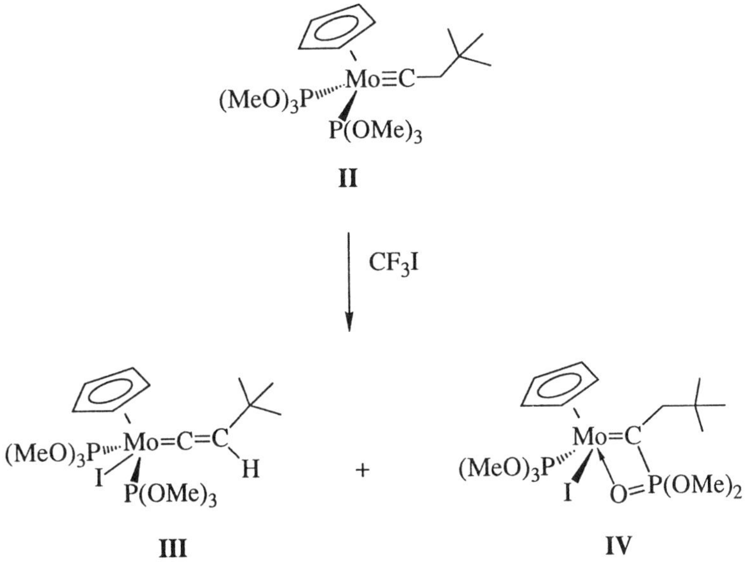

(MeO)$_3$P–Mo≡C–
P(OMe)$_3$

II

CF_3I

(MeO)$_3$P–Mo=C=C–H
I
P(OMe)$_3$

III

+

Mo=C
(MeO)$_3$P
I O=P(OMe)$_2$

IV

Scheme II. Reaction of **II** *with* CF_3I.

Scheme II). The common first step in these reactions was postulated to be electron transfer to form the 17-e$^-$ species, $[(\eta^5\text{-}C_5H_5)[P(OMe)_3]_2\text{-}Mo{\equiv}CCH_2\text{-}t\text{-}Bu]^+$.

Subsequent ligand exchange, proton abstraction, or both would then give the vinylidene products. The pathway for formation of chelate complex **IV** from **II** was suggested to involve $\cdot CH_3$ abstraction from a coordinated phosphite of the 17-e$^-$ species. Subsequent charge collapse and migration of the σ-bonded phosphonate group to the carbyne carbon would then afford carbene **IV**. These reactions differ from those in Scheme I in that no free organic products were reported. However, these processes did result in modification of the carbyne ligand, which was encouraging for the prospect of generating organic products through 1-e$^-$ oxidation of carbyne complexes.

Photooxidation of Metal Carbynes: "Inorganic" Reactivity

Photolysis of the carbyne complex $(\eta^5\text{-}C_5H_5)[P(OMe)_3]_2Mo{\equiv}CPh$ (**Va**, Scheme III) in the presence of $CHCl_3$ and PMe_3 resulted in photooxidation of **Va** followed by ligand exchange and halogen-atom abstraction to give the net 2-e$^-$ oxidation product $[(\eta^5\text{-}C_5H_5)Cl(PMe_3)_2Mo{\equiv}CPh)^+Cl^-$ (**VIIa**) (Scheme III) (*28*). Further work (*29*) demonstrated this to be a general reaction for the complexes $(\eta^5\text{-}C_5H_5)L_1L_2M{\equiv}CR)$ (**Vb–Vf**).

The primary photoprocess was postulated to be irreversible single-electron transfer to the halogenated solvent, yielding the radical cation $\mathbf{V}^{\cdot +}$ along with the Cl^- and $\cdot CHCl_3$ resulting from reduction of the solvent. Rapid ligand exchange of the more strongly donating phosphines for the phosphites or carbonyls in $\mathbf{V}^{\cdot +}$ would then give the next intermediate, the 17-e$^-$ complex $\mathbf{VI}^{\cdot +}$. The final isolated product, **VII**, would arise via halogen abstraction from the solvent by the metal atom of **VI**.

The steps in this mechanistic scheme are among the characteristic reactions of metal radicals. They are the rapid ligand-exchange and atom-abstraction processes discussed in conjunction with formation of odd electron organometallic species by photooxidation, chemical oxidation, or electrochemistry. For the photooxidation of **Va–Vf** in the presence of PMe_3, ligand exchange is the most rapid process available in the system and the chemistry is dominated by reactions at the metal atom. In contrast to the chemical oxidations of metal carbynes described in the previous section, under these conditions, the carbyne ligand is unchanged in the products **VIIa–VIIf**, and no free organic products derived from that ligand are observed.

$$\mathbf{V} \xrightarrow[\text{CHCl}_3]{h\nu} \mathbf{V}^{\cdot +} + \text{Cl}^- + {}^{\cdot}\text{CHCl}_2$$

$$\mathbf{V}^{\cdot +} \xrightarrow{\text{PMe}_3} [(\eta^5\text{-C}_5\text{H}_5)(\text{Me}_3\text{P})_2\text{M}{\equiv}\text{CR}]^{\cdot +}\ \text{Cl}^- \quad \mathbf{VI}^{\cdot +}$$

$$\mathbf{VI}^{\cdot +} \xrightarrow{\text{CHCl}_3} [(\eta^5\text{-C}_5\text{H}_5)\text{Cl}(\text{Me}_3\text{P})_2\text{M}{\equiv}\text{CR}]^{+}\ \text{Cl}^- + {}^{\cdot}\text{CHCl}_2 \quad \mathbf{VII}$$

Compound	M	L_1	L_2	R
Va	Mo	$P(OMe)_3$	$P(OMe)_3$	Ph
Vb	Mo	$P(OMe)_3$	CO	Ph
Vc	W	$P(OMe)_3$	$P(OMe)_3$	Ph
Vd	W	$P(OMe)_3$	CO	Ph
Ve	W	$P(OMe)_3$	CO	Me
Vf	W	$P(OMe)_3$	CO	$c\text{-}C_3H_5$

Scheme III.

Electronic Structure and Spectroscopy of Metal Carbynes

As part of our efforts to understand the photooxidation of complexes **Va–Vf**, we carried out a UV–visible spectroscopic study of the various compounds (*29*). Figure 1 shows the UV–visible spectrum of $(\eta^5\text{-}C_5H_5)[P(OMe)_3]_2Mo{\equiv}CPh$ (**Va**). The phenylcarbyne complexes **Va–Vd** give similar spectra with two main features: a strong band in the 320–330-nm range (molar absorptivity $\epsilon = 5 \times 10^3$–1×10^4) and a much weaker absorption in the 480–500-nm range (ϵ = 50–200). These transitions were assigned by analogy to the spectra of $X(CO)_2L_2W{\equiv}CR$ [R is Ph or *t*-Bu; X is Cl or Br; L_2 = tetramethylethylenediamine (tmeda), dipyridine (py_2), or 1,2-bis(diphenylphosphino)ethane (dppe)] described by

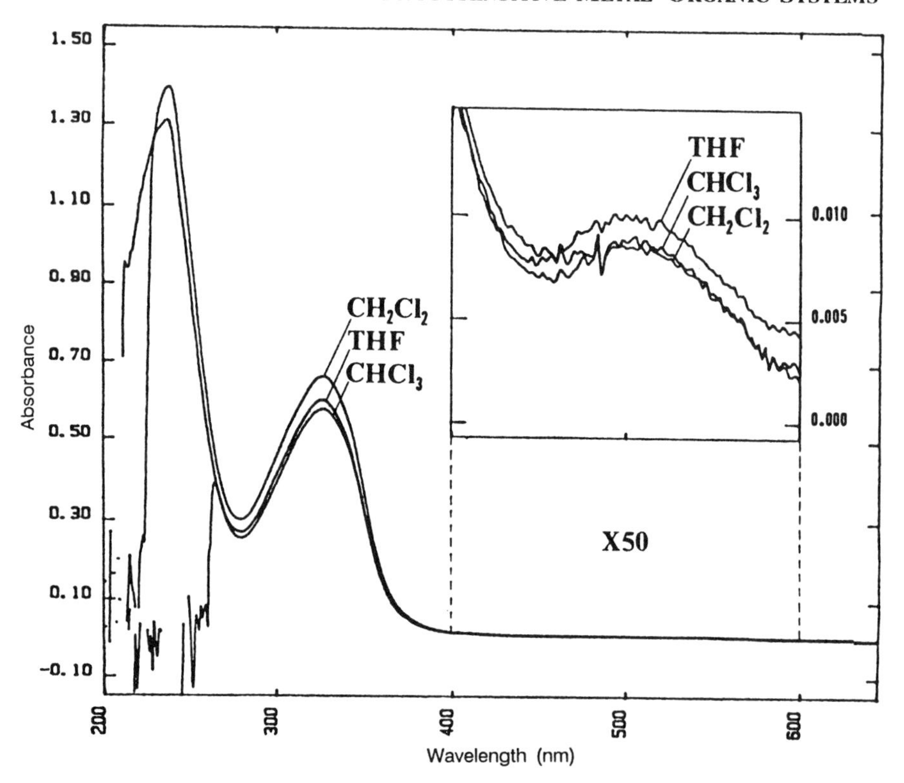

Figure 1. Absorption spectra of $(\eta^5\text{-}C_5H_5)[P(OMe)_3]_2Mo{\equiv}CPh$ *in solution at room temperature. The solutions are* 5×10^{-5} *M in* $CHCl_3$*,* CH_2Cl_2*, or THF. (Reproduced from reference 29. Copyright 1991 American Chemical Society.)*

Bocarsly and co-workers (*30, 31*). The 320-nm peaks are taken to arise from intraligand transitions in the phenyl substituents of arylcarbyne ligands. The 490-nm absorption is assigned as d–π* and is identified with the band referred to as MLCT in discussions of other carbyne complexes (*32, 33*). For carbynes with alkyl substituents (**Ve** and **Vf**), the strong transition at 320 nm is missing, and the d–π* transitions are blue-shifted.

Studies of the wavelength dependence of the photooxidation reaction demonstrate that the lowest lying d–π* state is responsible for electron transfer to halogenated hydrocarbons. Careful examination of the solvent dependence of the spectra reveals that the position of the photoactive transition is unchanged in $CHCl_3$, CH_2Cl_2, and tetrahydrofuran (THF), allowing the intervention of direct CTTS states to be ruled out.

Of the molecules that we have prepared, several with aryl substituents are luminescent in solution at room temperature, and the tungsten complexes exhibit reasonably long-lived excited states. For example, (η^5-C_5H_5)[$P(OMe)_3$](CO)W≡CPh (**Vd**) has a lifetime of 140 ns in solution at room temperature (*29*). Bocarsly et al. (*31*) reported that luminescence of the emissive complexes $X(CO)_2L_2W$≡CPh was quenched by a variety of organic compounds at rates dependent on the triplet energy of the quencher. Likewise, the emission of compounds **Vb** and **Vd** is quenched by organic compounds with low-lying triplet states, consistent with substantial triplet character in the excited state. Excitation spectra showed that the emission from complexes **Vb** and **Vd** arises from the same low-lying MLCT state that transfers the electron in the photooxidation process. In addition, these spectra demonstrated that the higher lying excited states are in part converted to the photoactive one.

Extended Hückel calculations on metal carbynes of the type (η^5-C_5H_5)L_1L_2Mo≡CR (*29, 34*) suggest that the highest occupied molecular orbital (HOMO) of the complexes should be primarily metal d in character, and that the lowest unoccupied molecular orbital (LUMO) is composed of one of the metal–carbon π^* orbitals. Figure 2 shows a partial molecular orbital diagram for the formation of the model compound (η^5-C_5H_5)[$P(OH)_3$]$_2$Mo≡CPh from the fragments (η^5-C_5H_5)[$P(OH)_3$]$_2Mo^-$ and CPh^+. The HOMO of the complex is composed largely of Mo $d_{x^2-y^2}$ and below it are the two metal–carbon π bonds. The LUMO and NLUMO (next lowest unoccupied molecular orbital) are the π^* orbitals, with the one that is conjugated into the phenyl π-system lying lower in energy.

Assignment of the HOMO as a nonbonding metal d orbital is consistent with structural information obtained by X-ray crystallography on $Br(dmpe)_2W$≡CPh and its 1-e^- oxidized congener [$Br(dmpe)_2W$≡CPh]-[PF_6] (dmpe is 1,2-bis(dimethylphosphino)ethane) (*35*). Oxidation results in only a slight shortening of the W≡C and W–Br bonds (0.024 and 0.042 Å), but the W–P bonds are lengthened somewhat. Upon this evidence, the HOMO is assigned as a nonbonding orbital that is primarily metal d_{xy} in character. Similar observations were made for $Cl(dppe)_2$Mo≡C(*p*-Tol)] and [$Cl(dppe)_2$Mo≡C(*p*-Tol)][PF_6] (*36*).

This picture of the electronic structure of (η^5-C_5H_5)[$P(OMe)_3$]$_2$-Mo≡CPh is consistent with the assignment of the lowest energy feature of the UV–visible spectra of carbyne complexes **Va–Vf** as a d–π^* transition. Although the extended Hückel calculations suggest the lowest energy excitations should show a significant amount of charge transfer to the carbyne ligand, all of these orbitals are highly delocalized, and there is still a reasonable contribution of Mo d atomic orbitals (AOs) to the virtual orbitals. The low-energy transitions would thus still have some residual d–d character. This result is consistent with the lack of solvent dependence and low extinction coefficients seen in the excitation to the lowest excited states.

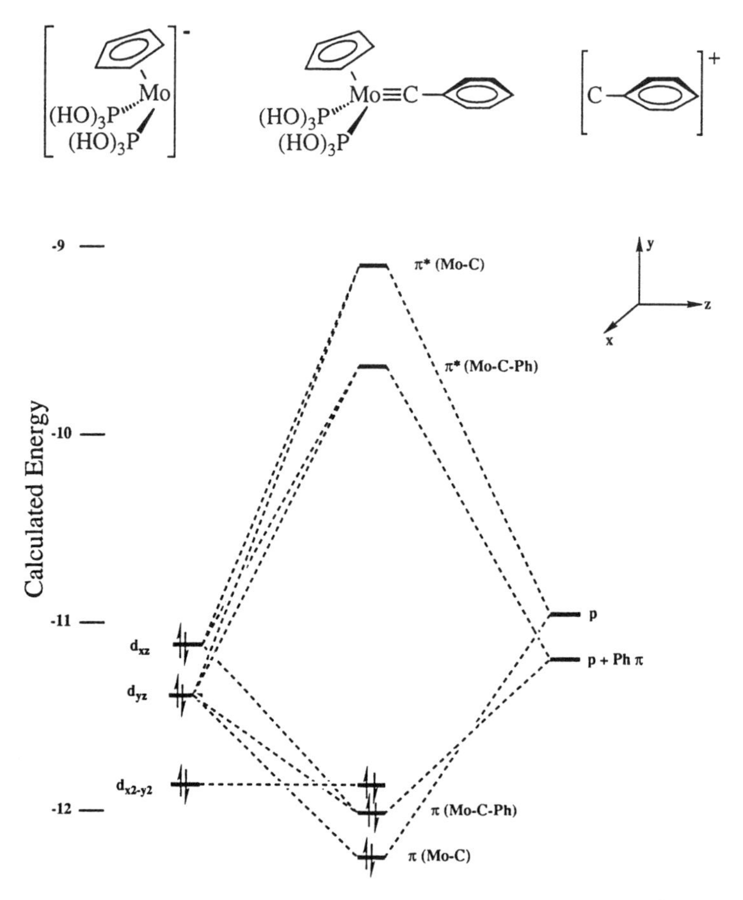

Figure 2. Orbital mixing diagram for the formation of $(\eta^5\text{-}C_5H_5)$-$[P(OH)_3]_2Mo{\equiv}CPh$ from the fragments $(\eta^5\text{-}C_5H_5)[P(OH)_3]_2Mo^-$ and CPh^+. (Reproduced from reference 29. Copyright 1991 American Chemical Society.)

Photooxidation of Metal Carbynes: "Organic" Rearrangements

Although the formation of the 2-e$^-$ oxidation products **VIIa–VIIf** upon photooxidation of the carbyne complexes **Va–Vf** in the presence of PMe_3 (Scheme III) is a general reaction, in the absence of PMe_3 the reactivity is quite different. When ligand substitution is not possible, the predominant process is rearrangement of the carbyne ligand and formation of free organic products. Although the product types vary depending on the carbyne substituents, for each of the alkyl-substituted carbynes we prepared, photooxidation in $CHCl_3$ in the absence of PMe_3 yields organic products derived from the original carbyne ligand. A few selected examples follow.

Although photolysis of the cyclopropylcarbyne $(\eta^5\text{-}C_5H_5)[P(OMe)_3]$-$(CO)W{\equiv}C(c\text{-}C_3H_5)$ (**Vf**) in $CHCl_3$–PMe_3 gives the chemistry summarized in Scheme III, photooxidation by $CHCl_3$ (no PMe_3) results in different reactivity (*37*). In the absence of phosphine, the inorganic product is trichloride complex **VIII** (Scheme IV), in which the carbyne ligand has been lost. The original cyclopropylcarbyne ligand has undergone rearrangement and carbonylation to produce cyclopentenone (Scheme IV).

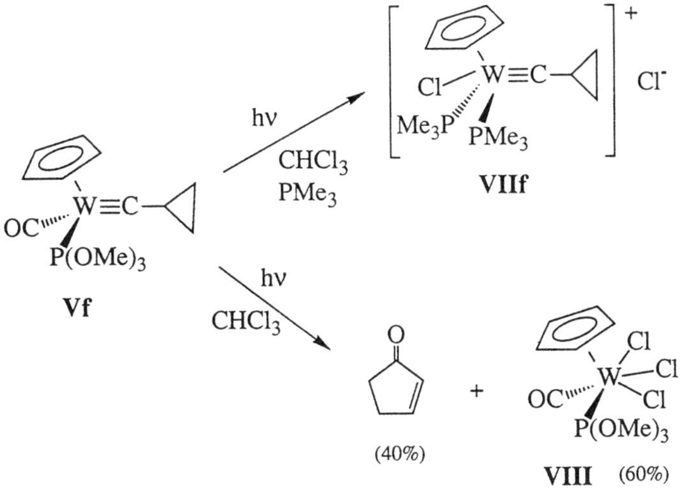

hν
$CHCl_3$
PMe_3
Cl^-
Vf
VIIf
hν
$CHCl_3$
(40%)
VIII (60%)

Scheme IV. (Reproduced from reference 37.)

Although mechanistic studies of this conversion are still in progress, the reaction must involve (at least formally) ring expansion, carbonylation, and hydrogen abstraction in the original carbyne ligand. Given that the ligand-centered reactions of metal radicals are generally simple processes such as dimerization through the ligand, the multistep reaction pathway necessary to produce cyclopentenone upon photooxidation of **Vf** is remarkable.

Given that the formation of cyclopentenone goes through highly reactive metal radical species, rearrangement of complexes with substituents on the cyclopropyl rings might be expected to be rather unselective. However, photooxidation of either dimethylcyclopropylcarbyne **Vg** or **Vh** results in formation of only *trans*-dimethylcyclopentenone (Scheme V). Photochemical equilibration of **Vg** and **Vh** is followed by rapid conversion of **Vh** to product, providing a means of obtaining only one stereochemistry in the substituted cyclopentenone.

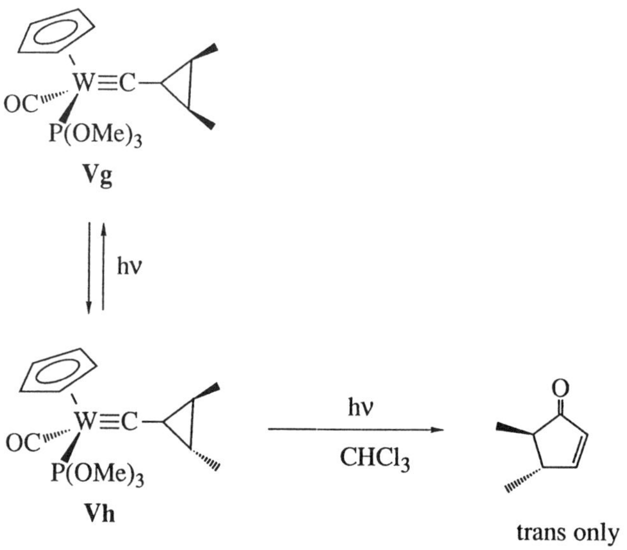

Vg

$h\nu$

Vh $\xrightarrow[\text{CHCl}_3]{h\nu}$ trans only

Scheme V.

In complexes with a single substituent on the cyclopropyl ring, the reaction is regiospecific as well. Photooxidation of **Vi** also results in formation of only one cyclopentenone (eq 1).

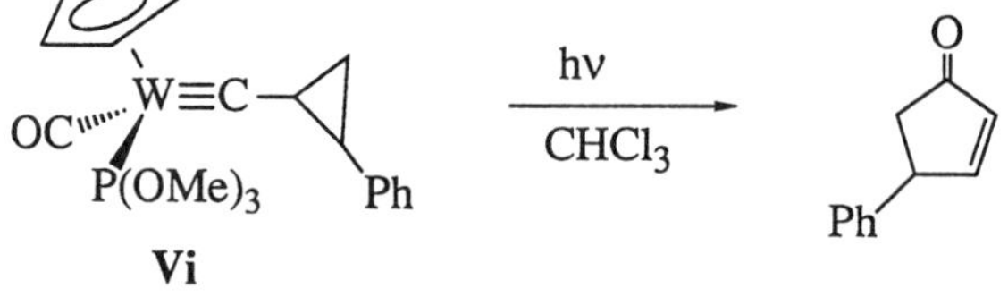

$$\textbf{Vi} \xrightarrow[\text{CHCl}_3]{h\nu} \quad (1)$$

Formation of cyclopentenones from substituted cyclopropylcarbyne complexes appears to be general when aryl or alkyl substituents are placed on C-2 and/or C-3 of the cyclopropyl ring. However, the chemistry of these systems is complex. Scheme VI illustrates a change in reaction manifold for **Vj**, which contains an acyl substituent at C-1 (*38*). Although the expected 3-substituted cyclopentenone is indeed formed, it is in low yield. In competition with the production of cyclopentenone, nucleophilic attack of Cl^- on the cyclopropyl ring results in formation of chelated σ-vinyl complex **IX**, which releases α,β-unsaturated ketone upon further reaction.

The ring opening–carbonylation sequence appears to be restricted to three-membered rings. The cyclobutyl compound **Vj** cleanly produces methylenecyclobutane upon photooxidation (eq 2) presumably via a

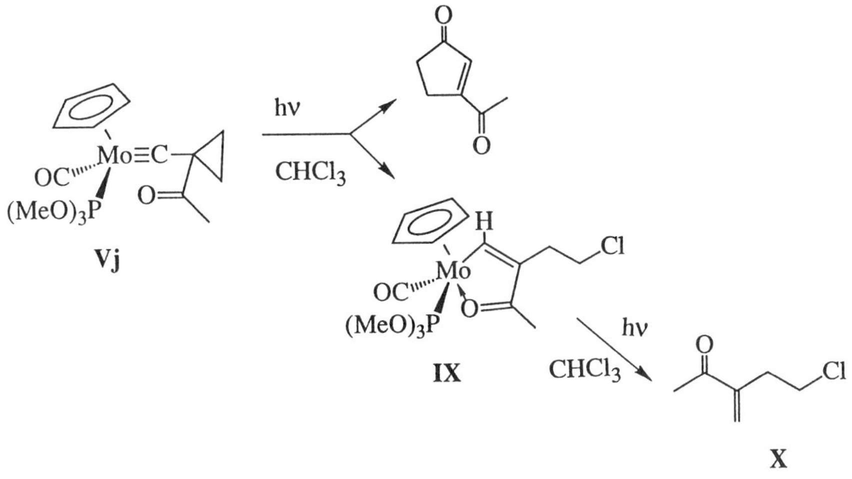

Scheme VI.

mechanism that involves H-shift to form a vinyl complex. Olefins also were obtained from complexes with acyclic substituents (eq 3). However, no matter what the original alkyl substituent on the carbyne complex, photooxidation results in rearrangement of the ligand and release of an organic product.

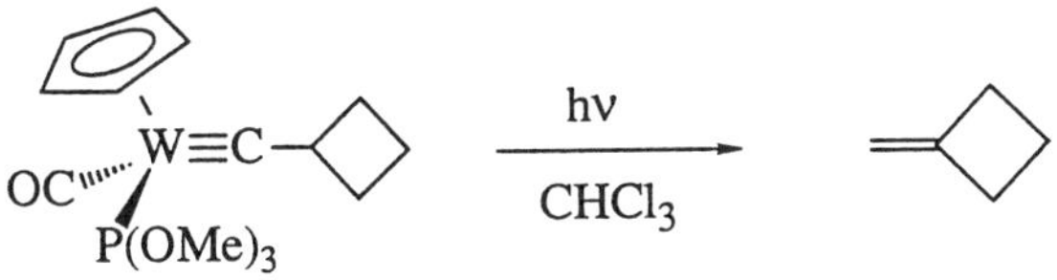

(2)

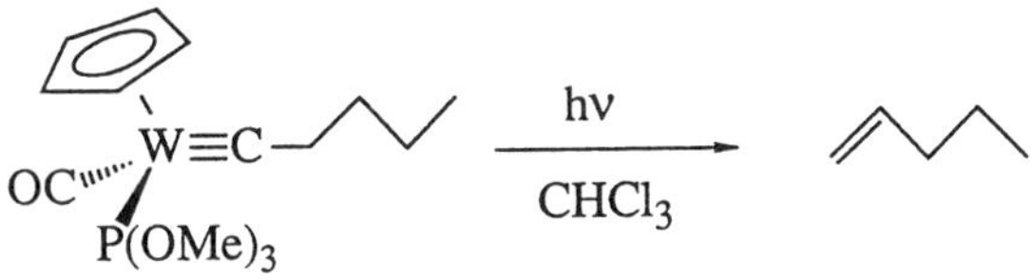

(3)

This system provides an extremely rare example (**Vf**) of a metal radical observed to undergo both the characteristic metal radical reactions (Scheme III) and organic reactions on a ligand (Scheme IV). By changing reaction conditions, the chemistry can be switched entirely from one reaction manifold to the other. These carbyne complexes thus provide a system for studying the types of chemical reactivity that can be induced by photochemical electron transfer from organometallic compounds. In addition, the conversion of more elaborate carbyne ligands to organic products with regiochemical and stereochemical control makes this an interesting model system for exploration of the possible use of organometallic radicals in organic synthesis.

Acknowledgment

We thank the National Science Foundation for support of this work.

References

1. Dötz, K. H. *Angew. Chem. Int. Ed. Engl.* **1984,** *23,* 587–608.
2. Fischer, H.; Hofmann, P.; Kreissl, F. R.; Schrock, R. R.; Schubert, U.; Weiss, K. *Carbyne Complexes;* VCH Publishers: New York, 1988.
3. *Organometallic Radical Processes;* Trogler, W. C., Ed.; Elsevier: Amsterdam, Netherlands, 1990.
4. Astruc, D. *Acc. Chem. Res.* **1991,** *24,* 36–42.
5. Baird, M. *Chem. Rev.* **1988,** *88,* 1217–1227.
6. Giannotti, C.; Gaspard, S.; Kramer, P. In *Photoinduced Electron Transfer, Part D. Photoinduced Electron Transfer Reactions: Inorganic Substrates and Applications;* Fox, M. A.; Chanon, M., Eds.; Elsevier: Amsterdam, Netherlands, 1988; pp 200–240.
7. Buehler, R. E. In *The Chemistry of the Carbon-Halogen Bond*; Patai, S., Ed.; Wiley: London, 1973; pp 795–864.
8. Mann, K. R.; Gray, H. B.; Hammond, G. S. *J. Am. Chem. Soc.* **1977,** *99,* 306–307.
9. Bergamini, P.; DiMartino, S.; Maldotti, A.; Sostero, S.; Traverso, O. *J. Organomet. Chem.* **1989,** *365,* 341–346.
10. Hershberger, J. W.; Klingler, R. J.; Kochi, J. K. *J. Am. Chem. Soc.* **1982,** *104,* 3034–3043.
11. Vogler, A.; Kunkely, H. *Angew. Chem. Intl. Ed. Engl.* **1981,** *20,* 386–387.
12. Wrighton, M. S.; Ginley, D. S. *J. Am. Chem. Soc.* **1975,** *97,* 2065–2072.
13. Tyler, D. R.; Schmidt, M. A.; Gray, H. B. *J. Am. Chem. Soc.* **1983,** *105,* 6016–6021.
14. Sandrini, D.; Maestri, M.; Balzani, V.; Chassot, L.; von Zelewsky, A. *J. Am. Chem. Soc.* **1987,** *109,* 7720–7724.
15. Summers, D. P.; Luong, J. C.; Wrighton, M. S. *J. Am. Chem. Soc.* **1981,** *103,* 5238–5241.
16. Geiger, W. E.; Gennett, T.; Lane, G. A.; Salzer, A.; Rheingold, A. L. *Organometallics,* **1986,** *5,* 1352–1359.
17. Ruiz, J.; Lacoste, M.; Astruc, D. *J. Am. Chem. Soc.* **1990,** *112,* 5471–5483.

18. Iyer, R. S.; Selegue, J. P. *J. Am. Chem. Soc.* **1987,** *109,* 910–911.
19. Beevor, R. G.; Freeman, M. J.; Green, M.; Morton, C. E.; Orpen, A. G. *J. Chem. Soc. Chem. Commun.* **1985,** 68–70.
20. (a) Sugimori, A.; Akiyama, T. *Sci. Papers I. P. C. R. Jpn.* **1984,** *78,* 172–177. (b) Akiyama, T.; Hoshi, Y.; Goto, S.; Sugimori, A. *Bull. Soc. Chem. Jpn.* **1973,** *46,* 1851–1854. (c) Akiyama, T.; Sugimori, A.; Hermann, H. *Bull. Soc. Chem. Jpn.* **1973,** *46,* 1855–1859.
21. Blaha, J. P.; Wrighton, M. S. *J. Am. Chem. Soc.* **1985,** *107,* 2694–2702.
22. Freeman, M. J.; Orpen, A. G.; Connelly, N. G.; Manners, I.; Raven, S. J. *J. Chem. Soc. Dalton Trans.* **1985,** 2283–2289.
23. Zou, C.; Ahmed, K. J.; Wrighton, M. S. *J. Am. Chem. Soc.* **1989,** *111,* 1133–1135.
24. Fonseca, E.; Geiger, W. E.; Bitterwolf, T. E.; Rheingold, A. L. *Organometallics,* **1988,** *7,* 567–568.
25. Seitz, F.; Fischer, H.; Riede, J.; Schottle, T.; Kaim, W. *Angew. Chem. Intl. Ed. Engl.* **1986,** *25,* 744–746.
26. Fischer, E. O.; Ruhs, A.; Kalder, H. J. *Z. Naturforsch., B: Anorg. Chem., Org. Chem.* **1977,** *32b,* 473–475.
27. Baker, P. K.; Barker, G. K.; Gill, D. S.; Green, M.; Orpen, A. G.; Williams, I. D. *J. Chem. Soc. Dalton Trans.* **1989,** 1321–1331.
28. Leep, C. J.; Kingsbury, K. B.; McElwee-White, L. *J. Am. Chem. Soc.* **1988,** *110,* 7535–7536.
29. Carter, J. D.; Kingsbury, K. B.; Wilde, A.; Schoch, T. K.; Leep, C. J.; Pham, E. K.; McElwee-White, L. *J. Am. Chem. Soc.* **1991,** *113,* 2947–2954.
30. Bocarsly, A. B.; Cameron, R. E.; Rubin, H. D.; McDermott, G. A.; Wolff, C. R.; Mayr, A. *Inorg. Chem.* **1985,** *24,* 3976–3978.
31. Bocarsly, A. B.; Cameron, R. E.; Mayr, A.; McDermott, G. A. In *Photochemistry and Photophysics of Coordination Compounds;* Yersin, H.; Vogler, A., Eds.; Springer-Verlag: Berlin, Germany, 1987; pp 213–216.
32. Vogler, A.; Kisslinger, J.; Roper, W. R. *Z. Naturforsch., B: Anorg. Chem., Org. Chem.* **1983,** *38b,* 1506–1509.
33. Sheridan, J. B.; Pourreau, D. B.; Geoffroy, G. L.; Rheingold, A. L. *Organometallics,* **1988,** *7,* 289–294.
34. Bottrill, M.; Green, M.; Orpen, A. G.; Saunders, D. R.; Williams, I. D. *J. Chem. Soc. Dalton Trans.* **1989,** 511–518.
35. Manna, J.; Gilbert, T. M.; Dallinger, R. F.; Geib, S. J.; Hopkins, M. D. *J. Am. Chem. Soc.* **1992,** *114,* 5870–5872.
36. Mortimer, M. D., Ph. D. Thesis, University of Bristol, Bristol, England, 1991.
37. Kingsbury, K. B.; Carter, J. D.; McElwee-White, L. *J. Chem. Soc. Chem. Commun.* **1990,** 624–625.
38. Carter, J. D.; Schoch, T. K.; McElwee-White, L. *Organometallics,* **1992,** *11,* 3571–3578.

RECEIVED for review November 7, 1991. ACCEPTED revised manuscript May 1, 1992.

18

Photocatalysis Induced by Light-Sensitive Coordination Compounds

Horst Hennig, Lutz Weber, and Detlef Rehorek

Department of Chemistry, University of Leipzig, Talstrasse 35, O–7010 Leipzig, Germany

Light-sensitive transition metal complexes and organometallic compounds are attractive precursors for light-induced catalytic processes because electronic excitation may lead to species that exhibit catalytic activity. Examples are coordinatively unsaturated species, complexes with changed formal oxidation numbers, free ligands, and ligand redox products. Photocatalytic reactions in organic synthesis based on the photogeneration of electron-rich metallofragments and photooxidations in the presence of metalloporphyrins and metal acetylacetonates, respectively, are discussed. Electron-rich metallofragments produced by photochemical homolytic metal–ligand bond cleavage of mixed-ligand azido complexes catalyze cyclization reactions of acetylene and alkyne derivatives, whereas metalloporphyrins and metal acetylacetonates are discussed in terms of their photocatalytic reactivity in oxygenation reactions of olefins, particularly of terpenes such as α-pinene and others.

THE PHOTOCHEMISTRY OF ORGANOMETALLIC COMPOUNDS and Werner-type transition metal complexes has been the subject of several books and monographs published during the past few decades (*1–6*). One of the main reasons for the increasing interest in the photochemistry (and also photophysics) of these compounds is the broader diversity of reaction pathways induced by light as compared with organic compounds. Usually, in organic photochemistry the general features of the electronic excitation and the resulting photochemical reactions may be satisfactorily described on the basis of three electronic states to be involved: singlet ground state,

0065–2393/93/0238–0351$07.50/0

first excited singlet state, and the corresponding triplet excited state formed by intersystem crossing. The excited states may lead to photoreactions or undergo fast deactivation back to the singlet ground state.

Transition metal complexes differ from organic compounds with respect to both the number and the spin multiplicity of accessible electronically excited states that then undergo very fast relaxation to thermally equilibrated electronically excited (thexi) states. Thus, depending on the wavelength of irradiation, various electronic states that are excited eventually result in the population of thexi states of different reactivity. In some favorable instances, this property allows tuning of photochemical reactivity and switching between various pathways, such as electron transfer, dissociation–substitution–rearrangement reactions, and ligand-centered reactivity. For example, the greater variety of available electronically excited states may be used for the photogeneration of coordinatively unsaturated species, transition metal compounds with changed formal oxidation numbers, as well as free ligands and ligand redox products. Most of these species are highly reactive and therefore of practical importance with respect to catalytic processes induced by light.

Thus, it is not surprising that photocatalysis based on light-sensitive transition metal complexes and organometallic compounds has attained considerable attention (*7–10*). Photoimaging processes (*11*), wastewater recycling and other environment-protecting processes (*12*), storage and conversion of solar energy (*13*), simulation and modeling of light-sensitive metalloenzymes (*14*), and organic synthesis (*15–19*) have attracted particular consideration in the application of photocatalysis.

This chapter focuses on some of our results obtained in homogeneous photocatalysis and their implications for organic syntheses. Therefore, it is not intended to cover the subject of photocatalysis in organic synthesis altogether. A more general and very comprehensive review on this subject was written by Salomon (*15*), and we have also reviewed this field in photocatalysis very recently (*20*).

Homogeneous Photocatalysis Induced by Transition Metal Complexes and Organometallic Compounds

Recently, Serpone and Pelizzetti (*10*) edited an excellent collection of reviews illustrating the recent state of the art in photocatalysis. This collection provides a comprehensive survey on "what is meant by photocatalysis", and it reflects the specific points of view of many authors.

Figure 1 displays a scheme of the generally accepted main principles regarding the current discussions on homogeneous photocatalysis induced by light-sensitive transition metal compounds. Figure 1 also sums up our

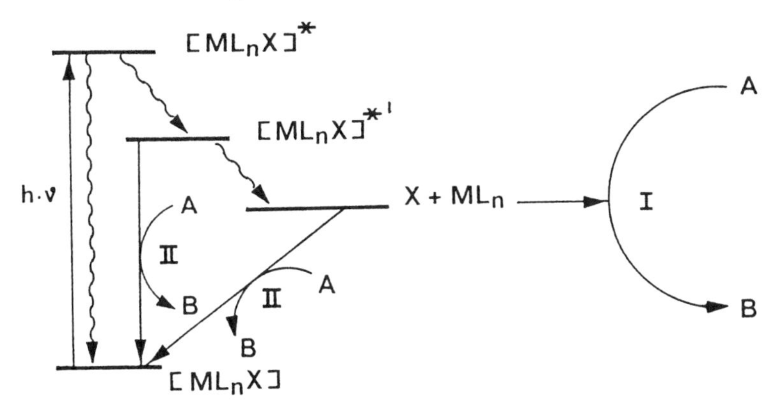

Figure 1. Simplified Jablonski diagram illustrating photoinduced catalytic (I) and photoassisted (II) conversions of a compound A into product B. (Reproduced with permission from reference 20. Copyright 1991.)

earlier contributions to photocatalytic reactions (*21–26*); it is not the aim of this chapter to reflect on fundamental contributions offered by other authors (*9, 10, 27–31*). A critical survey is given in a review by Chanon and Chanon (*18*).

Generally, it seems to be reasonable to distinguish between two limiting cases of reactions that are both light-induced and catalytic with respect to any transformations of a substrate A to a product B and reacting by a pathway not accessible thermally:

Case 1. The number of moles of product B formed photocatalytically exceeds the number of moles of absorbed (or strictly speaking, catalyst-generating) photons, and therefore, the overall product quantum yield is greater than the quantum yield of the primary photoreaction. That result implies, in terms of turnover numbers, that the overall product quantum yield may still be well below 1, even if the turnover number is >1. Quantum yield values of product formation considerably higher than unity may be achieved, and this result is of particular practical importance.

Case 2. The number of moles of product B produced photocatalytically is equal to or less than the number of moles of absorbed (catalyst-generating) photons. The product quantum yield is equal to or even lower than that of the primary photoreaction and cannot exceed unity.

This scheme of classification appears to be advantageous for designing photocatalytic systems. In most cases it might be of interest to search for reactions with quantum yields higher than unity. However, the undoubtedly most important photocatalytic process, that is, the photosynthesis of green plants, is efficient enough despite the fact that its quantum yield is well below unity.

Summarizing the current discussions, there appears to be a general agreement that reactions of Case 1 should be defined as photoinduced catalytic reactions. However, Wrighton et al.'s (*31*) and our (*21*) proposal to define reactions of Case 2 as photoassisted reactions is not generally accepted because of different mechanistic points of view (*18*).

Photoinduced catalytic reactions based on light-sensitive transition metal complexes and organometallic compounds are distinguished by the photochemical generation of a catalyst. Coordinatively unsaturated metal complexes as well as complex compounds with changed formal oxidation number (often associated with a change of kinetic lability or complex geometry) have been identified as photochemically generated catalysts. Free ligands and ligand redox products [for example, free radicals (*32*)] may also serve as catalysts in a subsequent dark reaction, resulting in a catalytic conversion of substrate A to product B. Most of these catalytically active species are well-known as catalysts in thermal homogeneous catalytic reactions.

As illustrated in Figure 1, photons are required only to generate the catalyst. The catalytic conversion of A to B proceeds as a dark reaction, and thus the overall quantum yield values of product formation may exceed unity. Quantum yield values considerably higher than unity are of particular interest for industrial applications when correlating the price of the photons necessary to make a photocatalytic reaction economically feasible. Moreover, both processes, the photochemical generation of a catalyst and the catalytic substrate conversion, may be carried out separately, as in the well-known example of the classic silver halide photographic process. Therefore, this class of reactions is of great interest with respect to unconventional information-recording processes (*11*).

Photoassisted reactions, the other limiting case of photocatalysis, involve catalytic interactions of short-lived intermediates generated under the influence of light. As possible candidates for those short-lived intermediates, both transition metal complexes in electronically excited states and short-lived coordinatively unsaturated species may be regarded. Short-lived coordinatively unsaturated species are preferably derived from organometallic compounds and metal carbonyls. Because of the short lifetime of these species, a catalytic cycle of this kind requires continuous irradiation to be effective with respect to substrate conversion.

To illustrate substantially these limiting cases of photocatalysis, we arbitrarily selected three examples of photoinduced catalytic and photoassisted reactions from our results in dealing with photocatalytic systems.

Figure 2 displays the photoinduced catalytic dimerization of heterocyclic aldehydes to the corresponding 1,2-enediols. Photogenerated cyanide ions catalyze this acyloin condensation (*33*). This photocatalytic system is of some practical consequence with respect to photoimaging processes, because we have shown that it also operates in polymer films. Hence, photocatalytically produced heterocyclic enediols may be used as strong

$$[Mo(CN)_8]^{4-} \underset{}{\overset{h\nu}{\rightleftharpoons}} \cdots \xrightarrow[H_2O]{\Delta T} 4CN^- + [Mo(CN)_4O(OH)]^{3-}$$

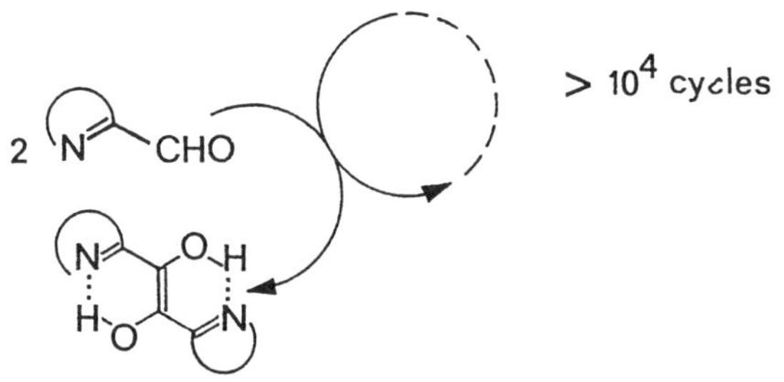

$> 10^4$ cycles

Figure 2. Photoinduced catalytic formation of 1,2-enediols from heterocyclic aldehydes. Here, the catalyst (CN^-) is generated by photosubstitution of $[Mo(CN)_8]^{4-}$.

reductants in physical development processes or may function as indigo-like dyestuffs (*34*), which allows the designing of photoimaging processes.

Figure 3 shows the photoassisted reduction of selected iron(III), cobalt(III), and copper(II) complexes in the presence of alkyl chromates, known as strong oxidants under thermal and usually under photochemical conditions, too. This rather surprising reaction proceeds via photochemical formation of short-lived and strongly reducing chromium(V) intermediates. However, thermal side reactions between chromic acid esters and radicals formed in the course of the photoreaction lead to the termination of the photoassisted cycle (*35–37*).

Finally, Figure 4 illustrates the photooxidation of molybdenum(IV) to molybdenum(V) in the presence of alcohols and the concomitant formation of the corresponding aldehydes and ketones. This photoinduced

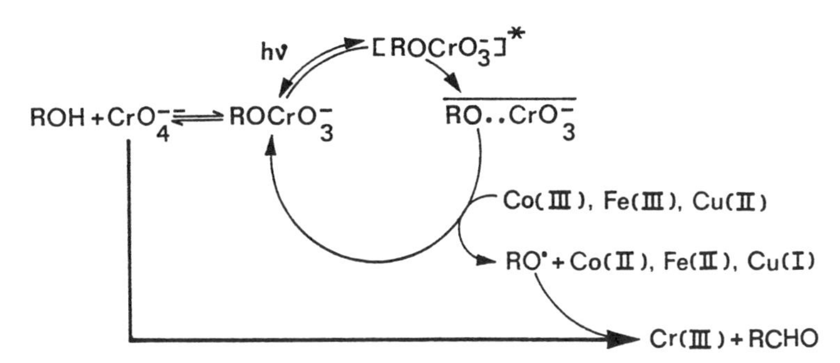

Figure 3. Photoassisted catalytic reduction of selected transition metal complexes in the presence of alkyl chromates.

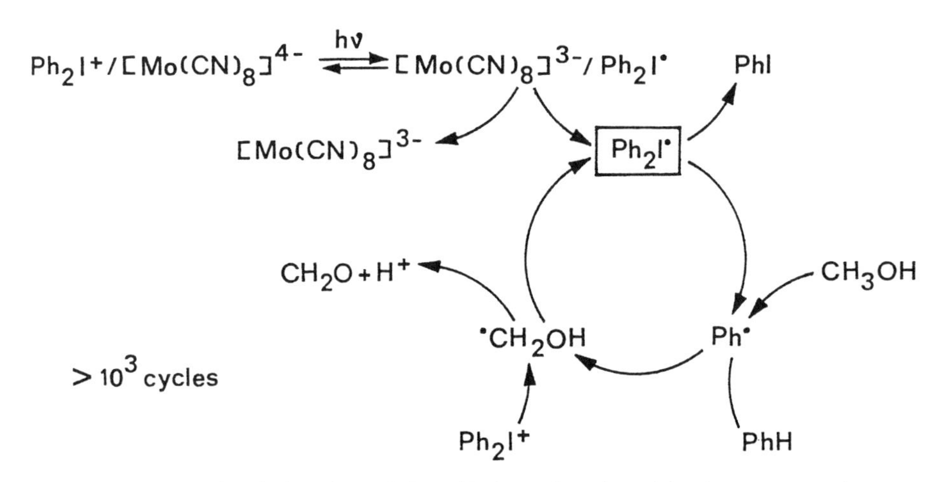

Figure 4. Photoinduced catalytic oxidation of methanol in the presence of diphenyliodonium octacyanomolybdate(IV).

catalytic cycle is based upon very efficient photoredox reactions due to second-sphere effects of ion pairs consisting of diphenyliodonium cations and cyanomolybdate(IV) anions (*38, 39*).

The discussion of the advantages of using photocatalytic routes in organic synthesis will be the main aspect of the following section. Because details of this rapidly developing field of photocatalysis have already been reviewed (*15, 18, 20*), we will focus here on some of our own results obtained in photocatalytic cyclotrimerization reactions of acetylene and alkyne derivatives as well as oxygenation–oxidation reactions of olefins.

Photocatalysis in Cyclization of Alkynes and Oxygenation–Oxidation Reactions of Olefins

Coordinatively unsaturated species, particularly in low formal oxidation states, have attracted considerable attention because they are the actual catalysts in homogeneous complex catalysis. Therefore, the convenient way of light-induced generation of such catalytically reactive species has become increasingly attractive also for organic synthesis. Our attention has been directed toward the photogeneration of electron-rich metallofragments and their use in catalytic cyclotrimerization reactions of acetylene and alkyne derivatives. The other aim of our investigations concerns oxygenation and oxidation reactions of olefins initiated by photochemically produced coordinatively unsaturated species derived from suitable metalloporphyrin derivatives and metal acetylacetonates, respectively.

Cyclotrimerization of Acetylene Induced by Photochemically Generated Electron-Rich Metallofragments. Since the pioneering work of Reppe (*40*), the functionalization of acetylene and alkyne derivatives, particularly by cyclization reactions, has captured considerable interest among chemists. However, most of these cyclization reactions of acetylene and its derivatives are governed by strict limitations with respect to pressure and temperature because of the explosiveness of acetylene, as illustrated recently by Bönnemann (*41*).

It is noteworthy, therefore, that Biagini et al. (*42*) reported on cyclotrimerization reactions of acetylene in the presence of $\{[Co(pyr)_6]^{2+}; BPh_4^-\}$ (pyr is pyridine) ion pairs which, upon photolysis, form low-valent and coordinatively unsaturated cobalt species acting as catalysts needed for the catalytic activation of acetylene.

Moreover, Schulz and co-workers (*43, 44*) showed that photocatalytic cyclotrimerization of acetylene in the presence of organic nitriles yields 2-substituted pyridine derivatives, using bis(olefin)cobalt(I)cyclopentadienyl derivatives as light-sensitive precursors for the generation of the catalyst.

Trogler (*45*) indicated that electron-rich metallofragments may be photogenerated from light-sensitive mixed-ligand oxalato (ox) or dithiooxalato (dto) complexes of platinum(II) of the type $[Pt(ox/dto)(PR_3)_2]$ where PR_3 is a stabilizing phosphine ligand. The resulting platinum(0) complex fragments have proved to be very reactive with respect to various kinds of oxidative addition reactions and other reactions of olefins such as hydrogenation, hydrosilation, and deuterium exchange.

In our studies, light-sensitive mixed-ligand azido complexes of nickel(II), palladium(II), and platinum(II) are employed. The motivation of dealing with these mixed-ligand complexes originates from the possibility of combining photooxidizable ligands (here N_3^-) with additional ligands that are able to stabilize central metal ions in low formal oxidation states. Examples of the stabilizing ligands are π-acceptor ligands like phosphines, diphosphines, and α-diimines. Upon irradiation of these mixed-ligand complexes, central ions like nickel(II), palladium(II), and platinum(II) were expected to be reduced photochemically to lower-valent species, which are coordinatively unsaturated but are stabilized in the presence of additional π-acceptor ligands. The general scheme of the photogeneration of such electron-rich metallofragments is depicted in equation 1.

$$\text{(X X)}M^{II}(N_3)_2 \xrightarrow{h\nu} \text{(X X)}M(0) + 3N_2 \quad (1)$$

$\overset{\frown}{X\ X}$: π-acceptor ligands of phosphine, diphosphine, and α-diimine type, such as triphenylphosphine, bis(diphenyl)phosphinoethane, and 1,1′-bipyridine.

M^{II}: Ni^{2+}, Pd^{2+}, or Pt^{2+}

N_3^-: Besides azido mixed-ligand complexes, oxalato, dithiooxalato, and malonato complexes have also been prepared.

A number of mixed-ligand azido complexes of nickel(II), palladium(II), and platinum(II) have been prepared with both phosphine and diphosphine ligands as well as with aliphatic and aromatic diimine ligands (*45–49*). All of these mixed-ligand complexes exhibit a very photosensitive behavior.

We will now summarize some general aspects of photochemical reactions of mixed-ligand azido metal(II) complexes. Surprisingly, photodecomposition reactions of mixed-ligand complexes of nickel(II), palladium(II), and platinum(II) have attracted only little attention in the recent years.

Reed and co-workers (*50*) demonstrated the intermediate formation of singlet nitrene by ultraviolet irradiation of $[Ni(tet)(N_3)_2]$ (here tet designates a tetradentate *N*-macrocyclic ligand), which is in accordance with previously reported mechanistic studies by Basolo and co-workers (*51*).

Fehlhammer et al. (*52*) reported that irradiation of $[Pd(N_3)_2(PPh_3)_2]$ results in the formation of the dinuclear species $[Pd_2(N_3)_4(PPh_3)_2]$ as a result of the photodissociation of phosphine ligands.

Nelson and co-workers (*53*) demonstrated the photochemically induced reversible *cis–trans* isomerization of palladium(II) mixed-ligand azido complexes with various benzylphosphines as additional ligands.

Beck and Schorpp (*54, 55*) pointed out that platinum(II) mixed-ligand azido complexes are photosensitive and undergo photoredox reactions that, however, were not further characterized.

Bartocci and Scandola (*56*) reported for the first time the photochemical formation of platinum(I) complexes and azidyl radicals upon irradiation into the $N_3 \rightarrow Pt$ charge-transfer region of azido(diethylenetriamine)platinum(II) nitrate.

Detailed investigations by Vogler and co-workers (*57–60*) indicated the formation of platinum(0) species together with azidyl radicals upon irradiation into the charge-transfer region of various mixed-ligand azido platinum complexes.

Finally, Kamath and Srivastava (*61*) recently showed that mixed-ligand complexes of palladium(II) and platinum(II) with azide and α-diimines as ligands behave as very efficient photosensitizers with respect to the generation of singlet oxygen.

Summarizing these results on the photochemistry of mixed-ligand azido complexes of nickel(II), palladium(II), and platinum(II), it appears that detailed mechanistic investigations of this photochemically highly attractive group of mixed-ligand complexes are very scarce.

The following reaction pathways, which partially depend on the wavelength of irradiation, have to be considered when dealing with the photochemistry of these compounds:

1. photosubstitution as well as photodissociation processes
2. photoredox reactions leading to intermediate nitrene derivatives and/or azidyl radicals together with the production of reduced metal centers
3. photochemically induced *cis–trans* isomerization
4. photosensitization reactions leading to singlet oxygen

Photochemistry of Platinum(II), Palladium(II), and Nickel(II) Mixed-Ligand Complexes. This section will focus on some mechanistic details of the photochemistry of these mixed-ligand azido complexes, particularly of palladium(II) and platinum(II) complexes. Additionally, preliminary results of photochemical reactions of the corresponding nickel(II) complexes will be discussed. Finally, some photocatalytic reactions based on these light-sensitive mixed-ligand azido complexes will be described.

Stationary photolysis of $[Pt(N_3)_2(PPh_3)_2]$ at 280 nm under argon results in the appearance of a shoulder in the UV–visible spectrum at 430 nm (Figure 5) and the disappearance of the platinum(II) complex with a quantum yield value of $\Phi = 0.8$ in $CHCl_3$. Finally photolysis under these experimental conditions leads to the formation of $[Pt_2(PPh_3)_4]$ and nitrogen (*46*). Irradiation in the presence of oxygen yields the very efficient formation of the bis(triphenylphosphine)platinum(II) peroxo complex and concomitant generation of nitrogen.

In $CHCl_3$, the same final product was observed when the stationary photolysis of $[Pd(N_3)_2(PPh_3)_2]$ is carried out under oxygen or argon. In both cases there is a very efficient formation ($\Phi_{315} = 0.6$) of $[PdCl_2(PPh_3)_2]$, apparently due to oxidative addition reactions of $CHCl_3$ to metal(0) species formed by charge-transfer excitation of the mixed-ligand complex. The use of more concentrated solutions leads to the pho-

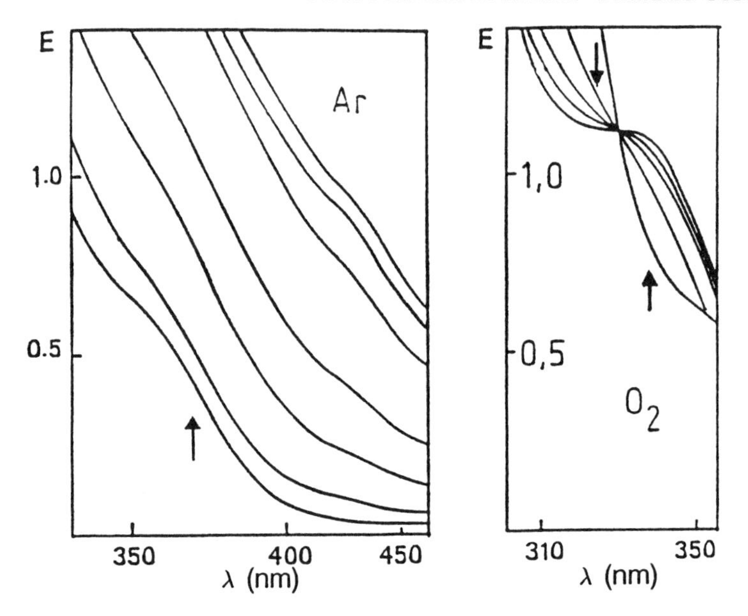

Figure 5. Spectral changes upon 280-nm photolysis of $[Pt(N_3)_2(PPh_3)_2]$ *in* $CHCl_3$ *under argon and dioxygen. Spectra marked with an arrow represent absorption spectra prior to photolysis. Spectral changes are due to the formation of* $[Pt_2(PPh_3)_4]$ *and* $[PtO_2(PPh_3)_2]$. *(Reproduced with permission from reference 20. Copyright 1991.)*

tochemical formation of $[Pd_2(N_3)_4(PPh_3)_2]$, as already reported by Fehlhammer et al. (*52*).

Flash photolysis of mixed-ligand complexes of palladium(II) and platinum(II) under argon in the microsecond time scale has revealed the intermediate formation of a solvent-stabilized metal(0) complex fragment.

The same experiment carried out under oxygen shows that, with $[Pt(N_3)_2(PPh_3)_2]$, a short-lived intermediate has already reacted to a peroxo complex within the microsecond time scale. Because flash photolysis in the microsecond time scale has confirmed that primary products formed immediately after the photophysical deactivation have escaped detection, nanosecond-laser flash experiments were carried out to detect these short-lived intermediates. The difference spectra (Figure 6) exhibit maxima at 380 and 460 nm. The 460-nm transient disappears with a lifetime of 35 to 210 μs, depending on the solvent, following a first-order decay. The band at 460 nm is assigned to the thermally unstable intermediate, $[PtN_3(PPh_3)_2]$.

In attempting to stabilize this platinum(I) intermediate, low-temperature UV–visible and IR spectroscopic investigations at 133 and 183 K were conducted, and IR spectroscopic studies in a poly(vinyl chloride) (PVC) matrix were performed at 10 K. However, the mixed-

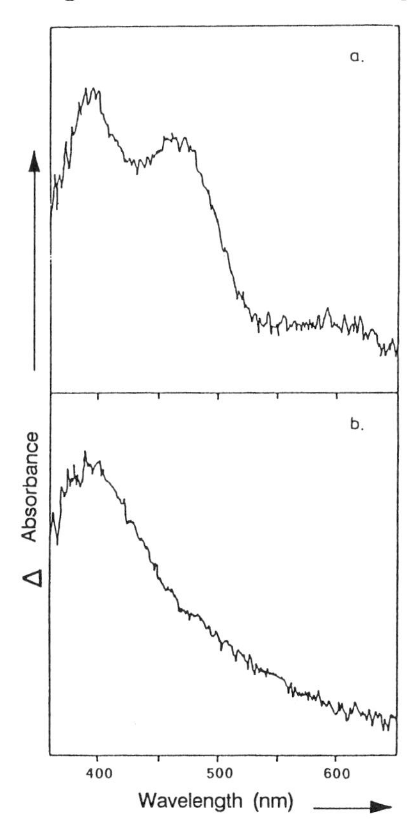

Figure 6. Difference spectra recorded 20 ns (top) and 1 ms (bottom) after laser flash excitation of [Pt(N_3)$_2$(PPh_3)$_2$] in CH_2Cl_2 at room temperature and excitation wavelength of 300 nm. (Reproduced with permission from reference 20. Copyright 1991.)

ligand platinum(I) complex could not be stabilized under these conditions. Instead, the solvent-stabilized intermediate [Pt(PPh_3)$_2$(2-MeTHF)$_2$] (THF is tetrahydrofuran) was observed for the first time by low-temperature UV–visible spectroscopy (*47*).

Low-temperature and PVC-matrix IR spectroscopy point to significant changes of the asymmetric stretching band of coordinated azide, resulting from a photochemically induced *cis–trans* isomerization as a minor pathway of the predominant photoredox reaction of [Pt(N_3)$_2$-(PPh_3)$_2$]. However, metal(I) mixed-ligand complexes like [PdN_3(PPh_3)$_2$] and azidyl radicals have been detected unambiguously by electron paramagnetic resonance (EPR) spin trapping (*62*).

The main results of our research into photochemical reactions of mixed-ligand azido complexes of platinum(II) and palladium(II) can be summarized as follows.

Charge-transfer excitation of these complexes leads to a short-lived metal(I) complex fragment as a result of primary photoredox processes between the metal(II) center and the azide ligand. The primary oxidation product of the azide ligands are azidyl radicals that decompose quickly to yield molecular nitrogen:

$$[M(N_3)_2(PR_3)_2] \overset{h\nu}{\rightleftarrows} [M(N_3)_2(PR_3)_2]^* \quad (2)$$

$$[M(N_3)_2(PR_3)_2] \longrightarrow [M(N_3)(PR_3)_2] + {}^{\bullet}N_3 \quad (3)$$

$${}^{\bullet}N_3 \longrightarrow 1.5N_2 \quad (4)$$

The metal(I) intermediates decay thermally with first-order kinetics, forming coordinatively unsaturated electron-rich metallofragments:

$$[M(N_3)(PR_3)_2] \longrightarrow [M(PR_3)_2] + {}^{\bullet}N_3 \quad (5)$$

The coordinatively unsaturated species can be stabilized by solvents (S) (e.g., S = THF or 2-MeTHF), but slowly dimerize in a dark reaction to form the well-known metal–metal bonded dinuclear complexes:

$$[M(PR_3)_2] + 2S \longrightarrow [M(PR_3)_2S_2] \quad (6)$$

$$2[M(PR_3)_2S_2] \longrightarrow [M_2(PR_3)_4] + 4S \quad (7)$$

Photochemically induced *cis–trans* isomerization (eq 8) is also observed and leads to the corresponding *trans* complexes, which photodecompose to the same final products as already discussed:

$$cis\text{-}[M(N_3)_2(PR_3)_2] \xrightarrow{h\nu} trans\text{-}[M(N_3)_2(PR_3)_2] \longrightarrow \cdots \quad (8)$$

Synproportionation reactions of metal(II) complexes with metal(0) intermediates leading to metal(I) species and disproportionation of metal(I) species to metal(0) and metal(II) complexes have been ruled out by detailed kinetic investigations.

However, some unusual photoreactions were observed in addition to photoredox processes and light-induced *cis–trans* isomerization (unpublished results). Thus, photosubstitution of phosphine ligands occurs upon low-frequency excitation of $[Pd(N_3)_2(PPh_3)_2]$ in the presence of *p*-nitrosodimethylaniline (*p*-NDMA) and other weak donor ligands:

$$[Pd(N_3)_2(PPh_3)_2] + p\text{-NDMA} \xrightarrow{h\nu} [Pd(N_3)_2(PPh_3)(p\text{-NDMA})] + PPh_3 \quad (9)$$

No photoredox reactions have been observed under these conditions, a finding that confirms earlier results by Fehlhammer et al. (*52*).

Preliminary results of our investigations show that the photochemical behavior of mixed-ligand azido complexes of nickel(II) is quite different when compared with the corresponding palladium(II) and platinum(II) complexes.

Photolysis of solutions of $[Ni(N_3)_2(PR_3)_2]$ in hexane under argon leads to the efficient photodissociation of phosphine ligands, whereas in halocarbon solvents the formation of $[NiCl_2(PR_3)_2]$ is observed. Prolonged irradiation in CCl_4 gives the corresponding nickel(III) complexes.

The generation of azidyl radicals may be ruled out, as shown by EPR spin-trapping experiments (Hennig, H.; Stich, R.; Rehorek, D.; Kemp, T. J., unpublished results). Instead, the spin adduct of radicals showing either β-hydrogen or ^{31}P-hyperfine splitting is observed by using nitrosodurene as a spin trap (Figure 7).

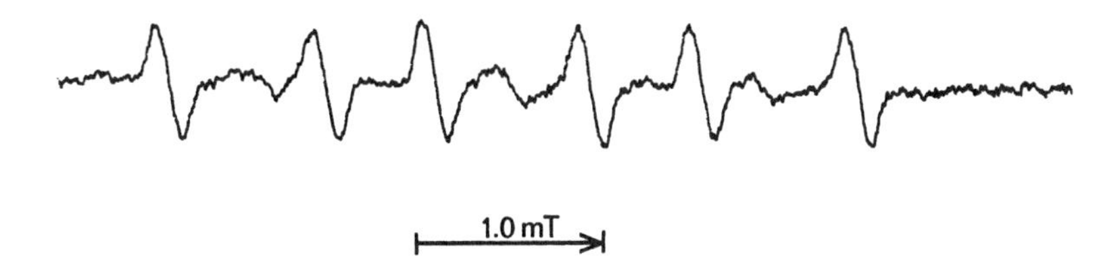

Figure 7. EPR spectrum of the spin adduct of a ligand radical generated by photolysis (350–600 nm of $[Ni(dppe)(N_3)_2]$ in the presence of nitrosodurene as spin trap. The solvent was CH_2Cl_2–CH_3CN (3:1); ^{14}N-hyperfine splitting constant due to the interaction with nitrosodurene (a_N) is 1.416 ± 0.010 mT, and additional hyperfine coupling due to interactions of ^{1}H or ^{31}P nucleus of dppe (a_H or a_P) is 0.805 ± 0.010 mT.

However, EPR signals of a lower-valent nickel complex were not detected, and intermediate nitrene formation has yet to be observed. This rather surprising result seems to indicate a photoinduced electron transfer between the metal center and the phosphine ligand:

$$[Ni(N_3)_2(R_2P\text{–}PR_2)] \xrightarrow{h\nu} \left[\begin{array}{l} R_2P^{+\bullet} \\ \quad | \quad Ni(N_3)_2 \\ \quad / \\ R_2P \end{array} \right] \cdots\!> \text{spin adduct} \quad (10)$$

where $R_2P{-}PR_2$ is bis(diphenylphosphino)ethane.

A decrease of the asymmetric stretching band of azide ligands was observed by IR spectroscopy together with an intermediate appearance of a new band at 2050 cm^{-1} that was assigned to a labile intermediate (Hennig, H.; Knoll, H.; Stufkens, D.J., unpublished results). However, further investigations are needed to elucidate the rather complicated mechanism of the photodecomposition of nickel(II) mixed-ligand azido complexes.

In summary, electron-rich metallofragments of platinum(0) and palladium(0) may be obtained very conveniently from thermally stable and catalytically inactive precursor complexes. Irradiation of the corresponding nickel(II) mixed-ligand azido complexes also leads to the generation of catalytically active species with respect to cyclotrimerization reactions of acetylene, although their constitution has not yet been elucidated so far.

Cyclotrimerization Reactions of Acetylene Induced by Photochemically Generated Electron-Rich Metallofragments. As mentioned previously, metal(II) mixed-ligand azido complexes provide good sources for the photogeneration of catalytically active species. We demonstrated their high reactivity and catalytic activity with respect to various reaction pathways. Thus, hydrosilation and hydrogenation reactions of olefins as well as oxidative addition of halocarbons were observed. Oxidative addition leads to metal(II) mixed-ligand halide complexes and short-lived halocarbon radicals as detected by EPR spin trapping (*63, 64*). However, cyclooligomerization reactions of acetylene and alkyne derivatives in the presence of photochemically generated electron-rich metallofragments are of much greater interest. In particular, photocatalytic cyclotrimerization reactions leading to benzene and pyridine derivatives, in which pyridine derivatives may be formed from organic nitriles or cyanates, appear to be an attractive alternative for the corresponding thermal reactions.

Although thermal cyclotrimerization reactions of acetylene and its derivatives are of considerable importance, the reactions require high pressures and temperatures, and these requirements, in view of the explosiveness of acetylene, limit the industrial application of these homogeneous catalytic reactions (*41*). Therefore, photocatalysis should be of advantage because only ambient temperature and low pressure are required for initiating the catalytic cycle; hence, the danger of explosion is reduced. The nickel(II) mixed-ligand azido complexes are efficient precursor complexes for the photocatalytic trimerization of acetylene to benzene using dimethyl sulfoxide (DMSO) as solvent and a slightly increased temperature of 50 °C. Photocatalytic turnovers greater than 100 have been observed under these mild conditions (*17*). However, a decrease of the catalytic activity (2-MeTHF > THF > DMSO) with increasing donor strength of the solvents was observed. Thus, in DMSO traces of benzene are formed at ambient temperature only by using [$Ni(N_3)_2$(dppp)] (dppp is bis(diphenylphosphino)propane), whereas all other complexes are inactive under these conditions. However, with 2-MeTHF and THF, catalytic turnovers of about 100 were observed at ambient temperature and normal pressure using nickel(II), palladium(II), and platinum(II) complexes as catalyst precursors.

Gas chromatographic–mass spectrometric (GC–MS) analysis of the oligomerization products obtained photocatalytically showed, however, that side reactions occur at ambient temperatures when using THF or 2-MeTHF as a solvent. Thus, isoalkanes and cycloalkanes of chain lengths between 8 and 12 were detected as side products in addition to benzene as the major product. This result may be explained by assuming a reaction of THF and 2-MeTHF as a hydrogen source. The role of THF as a source of hydrogen was recently demonstrated by Gstach and Kisch (*65*). However, the mechanistic details of the photocatalytic cyclotrimerization of acetylene and alkyne derivatives, as well as the influence of the solvent on the formation of side products, has still to be elucidated.

Finally, preliminary results seem to confirm the photocatalytic synthesis of benzene derivatives by using appropriate alkyne derivatives. Furthermore, the photocatalytic synthesis of 2-substituted pyridine derivatives by reaction of acetylene in the presence of nitriles and organic cyanates also appears to be promising.

Photocatalytic Oxygenation–Oxidation Reactions of Olefins Induced by Light-Sensitive Metal Acetylacetonates and Metalloporphyrins

The selective oxygenation of hydrocarbons with molecular oxygen is an attractive goal for the application of transition metal catalysts and has captured much attention over the past decades. Unlike thermal oxygenation reactions, however, the photocatalytic activation of metal complexes in oxygenation reactions has been studied with less intensity. In most cases, the mechanism of transition-metal-catalyzed oxygenation reactions involves the complexation of oxygen. This step is necessary to overcome the energy barrier in the spin-forbidden reaction of triplet oxygen with the singlet spin-state hydrocarbons. The ability of metal complexes to coordinate molecular oxygen is therefore an essential requirement to be met by oxygenation catalysts.

The generation of low-valent coordinatively unsaturated metal species capable of complexing dioxygen by irradiation of transition metal complexes may provide a basis for a general concept of photocatalytic oxygen activation. In this section, we describe our recent results obtained by using transition metal acetylacetonates or porphyrinates in photocatalytic oxygenation reactions with molecular oxygen.

Before beginning further discussions of the observed product selectivities and reaction mechanisms involved we first summarize the various classes in transition-metal-catalyzed oxygenation mechanisms. For this purpose, we chose α-pinene (structure **1**) as substrate for which the thermal oxygenation reactions have been thoroughly studied and which is known to give different products depending on the oxygenation mechanism.

OOH OH O

1 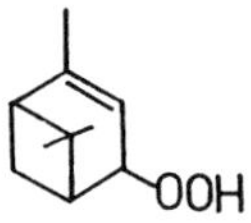 **2** 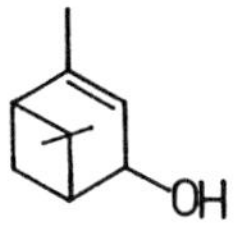 **3** 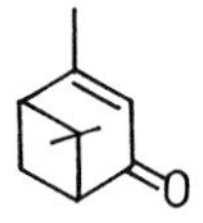 **4**

OOH OH O

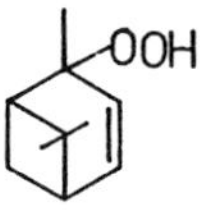

5 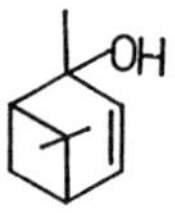 **6**  **7**

Class 1. A radical-chain oxygenation mechanism may be induced by abstraction of a hydrogen atom from the substrate and its diffusion from the reaction center. The alkyl radicals thus formed then react with dissolved oxygen to give peroxyl radicals. The alkyl peroxyl radicals may also abstract hydrogen from the substrate and thus promote the autoxidation. As the result of this mechanism, an unselective product mixture is expected.

Class 2. Abstraction of hydrogen from the substrate by the metal–oxygen species and fast recombination of a hydroxyl radical within an alkyl radical–metal complex cage, which is referred to as the oxygen-rebound mechanism, leads to selective formation of hydroxylated products, for example, *trans*-verbenol (**3**), verbenone (**4**), and *trans*-pin-3-en-2-ol (**6**) when α-pinene is the substrate.

Class 3. With unsaturated hydrocarbons, direct oxygen-atom-transfer processes with high-valent oxometal species yield epoxides.

As can be seen from Table I, generally a large variety of products is formed by photocatalytic oxygenation of α-pinene (**1**) in the presence of metal acetylacetonates (*66*). Apparently, more than one of the mechanisms just described is involved in the overall reaction. The reaction of α-pinene with *tert*-butyl hydroperoxide in the absence of metal complexes gives a product distribution characterized by relatively high yields of

Table I. Photocatalytic Oxygenation of α-Pinene with Transition Metal Acetylacetonato Complexes

		Product (mol%)					
Catalyst[a]	*Turnover*[b]	**2**	**3**	**4**	**5**	**6**	**7**
$TiO(acac)_2$	13	–	27	–	–	10	63
$Ti(acac)_2Cl_2$	26	–	36	–	–	19	45
$VO(acac)_2$	0						
$Cr(acac)_3$	92	8	16	27	8	9	32
$Cr(acac)_3$[c]	158	–	12	47	–	14	27
$Cr(dbm)_3$	66	15	14	22	15	–	34
$Cr(bzac)_3$	106	16	15	30	8	9	22
$Mn(acac)_2$	73	–	20	22	–	12	46
$Mn(acac)_3$	66	7	26	18	10	5	34
$Fe(acac)_2(py)_2$	53	24	25	10	12	–	29
$Fe(acac)_3$	20	32	–	8	16	–	44
$Co(acac)_2$	79	–	22	34	–	10	34
$Co(acac)_2$[c]	231	–	25	29	–	10	36
$Co(acac)_3$	79	–	32	36	–	–	32
$Co(acac)_3$[c]	238	–	25	33	–	7	35
$Co(dbm)_3$	125	–	21	32	–	8	39
$Ni(acac)_2$	13	–	32	–	31	–	37
$Cu(acac)_2$	53	13	15	12	23	13	24
$MoO_2(acac)_2$	0						
$Ru(acac)_3$[d]	7	–	22	11	–	–	44
$Rh(acac)_3$	40	13	20	19	–	13	35
$Pd(acac)_2$	0						
Hacac	0						
t-BuOOH[e]	46	16	18	16	16	9	23

NOTE: 15 mL of α-pinene (0.01 mol) in 15 mL of dry toluene was irradiated in the presence of 0.005 mol of the acetylacetonato complex in a thermostatically controlled (40 °C) photoreactor with a 55-W tungsten halogen lamp. Over the reaction period of 450 min, a stream (2 L/h) of dry air was bubbled through the solution.

[a]Ligands are as follows: acac, acetylacetonate; dbm, dibenzoylmethanate; and bzac, benzoylacetonate.

[b]Turnover numbers are moles of product formed per mole of catalyst.

[c]The reaction time was 24 h.

[d]22 mol% of *trans*-pinocarveol hydroperoxide was also formed.

[e]19 mol% of of *trans*-pinocarveol hydroperoxide and 6 mol% of myrtenol hydroperoxide were also formed.

SOURCE: Data are taken from reference 69.

hydroperoxides **2** and **5** typical for radical-chain autoxidation reactions (*67–69*). The reaction in the presence of metal complexes leads to different products, despite the fact that a radical mechanism is operating in most cases. Metal-catalyzed secondary reactions, for example, decomposition of hydroperoxides, seem to account for the discrepancy.

According to the product spectrum, the metals can be divided into three classes:

1. No photocatalysis was observed for the acetylacetonates of vanadium, molybdenum, and palladium.
2. A broad spectrum of products including hydroperoxides is produced in the presence of Cr, Mn(III), Fe(III), Ru, and Rh catalysts.
3. Ti, Co, Mn(II), and Ni complexes lead selectively to the thermodynamically most-stable allylic oxidation products in addition to large amounts of epoxide **7**. High yields of **7** were also observed for Fe(II) and Ru catalysts.

The third group of photocatalysts is of particular interest because they catalyze the formation of products that can easily be converted to *cis*-verbenol, which is known to be an attractant pheromone of the bark beetle (*Ips typographicus*), one of the most destructive insect pests in the forests of central Europe.

Considerably higher turnover numbers with respect to oxygenated derivatives of α-pinene and, in some cases, good selectivities were found for photocatalytic reactions in the presence of metalloporphyrins (*17, 70*) (Table II).

Table II. Products Formed During the Photocatalytic Oxygenation of α-Pinene in the Presence of Metalloporphyrins

		Products (%)				
Catalyst	*Turnover*[a]	**3**	**4**	**6**	**7**	*Other*
Mn(tpp)Cl	250	14	–	86	–	–
Cr(tpp)OH·$2H_2O$	1424	19	6	33	18	24
Fe(tpp)Cl	480	16	30	25	29	–
$[Nb(tpp)]_2O_3$	524	12	3	48	30	7
$MoO(tpp)OCH_3$	72	23	–	30	47	–

NOTE: 0.1 mol of α-pinene and 10^{-5} mol of complex were dissolved in 15 mL of dry benzene and irradiated for 8 h with a 500-W mercury lamp.

[a]Turnover numbers are moles of product formed per mole of catalyst.

SOURCE: Data are taken from references 17 and 71.

The results compiled in Table II deserve some comments. First, no photocatalysis was observed with [Co(tpp)Cl] (where tpp is tetraphenylporphyrin; not listed in the table) as a catalyst precursor. Second, for the corresponding tungsten complex (also not listed), demetallation occurs, and typical products known from attack of singlet oxygen (here generated by sensitization by metal-free porphyrin) at α-pinene (*67*) were isolated. Finally, the high selectivity found for the corresponding manganese complex leading exclusively to allylic alcohols *trans*-verbenol and *trans*-pin-3-en-2-ol (which may be converted very easily to the pheromone **3**) may be explained by the mechanism illustrated by equations 11–15 (*71*).

$$[\mathrm{Mn(TPP)Cl}] \xrightarrow{h\nu} [\mathrm{Mn^{II}(TPP)}] + \mathrm{Cl^{\bullet}} \quad (11)$$

$$2[\mathrm{Mn^{II}(TPP)}] + \mathrm{O_2} \longrightarrow [\mathrm{Mn^{III}(TPP){-}O{-}O{-}Mn^{III}(TPP)}] \quad (12)$$

$$[\mathrm{Mn^{III}(TPP){-}O{-}O{-}Mn^{III}(TPP)}] \longrightarrow 2[\mathrm{O{=}Mn^{IV}(TPP)}] \quad (13)$$

$$\mathrm{RH} + [\mathrm{O{=}Mn^{IV}(TPP)}] \longrightarrow [\mathrm{R^{\bullet}\,HO{-}Mn^{III}(TPP)}] \quad (14)$$

$$[\mathrm{R^{\bullet}\,HO{-}Mn^{III}(TPP)}] \longrightarrow \mathrm{ROH} + [\mathrm{Mn^{II}(TPP)}] \quad (15)$$

Selective hydrogen abstraction from α-pinene (eq 15) to produce a manganese(III) complex together with a pinenyl radical appears to be the key step in this mechanism. Fast recombination of this radical with a hydroxyl radical bound to Mn(III) within the cage then yields the allylic alcohol and manganese(II), which may reenter the catalytic cycle. Slower rates of recombination and diffusional escape of the radical from the metalloporphyrin result in reactions with dissolved oxygen and, hence, product distribution similar to the free radical autoxidation process.

Both free radical and in-cage oxygen-transfer mechanisms seem to be responsible for the formation of allylic oxidation products and pinene oxide (**7**) in the presence of iron porphyrins. Regardless of the kind of initial iron porphyrin, it is converted into the thermally inert μ-oxobis-[(tetraphenylporphyrinato)iron(III)] upon photolysis in the presence of dioxygen. This complex undergoes photoinduced disproportionation to give iron(II) and oxoiron(IV) porphyrins upon visible-light irradiation (*72–74*):

$$[\mathrm{Fe^{III}(tpp){-}O{-}Fe^{III}(tpp)}] \xrightarrow{h\nu} [\mathrm{O{=}Fe^{IV}(tpp)}] + [\mathrm{Fe^{II}(tpp)}] \quad (16)$$

Unlike the well-characterized oxoiron(V) porphyrin complex, the oxoiron(IV) species was reported to induce only radical-chain autoxidation reactions in the presence of hydrocarbons, leading to alcohols in the case of saturated hydrocarbons and to a mixture of allylic oxygenation products and epoxides in the case of alkenes by a mechanism similar to the one discussed for manganese porphyrins (*73–76*)

$$RH + [O{=}Fe^{IV}(tpp)] \longrightarrow [R\cdot HO{-}Fe^{III}(tpp)] \quad (17)$$

$$[R\cdot HO{-}Fe^{III}(tpp)] \longrightarrow ROH + [Fe^{III}(tpp)] \quad (18)$$

However, in the photocatalytic oxygenation of strained alicyclic alkenes, epoxides, allylic oxygenation products, or both have been found, depending on the structure of the alkene (*77*). In addition to in-cage reactions (17 and 18), direct oxygen transfer to the carbon–carbon double bond as well as diffusional escape of the radical $R^{\bullet}$ and its possible reaction with dioxygen to form hydroperoxides should also be considered. Hydroperoxides may then act as a source of oxygen in a Fe^{III}(tpp)-catalyzed thermal reaction (Scheme I).

Sterically strained olefins gave higher yields of epoxides; for example, with cyclooctene derivatives only epoxides are formed. Turnover numbers up to 2400 were estimated. Similar reactions were observed for the photocatalytic oxygenation of hydrocarbons with dioxygen using a manganese porphyrin catalyst (*71*).

More recently, we showed the enantioselective photocatalytic oxygenation of α-pinene enantiomers using a 2,6-permethylated cyclodextrin-linked iron porphyrin, **8** (*78*).

Matsuda and co-workers (*79, 80*) reported the selective photocatalytic formation of epoxides of cyclohexene, 1-hexene, 2-hexene, and 2,3-dimethyl-2-butene with dioxygen in the presence of molybdenum und niobium porphyrin catalysts. In contrast to these results, we obtained the corresponding allylic alcohols and ketones (together with the epoxides) as a result of a free radical chain process (*70*). Intermediate radicals as well as $[O{=}M{-}O{-}O^{\bullet}(tpp)]$ species were identified by EPR and spin-trapping experiments (*70, 81*).

Concluding Remarks

Coordination compounds play an important role in the catalytic conversion of both organic and inorganic substrates. The activation of transition metal complexes by visible and ultraviolet light provides definite advantages when compared with the usual thermal activation.

The very convenient photochemical generation of catalytically reactive species should be stressed in particular. Compared with thermal reac-

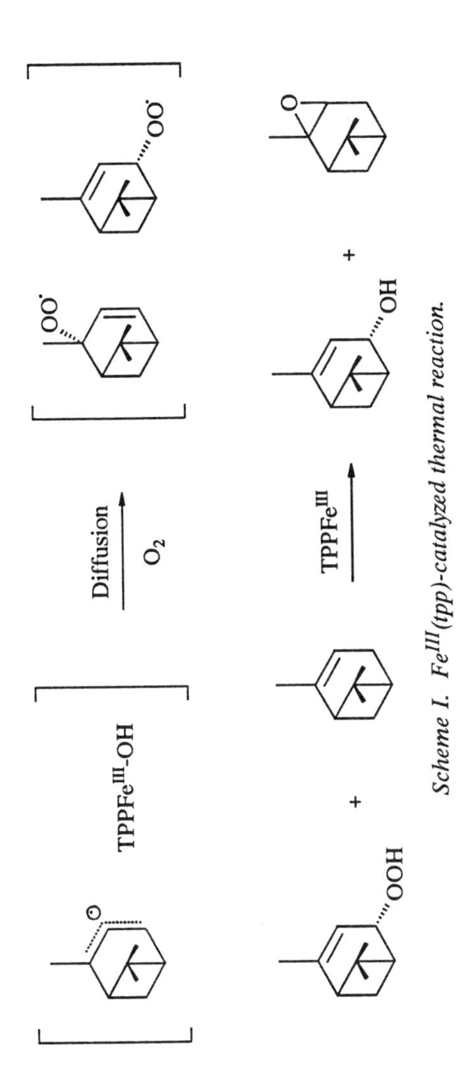

Scheme I. Fe^{III}(tpp)-catalyzed thermal reaction.

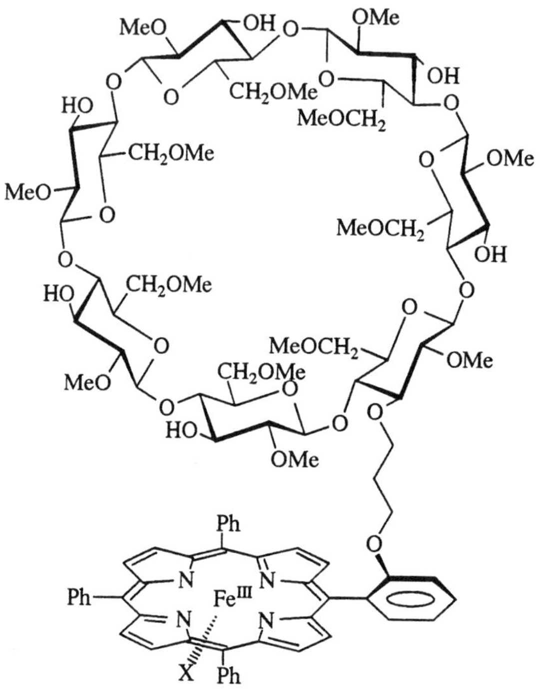

8

tion pathways, photochemically generated catalysts can be identified very conveniently with respect to their constitution, which contributes to a better understanding of mechanistic details of catalytic reactions. Furthermore, the excitation of different electronically excited states, depending on the wavelength of irradiation, provides a tuning between different reaction pathways. A further function of tuning by light is the ability to design photoinduced catalytic or photoassisted reactions, depending on the application of the photocatalysis. Photocatalytic reactions may be carried out under very mild conditions (temperature and pressure), which can be very advantageous when considering systems that are heat sensitive. Photocatalysis accomplished at low temperatures often allows the isolation and identification of reactive intermediates and thus provides further insight into the mechanisms of various reaction pathways.

However, fast back-electron-transfer processes and recombination reactions may considerably reduce the efficiency of photocatalytic reactions, and photochemical decomposition of the catalyst may lead to fast termination of photocatalytic cycles. Finally, spectral sensitization, which is essential with respect to using a broad range of photonic energy (for example, solar energy), has to be considered when practical aspects are concerned. These disadvantages should also be taken into consideration when dealing with photocatalysis.

References

1. Balzani, V.; Carassiti, V. *Photochemistry of Coordination Compounds;* Academic: Orlando, FL, 1977.
2. Geoffroy, L. L.; Wrighton, M. S. *Organometallic Photochemistry;* Academic: Orlando, FL, 1979.
3. Adamson, A. W.; Fleischauer, P. D. *Concepts of Inorganic Photochemistry;* Wiley: New York, 1975.
4. Ferraudi, G. J. *Elements of Inorganic Photochemistry;* Wiley: New York, 1987.
5. Sykora, J.; Sima, J. *Coord. Chem. Rev.* **1990,** *107,* 1.
6. Balzani, V.; Scandola, F. *Supramolecular Photochemistry;* Ellis Horwood: New York, 1991.
7. Hennig, H.; Rehorek, D. *Photochemische und Photokatalytische Reaktionen von Koordinationsverbindungen;* Teubner: Stuttgart, Germany, 1988.
8. *Energy Resources through Photochemistry and Catalysis;* Grätzel, M., Ed.; Academic: Orlando, FL, 1983.
9. *Homogeneous and Heterogeneous Photocatalysis;* Pelizzetti, E.; Serpone, N., Eds.; Reidel: Dordrecht, Netherlands, 1986.
10. *Photocatalysis;* Serpone, N.; Pelizzetti, E., Eds.; Wiley: New York, 1989.
11. Böttcher, H.; Epperlein, J. *Moderne Photographische Systeme;* 2nd ed.; Verlag Grundstoffind: Leipzig, Germany, 1988.
12. Ollis, D. F.; Pelizzetti, E.; Serpone, N. In *Photocatalysis;* Serpone, N.; Pelizzetti, E., Eds.; Wiley: New York, 1989; p 603.
13. Parmon, V. N.; Zamarajev, K. I. In *Photocatalysis;* Serpone, N.; Pelizzetti, E., Eds.; Wiley: New York, 1989; p 565.
14. Weber, L.; Haufe, G.; Rehorek, D.; Hennig, H. *J. Chem. Soc. Chem. Commun.* **1991,** 502.
15. Salomon, R. G. *Tetrahedron* **1983,** *39,* 485.
16. Kutal, C. *Coord. Chem. Rev.* **1985,** *64,* 191.
17. Hennig, H.; Rehorek, D.; Stich, R.; Weber, L. *Pure Appl. Chem.* **1990,** *62,* 1489.
18. Chanon, M.; Chanon, F. *Photocatalysis;* Serpone, N.; Pelizzetti, E., Eds.; Wiley: New York, 1989; p 489.
19. Hennig, H.; Weber, L.; Stich, R.; Grosche, M.; Rehorek, D. *Progress in Photochemistry and Photophysics;* Rabek, G. F., Ed.; CRC Press: Bota Raton, FL, 1992, Vol. VI, p 167.
20. Hennig, H.; Knoll, H.; Stich, R.; Rehorek, D.; Stufkens, D. J. *Coord. Chem. Rev.* **1991,** *111,* 131.
21. Hennig, H.; Thomas, P.; Wagener, R.; Rehorek, D.; Jurdeczka, K. *Z. Chem.* **1977,** *17,* 241.
22. Hennig, H.; Rehorek, D.; Archer, R. D. *Coord. Chem. Rev.* **1985,** *61,* 1.
23. Hennig, H.; Billing, R.; Rehorek, D. *J. Inf. Rec. Mater.* **1987,** *15,* 423.
24. Hennig, H.; Rehorek, D. In *Excited States and Reactive Intermediates;* Lever, A. B. P., Ed.; ACS Symposium Series 307; American Chemical Society: Washington, DC, 1985; p 104.
25. Kisch, H.; Hennig, H. *EPA Newsletter* **1983,** *19,* 23.
26. Hennig, H.; Rehorek, D.; Billing, R. In *Coordination Chemistry and Catalysis;* Ziolkowski, J. J., Ed.; World Scientific: Singapore, 1988; p 421.
27. Carassiti, V. *EPA Newsletter* **1984,** *21,* 12.

28. Mirbach, F. *EPA Newsletter* **1984,** *20,* 16.
29. Kisch, H. In *Photocatalysis;* Serpone, N.; Pelizzetti, E., Eds.; Wiley: New York, 1989; p 1.
30. Moggi, L.; Juris, A.; Sandrini, D.; Manfrin, M. F. *Rev. Chem. Intermed.* **1981,** *4,* 171.
31. Wrighton, M. S.; Ginley, D. S.; Schroeder, M. A.; Morse, D. L. *Pure Appl. Chem.* **1975,** *41,* 671.
32. Rehorek, D. *Chem. Soc. Rev.* **1991,** *20,* 341.
33. Weissenfels, M.; Punkt, J. *Tetrahedron* **1978,** *34,* 311.
34. Hennig, H.; Hoyer, E.; Lippmann, E.; Nagorsnik, E.; Thomas, P.; Weissenfels, M. *J. Inf. Rec. Mater.* **1978,** *6,* 39.
35. Hennig, H.; Scheibler, P.; Wagener, R.; Thomas, P.; Rehorek, D. *J. Prakt. Chem.* **1982,** *324,* 279.
36. Hennig, H.; Scheibler, P.; Wagener, R.; Rehorek, D. *J. Inf. Rec. Mater.* **1980,** *88,* 383.
37. Hennig, H.; Scheibler, P.; Wagener, R.; Rehorek, D. *Inorg. Chim. Acta,* **1980,** *44,* L231.
38. Hennig, H.; Salvetter, J.; Rehorek, D.; Billing, R. *Z. Chem.* **1986,** *26,* 137.
39. Billing, R.; Rehorek, D.; Salvetter, J.; Hennig, H. *Z. Anorg. Allg. Chem.* **1988,** *557,* 234.
40. Reppe, W. *Chemie und Technik der Acetylen-Druck-Reaktionen;* Verlag Chemie: Weinheim, Germany, 1952.
41. Bönnemann, H. *Angew. Chem.* **1985,** *97,* 264.
42. Biagini, P.; Funaidi, F.; Juris, A.; Fachinetti, G. *J. Organomet. Chem.* **1990,** *390,* C61.
43. Schulz, W.; Pracejus, H.; Oehme, G. *Tetrahedron Lett.* **1989,** *30,* 1229.
44. Rosenthal, U.; Schulz, W. *J. Organomet. Chem.* **1987,** *321,* 103.
45. Trogler, W. C. In *Excited States and Reactive Intermediates;* Lever, A. B. P., Ed.; ACS Symposium Series 307; American Chemical Society: Washington, DC, 1985; p 177.
46. Hennig, H.; Stich, R.; Knoll, H.; Rehorek, D. *Z. Anorg. Allg. Chem.* **1989,** *567,* 139.
47. Knoll, H.; Stich, R.; Hennig, H.; Stufkens, D. *J. Inorg. Chim. Acta* **1990,** *178,* 71.
48. Stich, R., Ph.D. Thesis, University of Leipzig, Leipzig, Germany, 1988.
49. Sachsinger, N., Diploma Thesis, University of Leipzig, Leipzig, Germany, 1991.
50. Ngai, R.; Wang, Y.-H. L.; Reed, J. L. *Inorg. Chem.* **1985,** *24,* 3803.
51. Reed, J. L.; Wang, Y.-H. L.; Basolo, F. *J. Am. Chem. Soc.* **1972,** *94,* 7173.
52. Fehlhammer, W. P.; Beck, W.; Pollmann, P. *Chem. Ber.* **1969,** *102,* 3903.
53. Verstuyft, A. W.; Redfield, D. A.; Cary, L. W.; Nelson, J. N. *Inorg. Chem.* **1977,** *16,* 2776.
54. Kreutzer, P. H.; Schorpp, K. T.; Beck, W. *Z. Naturforsch. B* **1975,** *30,* 544.
55. Beck, W.; Schorpp, K. *Chem. Ber.* **1975,** *108,* 3317.
56. Bartocci, C.; Scandola, F. *J. Chem. Soc. Chem. Commun.* **1970,** 531.
57. Vogler, A.; Kern, A.; Fusseder, A.; Huttermann, J. *Z. Naturforsch. B* **1978,** *33,* 1352.
58. Vogler, A.; Wright, R. E.; Kunkely, H. *Angew. Chem.* **1980,** *92,* 745.

59. Vogler, A.; Hlavatsch, J. *Angew. Chem.* **1983,** *95,* 153.
60. Vogler, A.; Quett, C.; Kunkely, H. *Ber. Bunsen-Ges. Phys. Chem.* **1988,** *92,* 1486.
61. Kamath, S. S.; Srivastava, T. S. *J. Photochem. Photobiol. A* **1990,** *52,* 83.
62. Hennig, H.; Stich, R.; Rehorek, D.; Thomas, P.; Kemp, T. J. *Inorg. Chim. Acta* **1988,** *143,* L7.
63. Rehorek, D.; Hennig, H. *Can. J. Chem.* **1982,** *60,* 1565.
64. Rehorek, D.; Hennig, H.; DuBose, C. M.; Kemp, T. J.; Janzen, E. G. *Free Radical Res. Commun.* **1990,** *10,* 75.
65. Gstach, H.; Kisch, H. *Z. Naturforsch. B* **1983,** *38,* 251.
66. Stich, R.; Weber, L.; Rehorek, D.; Hennig, H. *Z. Anorg. Allg. Chem.* **1991,** *600,* 211.
67. Schenck, G. O.; Eggert, H.; Denk, W. *Liebigs Ann. Chem.* **1953,** *584,* 177.
68. Pritzkow, W.; Van Trien, V.; Schmidt-Renner, W. *Miltitzer Ber.* **1982,** 17.
69. de Carvalho, M.-E.; Meunier, B. *Tetrahedron Lett.* **1983,** *24,* 3621.
70. Weber, L.; Haufe, G.; Rehorek, D.; Hennig, H. *J. Mol. Catal.* **1990,** *60,* 267.
71. Weber, L.; Behling, J.; Haufe, G.; Hennig, H. *J. Prakt. Chem.* **1992,** *334,* 265.
72. Peterson, M. W.; Rivers, D. S.; Richman, R. M. *J. Am. Chem. Soc.* **1985,** *107,* 2907.
73. Hendrickson, D. N.; Kinnaird, M. G.; Suslick, K. S. *J. Am. Chem. Soc.* **1987,** *109,* 1243.
74. Berthold, T.; Rehorek, D.; Hennig, H. *Z. Chem.* **1986,** *26,* 183.
75. Suslick, K. S.; Acholla, F. V.; Cook, B. R.; Kinnaird, M. G. *Rec. Trav. Chim. Pays-Bas.* **1987,** *106,* 329.
76. Shelnutt, J. A.; Trudell, D. E. *Tetrahedron Lett.* **1989,** *30,* 5231.
77. Weber, L.; Haufe, G.; Rehorek, D.; Hennig, H. *J. Chem. Soc. Chem. Commun.* **1991,** 502.
78. Weber, L.; Imiolzcyk, I.; Haufe, G.; Rehorek, D.; Hennig, H. *J. Chem. Soc. Chem. Commun.* **1992,** 301.
79. Matsuda, Y.; Sakamoto, S.; Koshima, H.; Murakami, Y. *J. Am. Chem. Soc.* **1985,** *107,* 6415.
80. Matsuda, Y.; Koshima, H.; Nakamura, K.; Murakami, Y. *Chem. Lett.* **1988,** 625.
81. Rehorek, D.; Berthold, T.; Behling, J.; Weber, L.; Hennig, H.; Kemp, T. J. *Z. Chem.* **1989,** *29,* 331.

RECEIVED for review November 7, 1991. ACCEPTED revised manuscript May 21, 1992.

19

Photoredox Reactivity of Copper Complexes and Photooxidation of Organic Substrates

Ján Sýkora[1], Eva Brandšteterová[2], and Adriana Jabconová[1]

[1]Department of Inorganic Chemistry and [2]Department of Analytical Chemistry, Slovak Technical University, Radlinského 9, 81237, Bratislava, Czechoslovakia

Following a brief overview of the thermal and photochemical properties of copper complexes in condensed media, the possibilities for using copper coordination compounds for the photooxidation of organic substrates are discussed. As part of the classification of such photooxidation reaction types, the Cu(I)–Cu(II) photoredox cycle that we developed is treated in detail. The systems examined include mainly organic substrates containing hydroxyl groups, such as aliphatic alcohols and phenols, in the presence of copper complexes possessing ligands such as halide, nitrite, water, 1,10-phenanthroline, and 2,2-bipyridine. An attempt is made to evaluate the relationship between the composition of the coordination sphere of the copper complex and the types and yields of oxidized photoproducts formed. Possible reaction schemes are discussed, and the practical importance of the results is documented.

THE PHOTOCHEMISTRY OF COORDINATION COMPOUNDS has been studied extensively in the past few years (*1–5*), and progress has also been made in the field of copper photochemistry. The photochemistry of copper complexes has been reviewed (*6*), and their reactivity has been classified (*7*) in terms of their catalytic role (*8*) in phototransformations (*9, 10*) or photooxidations (*11*) of organic substrates.

0065–2393/93/0238–0377$06.50/0

The photosensitivity of copper, or more generally, metal complex organic systems, may lead to many interesting phototransformations of organic molecules, for example, isomerizations, substitutions, oxidations, and double-bond migrations. This chapter focuses only on one class of photoreactions, the photooxidations of organic substrates in the presence of copper complexes. Along with the characterization of the approach used, we present a brief overview of redox photoreactivity and classification of photooxidations only to the extent necessary for discussing the topic treated in more detail later. We do not intend to give a complete review of photooxidations in the presence of coordination compounds of copper. The emphasis is on the photooxidations of hydroxylic organic compounds within the Cu(I)–Cu(II) photocatalytic cycle performed in our laboratories. The structure of the chapter reflects the various stages of research done, its sequence, and background: thermal and photoredox reactivity and photooxidations, which are discussed in terms of their mechanistic principles and applications.

Thermal and Photochemical Reactivity of Copper Complexes

The research in our laboratory was based on the concept of mutual influence of ligands (MIL) through the central atom (*12*). Within this approach a redox process is regarded as a consequence of interactions that depend on the ligand's bonding properties. The mutual influence of ligands of peculiar bonding properties (σ-donor, π-acceptor like acetone and σ,π-donor like Cl^-) creates good conditions for a spontaneous intramolecular redox process, for example, reduction of Cu(II) to Cu(I) and oxidation of Cl to radical $Cl^{\bullet}$. These dark reactions, explained as a consequence of MIL, were accelerated by irradiation. On the other hand, on irradiation the thermally stable complexes underwent changes analogous to those observed as a consequence of MIL. Therefore, the nature of electron redistribution caused by the entrance of ligand-specific properties (e.g., σ-donor and π-acceptor) and that caused by photon absorption are expected to be the same or at least very similar.

Classification of Photochemical Reactivity. The photochemistry of copper complexes [Cu(I), Cu(II), and Cu(III)] has been classified (*6, 7, 9, 10*) according to the nature of the photoreactive excited state and product(s) formed. Coordination compounds of copper can participate in many types of redox reactions arising from the population of metal-to-ligand charge-transfer (MLCT), ligand-to-metal charge-transfer (LMCT), charge-transfer-to-solvent (CTTS), intraligand (IL), interconfigurational metal-centered (ICMC), and ligand-to-ligand charge-transfer (LLCT) excited states (*13*). When taking into consideration the changes of the oxidation state of the metal center, three principal types of Cu(II) photochemical reactivity are distinguished (*7*): Cu(II) to Cu(I) photoreduction, Cu(II) to Cu(III) photooxidation, and Cu(II) photoredox reaction [Cu(II) → Cu(I) → Cu(II)] without irreversible photoreduction to Cu(I).

The Cu(I) photochemistry was classified into at least four various classes (details are given in ref. 7). Those photoredox reactions that are interesting from the point of view of photooxidation will be discussed in more detail in the next section.

Primary Photoprocesses. Our research was concentrated on copper photochemistry in various organic solvents. Chlorocopper(II) complexes in acetonitrile (ACN), a strong π-acceptor, represent an interesting photocatalytic system (*14*) working under visible light. Irradiation of the system in the region of the lowest spin-allowed charge-transfer (CT) excited state results in the photoreduction of Cu(II) metal center to Cu(I) (*15*) and $Cl^{\bullet}$ radical formation evidenced by pulsed-laser flash photolysis (*16, 17*) and an electron spin resonance (ESR) spin-trapping technique (*18*).

In the presence of free Cl^- ions, the radical $Cl^{\bullet}$ forms a transient that corresponds to the $Cl^{\bullet -}$ radical anion (ϵ_{366} = 14,400 ± 10% dm^3/mol·cm, τ = 4.0 μs) (Figures 1 and 2). This spectrally detectable (Figure 1) transient (λ_{max} = 366 nm) has a first-order decay (Figure 2), and its lifetime is unaffected by oxygen. Moreover, the gas–liquid chromatographic (GLC) analyses of the solutions after prolonged continuous irradiation show that the amounts of acetonitrile chloro derivatives obtained are less than 1% of the amount of Cu(II) that is photoreduced (*16*).

Possible scavenging processes were demonstrated experimentally (Figure 3). The $-Cl_2^{\bullet -}$ is quenched only by Cu(I) (k_1 = 2.8 × 10^8 $mol^{-1}dm^3s^{-1}$) and Cu(II) (k_2 = 1.7 × 10^8 $mol^{-1}dm^3s^{-1}$) and definitely not by Li^+, Cl^-, O_2, or solvent. The $Cl^{\bullet}$ radical is trapped by the *N-tert*-butyl-α-phenylnitrone (PBN) spin-trap and the hyperfine constants (A_N, $A_{Cl}35$, $A_{Cl}37$, and A_H) of PBN·Cl spin-adduct (Figure 4) prove (*18*) its identity (A_N = 12.10, $A_{Cl}35$ = 6.25, $A_{Cl}37$ = 5.15, and A_H = 7.5 G).

The experimentally observed photosensitivity of chlorocopper(II) complexes led to the idea of using this system for photooxidations of more reactive (*20*) organic substrates in ACN.

$$CuCl_x^{2-x} \xrightarrow[ACN\,(Cl^-)]{h\nu} CuCl_{x-1}^{2-x} + Cl^{\bullet}(Cl_2^{\bullet -}) \quad (1)$$

[$Cl_2^{\bullet -}$ is a strong oxidizing agent (*19*) in ACN (standard oxidation potential E^o = 2.3 V); ACN is a solvent that is practically inert toward chlorination.]

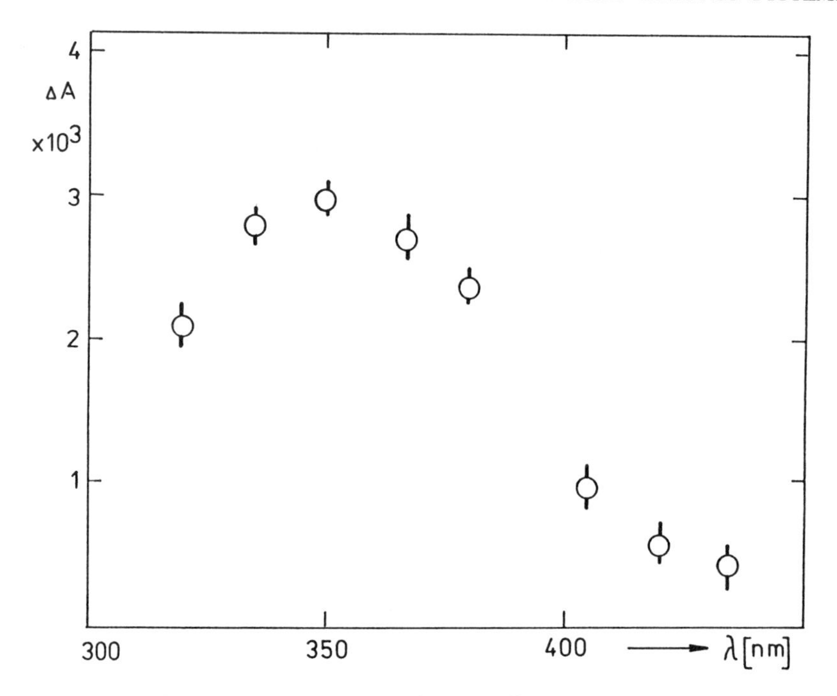

Figure 1. The absorption spectrum of the $Cl_2^{\bullet -}$ transient in an acetonitrile solution of $[CuCl_4]^{2-}$; $[Cu(II)] = 1.5 \times 10^{-3}$ mol/dm^3 (Nd pulsed-laser intensity ca. 1×10^{15} photons, pulse width 250 ns, and sample thickness 2 mm). (Reproduced with permission from reference 16. Copyright 1978, Royal Society of Chemistry.)

Photooxidations of Organic Substrates in the Presence of Copper Complexes

The term photochemical oxidation (photooxidation) in the presence of copper complexes is defined here as the photoformation of oxidized product OS′ from the original OS (OS is organic substrate) molecule in the presence of CuL_x (copper coordination compound) as represented by equation 2:

$$OS \xrightarrow[CuL_x]{h\nu} OS' \tag{2}$$

The presence of CuL_x is an essential condition for OS phototransformation (the OS′ formation does not proceed without CuL_x). The CuL_x is a copper complex with the central atom in any suitable oxidation state and any suitable ligand L (L is or is not OS).

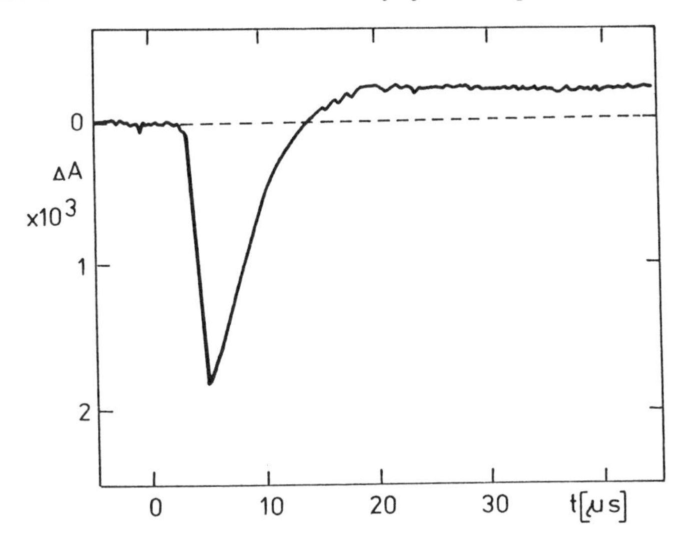

Figure 2. The $Cl_2^{\bullet -}$ transient decay curve monitored at 366 nm (λ_{irr} = 473 nm) in an acetonitrile solution of $[CuCl_4]^{2-}$; [Cu(II)] = 1.5 × 10^{-3} mol/dm^3 (Nd pulsed-laser intensity ca. 1 × 10^{15} photons, pulse width 250 ns, and sample thickness 2 mm). (Reproduced with permission from reference 16. Copyright 1978, Royal Society of Chemistry.)

According to the suggested classification of photooxidations in the presence of copper complexes (*11*), two main classes can be distinguished (only representative examples of particular type of systems are given): photosensitized and photoassisted.

Representative examples of photosensitized photooxidations are as follows.

- dehydrogenation: photooxidation of alcohol to aldehyde sensitized by $Cu(acac)_2$ (acac is acetylacetonate) and aryl ketones (*21*) under anaerobic conditions or by $[CuL(PPh_3)_2]^+$ (PPh_3 is triphenylphosphine) (*22*) in the cyclic Co(III) complex photoreduction, where L is methyl-substituted 2,2′-bipyridine (bpy) or 1,10-phenanthroline (phen)
- oxidative addition: 1,2-shift of the cyclopropyl group as a consequence of the photooxidative addition of methanol (*23*) in the system containing Cu(II), $C_{14}H_8(CN)_2$, and methanol
- ligand oxidation: "photosensitized" oxidation of the malonato, acetato, and oxalato ligand to CO_2 in the Cu(II) complex without (*24*) irreversible photoreduction to Cu(I)

Representative examples of photoassisted photooxidations are as follows.

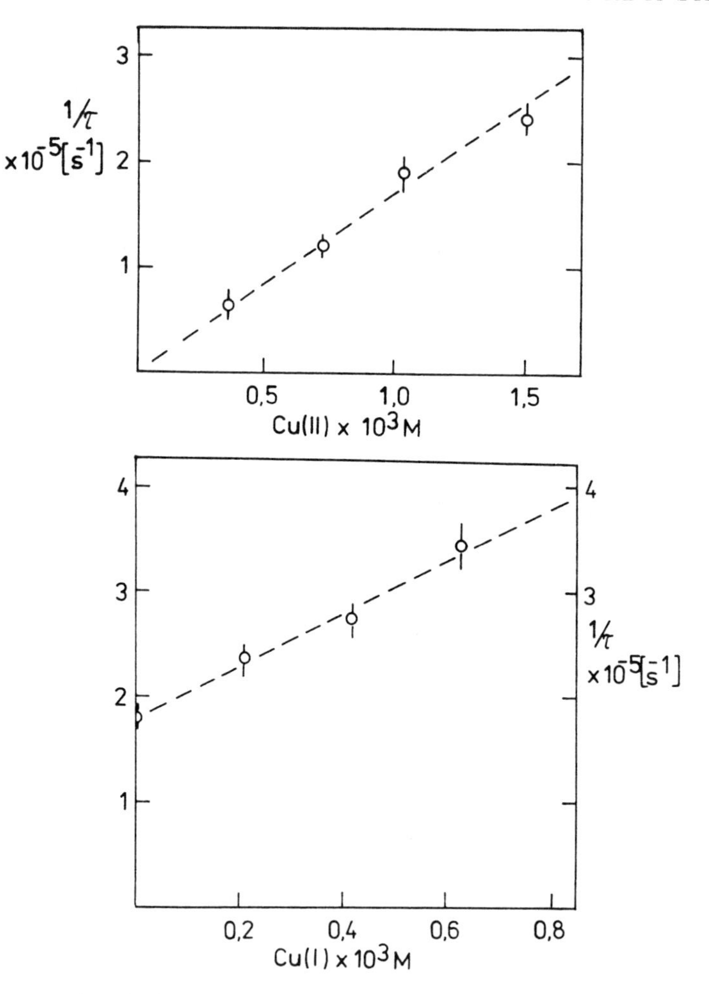

Figure 3. The $Cl_2^{\bullet -}$ reciprocal lifetime τ vs. [Cu(II)] at a constant ratio of [Cu(II)]:[Cl^-] = 1:8 and [Cu(I)] at a constant [Cu(II)] of 1 × 10^{-3} mol/dm^{-3}. (CuCl was added to keep the ratio of [Cu(II)]:[Cl^-] = 1:8.) (Reproduced with permission from reference 17. Copyright 1979.)

- dehydrogenation: anaerobic aldehyde formation (*25*) from aliphatic alcohols in the presence of cupric halogeno complexes (discussed later)
- oxidation: phenols are photooxygenated in the presence of dioxygen in solutions of CuL complexes, where L is Cl^-, Br^-, I^-, NO_3^- (*26*), or L is bpy or phen (*27*) (discussed later)
- oxidative photodecarboxylation: CO_2 elimination is observed (*28*) during photooxidation of 2-hydroxy acids to α-keto acids in the presence of Cu(II)

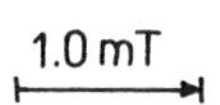

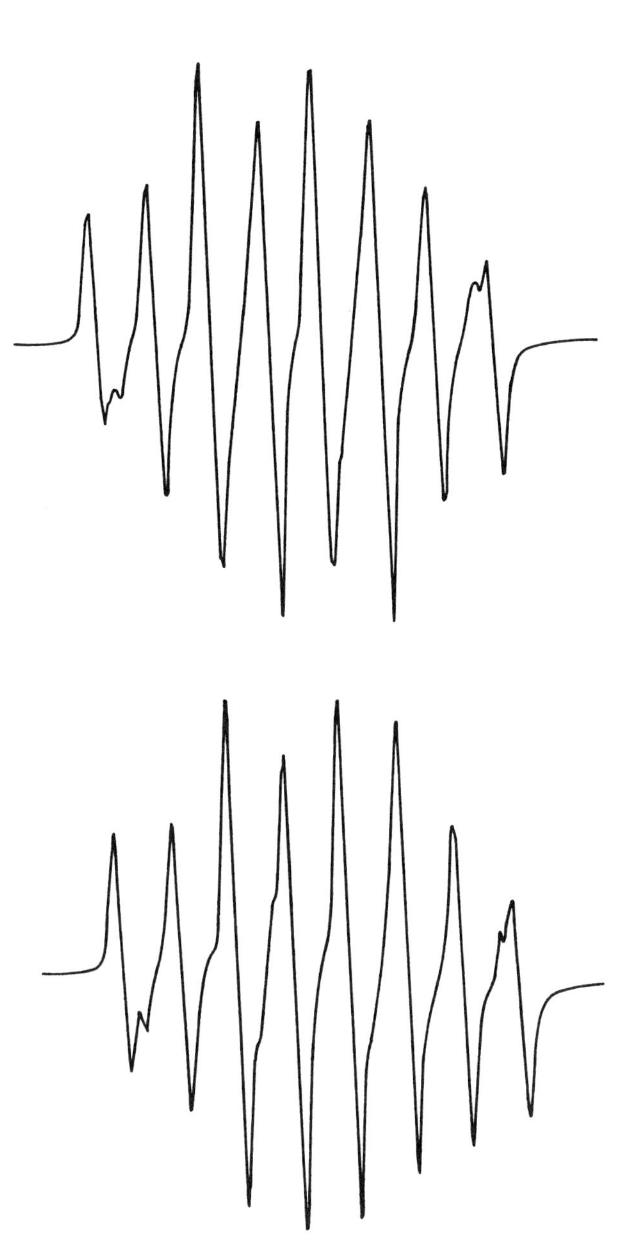

Figure 4. Experimental (top) and computer-simulated (bottom) ESR spectra of the PBN·Cl spin adduct in acetonitrile (Bruker ER 200D).

Anaerobic Systems. The photooxidizing ability of our chlorocopper(II) system in acetonitrile can be exemplified using alcohols as organic substrates (*25*). Alcohols were oxidized to corresponding aldehydes (*25*) with the quantum yield values (for product formation) of 0.01–0.001 mol/einstein (under anaerobic conditions). The overall reaction pathway of the photooxidizing effect of copper complexes on the aliphatic alcohols in acetonitrile has been suggested (*25*). With methanol or ethanol, methanal or ethanal is formed:

$$2[CuCl_x]^{2-x} + R{-}CH_2OH \xrightarrow{h\nu} R{-}CHO + 2[CuCl_{x-1}]^{2-x} + 2HCl \quad (3)$$

(R is H or CH_3), whereas 2-propanol is converted into acetone:

$$2[CuCl_x]^{2-x} + (CH_3)_2CH{-}OH \xrightarrow{h\nu} (CH_3)_2C{=}O + 2[CuCl_{x-1}]^{2-x} + 2HCl \quad (4)$$

The formation of aldehydes from 1-propanol or 1-butanol is followed by additional formation of their chloro derivatives:

$$2[CuCl_x]^{2-x} + R{-}CHO \xrightarrow{h\nu} R'{-}CHO + 2[CuCl_{x-1}]^{2-x} + HCl \quad (5)$$

where R′ is C_2H_4Cl or C_3H_6Cl.

From the mechanistic point view, it was of interest to study the nature of chloro-radical secondary processes and their role in the overall mechanism of photooxidation reactions of various organic substrates as scavengers. These processes may proceed according to the Noyes geminate-pair-scavenging mechanism (*29, 30*) and were tested for such behavior very recently (*31–33*) in the systems containing organic substrate scavengers of different structure and chemical nature (aliphatic alcohol, ketone, and carbohydrate).

The tested Noyes geminate-pair-scavenging mechanism in the system under study can be schematized as follows:

Primary pair formation

$$[CuCl_4]^{2-} \xrightarrow{h\nu} \{[CuCl_3]^{2-} \cdot Cl^{\bullet}\} \quad (6)$$

Primary recombination

$$\{[CuCl_3]^{2-} \cdot Cl^{\bullet}\} \longrightarrow [CuCl_4]^{2-} \quad (7)$$

Secondary pair formation

$$\{[CuCl_3]^{2-} \cdot Cl^{\bullet} \longrightarrow [CuCl_3]^{2-} + Cl^{\bullet} \quad (8)$$

Secondary recombination

$$[CuCl_3]^{2-} + Cl^{\bullet} \longrightarrow [CuCl_4]^{2-} \quad (9)$$

Reactive scavenging

$$[CuCl_3]^{2-} + Cl^{\bullet} + Q \xrightarrow{k_Q} [CuCl_3]^{2-} + \text{products} \quad (10)$$

When plotting the dependence of $\Phi_{Cu(I)}$ (defined as moles of Cu(I) photochemically produced per mole of photons adsorbed) versus the square root of scavenger concentration according to the Noyes equation (eq 11)

$$\frac{d\Phi_{Cu(I)}}{dc_Q{}^{1/2}} = 2a\,(\pi k_Q)^{1/2} \quad (11)$$

where a is the diffusion parameter *(29)* ($a = 1.6 \times 10^{-6}\ s^{1/2}$) and c_Q is concentration of scavenger, the scavenging rate constant k_Q can be evaluated from the slopes of equation 11 as given by equation 12:

$$k_Q = \frac{b^2}{4\pi a^2} \quad (12)$$

where

$$b = \frac{d\Phi_{Cu(I)}}{dc_Q{}^{1/2}} \quad (13)$$

The values of c_{sat} (the lowest concentration of c_Q for which $\Phi_{Cu(I)} = \Phi_{max}$), b, and k_Q experimentally obtained are summarized in Table I.

The experimentally obtained linear dependence of the quantum yield for Cu(I) formation on the square root of scavenger concentration within the range where the scavenging reaction competes with the secondary recombination supports the Noyes mechanism. The estimated values for scavenging rate constants *(32, 33)* are in good agreement with those previously reported and obtained by using different experimental methods.

The typical electronic absorption spectral changes during irradiation are shown in Figure 5 (the Cu(II) concentration decrease and the Cu(I) concentration increase as a product of photolysis are reflected in the appearance of the isosbestic point at 41,670 cm^{-1}).

Table I. Characteristics of the Φ versus c_{sat} Plots and Estimated Rate Constants for Scavenging of Chlorine Radicals

Scavenger	c_{sat} *(M)*	b *(10^{-2} $M^{1/2}$)*	k_Q *(10^8 $M^{-1}s^{-1}$)*	*Ref.*
Benzene	0.59	2.06	0.13	33
Ethanol	0.12	8.57	2.3	31
1-Propanol	0.068	10.9	3.7	31
Methanol	0.40	11.14	3.86	33
1-Butanol	0.048	12.4	4.8	31
1-Octanol	0.048	13.1	5.3	31
2-Propanol	0.49	16.63	8.60	33
Acetone	0.54	16.99	8.98	33

NOTE: In acetonitrile; λ_{irr} is 451 nm for benzene, methanol, 2-propanol, and acetone and 313 nm for the other scavengers.

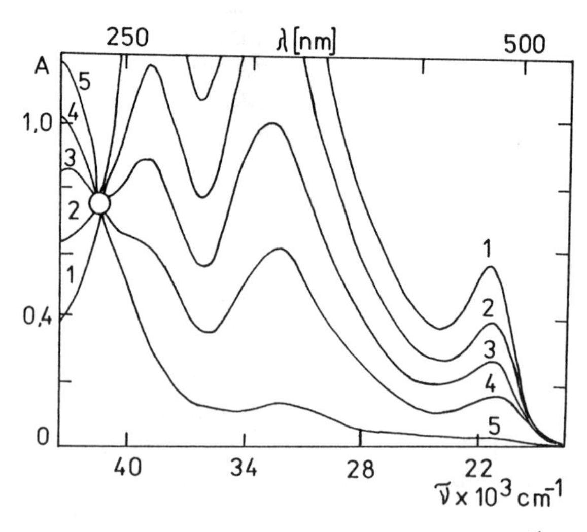

Figure 5. Electronic absorption spectral changes of the system Cu(II)–Cl⁻–butanol (10% v/v)–ACN during irradiation. [Cu(II)]:[Cl⁻] = 1:8; [Cu(II)] = 1 × 10^{-3} mol/dm³; λ_{irr} = 313 nm; and irradiation periods of 0–1, 2–25, 3–33, 4–39, and 5–51 min. (Reproduced with permission from reference 25. Copyright 1982.)

The yield of oxidation products was strongly dependent on the composition of the chlorocopper(II) complexes present in the system *(25)*. Apparently the reason was in the different photoredox reactivity of individual kinetically labile chlorocopper(II) species whose concentration distribution depends on the molar ratio of Cl^- to Cu(II) in ACN. Thus, considering both kinetic and thermodynamic aspects (Table II) of the redox reactivity of individual $CuCl_x$ species [the $\Phi_{Cu(I)}$ values calculated *(34, 35)* from $\Phi_{Cu(I)}$ overall *(17)* quantum yields and $E_{1/2}$* (excited-state half-wave oxidation potential) *(35)* from $E_{1/2}$ (ground-state half-wave oxidation potential) values *(36)* using a published procedure *(37, 38)* and known spectral characteristics of individual Cu(II) complex particles], we concluded that the oxidation driving force decreases with the increasing number of chloro ligands in the coordination sphere both in the ground and excited states. The same trend is observed also for the photoredox efficiency ($\Phi_{Cu(I)}$ decrease obtained).

Finally, if the OS oxidation depends on the ability of the Cu(II) metal center to be photoreduced, this relationship might be of great importance in tailoring the most efficient system suitable for photooxidation of various organic substrates.

Table II. Dependence of Oxidation Efficiency of Chlorocopper Complexes ($CuCl_x$) on Composition of Coordination Sphere

Complex Particle	$E_{1/2}$ *(V)*	$E_{1/2}$*[a] *(V)*	$\Phi_{Cu(I)}$[b]
$CuCl_2$	0.418	3.4	0.25
$[CuCl_3]^-$	0.342	2.8	0.09
$[CuCl_4]^{2-}$	0.208	2.6	0.05

[a]Calculated according to refs. 34, 35, 37, and 38.

[b]$\Phi_{Cu(I)} = \phi_{t=0}$ values at λ_{irr} = 471 nm (extrapolated to t_{irr} = 0).

Aerobic Systems. *Alcohols.* The rate of Cu(II) photoreduction is also markedly influenced by the content of oxygen in the system (Figure 6). Spectrophotometric investigations showed that photoproduced Cu(I) is reoxidized by oxygen, and, in fact (Figure 6), the actual rate of Cu(II) photoreduction is lowered. This observation, often regarded as a drawback in photochemical studies, is of importance in possibly raising the yield of oxidation products and led us to the suggestion of a photoassisted Cu(II)–Cu(I) redox cycle *(14)*. We were able, indeed, to realize several cycles (Figure 7) and to raise the yield of alcohol oxidation products.

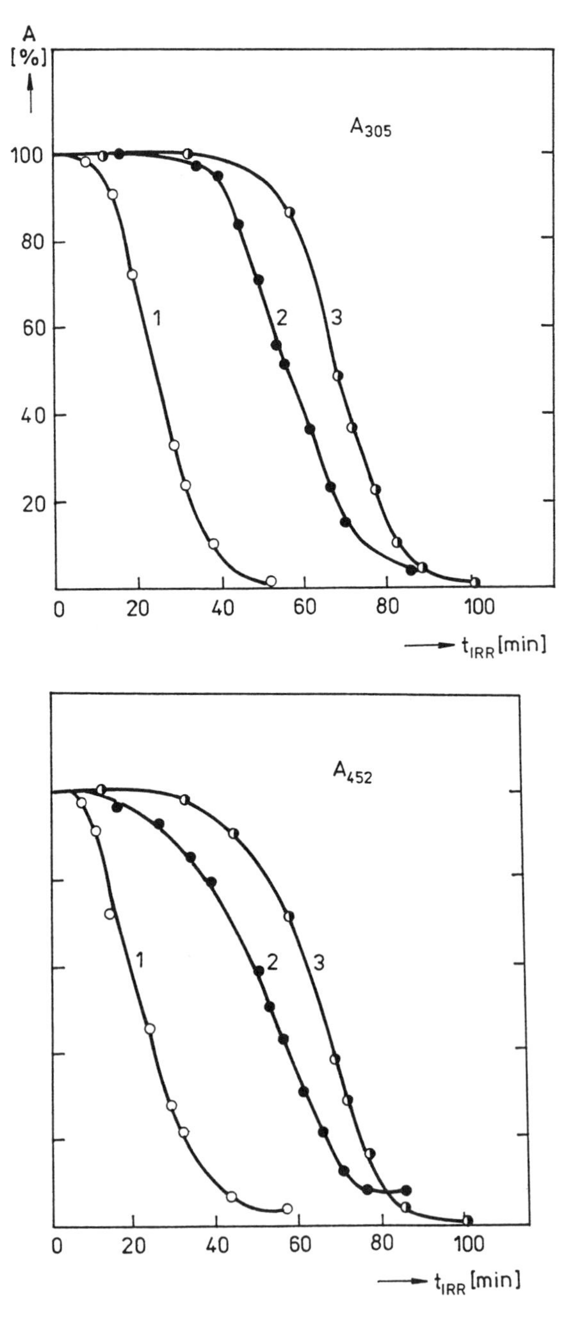

Figure 6. The influence of oxygen on the photoreduction of Cu(II) to Cu(I) in the system Cu(II)–Cl^-–methanol (10% v/v)–ACN; [Cu(II)]:[Cl^-] = 1:8; [Cu(II)] = 1×10^{-3} mol/dm^3; λ_{irr} = 313 nm (absorbance changes at 305 and 452 nm); and (1) 15 min bubbled with argon, (2) 60 min closed, and (3) 90 min opened to air.

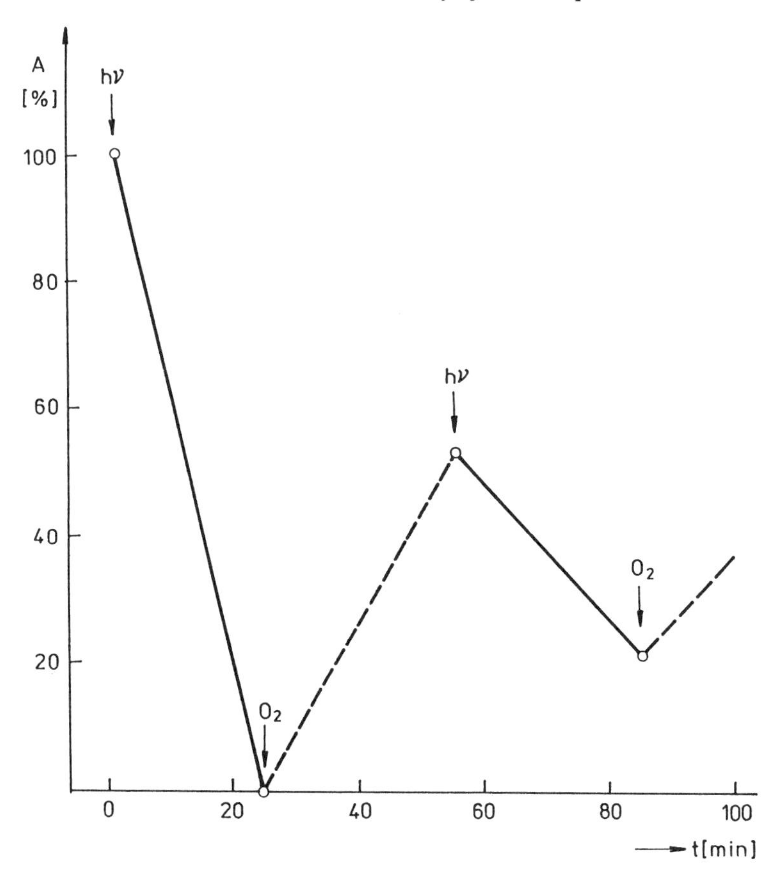

Figure 7. The influence of oxygen and irradiation on the redox changes in the system Cu(II)–Cl^-–ClO_4^-–propanol (15% v/v)–ACN, where [Cu(II)]:[Cl^-]:[ClO_4^-] = 1:8:4; [Cu(II)] = 1 × 10^{-3} mol/dm^3; λ_{irr} = 313 nm (absorbance changes at 305 nm).

Primary and secondary alcohols can be converted to the corresponding aldehydes or ketones without overoxidation to the carboxylic acids by using visible light in the presence of H_2PtCl_6, $CuCl_2$, and dioxygen (*39, 40*). Interestingly, neither $CuCl_2$–O_2 alone nor $[PtCl_6]^{2-}$–O_2 alone produces a catalytic photooxidation of alcohols. The requirement that $CuCl_2$ be present is explained (*40*) in terms of its dual role as an oxidant of Pt(III) and a radical trap for RRHCO• radicals. A binuclear (possibly chloro-bridged) platinum copper complex is thought to be the active catalyst.

Phenols. Copper complexes are certainly among the most thoroughly studied and most versatile redox catalysts, and they are useful, for example, in the oxidation of organic substrates by molecular oxygen in both biological and nonbiological systems. In addition to considerable interest in the development of chemical copper model systems mimicking the biological oxygenases, great efforts can be seen also in the development of some commercially interesting products obtainable by oxidative coupling of phenols catalyzed by copper complexes.

The oxidation of substituted phenols to quinones by dioxygen in the presence of copper complexes as catalysts has been the object of several studies. The reagents most extensively employed are morpholine–Cu(II) (*41*) and ammine–Cu(I) (*42*) complexes, $CuCl_2$ in pyridine (*43*), copper salts in dimethylformamide (*44*), Cu(II) nitrate (*45*), copper halides (*46*), carboxylates, phenoxides (*47*), sulfate (*48*), and acetate (*49*). These reagents usually give a mixture of quinones, dimers, and polymers (e.g., polyphenylene oxides).

The influence of various factors (e.g., solvent and experimental conditions) and the role of the +1 and +2 oxidation states of copper toward dioxygen reactivity have been discussed (*50*). The detailed investigation on the reactivity of the copper(0) oxidation state (metallic copper) with O_2 in the presence of various organic substrates (e.g., phenol, methanol, nitromethane, and benzoic acid) started only recently (*51, 52*). Various types of heterogeneous copper catalysts for the oxidation of phenols like Cu–Al_2O_3, Cu–Cr_2O_3 (*51*), or combination of Cu(0) with catalytic amounts of CuCl (*53*) instead of the homogeneous Cu(II) systems were used. The interaction of phenol with oxygen in the presence of metallic copper has been indicated (*51*) to be very similar to that observed in the presence of homogeneous Cu(I) and Cu(II) catalysts.

The majority of such systems have been studied only thermally (in the dark), and only in some cases has the effect of light also been established (*54*). Similarly, no attention has been paid to the photochemistry of phenol oxidations in the presence of copper catalysts.

In one study (*55*) *p*-benzoquinone (pbq) was obtained as a product of the liquid-phase oxidation of phenol catalyzed by $CuCl_2$ in dimethylformamide (DMF), dimethyl sulfoxide (DMSO), 1,4-dioxane, and ethylene glycol. By investigating the influence of solvent on the overall conversion and selectivity of phenol (O_2 at 35 kg/cm^2, 50–110 °C), solvents of dipolar aprotic character, particularly DMF, were found to be convenient. Copper halides in acetonitrile were claimed (*56*) to be particularly effective for pbq production from phenol in good yields, but only at pressures greater than 10,100 kPa (100 atm) of dioxygen (*57*) or elevated temperatures (*55*).

In seeking to extend the application of copper complexes (*58*) as catalysts in oxidations of organic substrates (*11*), we undertook a study of a novel route to phenol oxidation by dioxygen as concerns the copper catalytic (*26*) system used: Cu(I), X, phenol, O_2, and acetonitrile (typical experimental conditions were [Cu(I)] = 1×10^{-3} mol/dm^3, phenol in large excess, medium-pressure Hg lamp, and cutoff λ >315 nm). The various effects on the pbq yield in the system just described (composition of the catalytically active copper complexes; presence of various anions as X = Cl^-, Br^-, I^-, NO_3^-; addition of H_2O_2 and solvents like H_2O, CH_3OH, and CCl_4; and dark reactivity or irradiation with visible light) were also reported (*59*).

In line with the redox properties of the anions, a decrease of pbq yield was observed according to the sequence $Cl^- > Br^- > I^-$. The increase of oxidation ability in the presence of NO_3^- (higher pbq yield found) corresponds well to the known redox potentials in ACN (*60*). The addition of H_2O_2 (10%) shortens the induction period of the pbq formation and causes a decrease of the pbq yield, as was the case in similar copper–pyridine–O_2 systems (*61*) (ascribed to the effect of peroxide-like species formed during oxidation). The presence of other solvents (10% v/v) in the ACN–copper system led to the decrease (CCl_4 had no influence) of pbq yield, probably due to the destructive effect toward the catalytically active copper species of particular composition.

Light enhanced the pbq yield markedly, an effect that is strongly dependent on the composition of the system, and there is a possibility of useful control and tailoring of copper catalyst properties. The overall reaction scheme suggested (Figure 8) is very similar to that of an analogous phen and bpy systems (Figure 9) and will be discussed later.

Although the importance of the presence of chloride in the copper catalyst (*25, 26*) as well as in the chlorination mechanism [via chlorination of phenol, which is known to occur in DMF in the absence of oxygen (*62*)] was emphasized, formation of pbq from phenol has also been observed in

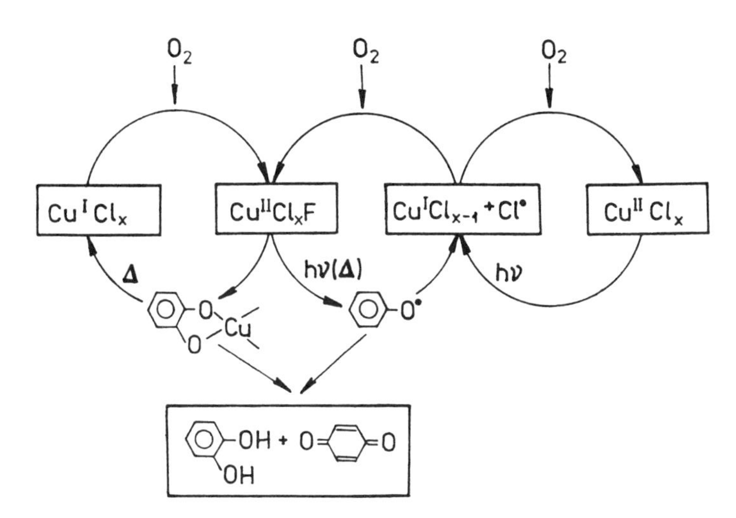

Figure 8. (Photo)transformation of phenol in the presence of chlorocopper complexes; λ_{irr} > 315 nm (F = phenol).

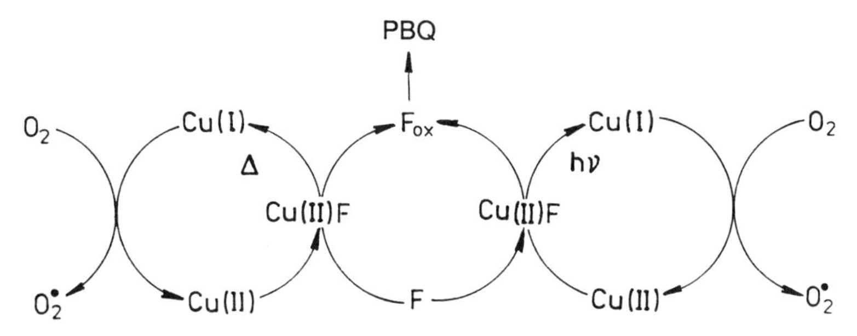

*Figure 9. Phenol-to-*p*-benzoquinone (photo)transformation in the presence of CuL_2 complexes; λ_{irr} > 315 nm (F is phenol, F_{ox} is phenoxy radical, PBQ is parabenzoquinone, and L is phen or bpy).*

the absence of chloride using, for example, Cu(II) sulfate or bpy or phen copper complexes (*63–65*).

Continuous irradiation of a mixture of cuprous bpy (or phen) complex and phenol in ACN solution in the presence of dioxygen with visible light led to the formation of pbq according to equation 14:

$$\text{HO–C}_6\text{H}_5 \xrightarrow[\Delta/h\nu;\ O_2]{CuL_2\ -\ ACN} \text{O=C}_6\text{H}_4\text{=O} \qquad (14)$$

The amount of pbq formed was estimated to be some 10 times higher than the corresponding stoichiometric concentration of the Cu(I) catalyst used even under dark or photolysis conditions. The enhancement of pbq formation (*66, 67*) under photolysis (with respect to dark conditions) was found to be 161% and 214% when using $[Cu(bpy)_2]^+$ and $[Cu(phen)_2]^+$ complexes, respectively. The mechanism of phenol-to-quinone aerobic oxidation catalyzed by Cu(I) bpy and phen complexes involves dioxygen activation, phenoxy radical formation, and at least one photochemical step (Figure 9).

The continuous irradiation of the acetonitrile solution of cuprous (*68, 69*) chloro complexes in the presence of *o*-cresol, *m*-cresol, *p*-cresol, 2,4-dimethylphenol, and 2,5-dimethylphenol led to the very effective (98% yield, 98% selectivity, and 100% conversion) phototransformation of phenol to monomeric quinones (eq 15):

$$\text{HO–C}_6\text{H}_3(R_1)(R_2) \xrightarrow[\Delta/h\nu;\ O_2]{CuX\ -\ ACN} \text{monomeric benzoquinones} \qquad (15)$$

where X is Cl^- and R_1 and R_2 are H or CH_3.

The overall reaction scheme covers first of all the phenoxy radical formation due to Cu(II)–phenol bond breaking [Cu(II) being reduced to Cu(I)] followed by interaction of this radical with "activated dioxygen" [coming probably from the Cu(I)–O_2 moiety, as observed in analogous cobalt systems (*70*)] and reoxidation of the Cu(I) metal center by molecular dioxygen, thus closing the cycle. The observed enhancement of phenol-to-quinone photooxidation under irradiation (by factor of 4 with respect to the dark reaction) is explained by the enhanced rate of Cu(II) to Cu(I) photoreduction and at the same time by the photoformation of phenoxy radical as a consequence of the photoredox reaction of an excited Cu(II)-phenolate complex. The second possible influence of such photochemical behavior is expected to be the Cl radical formation from $CuCl_x$ cupric species (pure photochemical step), which, in turn, abstracts the hydrogen H atom from a phenol molecule outside the coordination sphere and thus leads to the phenoxy radical concentration increase needed for increasing quinone production.

A very elegant quantitative structure–activity relationship (QSAR) study (*71*) of phenol oxidation by singlet oxygen in the model system containing rose bengal as sensitizer revealed that the observed second-order rate constant of phenol disappearance correlates with the $E_{1/2}$ and σ-constant of both the undissociated phenols and the phenolate anion. Although electron transfer with $^1O_2$ is thermodynamically favorable for undissociated phenols with electron-donating substituents in aqueous solution, this electron transfer is not likely to be competitive with other environmental transformation pathways for undissociated phenols with electron-withdrawing substituents. Unfortunately, the estimated QSARs were relevant only to note all possible reaction pathways, the so-called environmental fate of phenols (i.e., oxidation of phenols by $^1O_2$), and the actual product formed (e.g., quinones, coupling products, or polymers) was not specified. The observed yields and selectivity of substituted quinones formed (*68, 69*) do not correlate with the redox potentials of substituted phenols (*72*).

The reactivity of phenols toward oxidation governed by a copper complex catalyst might therefore be profoundly influenced by steric and electronic factors. Moreover, when taking into consideration that there exists more than one phenol oxidation mode (e.g., various oxidants), the observed failure of redox potential versus yield and selectivity correlations is not so surprising.

Future Prospects

The role of copper ions (salts and complexes) seems to be quite interesting in many processes connected with phototransformation, sometimes decomposition of various organic substrates, and very often pollutants. For example, the complete oxidation of phenol using sulfite oxidant was observed (*73*) in the presence of copper sulfate. On the photochemical level, the rate of acetic acid (*74*) and formic acid (*75*) photooxidation in the presence of UV light, with TiO_2, and dissolved copper ions as catalysts, was considerably enhanced (*76, 77*) by the copper ions. Due to the richness of the photoredox pathways, interest in copper photoreactivity will no doubt remain high. In particular, photooxidations based on the photoredox reactivity of copper complexes will likely play a very important and promising role in nonbiological as well as biological systems.

Acknowledgments

J. Sýkora acknowledges scientific collaboration with the groups of Horst Hennig, University of Leipzig, Leipzig, Germany; Orazio Traverso, University of Ferrara, Ferrara, Italy; and Claudio Furlani, University of Rome, Rome, Italy. We are grateful for the contributions of our coworkers, cited in the literature, to our project of systematic research on the photochemical reactivity of copper complexes.

References

1. Sýkora, J.; Šima, J. *Photochemistry of Coordination Compounds;* Veda: Bratislava, Czechoslovakia, 1986.
2. Hennig, H.; Rehorek, D. *Photochemische und Photokatalytische Reaktionen von Koordinationsverbindungen;* Teubner: Stuttgart, Germany, 1988.
3. *Photoinduced Electron Transfer;* Fox, M. A.; Chanon, M., Eds.; Elsevier: Amsterdam, Netherlands, 1988.
4. Sýkora, J.; Šima, J. *Photochemistry of Coordination Compounds;* Elsevier-Veda: Amsterdam, Netherlands, 1990.
5. Balzani, V.; Scandola, F. *Supramolecular Photochemistry;* Ellis Horwood: Chichester, England, 1991.
6. Ferraudi, G.; Muralidharan, S. *Coord. Chem. Revs.* **1981,** *336,* 45.
7. Sýkora, J. *Chem. Listy* **1982,** *376,* 1047.
8. Sýkora, J. In *Photochemistry and Photophysics of Coordination Compounds;* Yersin, H.; Vogler, A., Eds.; Springer Verlag: Heidelberg, Germany, 1987; p 193.
9. Kutal, C. *Coord. Chem. Rev.* **1985,** *364,* 191.
10. Salomon, R. G. *Tetrahedron* **1983,** *339,* 485.
11. Sýkora, J. *J. Inf. Rec. Mat.* **1989,** *317,* 415.
12. Gažo, J. *Zh. Neorg. Khim.* **1977,** *322,* 2936.
13. Kutal, C. *Coord. Chem. Revs.* **1990,** *399,* 213.

14. Sýkora, J.; Horváth, E.; Gažo, J. In *Proceedings of the 4th Symposium on Metal Complexes in Catalytic Oxidation Reactions;* Polish Chemical Society: Wroclaw, Karpacz-Bierutovce, Poland, 1979; p 86.
15. Sýkora, J.; Horváth, E.; Gažo, J. *Z. Anorg. Allg. Chem.* **1978,** *3442,* 245.
16. Sýkora, J.; Giannini, I.; Diomedi-Camassei, F. *J. Chem. Soc. Chem. Commun.* **1978,** 207.
17. Cervone, E.; Diomedi-Camassei, F.; Giannini, I.; Sýkora, J. *J. Photochem.* **1979,** *311,* 321.
18. Bergamini, P.; Maldotti, A.; Sostero, S.; Traverso, O.; Sýkora, J. *Inorg. Chim. Acta* **1984,** *385,* L15.
19. Laurence, G. S.; Thornton, A. T. *J. Chem. Soc. Dalton Trans.* **1973,** 1637.
20. Kochi, J. K. *J. Am. Chem. Soc.* **1955,** *377,* 5274.
21. Chow, Y. L.; Buono-Core, G. E. *J. Am. Chem. Soc.* **1986,** *3108,* 1234.
22. Sakaki, S.; Koga, K.; Ohkubo, K. *Inorg. Chem.* **1986,** *325,* 2330.
23. Mizuno, K.; Yoshioka, K.; Otsuji, Y. *J. Chem. Soc. Chem. Commun.* **1984,** 1665.
24. Morimoto, J. Y.; DeGraff, B. A. *J. Phys. Chem.* **1975,** *79,* 326.
25. Sýkora, J.; Jakubčová, M.; Cvengrošová, Z. Collect. Czech. Chem. Commun. **1982,** *47,* 2061.
26. Engelbrecht, P.; Thomas, Ph.; Hennig, H.; Sýkora, J. *Z. Chem.* **1986,** *26,* 137.
27. Sýkora, J.; Brandšteterová, E.; Jabconová, A. *Proceedings of the 3rd Meeting on Photochemistry;* Czechoslovak Chemical Society: Liblice, Czechoslovakia, 1991; p 7.
28. Matsushima, R.; Ichikawa, Y.; Kuwabara, K. *Bull. Chem. Soc. Jpn.* **1980,** *53,* 1902.
29. Noyes, R. M. *J. Am. Chem. Soc.* **1955,** *77,* 2042.
30. Noyes, R. M. *J. Am. Chem. Soc.* **1956,** *78,* 5486.
31. Horváth, O. *J. Photochem. Photobiol. A: Chemistry* **1989,** *48,* 243.
32. Sýkora, J.; Molčan, M. *Proceedings of the 13th Conference on Coordination Chemistry;* Slovak Chemical Society: Smolenice-Bratislava, Czechoslovakia,, 1991; p 291.
33. Sýkora, J.; Molčan, M. *Bull. Soc. Chim. Belg.* **1992,** *101,* 775.
34. Sýkora, J.; Šima, J.; Valigura, D. *Chem. Zvesti* **1981,** *35,* 345.
35. Sýkora, J.; Šima, J.; Valigura, D.; Horváth, E.; Gažo, J. *Proceedings of the 3rd Symposium on Photochemical and Thermal Reactions of Coordination compounds;* Polish Chemical Society: Mogilany-Kraków, Poland, 1980; p 12.
36. Sestili, L.; Furlani, C.; Ciana, A.; Garbassi, F. *Electrochim. Acta,* **1970,** *15,* 225.
37. Balzani, V.; Boletta, F.; Gandolfi, M. T.; Maestri, M. *Top. Curr. Chem.* **1978,** *75,* 1.
38. Grabowski, Z. R.; Rubaszewska, W. *J. Chem. Soc. Faraday Trans. 1* **1977,** *73,* 11.
39. Cameron, R. E.; Bocarsly, A. B. *J. Am. Chem. Soc.* **1985,** *107,* 6116.
40. Bocarsly, A. B.; Cameron, R. E.; Zhou, M. In *Photochemistry and Photophysics of Coordination Compounds;* Yersin, H.; Vogler, A., Eds.; Springer-Verlag: Heidelberg, Germany, 1987; p 177.
41. Brackman, W.; Havinga, E. *Rec. Trav. Chim.* **1955,** *74,* 37, 1021, 1070, 1100, 1107.

42. Hay, A. S.; Blanchard, H. S.; Enders, G. F.; Eustance, J. W. *J. Am. Chem. Soc.* **1959,** *81,* 6355.
43. Nishida, N.; Minakata, T.; Tsuchida, G. *Makromol. Chem.* **1982,** *183,* 1889.
44. Hutchins, D. A. U.S. Patent 3,859,317, 1975; U.S. Patent 3,870,731, 1975.
45. Tsuruya, S.; Kishikawa, J.; Tanaka, R.; Kuse, T. *J. Catal.* **1977,** *49,* 254.
46. Hsu, Ch. Y.; Lyons, J. E. U.S. Patent 4,442,036, 1984.; E.P. Patent 423,985, 1982.
47. Costantini, M.; Jouffret, M. E.P. Patent 15,221,03, 1981.
48. Morisaki, S.; Komamiya, K.; Naito, M. *Nippon Kagaku Kaishi* **1980,** *7,* 1191.
49. Beltrame, P.; Beltrame, L.; Bussola, M. *Gazz. Chim. Ital.* **1980,** *110,* 141.
50. Tyeklár, Z.; Karlin, K. D. *Acc. Chem. Res.* **1989,** *22,* 241.
51. Gargano, M.; Ravasio, N.; Rossi, M.; Tiripicchio, A.; Camellini, M. T. *J. Chem. Soc. Dalton Trans.* **1989,** 921.
52. Ravasio, N.; Gargano, M.; Rossi, M. In *New Developments in Selective Oxidation;* Centi, G.; Trifiro, F., Eds.; Elsevier: Amsterdam, Netherlands, 1990; p 139.
53. Capdevielle, P.; Maumy, M. *Tetrahedron Lett.* **1982,** *23,* 1577.
54. Sýkora, J. *Proceedings of the 10th Conference on Coordination Chemistry;* Slovak Chemical Society: Smolenice-Bratislava, Czechoslovakia, 1985; p 387.
55. Beltrame, P.; Beltrame, P. L.; Carniti, P. *Ind. Chem. Prod. Res. Dev.* **1979,** *18,* 208.
56. Lyons, J. E.; Hsu, Ch. Y. In *Biological and Inorganic Copper Chemistry;* Karlin, K. D.; Zubieta, J., Eds.; Adenine Press: New York, 1985; p 57.
57. Reilly, E. L. U.S. Patent 3,987,068, 1976.
58. Sýkora, J.; Šima, J. *Coord. Chem. Revs.* **1990,** *107,* 1.
59. Kureková, M., Diploma Thesis, Slovak Technical University, Bratislava, Czechoslovakia, 1985.
60. Mruthyunjaya, H. C.; Murthy, A. R. V. *Indian J. Chem.* **1973,** *11,* 481.
61. Demmin, T. R.; Swerdloff, M. D.; Rogic, M. M. *J. Am. Chem. Soc.* **1981,** *103,* 5795.
62. Kosower, E. M.; Wu, A. S. *J. Org. Chem.* **1963,** *28,* 633.
63. Brenner, W. Ger. Patent 2,221,624, 1972.
64. Sýkora, J.; Lopatová, A.; Molčan, M. *Proceedings of the 6th Symposium on Photochemical and Thermal Reactions of Coordination Compounds;* Slovak Chemical Society: Smolenice-Bratislava, Czechoslovakia, 1988; p 157.
65. Sýkora, J.; Lopatová, A.; Molčan, M. *Proceedings of the 12th Conference on Coordination Chemistry;* Slovak Chemical Society: Smolenice-Bratislava, Czechoslovakia, 1989; p 377.
66. Sýkora, J.; Brandšteterová, E.; Jabconová, A. *Proceedings of the 9th International Symposium on Photochemistry and Photophysics of Coordination Compounds;* University of Fribourg: Fribourg, Switzerland, 1991; p 47.
67. Sýkora, J.; Brandšteterová, E.; Jabconová, A. *Bull Soc. Chim. Belg.* **1992,** *101,* 821.
68. Sýkora, J.; Engelbrecht, P.; Hennig, H.; Thomas, Ph.; Gliesing, H., Ger. Patent DD 256,323 Al, 1988.
69. Sýkora, J.; Engelbrecht, P.; Hennig, H.; Thomas, Ph.; Gliesing, H., Czech. Patent 255,399, 1987.
70. Bailey, C. L.; Drago, R. S. *Coord. Chem. Revs.* **1987,** *79,* 321.

71. Tratnyek, P. G.; Hoigné, J. *Envir. Sci. Technol.* **1991,** *25,* 1596.
72. Suatoni, J. C.; Snyder, R. E.; Clark, R. O. *Anal. Chem.* **1961,** *33,* 1894.
73. Kulkarni, U. S.; Dixit, S. G. *Ind. Eng. Chem. Res.* **1991,** *30,* 1916.
74. Bideau, M.; Claudel, B.; Faure, L.; Kazouan, H. *J. Photochem. Photobiol. A: Chem.* **1991,** *61,* 269.
75. Bideau, M.; Claudel, B.; Otterbein, M. *J. Photochem.* **1980,** *14,* 291.
76. Bideau, M.; Claudel, B.; Faure, L.; Rachimoellah, M. *J. Photochem.* **1987,** *39,* 107.
77. Bideau, M.; Claudel, B.; Faure, L.; Rachimoellah, M. *Chem. Eng. Commun.* **1990,** *93,* 167.

RECEIVED for review January 21, 1992. ACCEPTED revised manuscript May 14, 1992.

20

Light-Sensitive Organometallic Compounds in Photopolymerization

Achim Roloff

Polymer Division, Ciba-Geigy AG Research Center Marly, CH 1701 Fribourg, Switzerland

Photopolymerization processes initiated by organometallic compounds have found industrial applications recently. Titanium- and iron-based compounds have been found suitable for the initiation of radical as well as cationic polymerization reactions. The basic processes, industrial applications, and the underlying photochemistry of these compounds are discussed.

THE INTERACTION OF LIGHT WITH POLYMERIC MATERIALS was regarded as highly undesirable for many years. UV absorbers and light stabilizers were put into polymers to stabilize these materials. It has not been until fairly recently that the curing of polymers using light has been technically considered. Today radiation curing has applications in surface coating, lithography, the manufacture of three-dimensional objects, and adhesives as well as the manufacture of integrated circuits (*1*).

Industrial Applications

Areas in which photopolymerization is used on a commercial scale today are the following:

- surface coatings
- printing inks
- photoresist materials
- solder mask resists

0065–2393/93/0238–0399$06.00/0

- protective coatings of electronic parts
- adhesives
- holography
- stereolithography
- contact lenses
- high solid coatings
- can coatings
- binders for abrasives

The principal technique for imagewise photopolymerization as it is used with photoresist materials is still irradiation of a polymerizable material through a mask. Upon contact with light, the material polymerizes in the area not covered by the mask, and in a development step the unpolymerized material can be removed; the result is a relief image (Figure 1). These photopolymerization processes can be carried out with polymers that are either cationically or radically initiated. The first report of anionic photopolymerization has only recently appeared (*2*).

A photopolymerization reaction can be achieved by two fundamentally different processes. Direct photopolymerization applies to cases in which one photon is absorbed per bond formed (*3*). The other, more efficient process is photoinitiated polymerization, which requires the absorption of a photon only for the production of an active initiating species. Polymerization reactions occurring via this second process therefore require at least two components: a photoinitiator and a polymerizable material. Photoinitiators are compounds that, upon interaction with light, liberate a radical, a cation, or an anion to start the polymerization reaction, which then proceeds thermally.

Organometallic Photoinitiators

Organometallics are known to form radicals or cations when interacting with light (*4, 5*). If they absorb the desired wavelength they can be used as photoinitiators, provided they meet a number of additional requirements (*1*):

1. adequate thermal stability of the compounds as such, as well as their mixtures with the polymerizable material
2. suitable absorption characteristics in the UV–visible region
3. efficient pathways to yield radicals or cations

We found (*1, 7*) organometallic photoinitiators for cationic polymerization based on iron arene chemistry (structure **1**), and we carried out (*1, 24*) radical polymerization using titanocene compounds (structure **2**).

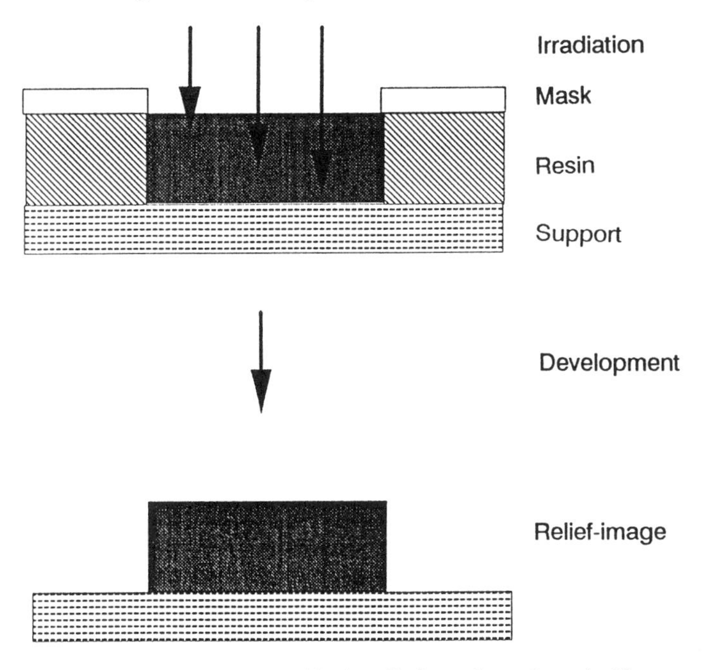

Figure 1. Relief image prepared by irradiation of a polymerizable material through a mask.

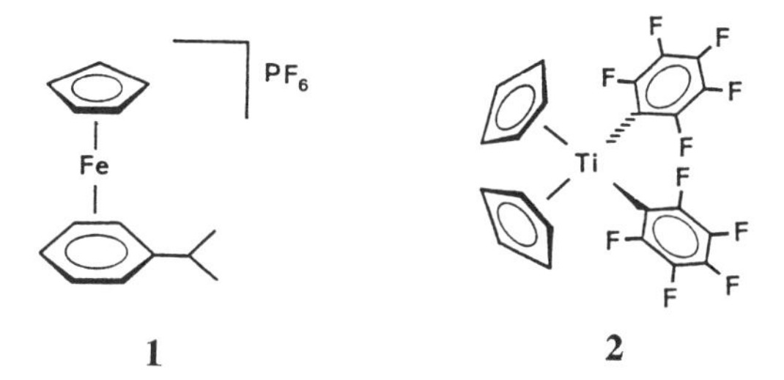

Iron Arene Compounds. The chemistry and application of the cationic iron arene complexes as initiators for cationic polymerization were described in a number of papers (*6–11*) and in Chapter 21 by Hendrickson and Palazzotto in this volume. Iron arene complexes were synthesized by Nesmeyanov et al. (*12*), who also studied their photochemistry (*13*). Gill and Mann (*14*) described photochemical ligand-exchange reactions, substituting the arene ligand by three monodentate ligands like CO, phosphines, or isonitriles.

The primary photochemical reaction, however is still unclear. Meier and Zweifel (*10, 11*) formulated a Lewis acid as the active intermediate in photopolymerization of epoxides. Kutal and co-workers (*15*) proposed a

$CpFeO^+$ intermediate (Cp is cyclopentadienyl) on the basis of observations in the gas phase.

Titanocene Photoinitiators. Titanium(IV) compounds have been proposed as photoinitiators for the polymerization of acrylic monomers as well as in some cases styrene, ethylene, and vinyl chloride (*16–18*). The photochemistry was investigated for the dimethyl (*19*) and the diphenyl (*20–22*) derivatives, but these compounds proved to be too unstable for technical application (*23*), in which thermal as well as oxidative stability are required. The dimethyl derivative (structure **3**) decomposed in acetic acid and above −30 °C autocatalytically and was oxygen-sensitive; the diphenyl derivative (structure **4**) decomposed in acetic acid and above 146 °C and was also oxygen-sensitive.

Thermal and oxidative stability were greatly increased when a fluorinated aryl ligand was introduced. Perfluorodiphenyltitanocene (structure **5**) is sufficiently stable thermally (decomposition at 230 °C) as well as toward oxygen and acetic acid and was therefore suitable as a photoinitiator for radical photopolymerization (*24*).

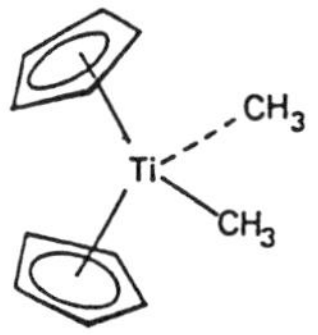

Ti CH3 CH3 **3** Ti **4** Ti F F F F F F F F F F **5**

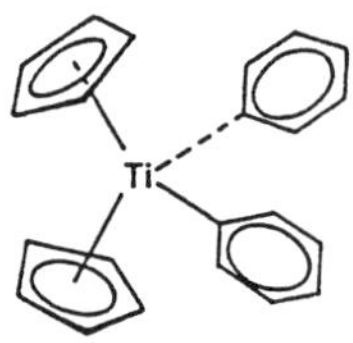

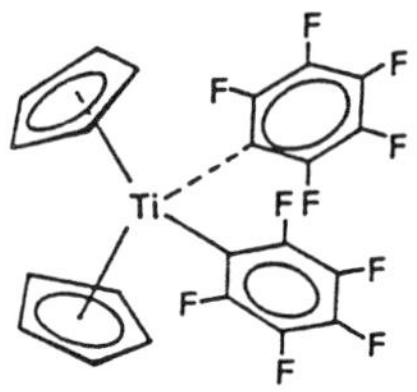

Mechanistic Aspects. We wished to have as much knowledge as possible about the primary photochemical steps in this initiation process. With radical scavenger experiments we wanted to establish the photochemical behavior of all three types of titanocene complexes. For the dimethyl derivative the literature described the formation of methyl radicals (*19*), and we were able to isolate and identify the scavenger adduct (Scheme I).

For the phenyl derivative Rausch (*22*) and Brubaker and co-workers (*20, 21*) postulated the formation of benzene and diphenyl (Scheme II). We trapped the phenyl radical with tetramethylpiperidine *N*-oxide (TEMPO). In trapping experiments with TEMPO as a radical scavenger we found, surprisingly, that in the perfluorinated compound, the cyclopentadienyl and perfluorophenyl ligands were replaced upon irradiation. No organic radicals were trapped by TEMPO in this experiment (*25*) (Scheme III).

The question was now whether this ligand-coupling process was inter- or intramolecular. To clarify this question, we synthesized two titanium

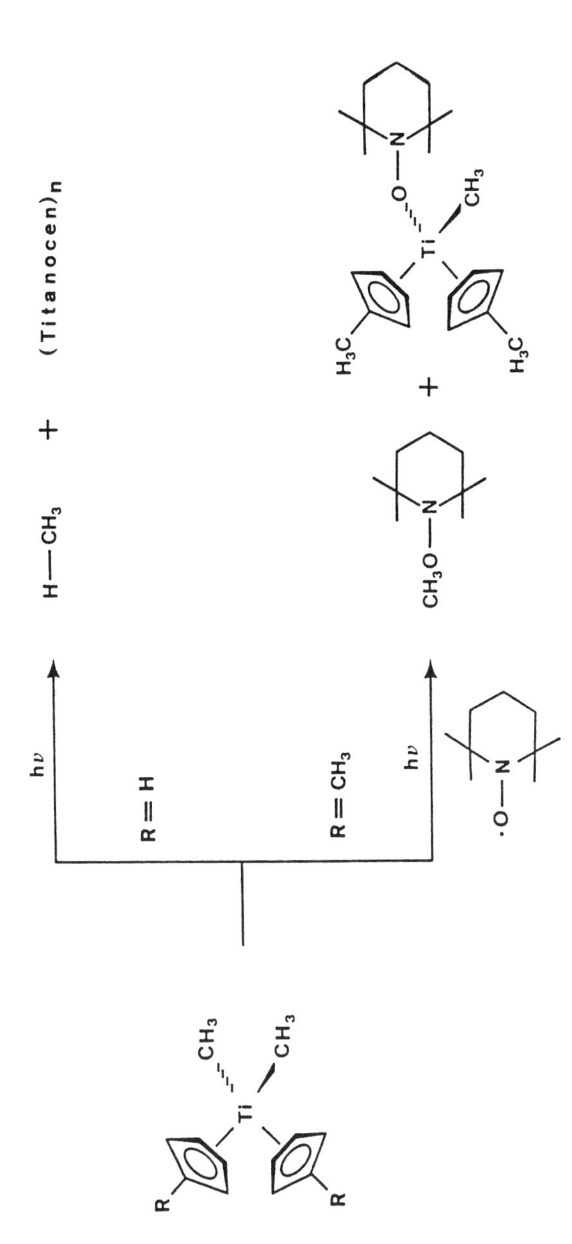

Scheme I. Photochemical behavior of the methyl derivative of the titanocene complex.

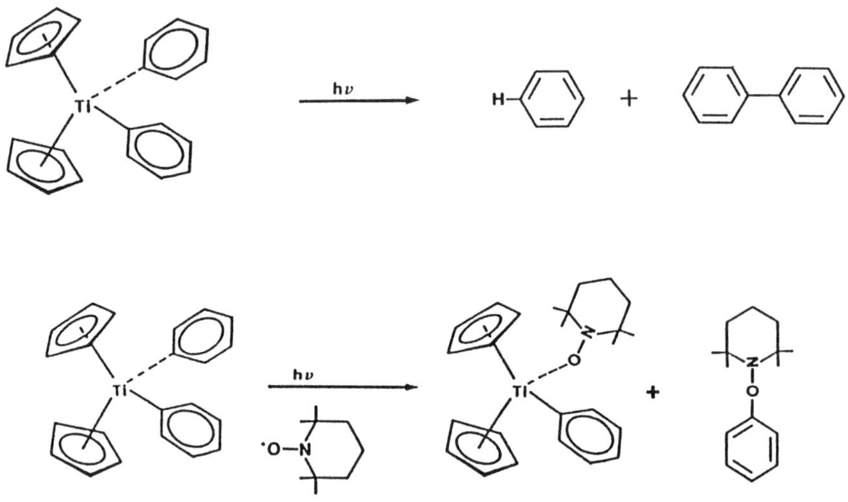

Scheme II. Photochemical behavior of the phenyl derivative of the titanocene complex.

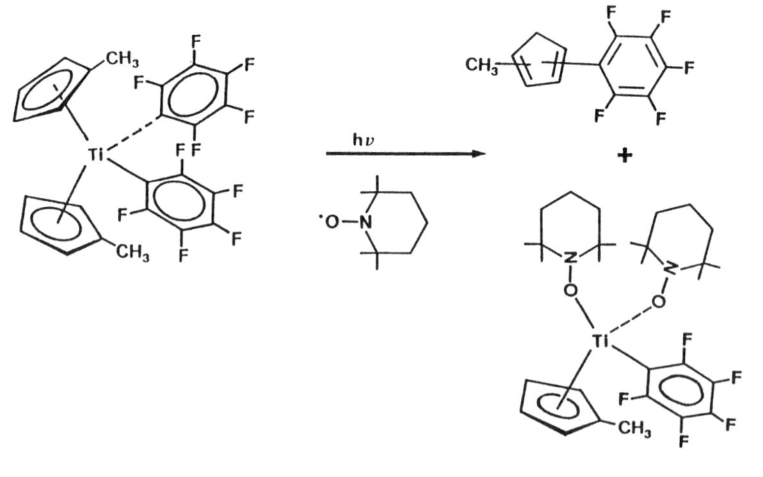

Scheme III. Trapping experiment of the perfluorinated phenyl derivative of the titanocene complex with TEMPO.

complexes having similar absorption characteristics and decay rates (Scheme IV). A methyl-substituted cyclopentadienyl ligand in the first compound and a methoxy substituent on the aryl ligand allowed easy detection spectroscopically. As no cross-coupling products were identified, the ligand-coupling reaction was established to be a purely intramolecular decay (Scheme V).

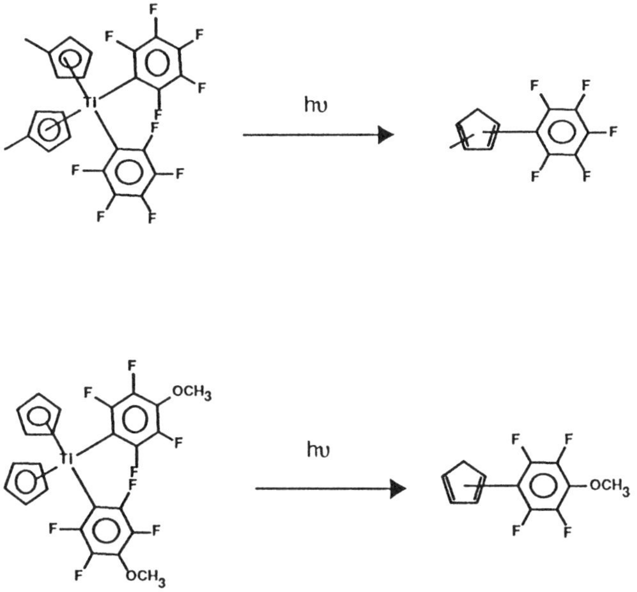

hυ

hυ

Scheme IV. Photochemical behavior of titanium complexes with similar absorption characteristics.

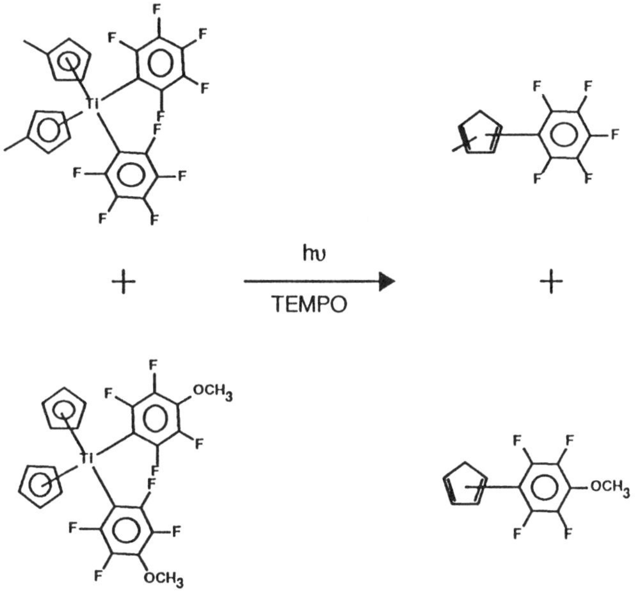

hυ

TEMPO

Scheme V. Ligand-coupling reaction of titanium complexes.

Laser flash photolysis experiments established that the primary photoreaction was purely first order. Therefore, the reaction product and the starting material had to be isomers (*26*). The primary photoisomerization is tentatively ascribed to a cyclopentadienyl ring slippage from η^5 to a lower coordination number, a process that allows coordinative unsaturation (Scheme VI). On the basis of IR spectroscopic information we showed that the primary photoproduct reacts with a number of potential ligands (*26*) (Scheme VII).

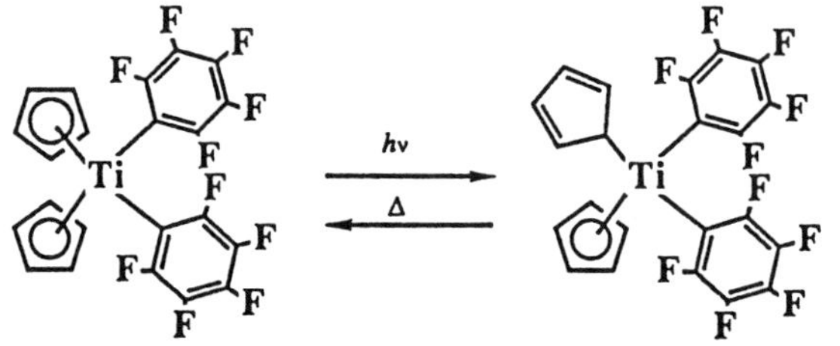

Scheme VI. Cyclopentadienyl ring slippage in the titanium complex.

Photobleaching. The organometallic photoinitiators have two unique features. The first is their ability to absorb in the visible range of the spectrum. For the titanocene compound the absorption extends up to 560 nm with pronounced absorption bands in the visible region at 405 and 480 nm. The second unique feature is the fact that they undergo photobleaching (*23*). When using conventional photoinitiators of the aromatic ketone type, the optical density of the formulation before and after irradiation stays more or less the same. When incorporating initiators that bleach, the optical density of the formulation containing the photoinitiator decreases with increasing exposure time. This phenomenon allows the curing of extremely thick coatings.

In order to study the extent of curing thick layers, we evaluated the dependence of the polymerization depth at a constant-volume cross section on irradiation and photoinitiator concentration.

An opaque tube sealed with a foil to form a base was filled with a formulation containing one part of an epoxyacrylate and one part of trimethylolpropane triacrylate (TMPTA) and various concentrations of our bis(pentafluorophenyl)titanocene. The tube was covered with a transparent polyester foil and placed in the irradiation unit. The tube axis was aligned with the projection axis of a UV lamp, and the upper tube surface was positioned at a distance of 75 cm from the lamp. Different exposure times were applied. After exposure the tube was removed from the unit and set on its hardened end, the base foil was peeled off, and the polymerization depth was measured with a calibrated ruler.

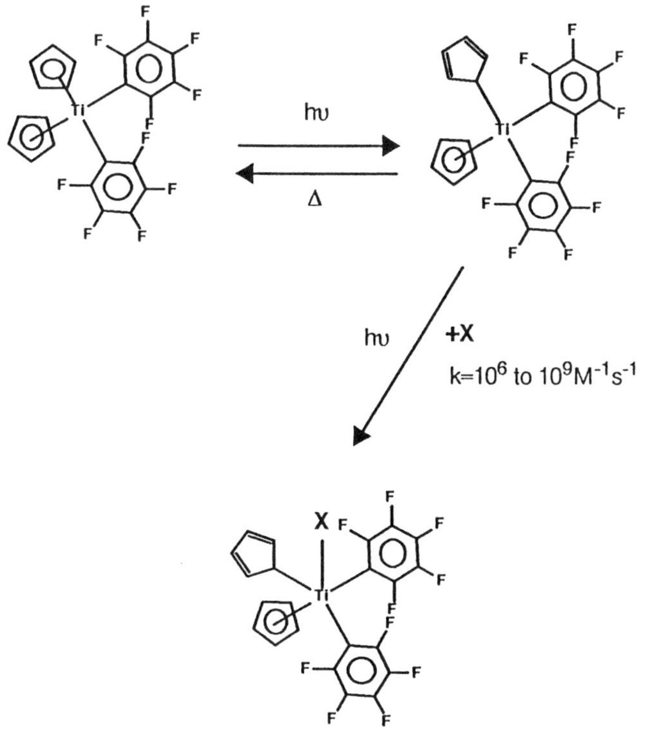

Scheme VII. Reaction of primary photoproduct of titanium complex. X is water, methanol, acetone, acetonitrile, nitromethane, 1,4-butanediol bisacrylate, molecular oxygen, molecular nitrogen, TEMPO, or carbon monoxide.

Figure 2 shows the dependence of curing depth on irradiation time for a series of initiator concentrations. The two main features shown in the general behavior of this model system are as follows:

1. At a constant photoinitiator concentration, the curing depth rapidly increases in the beginning and slows down with extended exposure times.
2. At a given length of exposure, the curing depth goes through a maximum with increasing photoinitiator concentrations.

The best result that we achieved in our model formulation was a curing depth of 18 mm after 16-min exposure time using a concentration of 0.18% photoinitiator. The efficiency of the curing process of course also depends on the resin formulation and the irradiation source used.

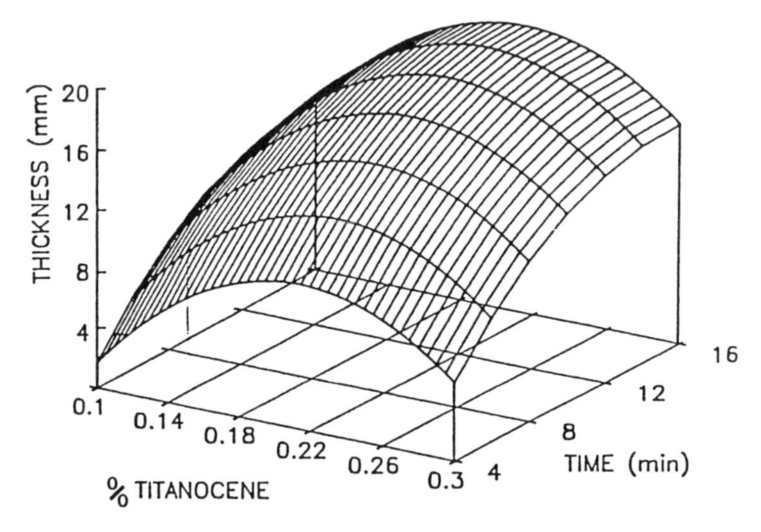

Figure 2. The dependence of curing depth on irradiation time for a series of titanocene initiator concentrations.

Laser Direct Imaging

Since lasers have become important tools in structured curing of photopolymers, maskless structuring such as laser direct imaging has been introduced (*27*). In this process the patterns are directly formed by scanning the laser beam over the photoresist layer; this process makes the troublesome image transfer by masks obsolete. Quick prototyping and fast design by link to a computer-aided design and manufacturing (CAD/CAM) system thus have been made possible. The sensitivity of the titanocenes to the argon laser emission lines makes them highly promising candidates for maskless photostructuring and for holographic experiments (*23, 28*).

Summary

Organometallic photoinitiators allow for the first time the possibility of curing thick layers because of their bleaching effect. Furthermore the curing of non-UV-transparent monomer formulations has become possible. The absorption bands of the iron-based as well as the titanium-based compounds extend into the visible region of the spectrum, and hence they are suitable initiators for laser-imaging processes.

References

1. Klingert, B.; Riediker, M.; Roloff, A. *Comments Inorg. Chem.* **1988,** *7,* 109–138.
2. Kutal, C.; Grutsch, P. A.; Yang, D. B. *Macromolecules* **1991,** *24,* 6872.
3. Curtis, H.; Irving, E.; Johnson, B. F. G. *Chem. Brit.* **1986,** *22,* 327.
4. Wrighton, M. S.; Graff, J. L.; Kazlauskas, R. J.; Mitchener, J. C.; Reichel, C. L. *Pure Appl. Chem.* **1982,** *54,* 161.
5. Hennig, H.; Rehorek, D.; Archer, R. D. *Coord. Chem. Rev.* **1985,** *61,* 11.
6. Lohse, F.; Zweifel, H. *Adv. Polym. Sci.* **1986,** *78,* 61.
7. Meier, K.; Bühler, N.; Zweifel, H.; Berner, G.; Lohse, F. Eur. Patent Appl. No. 094915, 1983.
8. Meier, K.; Eugster, G.; Schwarzenbach, F.; Zweifel, H. Eur. Patent 126712, 1986.
9. Palazzotto, M. C.; Hendrickson, A. Eur. Patent Appl. No. 109851, 1984.
10. Meier, K.; Zweifel, H. *Polym. Prepr.* **1985,** *26,* 347.
11. Meier, K.; Zweifel, H. *J. Imaging Sci.* **1986,** *30,* 174.
12. Nesmeyanov, A. N.; Vol'kenau, N. A.; Bolesova, I. N. *Dokl. Akad. Nauk. SSSR* **1963,** *19,* 3007.
13. Nesmeyanov, A. N.; Vol'kenau, N. A.; Shilovtseva, L. S. *Dokl. Akad. Nauk. SSSR* **1970,** *190,* 857.
14. Gill, T. P.; Mann, K. R. *Inorg. Chem.* **1980,** *19,* 3007.
15. Amster, I. J.; Gwathney, W. J.; Lin, L.; Kutal, C. *Abstracts of Papers, 9th International Symposium on the Photochemistry and Photophysics of Coordination Compounds;* Institute of Inorganic Chemistry, University of Fribourg: Fribourg, Switzerland, 1991; p 103.
16. Chien, J. C. W.; Wu, J. C.; Rausch, M. D. *J. Am. Chem. Soc.* **1981,** *103,* 1180.
17. Zucchini, U.; Albizzati, E.; Giannini, W. *J. Organomet. Chem.* **1971,** *26,* 357.
18. Ballard, D. H. G.; van Lienden, P. W. *Makromol. Chem.* **1972,** *154,* 177.
19. Bamford, C. H.; Puddephatt, R. J. Slater, D. M. *J. Organomet. Chem.* **1978,** *159,* C31.
20. Tung, H. S.; Brubaker, C. H., Jr. *Inorg. Chim. Acta* **1981,** *52,* 197.
21. Peng, M.; Brubaker, C. H., Jr. *Inorg. Chim. Acta* **1978,** *26,* 231.
22. Rausch, M. D. *J. Organomet. Chem.* **1978,** *160,* 81.
23. Angerer, H.; Desobry, V.; Riediker, M.; Spahni, H.; Rembold,M. *Proceedings of the Conference on Radiation Curing ASIA;* Society of Manufacturing Engineers: Dearborn, MI, 1988; Paper No. 3, M06, p 461.
24. Riediker, M.; Roth, M.; Bühler, N.; Berger. J. Eur. Patent 0122223, 1984.
25. Roloff, A.; Meier, K.; Riediker, M. *Pure Appl. Chem.* **1986,** *58,* 1267.
26. Klingert, B.; Roloff, A.; Urwyler, B.; Wirz, J. *Helv. Chim. Acta* **1988,** *71,* 1858.
27. Chen, G. Y. Y. *Printed Circuit Fabrication* **1986,** *1,* 41.
28. Finter, J.; Riediker, M.; Rohde, O.; Rotzinger, B. *Makromol. Chem. Macromol. Symp.* **1989,** *24,* 177–187.

RECEIVED for review November 7, 1991. ACCEPTED revised manuscript December 9, 1992.

21

Photoinitiator Activity, Electrochemistry, and Spectroscopy of Cationic Organometallic Compounds

W. A. Hendrickson and M. C. Palazzotto

Corporate Research Laboratories, 3M Company, St. Paul, MN 55144

Of particular interest as photoinitiators of cationic polymerization are the (η^5-cyclopentadienyl)(η^6-arene)iron(+) compounds. In an effort to produce compounds with longer wavelength sensitivity, we prepared a series of (η^6-fluorenyl)(η^5-cyclopentadienyl)iron(+) compounds along with the corresponding free fluorenyl ligands. The iron complexes of these ligands have transitions that are much more intense than the parent simple arene derivatives and electrochemistry more characteristic of the free ligand. A comparison of the electrochemical and spectroscopic data for metal complex and free ligand showed that the electronic transitions in the metal complex are intraligand in origin, yet still give rise to metal-centered photochemistry. The metal complexes of the fluorenyl ligands provide a route to easily prepare variable-wavelength photoinitiators.

CATIONIC ORGANOMETALLIC COMPOUNDS OF MANY TYPES are efficient photoinitiators of cationic polymerization. The activity of these compounds is primarily dependent on the nature of the counterion and to a lesser degree on the identity of the ancillary ligands remaining in the coordination sphere of the metal. The (η^5-cyclopentadienyl)(η^6-arene)iron(+) compounds are attractive for industrial applications because of their ease of preparation, efficiency of initiating polymerization, and visible light sensitivity. One of the most important industrial applications is in the solvent and coatings industry.

0065–2393/93/0238–0411$06.00/0

Solvents and Coatings

The coatings industry is a major user of solvents (consuming 33% of an estimated 12 billion tons in 1987) (*1*). Related industries consume an additional 15% of the U.S. solvent market. The trend is toward decreasing the use of solvent-borne systems (*see* Figure 1), but solvents still will represent a major fraction in the year 2010 (*1*).

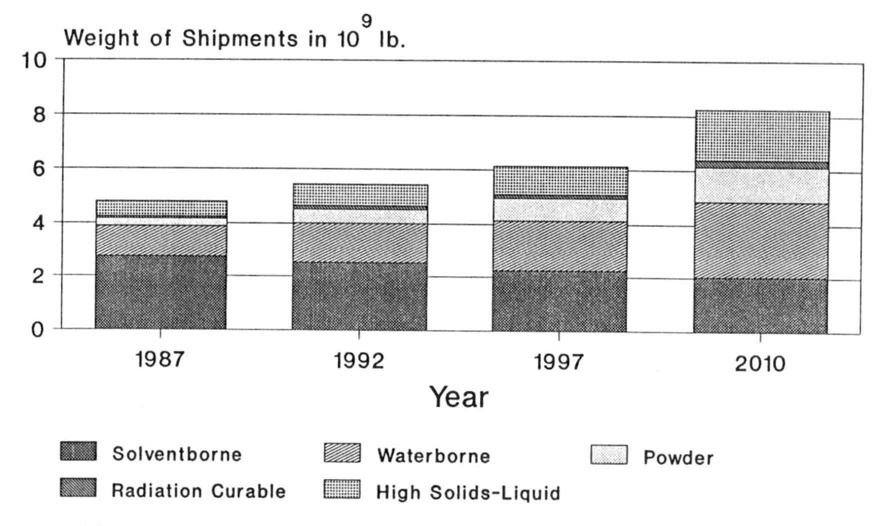

Figure 1. Trends in shipments of industrial coatings in the United States.

Continued use of solvents in the coatings and related industries (sealants, printing inks, electronics, and magnetic media) has led to some critical energy and environmental concerns. First, solvent-based manufacturing is energy intensive because of the high energy content of petroleum-based solvents and the energy required for processing and solvent emission control. Second, lower emission limits for volatile organic compounds (VOC) are forcing the coatings industry to invest in expensive solvent emissions control equipment with the added cost of disposal of recovered solvents.

In addition to these important economic and environmental issues, the coatings and related industries must meet these demands while maintaining or improving product standards. The ideal solution is to convert coating processes to 100% reactive systems. This conversion would eliminate both the energy and emission control requirements of solvent-based systems, but this must be done without sacrificing product quality.

Solventless coatings can be produced in many ways, but the focus of the work presented here is to develop methods based on radiation processing. This work is built around a new series of photocatalysts based on cationic organometallic compounds. The photochemical–thermal behavior of these catalysts combined with resultant physical properties of the cured compositions should make the production of 100% reactive systems more likely.

Description of the Photocatalyst Systems

For photoinitiated cationic polymerization, the technology that has received the most attention in recent years has been based on the onium salt systems (diazonium, iodonium, and sulfonium). These onium salts, when exposed to light, will decompose to generate both protonic acid and free radicals. A general mechanism for iodonium salt photochemistry is shown in Scheme I (*2*).

$$[Ph_2I^+]\,X^- \xrightarrow{h\nu} [Ph_2I^+]^*X^-$$

$$[Ph_2I^+]^* \longrightarrow Ph\text{-}I^{+\cdot} + Ph\cdot$$

$$Ph\text{-}I^{+\cdot} + S\text{-}H \longrightarrow Ph\text{-}I\text{-}H^+ + S\cdot$$

$$Ph\text{-}I\text{-}H^+ \longrightarrow Ph\text{-}I + H^+$$

Scheme I. General mechanism for onium salt photochemistry.

The protonic acid generated can be used to initiate the cationic polymerization of monomers such as epoxies. The counterion X^- plays a role in the propagation of the polymerization reaction, and for epoxies, the trend for the relative rate of polymerization is $BF_4^- < PF_6^- < AsF_6^- < SbF_6^-$ (*3*). The free radicals generated in these systems can be used to initiate the free radical polymerization of acrylates, for example.

Cationic organometallic compounds are capable of generating a species that can also initiate cationic polymerization (*4*). The compounds can be selected from a wide variety of complexes (Figure 2), a number of which are relatively easy to synthesize and are air-stable. These last two factors are important when considering the application of such catalysts to industrial processes. The focus of this chapter will be the iron-containing compounds.

Some of the characteristics of the compounds themselves will be described along with some of their curing chemistry, with a focus on epoxy compositions. A major portion of the chapter will be devoted to describing a particular series of iron-containing cationic organometallics and the effort expended to modify their absorption properties and understand the electronic and electrochemical behavior.

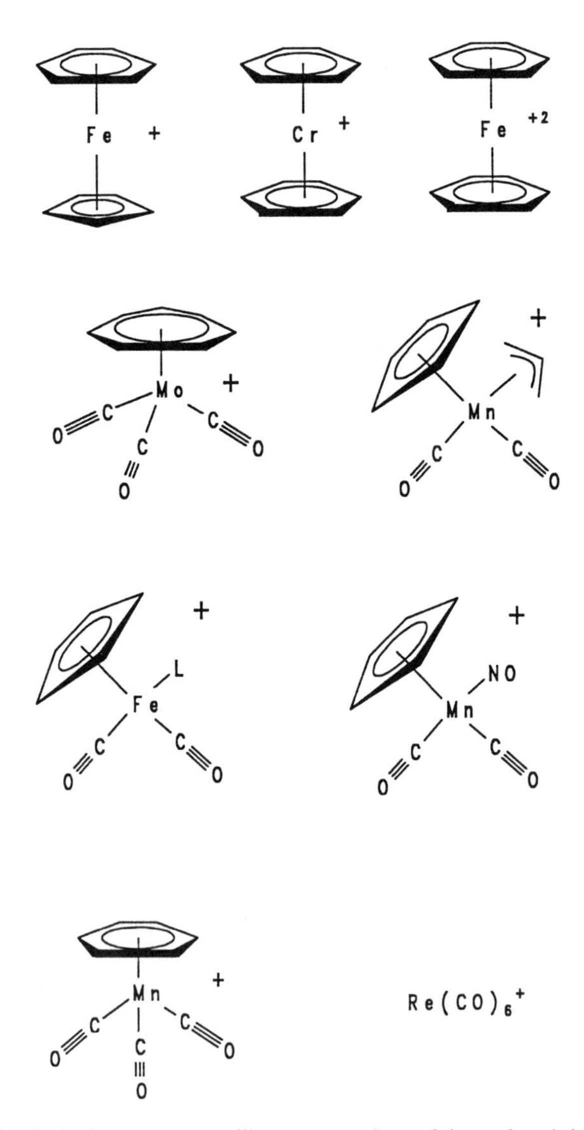

Figure 2. Cationic organometallic compounds useful as photoinitiators of cationic polymerization.

Photochemistry of the Catalyst Systems

In order to appreciate the capabilities of these organometallic-based photocatalysts, it is important to have some understanding of the photochemical reactions involved in the generation of the active catalysts. The action of light upon these compounds generates a Lewis acid. No evidence has been found for the initial generation of protonic acid. This property differentiates these compounds from the onium salts described. The Lewis acid can be used to initiate cationic polymerization. The Lewis acid is a thermally active catalyst. Cationic polymerization is generally slower than free radical, but it can continue after irradiation has ceased.

The two general types of cationic organometallic compounds that are active as photoinitiators of cationic polymerization are

1. arene-ring-only metal complexes
2. carbonyl-containing metal complexes

Some discussion of the photochemistry and curing chemistry of a representative example of each type of complex will be presented.

The basic photochemistry of the type 1 compounds can be understood by examining the activity of iron(+) ($CpFeArene^+$) upon exposure to light. An absorption spectrum characteristic of this compound is shown in Figure 3. The spectrum shows the long wavelength and low extinction coefficient ligand field absorption bands typical of 3d transition metal complexes.

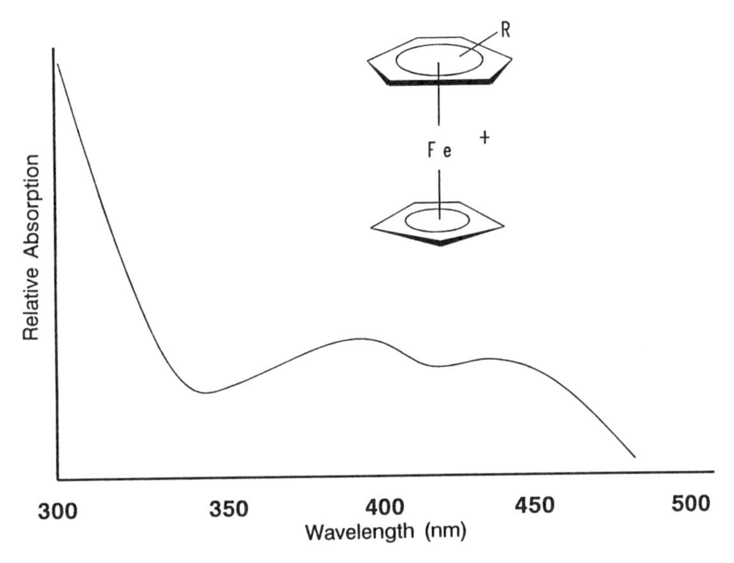

Figure 3. Absorption spectrum of a typical cyclopentadienylareneiron cationic organometallic compound.

Whether the compound is irradiated in the ultraviolet or visible region, the first action upon absorption of a photon is loss of the arene ligand (Scheme II) (5). This loss generates a coordinatively unsaturated metal center with potentially three sites available to bond ligands or reactive monomers. This initially generated species still has the cyclopentadienyl moiety attached (this fact has been verified by ligand-exchange studies). Subsequent thermal or photochemical steps may generate a free Fe^{2+} species. The actual active Lewis acid may be the $CpFe^+$ or the Fe^{2+} species. Free radicals are generated from the photochemistry of these compounds, although not very efficiently. The source of the free radicals is most likely a product of thermal reactions of the Cp^- moiety. If no suitable ligand is present, the intermediates go on to form ferrocene and Fe^{2+}.

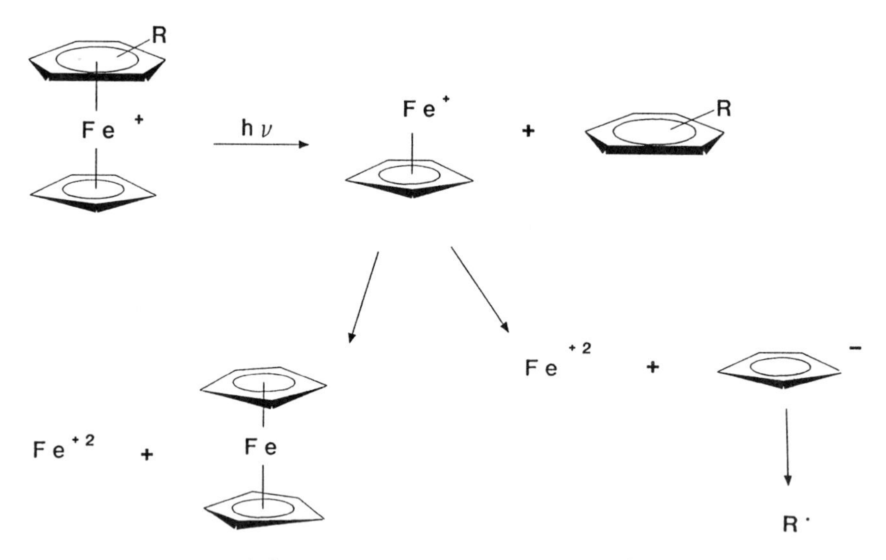

Scheme II. Irradiation of CpFeArene$^+$.

These CpFeArene$^+$ catalysts can be used to photoinitiate the cure of epoxy-containing monomers. The compositions can be made to cure quite rapidly in air, and the speed of polymerization can be controlled by the identity of the counterion, as the following table shows:

Counterion	*Time to Cure (s)*
SbF_6^-	30
AsF_6^-	45
PF_6^-	120
BF_4^-	>1800

This order of counterion dependence follows that of the onium salts.

η-Cyclopentadienyl)dicarbonyltriphenylphosphineiron(+) is a typical example of compounds of type 2. The absorption spectrum of $CpFe(CO)_2L^+PF_6^-$ is shown in Figure 4. When this compound is exposed to light, it loses a carbonyl (Scheme III). This is the typical reaction of a metal carbonyl-containing complex. This process has been verified by infrared studies, because the loss of the carbonyl group is easily detected. The coordinatively unsaturated species generated in this step is the active Lewis acid.

This series of carbonyl-containing compounds allows for subtle control of reactivity because of the ease of which ancillary ligands can be introduced into the coordination sphere of the metal (Table I). The same counterion dependence is seen in these compounds, but there is also a

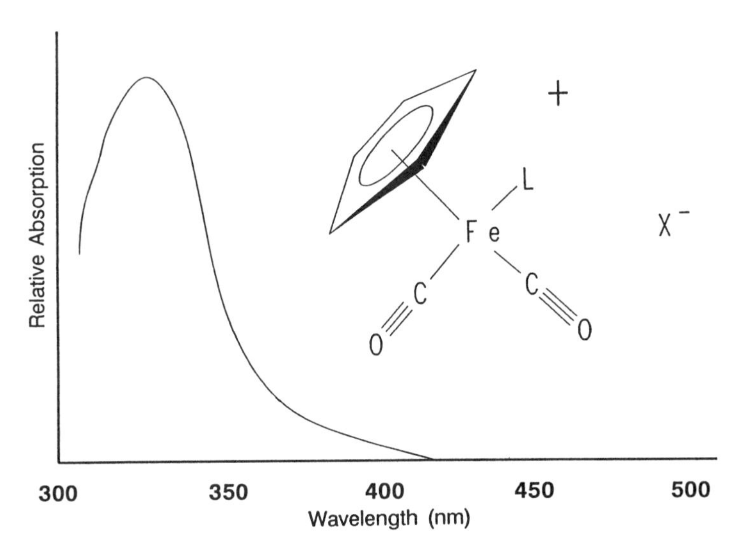

Figure 4. Absorption spectrum of a typical cyclopentadienyldicarbonyliron ligand cationic organometallic compound. L is triphenylphosphine and X^- is PF_6^-.

Scheme III. Irradiation of $CpFe(CO)_2L^+PF_6^-$.

Table I. Effect of Counterion and Ligand on Epoxy Cure Rate for $CpFe(CO)_2L^+$ Compounds

Counterion	*PPh_3*	*$SbPh_3$*
SbF_6^-	45	<5
AsF_6^-	45	10
PF_6^-	240	45
BF_4^-	<3600	<3600

All values are tack-free time in seconds.

marked dependence on the identity of L, the ligand remaining attached to the metal after photolysis. The Lewis acidity at the metal center can be tuned by these ancillary ligands.

The photogenerated thermally active species has been shown to activate reactive monomers such as epoxies. In the type 1 arene-only complexes, the identity of this species is not certain. The action of the active species in the epoxy case is believed to be as described in Scheme IV. The active species activates an epoxy to subsequent nucleophilic attack by another epoxy, following a typical acid-catalyzed scheme. The propagation efficiency or activity of the catalyst is strongly dependent on the counterion, with less coordinating anions being more active. With the same counterion, the efficiency in initiating polymerization can be controlled by varying the ligands on the metal center (Table I). The two effects can be combined, ligand and counterion, to achieve a wide range of activity from essentially the same compound (Table I).

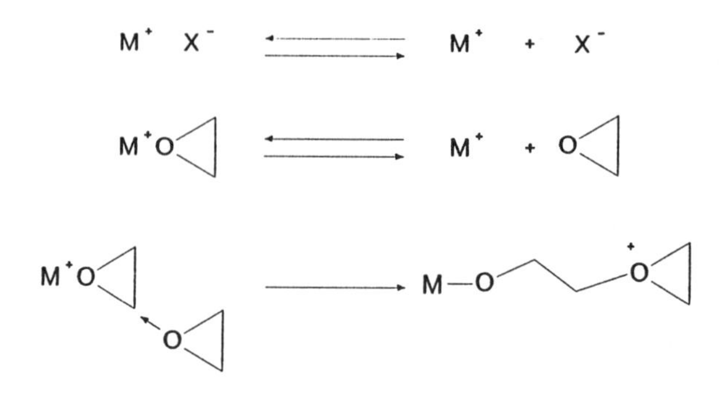

Scheme IV. Activation of reactive monomers in photogenerated thermally active epoxy species.

The Design and Synthesis of Long-Wavelength-Absorbing Cationic Complexes

The new photocatalysts are active in the UV and visible regions of the spectrum, but they do not extend into the visible region, and those that do have relatively weak (low extinction coefficients) transitions in the visible region. For other potential industrial uses, such as imaging applications, it would be useful to have greater flexibility in tuning a compound's spectral response and absorptivity. Specifically, it would be useful to have compounds with high extinction coefficients at long wavelengths.

For the purpose of studying compounds that are highly absorbing at long wavelengths, the series represented by structures **I** and **II**, where R is a phenyl or substituted phenyl group, were prepared. They were examined by spectroscopic and electrochemical techniques. This information will be used to design compounds of the desired absorption characteristics.

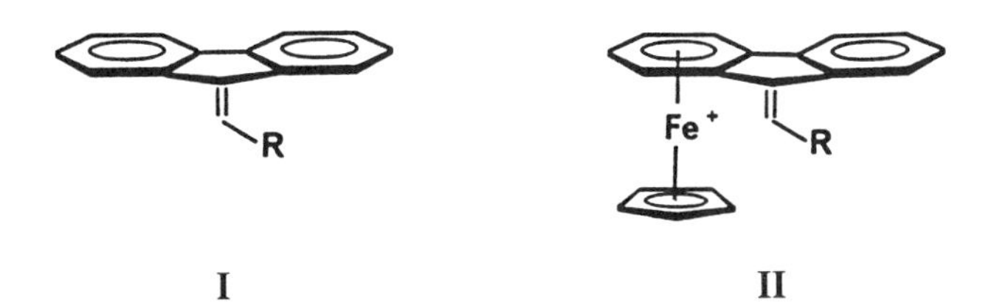

I II

The purpose of preparing series **II** was to examine the effect of changes in the absorption maximum and absorptivity on the photochemical activity of these compounds. The series of free ligands **I** was prepared to help understand the spectroscopy and electrochemistry for series **II**.

Spectroscopic and electrochemical techniques can be combined to yield information about the energy levels in a related series of compounds (*6*). The basis for the relationship between the electrochemistry and absorption spectrum of a compound can be understood when Figure 5 is examined. The values for the lowest unoccupied molecular orbital (LUMO) and highest occupied molecular orbital (HOMO) are related to E_{red}, reduction potential, and E_{ox}, oxidation potential. Their difference is related to $E_{h\nu}$, the energy of the optical transition, by the simple expression

$$E_{hu} = E_{ox} - E_{red} + \text{const} \quad (1)$$

where the terms are as just defined and const contains terms to account for the reorganization energy upon addition or removal of an electron.

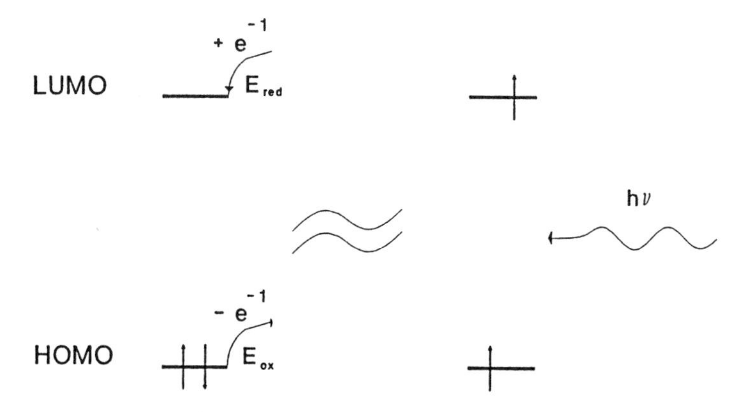

Figure 5. Equivalence of redox properties and the absorption process.

This type of correlation has been used extensively in the area of silver-film sensitization to select the best dye to use in a particular spectral region (*7*). Studies have investigated this relationship in organic (*8*) and inorganic (*9*) systems. We felt that such a study of series **I** and **II** would lead to a better understanding and would develop a predictive ability of the photochemical activity of **II**.

Experimental Details

Absorption Measurements. Absorption spectra were obtained from spectral grade acetonitrile solutions using a Perkin-Elmer model 330 UV–visible or an HP 8452A diode array spectrometer. Measurements of the iron compounds were done under subdued lights to eliminate any photochemical bleaching in solution.

Thin-Film Curing Studies. Cure times were measured on thin-film coatings. The compositions consisted of a 1:1 (w/w) vinylcyclohexene dioxide and 3,4-epoxycyclohexanemethylenecarboxylate-3,4-epoxycyclohexane. The coating compositions were prepared under red safelights. A solution containing 1% photocatalyst was prepared by weighing out the desired amount of photocatalyst, adding acetonitrile to dissolve it completely, and then adding the mixture of the epoxies. Coatings were made onto 3-mil polyvinylidene subbed polyester using a No. 3 coating rod from R.D. Associates. The samples were then exposed at 10 cm to a 275-W lamp from General Electric. The time to produce a nontacky surface was recorded.

Solution-Curing Studies. Photosensitive solutions were prepared as 1% by weight photocatalyst in cyclohexene oxide. For each of these solutions, a 3-mL aliquot was placed in a Pyrex test tube 5 cm from a 300-W quartz halogen source and irradiated for 5 min. After irradiation, the solution was quenched in methanol, and this quenching precipitated any polymer formed. The isolated polymer was dried in a vacuum oven. The percent conversion was calculated from the yield of isolated polymer. Three such runs were made for each solution. For irradiation at 440 nm or greater, a 440-nm cutoff filter was used.

Preparation of Cationic Organometallics for Curing Studies. The compounds used in the curing studies were prepared as described in U.S. Patent No. 5,089,536.

Preparation of 9-Substituted Fluorenyl Compounds. Stoichiometric amounts of the aldehyde and fluorene were refluxed in ethanol or ethanol–water in the presence of a catalytic amount of potassium hydroxide. The desired product precipitated from solution and often did not require further purification. The compounds could be recrystallized from methylene chloride–heptane as desired. Yields were typically in the 50–90% range. The compounds were identified by melting point, NMR and IR spectra, and elemental analysis.

Preparation of (η^6-9′-Substituted Fluorenyl)(η^5-cyclopentadienyl)iron(+) Hexafluorophosphate Compounds. Stoichiometric amounts of the (η^6-fluorenyl)(η^5-cyclopentadienyl)iron(+) hexafluorophosphate and the aldehyde were stirred overnight at room temperature in ethanol–water (75:25 mL) under nitrogen in the presence of a catalytic amount of potassium hydroxide. Upon completion of the reaction, a green product was present. This product could be eliminated by the addition of aqueous hydrogen hexafluorophosphate. The product could be filtered off and purified by recrystallization from acetone–ether or acetonitrile–ether. Yields were typically in the 50–75% range, and all compounds were characterized by melting point, NMR and IR spectra, and elemental analysis. For spectroscopic and electrochemical studies, the compounds were further purified by chromatography on silica gel.

Electrochemical Measurements. Electrochemical measurements were made with a Princeton Applied Research model 177/178 programmer, a model 179 potentiostat–galvanostat, and a Houston Instruments model 2000 *x–y* recorder. The cell was of the standard three-electrode configuration. The working electrode was a platinum wire 3 mm in diameter epoxied into the end of a glass capillary, the auxiliary electrode was a large platinum wire encircling the working electrode, and the reference electrode was 0.01 M $AgNO_3$–acetonitrile–Ag wire in a ceramic frit electrode from Sargent Welch. Ferrocene was used as an internal standard and appeared at +0.09 V versus this reference electrode. The working electrode was polished manually with 3- then 0.5-μm alumina before each scan. This polishing is especially important for the oxidation scans, which tend to leave a deposit on the electrode surface that distorts subsequent scans.

The supporting electrolyte was tetraethylammonium hexafluorophosphate from Southwest Analytical that had been dried thoroughly before use. Solvent was acetonitrile from MCB OmniSolv. Solvent was transferred under argon, and the cell was purged with argon before and during the measurements. This procedure produced an operating window of 4.5 V.

Results and Discussion

The series **I** compounds that were prepared are described and their absorption maxima (λ_{max}) and extinction coefficients (ϵ) are listed in Table II. The compounds all have relatively strong absorptions and extinction coefficients of 10^4 in the UV and visible portion of the spectrum. All of the compounds are electroactive; the position of E_{ox} is strongly dependent on the nature of R and the value of E_{red} is less so.

Table II. Absorption Spectra of the Free Ligands and Metal Complexes

Compound No.	*Phenyl Substituent*	*Free Ligand*	*Metal Complex*
1	4-nitro	354 (14.0)	338 (13.8)
2	4-cyano	333 (16.5)	325 (14.5)
3	4-chloro	330 (16.1)	328 (13.7)
4	3-nitro	322 (14.8)	316 (12.7)
5	3-chloro	324 (14.1)	316 (13.7)
6	–H	324 (14.5)	323 (13.4)
7	4-methyl	332 (16.6)	345 (14.1)
8	4-methoxy	344 (18.2)	368 (17.4)
9	3,4-dimethoxy	354 (17.4)	379 (17.0)
10	2.4-dimethoxy	353 (17.8)	378 (18.3)
11	2,3,4-trimethoxy	343 (12.9)	368 (14.0)
12	2,4,5-trimethoxy	375 (12.6)	402 (14.2)
13	4-dimethylamino	401 (23.6)	440 (21.5)
14	4-diethylamino	414 (28.6)	457 (25.7)

NOTE: All values for free ligands and metal complexes are λ_{max} in nanometers, and the numbers in parentheses are $\epsilon \times 10^{-3}$.

The shift to longer wavelength (lower energy) as R becomes more electron-donating can be understood when the molecular orbital calculation performed on this type of compound is examined (*10*). The calculations show that the excited states of compounds of type **I** have increased electron density on the fluorene portion of the molecule (Scheme V). This increased electron density indicates that the optical transition has charge-transfer character, from R to the fluorene portion of the molecule. As R becomes more electron-donating, this transfer becomes more facile, lowering the energy of the transition.

The electrochemistry of a series of type **I** compounds with electron-withdrawing groups was examined previously (*11*), and a correlation made between E_{red} and the Hammett sigma constants. Neither oxidation potentials nor absorption spectra were reported. A reasonably good correlation ($r = 0.976$) exists between $E_{h\nu}$ and $E_{ox} - E_{red}$, as can be seen in Figure 6.

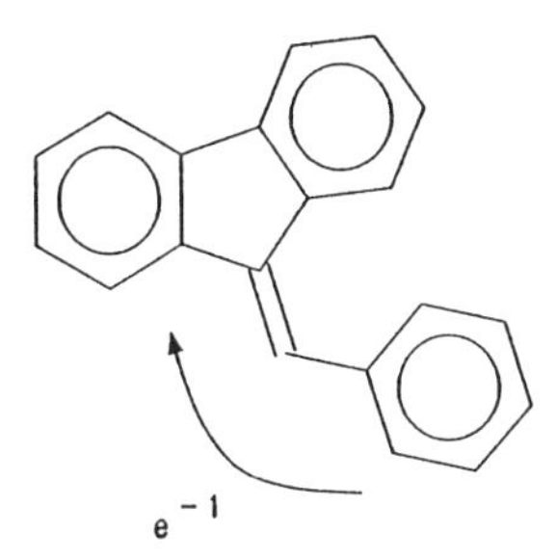

Scheme V. Increased electron density of the fluorene portion of the molecule.

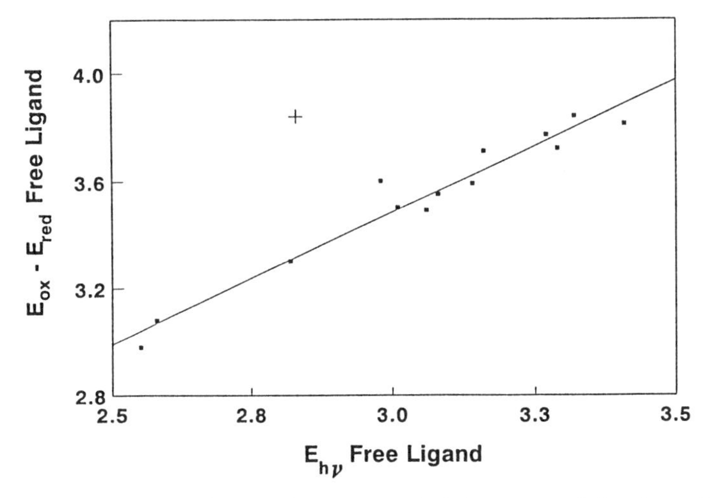

Figure 6. The correlation for the free ligand of the energy of the absorption maximum vs. $E_{ox} - E_{red}$. *The + indicates the 3-nitrophenyl derivative.*

The absorption maxima and extinction coefficients for the compounds of type **II** are listed in Table II. The absorption spectra of these compounds are characterized by relatively intense transitions (extinction coefficients ~10^4) in the UV and visible regions. (This feature compares to the relatively weak transitions, extinction coefficients ~10^2, for the η^6-arene compounds). These transitions are, for the most part, red-shifted with respect to the corresponding free ligand absorption, and their position depends on the nature of R. All of the compounds are electroactive; the value of E_{ox} is strongly dependent on the nature of R, and E_{red} is generally less so. As with the type **I** compounds, a correlation between $E_{h\nu}$ and $E_{ox} - E_{red}$ was made and is shown in Figure 7. Two lines are present, with a reasonably good correlation for electron-donating substituents (r = 0.975) and a much poored correlation for electron-withdrawing substituents (r = 0.642).

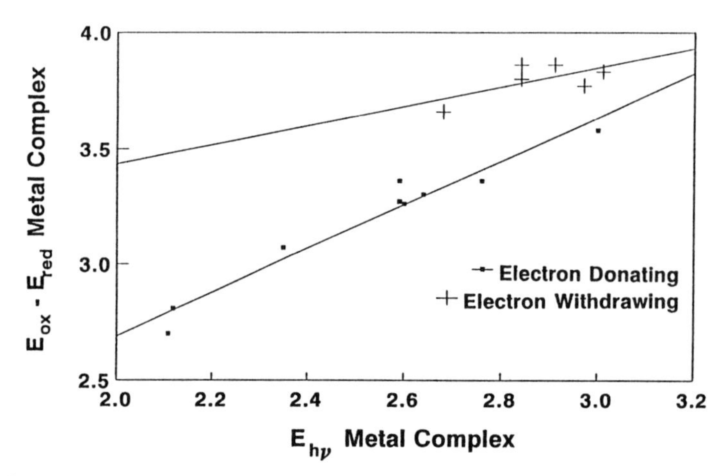

Figure 7. The correlation for the metal complex of the energy of the absorption maximum vs. $E_{ox} - E_{red}$.

The electrochemistry and spectroscopy of compounds related to series **II** were investigated in some detail (*12*). The arene ligand in these studies has generally been a benzene ring with simple alkyl group substitution. The lowest energy absorptions in these compounds are weak (extinction coefficient ~100) and have been attributed to metal-centered d–d transitions (*12b*). The observed electrochemical activity has been assigned to reversible reduction at the metal center, and the oxidation of these compounds has not been reported except to say that they are difficult to oxidize (*12c, 12d*).

The optical transitions and electrochemical behavior of the η^6-arene complexes can be understood from the simple molecular orbital diagram presented in Figure 8 (*13*). Figure 8 shows the highest occupied and lowest unoccupied orbitals associated with the metal center and the ligand.

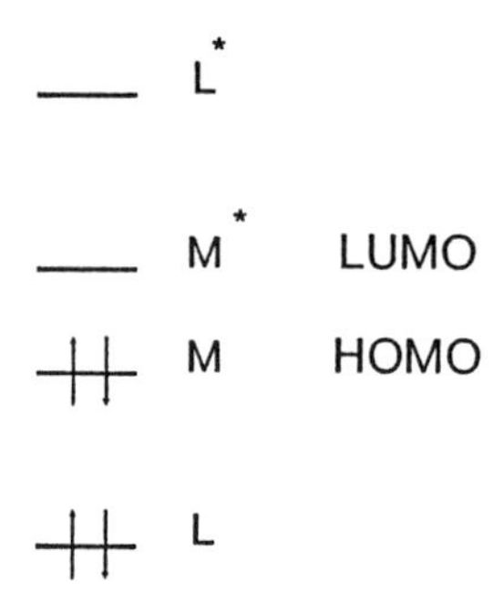

Figure 8. Simple molecular orbital representation for arene metal complexes.

The lowest energy transitions are between the metal orbitals with the metal-to-ligand charge-transfer transition being the next higher energy transition. The reduction behavior of these compounds can be understood as adding an electron to the lowest unoccupied metal-centered orbital.

Some major differences are seen between data for typical arene metal complexes and the related values for series **II**. The most obvious difference is in the optical spectrum. The simple arene complexes are characterized by relatively weak transitions whose position is not very dependent on the identity of the arene. The complexes of series **II** have intense transitions whose position is strongly dependent on the arene. The electrochemical behavior also shows major differences. The simple arene complexes exhibit reversible reduction and are quite difficult to oxidize, but series **II** shows irreversible reduction behavior and are reasonably easy to oxidize. This difference suggests that the energy levels in **II** are not adequately described by the orbital ordering in Figure 8.

An indication of the origin of the electrochemical and spectroscopic behavior of **II** can be obtained from a closer examination of the data presented in Table II. The absorptions for the metal complexes follow closely the pattern for the free ligand, only slightly red-shifted. The extinction coefficients for the related metal complex and free ligand are almost identical. This feature implies that the ligand orbitals play some role in the optical transitions of the metal complex and should be involved with their electrochemical behavior as well.

For both **I** and **II** a correlation exists between their electrochemical activity and absorption behavior. If the ligands play some role in the optical transitions of the metal complex, then there should be some relationship between the electrochemistry of the free ligand and that of the corresponding metal complex (*14*).

If a plot is made of E_{red} ligand versus E_{red} metal (Figure 9), a reasonably good correlation is obtained (r = 0.995). This correlation suggests that the reduction of the metal complex is really an addition of an electron to an orbital of mainly ligand character.

If a plot is made of E_{ox} ligand versus E_{ox} metal (Figure 10), the interpretation of the correlation is not as straightforward. A good correlation (r = 0.996) with straight-line behavior is seen for those compounds with electron-donating substituents. The compounds with electron-withdrawing substituents do not follow the behavior of the other compounds and level off at some upper limit. For the electron-donating substituents, the straight-line correlation suggests that oxidation in these compounds is from a mainly ligand-centered orbital.

If a plot is made of $E_{h\nu}$ ligand versus $E_{h\nu}$ metal (Figure 11), two groups of compounds are found. Straight-line correlations are seen for the compounds with electron-donating and electron-withdrawing substituents. The correlation is good (r = 0.998) for compounds with electron-

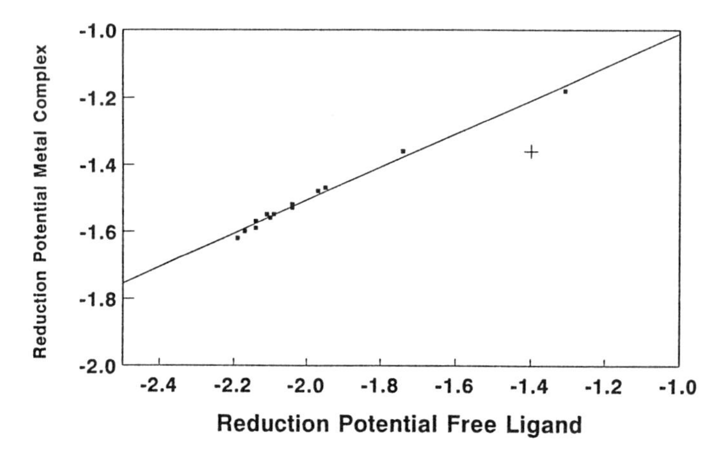

Figure 9. The correlation of the reduction potential of the free ligand vs. the metal complex. The + indicates the 3-nitrophenyl derivative.

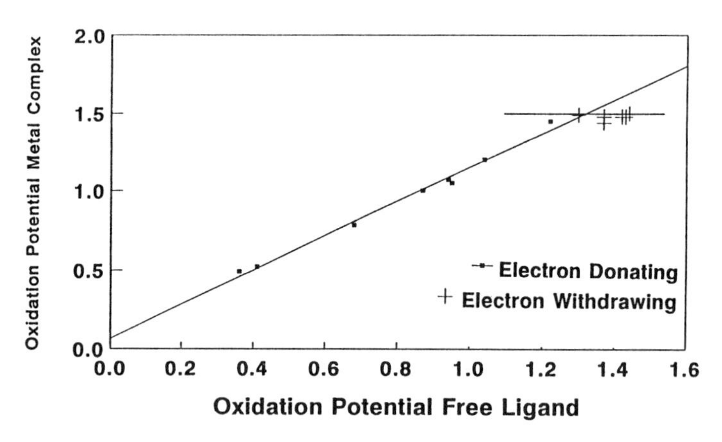

Figure 10. The correlation of the oxidation potential of the free ligand vs. the metal complex.

donating substituents and comparable (r = 0.956) for electron-withdrawing substituents.

If a plot is made of $E_{ox} - E_{red}$ ligand versus $E_{ox} - E_{red}$ metal (Figure 12), again two different correlations are found for the two types of substituents. The correlation is good for electron-donating (r = 0.990) but poor for electron-withdrawing (r = 0.883) substituents.

The electrochemical and spectroscopic data for the compounds of series **II** with electron-donating substituents suggests a new orbital ordering for these compounds (Figure 13). The electrochemical data suggest that the LUMO and HOMO are composed of predominately ligand-centered orbitals.

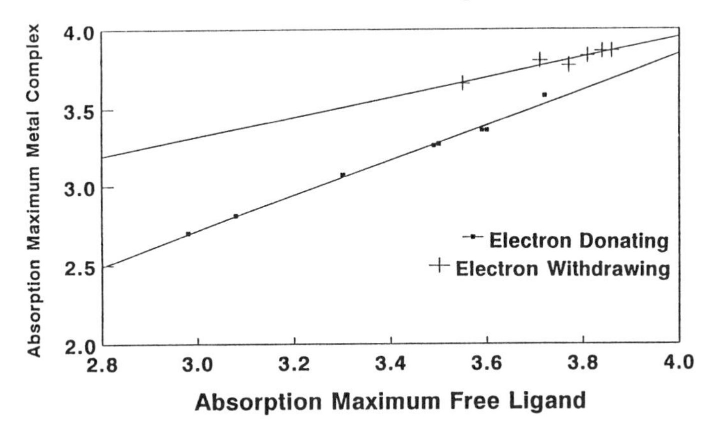

Figure 11. The correlation of the absorption maximum of the free ligand vs. the metal complex.

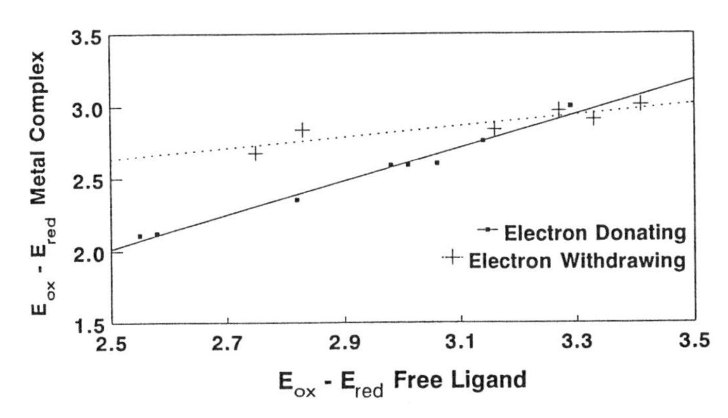

Figure 12. The correlation of $E_{ox} - E_{red}$ *for the free ligand vs. the metal complex.*

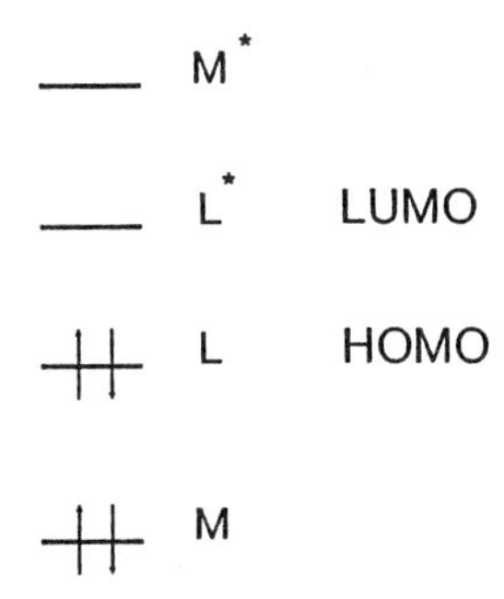

Figure 13. Simple molecular orbital representation of complexes with electron-donating substituents.

For compounds with electron-withdrawing substituents, the highest filled ligand orbital has moved beneath the highest filled metal-centered orbital. An ordering such as this would also explain the leveling off of the oxidation potentials of the metal complexes seen in Figure 10. As the ligand-based orbital moves lower in energy due to electron-withdrawing substituents, it moves lower than the metal-centered orbital. Oxidation now takes an electron out of a metal-centered orbital. This metal-centered orbital is relatively unaffected by changes in the ligand, and its position remains constant. Further lowering of the ligand-centered orbital makes no change in this level, so that the oxidation potential of the complex changes very little. The molecular orbital ordering in these compounds has been changed to something approximated by Figure 14.

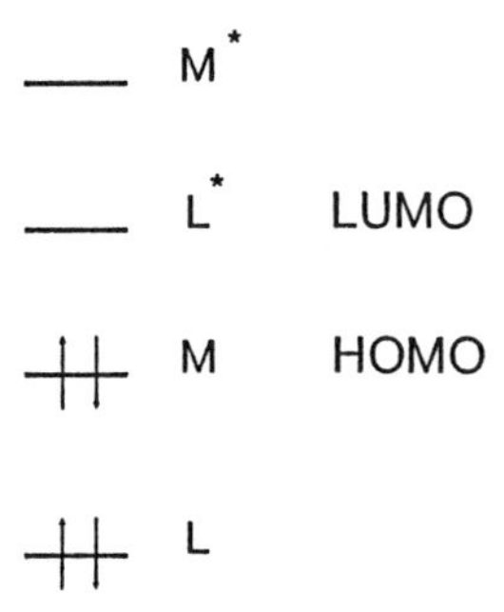

Figure 14. Simple molecular orbital representation of complexes with electron-withdrawing substituents.

The origin of the red shift for the series **II** relative to **I** can now be understood. The transition in **II** is mainly intraligand in character, so the electron density shift in the excited state can still be roughly described by Scheme V. Attaching the positive metal center to the fluorene portion of the molecule increases its electron-accepting ability. This increased ability will facilitate the transfer of electron density as per Scheme V, and thereby lower (red-shift) the energy of the transition.

Conclusion

The result of this study indicates that the most intense absorption in the metal complexes **II** is due to an intraligand transition for electron-donating substituents and MLCT for electron-withdrawing substituents. This finding is compared to the usual weak d–d absorptions found in this type of compound. Compounds **II** are known to photodissociate at least one ligand efficiently because they are active photoinitiators of cationic polymerization. In this case, intraligand or charge-transfer excitation readily produces metal-centered substitution.

That these compounds function as active photoinitiators of cationic polymerization can be seen in Table III. The series based on the η^6-fluorene compounds are very efficient initiators, more active than sensitized sulfonium salts in the near-UV and visible ranges.

Table III. Comparison of Photoinitiator Activity in Epoxy Polymerization

Photoinitiator	*No Filter*	*440-nm Filter*
Fe⁺ PF₆⁻	15	14
Fe⁺ PF₆⁻ MeO OMe	11	13
OMe OMe + [ ]₃ S⁺ SbF₆⁻	1	<1

NOTE: All values are yield of isolated polymer in percent.

References

1. Linak, E. *An Overview of the Paint and Coatings Industry;* Chemical and Economics Handbook; SRI: Menlo Park, CA, 1988; 592.5100A.
2. Crivello, J. V. *CHEMTECH* **1980,** *10,* 624.
3. Odian, G. *Principles of Polymerization;* McGraw-Hill: New York, 1970; p 321.
4. European Patents 94914, 94915, 109,851, and 126,712.
5. Gill, T. P.; Mann, K. R. *Inorg. Chem.* **1980,** *19,* 3007.
6. Loufty, R. O.; Loufty, R. O. *Can. J. Chem.* **1976,** *54,* 1454.
7. Tani, T.; Kikuchi, S.-I.; Honda, K.-J. *Photo. Sci. Eng.* **1968,** *12,* 80; Leubner, J. H. ibid. **1976,** *20,* 61; Carroll, B. H. ibid. **1972,** *21,* 151; Leubner, J. H. ibid. **1978,** *22,* 270; Dahne, S. ibid. **1979,** *23,* 219.
8. Peover, M. E. *Electrochim. Acta* **1968,** *13,* 1083.
9. Vleck, A. A. *Electrochim. Acta* **1968,** *13,* 1063; Sabi, T.; Adyagui, S. J. *Electroanal. Chem. Interfac. Electrochem.* **1975,** *60,* 1; Chum, H. L.; Rock, M. *Inorg. Chim. Acta* **1979,** *37,* 113; Smith, D. A.; Schultz, F. A. *Inorg. Chem.* **1982,** *21,* 3035.

10. Griffiths, J.; Lockwood, M. *J. Chem. Soc. Perkin Trans. I* **1976,** 48.
11. Archer, J. F.; Grimshaw, J. *J. Chem. Soc. B* **1969,** 266.
12. (a) Sohn, Y. S.; Hendrickson, D. N.; Gray, H. B. *J. Am. Chem. Soc.* **1971,** *93,* 6303. (b) Morrison, W. H.; Ho, E. Y.; Hendrickson, D. N. *Inorg. Chem.* **1975,** *14,* 500. (c) Hamon, J.-R.; Astruc, D.; Michaud, P. *J. Am. Chem. Soc.* **1981,** *103,* 758. (d) Green, J. C.; Kelly, M. R.; Payne, M. P.; Seddon, E. A. *Organometallics* **1982,** *2,* 211.
13. Muetterties, E. L.; Bleeke, J. R.; Wucherer, E. J.; Albright, T. A. *Chem. Rev.* **1982,** *82,* 499.
14. Vleck, A. A. *Coord. Chem. Rev.* **1982,** *43,* 39. Elliott, C. M.; Hershenhart, E. J. *J. Am. Chem. Soc.* **1982,** *104,* 7519.

RECEIVED for review November 7, 1991. ACCEPTED revised manuscript May 1, 1992.

AUTHOR INDEX

AFFILIATION INDEX

SUBJECT INDEX

A

F

G

H

I

M

P

Q

R

S

T

U

V

W

X

Z

Copy editing and indexing: Janet S. Dodd
Production assistance: Margaret J. Brown
Cover design: Ronna Hammer

Production Manager: Robin Giroux

Printed and bound by Maple Press, York, PA

ACS Books of Related Interest

Homogeneous Transition Metal Catalyzed Reactions
Edited by William R. Moser and Donald W. Slocum
Advances in Chemistry Series 230
625 pp; clothbound, ISBN 0–8412–2007–7

Catalytic Selective Oxidation
Edited by S. Ted Oyama and Joe W. Hightower
ACS Symposium Series 523
464 pp; clothbound, ISBN 0–8412–2637–7

Selectivity in Catalysis
Edited by Mark E. Davis and Steven L. Suib
ACS Symposium Series 517
410 pp; clothbound, ISBN 0–8412–2519–2

Biocatalyst Design for Stability and Specificity
Edited by Michael E. Himmel and George Georgiou
ACS Symposium Series 516
335 pp; clothbound, ISBN 0–8412–2518–4

Biocatalysis at Extreme Temperatures: Enzyme Systems Near and Above 100 °C
Edited by Michael W. W. Adams and Robert M. Kelly
ACS Symposium Series 498
215 pp; clothbound, ISBN 0–8412–2458–7

Catalysis in Polymer Synthesis
Edited by Edwin J. Vandenberg and Joseph C. Salamone
ACS Symposium Series 496
291 pp; clothbound, ISBN 0–8412–2456–0

Catalytic Control of Air Pollution: Mobile and Stationary Sources
Edited by Ronald G. Silver, John E. Sawyer, and Jerry C. Summers
ACS Symposium Series 495
175 pp; clothbound, ISBN 0–8412–2455–2